HANDBOOK OF RESEARCH ON FUNCTIONAL MATERIALS

Principles, Capabilities, and Limitations

HANDBOOK OF RESEARCH ON FUNCTIONAL MATERIALS

Principles, Capabilities, and Limitations

Edited by

Charles Wilkie, PhD, Georges Geuskens, PhD, and Victor Manuel de Matos Lobo, PhD

Gennady E. Zaikov, DSc, and A. K. Haghi, PhD
Reviewers and Advisory Board Members

Apple Academic Press
TORONTO NEW JERSEY

Apple Academic Press Inc.
3333 Mistwell Crescent
Oakville, ON L6L 0A2
Canada

Apple Academic Press Inc.
9 Spinnaker Way
Waretown, NJ 08758
USA

Printed in the United States of America on acid-free paper

International Standard Book Number-13: 978-1-926895-65-9 (Hardcover)

Library of Congress Control Number: 2013954918

Library and Archives Canada Cataloguing in Publication

Handbook of research on functional materials: principles, capabilities, and limitations/ edited by Charles Wilkie, PhD, Georges Geuskens, PhD, and Victor Manuel de Matos Lobo, PhD ; Gennady E. Zaikov, DSc, and A. K. Haghi, PhD, reviewers and advisory board members.

Includes bibliographical references and index.
ISBN 978-1-926895-65-9
1. Materials--Technological innovations--Handbooks, manuals, etc. I. Lobo, Victor M. M., writer of preface, editor of compilation II. Wilkie, C. A. (Charles A.), writer of preface, editor of compilation III. Geuskens, G. (Georges), writer of preface, editor of compilation IV. Title: Research on functional materials.

TA403.6.H35 2014 620.1'1 C2013-907752-9

Apple Academic Press also publishes its books in a variety of electronic formats. Some content that appears in print may not be available in electronic format. For information about Apple Academic Press products, visit our website at **www.appleacademicpress.com** and the CRC Press website at **www.crcpress.com**

ABOUT THE EDITORS

Charles Wilkie, PhD
Professor of Polymer and Organic Chemistry, Marquette University, Milwaukee, Wisconsin, USA

Charles Wilkie, PhD, is the Pfetschinger-Habermann Professor, Polymer and Organic Chemistry at Marquette University in Milwaukee, Wisconsin. He is an editor of *Polymers for Advanced Technologies* and serves on the editorial boards of *Polymer Degradation and Stability* and *Thermochimica Acta*. In 1992, Professor Wilkie was the recipient of the ACS Milwaukee Section Award. He was a Fulbright-Hays scholar at the Universite Libre de Bruxelles in 1991–1992. In 2007, he received the Marquette University Award for Excellence in Research. He is one of the organizers of the American Chemical Society Meeting on fire retardancy, held every four years, and he has taught a number of short courses on fire retardancy.

Georges Geuskens, PhD
Professor Emeritus, Department of Chemistry and Polymers, Universite de Libre de Bruxelles, Belgium

Georges Geuskens, PhD, is Professor Emeritus in the Department of Chemistry and Polymers at the Universite de Libre de Bruxelles in Belgium. Dr. Geuskens is world-renowned scientist in the field of photochemistry and particularly in the field of photodegradation and light stabilization of organic compounds (including oligomers, polymers, composites and nanocomposites). He has published several books and volumes in this area and in general, 200 original papers and reviews.

Victor Manuel de Matos Lobo, PhD
Professor, Coimbra University, Coimbra, Portugal

Victor Manuel de Matos Lobo, PhD, earned his PhD in 1966 from Cambridge University (UK), where he developed a new isothermal diffusion cell. He was the Head for the Department of Chemistry at the University of Coimbra in Portugal for many years. He has been a member of the Portuguese National Council for Education, representing the Portuguese scientific societies. He is a member of the council responsible for accepting scientific journals in SciELO (Scientific Electronic Library Online). He has published around 50 articles in the Portuguese daily press on matters concerning education. He has been the Editor of *Portugaliae Electrochimica Acta* from September 2002. Since 1991, he headed the Portuguese delegation at international meetings of the ISO (International Organization for Standardization) and the CEN (European Committee for Standardization).

REVIEWERS AND ADVISORY BOARD MEMBERS

Gennady E. Zaikov, DSc

Gennady E. Zaikov, DSc, is Head of the Polymer Division at the N. M. Emanuel Institute of Biochemical Physics, Russian Academy of Sciences, Moscow, Russia, and Professor at Moscow State Academy of Fine Chemical Technology, Russia, as well as Professor at Kazan National Research Technological University, Kazan, Russia. He is also a prolific author, researcher, and lecturer. He has received several awards for his work, including the the Russian Federation Scholarship for Outstanding Scientists. He has been a member of many professional organizations and on the editorial boards of many international science journals.

A. K. Haghi, PhD

A. K. Haghi, PhD, holds a BSc in urban and environmental engineering from University of North Carolina (USA); a MSc in mechanical engineering from North Carolina A&T State University (USA); a DEA in applied mechanics, acoustics and materials from Université de Technologie de Compiègne (France); and a PhD in engineering sciences from Université de Franche-Comté (France). He is the author and editor of 65 books as well as 1000 published papers in various journals and conference proceedings. Dr. Haghi has received several grants, consulted for a number of major corporations, and is a frequent speaker to national and international audiences. Since 1983, he served as a professor at several universities. He is currently Editor-in-Chief of the *International Journal of Chemoinformatics and Chemical Engineering* and *Polymers Research Journal* and on the editorial boards of many international journals. He is also a faculty member of University of Guilan (Iran) and a member of the Canadian Research and Development Center of Sciences and Cultures (CRDCSC), Montreal, Quebec, Canada.

CONTENTS

List of Contributions .... *xi*
List of Abbreviations .... *xv*
List of Symbols .... *xvii*
Preface .... *xix*

1. **A Study on Natural Hybrid Nanocomposites 1**
G. M. Magomedov, K. S. Dibirova, G. V. Kozlov, and G. E. Zaikov

2. **Green Nanofibers—Production and Limits 11**
A. K. Haghi and G. E. Zaikov

3. **Polymeric Nanosystems 105**
A. K. Haghi and G. E. Zaikov

4. **Carbon Nanotubes and Related Structures 147**
A. K. Haghi and G. E. Zaikov

5. **A Detailed Review on Fabrication and Characterization of the Metal Nano-Sized Branched Structures 255**
Guoqiang Xie, Minghui Song, kazuo Furuya, and Akihisa Inoue

6. **New Horizons in Nanotechnology 275**
Anamika Singh, Rajeev Singh, Kozlov Georgii Vladimirovich, and Yanovskii Yurii Grigor'evich

7. **Some New Aspects of Polymeric Nanocomposites 313**
A. K. Mikitaev, A. Y. Bedanokov, and M. A. Mikitaev

8. **Quantum – Chemical Modeling 343**
M. A. Chashkin, V. I. Kodolov, A. I. Zakharov, Yu. M. Vasilchenko, M. A. Vakhrushina, V. V. Trineeva, and G. E. Zaikov

9. **Novel Phase of Elemental Silver Nano-Particles Formation 361**
N. I. Naumkina, O. V. Mikhailov, and T. Z. Lygina

10. **A Study on Polycarbonate Modified with Cu/C Nanocomposite 377**
Yu. V. Pershin and V. I. Kodolov

11. Nanofibers and Solar Cells **385**
V. Mottaghitalab, M. Sajedi, M. S. Motlagh, M. Abbasi, and A. K. Haghi

12. New Generation of Nanomolecular Structures **413**
A. L. Iordanskii, S. Z. Rogovina, I. Afanasov, and A. A. Berlin

13. Influence of Various Metal/Carbon Nanocomposites on Changes in the Properties of Compositions on Liquid Glass Basis **441**
L. F. Akhmetshina, V. I. Kodolov, and G. E. Zaikov

Index **457**

LIST OF CONTRIBUTORS

M. Abbasi
Textile Engineering department, Faculty of Engineering, University of Guilan, Rasht P.O. BOX 3756, Guilan, Iran

I. Afanasov
EFTEK-Bio LTD. 123458, Technopark "Strogino" ul. Tvardovskogo 8, Moscow 123458, Russia

L. F. Akhmetshina
Izhevsk State Technical University, Studencheskaya St. 7, Izhevsk 426069, Russia; OJSC "Izhevsk Electromechanical Plant "Kupol", Pesochnaya St. 3, Izhevsk 426033, Russia, E-mail: techprour@mail.ru

A. Y. Bedanokov
D.I. Mendeleev Russian University for Chemical Technology, Moscow, Russia

A. A. Berlin
N.N. Semenov Institute of chemical Physics, Russian Academy of Sciences, 4 Kosygin str., Moscow 119991, Russia

M. A. Chashkin
Izhevsk State Technical University, Russian Academy of Sciences; Izhevsk Electromechanical Plant, Russian Academy of Sciences; BRHE Centre of Chemical Physics and Mesoscopy, Udmurt Scientific Centre, UD, RAS

K. S. Dibirova
Dagestan State Pedagogical University, Makhachkala 367003, Yaragskii st., 57, Russian Federation

Kazuo Furuya
High Voltage Electron Microscopy Station, National Institute for Materials Science, 3–13 Sakura, Tsukuba 305-0003, Japan

Yanovskii Yurii Grigor'evich
Institute of Applied Mechanics of Russian Academy of Sciences, Leninskii pr., 32 a, Moscow 119991, Russian Federation, E-mail: iam@ipsun.ras.ru

A. K. Haghi
Textile Engineering department, Faculty of Engineering, University of Guilan, Rasht P.O.BOX 3756, Guilan, Iran

Akihisa Inoue
WPI Advanced Institute for Materials Research; Institute for Materials Research, Tohoku University, Sendai 980-8577, Japan

Iordanskii, A. L.
N.N. Semenov Institute of chemical Physics, Russian Academy of Sciences, 4 Kosygin str., Moscow 119991, Russia, Tel. +7-495-939-7434. E mail: aljordan08@gmail.com

V. I. Kodolov
Izhevsk State Technical University, Russian Academy of Sciences; Izhevsk Electromechanical Plant, Russian Academy of Sciences; BRHE Centre of Chemical Physics and Mesoscopy, Udmurt Scientific Centre, UD, RAS

V. I. Kodolov
Studencheskaya St. 7, Izhevsk 426069, Russia; OJSC "Izhevsk Electromechanical Plant "Kupol", Pesochnaya St. 3, Izhevsk 426033, Russia; BRHE Centre of Chemical Physics and Mesoscopy, Udmurt Scientific Centre, UD, RAS

G. V. Kozlov
N.M. Emanuel Institute of Biochemical Physics of Russian Academy of Sciences, Moscow 119334, Kosygin st., 4, Russian Federation

T. Z. Lygina
Central Scientific-Research Institute of Geology Non-Ore Minerals, Zinin Street 4, 420097 Kazan, Russia, E-mail lygina@geolnerud.ru

G. M. Magomedov
Dagestan State Pedagogical University, Makhachkala 367003, Yaragskii st., 57, Russian Federation

O. V. Mikhailov
Kazan National Research Technological University, K. Marx Street 68, 420015 Kazan, Russia, E-mail ovm@kstu.ru

A. K. Mikitaev
Kabardino – Balkarian State University, Nalchik, Russia

M. A. Mikitaev
L.Ya.Karpov Research Institute, Moscow, Russia

M. S. Motlagh
Textile Engineering department, Faculty of Engineering, University of Guilan, Rasht P.O.BOX 3756, Guilan, Iran

V. Mottaghitalab
Textile Engineering department, Faculty of Engineering, University of Guilan, Rasht P.O.BOX 3756, Guilan, Iran

N. I. Naumkina
Central Scientific-Research Institute of Geology Non-Ore Minerals, Zinin Street 4, 420097 Kazan, Russia, E-mail lygina@geolnerud.ru

Yu. V.

Pershin
Izhevsk State Technical University, Russian Academy of Science, Izhevsk, Russia

S. Z. Rogovina
N.N. Semenov Institute of chemical Physics, Russian Academy of Sciences, 4 Kosygin str., Moscow 119991, Russia

M. Sajedi
Textile Engineering department, Faculty of Engineering, University of Guilan, Rasht P.O.BOX 3756, Guilan, Iran

Anamika Singh
Division of Reproductive and Child Health, Indian Council of Medical Research, New Delhi

Rajeev Singh
Division of Reproductive and Child Health, Indian Council of Medical Research, New Delhi

Minghui Song
High Voltage Electron Microscopy Station, National Institute for Materials Science, 3–13 Sakura, Tsukuba 305-0003, Japan

V. V. Trineeva
Izhevsk State Technical University, Russian Academy of Sciences; Izhevsk Electromechanical Plant, Russian Academy of Sciences; BRHE Centre of Chemical Physics and Mesoscopy, Udmurt Scientific Centre, UD, RAS; Institute of Applied Mechanics, Ural Division, Russian Academy of Science

M. A. Vakhrushina
Izhevsk Electromechanical Plant, Russian Academy of Sciences

Yu. M. Vasilchenko
Izhevsk State Technical University, Russian Academy of Sciences; Izhevsk Electromechanical Plant, Russian Academy of Sciences; BRHE Centre of Chemical Physics and Mesoscopy, Udmurt Scientific Centre, UD, RAS

Kozlov Georgii Vladimirovich
Kh.M. Berbekov Kabardino-Balkarian State University, Department of Chemistry, 360004, Nalchik, Chernyshevskogo, 173, Russia, E-mail: I_dolbin@mail.ru

Guoqiang Xie
Institute for Materials Research, Tohoku University, Sendai 980-8577, Japan, Tel.: +81-22-215-2492; E-mail: xiegq@imr.tohoku.ac.jp

G. E. Zaikov
N.M. Emanuel Institute of Biochemical Physics of Russian Academy of Sciences, Moscow 119334, Kosygin st., 4, Russian Federation, E-mail: gezaikov@yahoo.com

A. I. Zakharov
Izhevsk Electromechanical Plant, Russian Academy of Sciences

LIST OF ABBREVIATIONS

ACA	amino caproic acid
ANN	artificial neural network
ANOVA	analysis of variance
BSA	bovine serum albumin
CA	contact angle
CCD	central composite design
CHEC	cold-hardened epoxy composition
CHT	chitosan
CMC	cell membrane complex
CME	clathrin-mediated endocytosis
CNTs	carbon nanotubess
CvME	caveolae-mediated endocytosis
DCM	dichloromethane
DLS	dynamic light scattering
DSSC	dye-sensitized solar cells
EBID	electron-beam-induced deposition
ED	electron diffraction
EDS	energy dispersive spectroscopy
EMD	electron microdiffraction
ER	epoxy resin
ERM	effective reinforcing modulus
EWG	electron-withdrawing groups
HDPE	high-density polyethylene
LMC	low-molecular compound
MFD	mean fiber diameter
MNPs	magnetic nanoparticles
MNSs	magnetically targeted nanosystems
MWCNT	multiwalled carbon nanotube
MWNTs	multiwalled nanotubes
NPT	isothermal–isobaric
NVE	microcanonical
NVT	canonical
PAN	polyacrylonitrile
PCA	polycaproamide
PEO	polyethylene oxide

PEPA	polyethylene polyamine
PVA	polyvinyl alcohol
RDP	radial density profile
RES	reticuloendothelial system
RME	receptor-mediated endocytosis
RMSE	root mean square errors
RSM	response surface methodology
RVE	representative volume element
RVP	radial velocity profile
SAD	selected-area diffraction
SEM	scanning electron microscope
SLN	solid lipid nanoparticles
STM	scanning tunneling
SUSHI	Simulation Utilities for Soft and Hard Interfaces
SWCN	single walled carbon nanotube
SWNTs	single walled nanotubes
TDGL	time-dependent Ginsburg–Landau
TEM	transmission electron microscopy
TFA	triflouroacetic acid
TG	thermogravimetric
μVT	grand canonical

LIST OF SYMBOLS

x_i and x_j	independent variables
a	temperature conductivity coefficient
d	dimension of Euclidean space
D	optical density
F	work angle constant
Y	is the predicted response

GREEK VARIABLES

h	sample thickness
l_0	length of the main chain skeletal bond
M_e	molecular weight of polymer chain
N_A	Avogadro number
R_s	sample heat resistance
S	cross-sectional area of macromolecule
S	sample area
c	heat capacity
η	dynamic viscosity
κ	sphere constant according to the test certificate
λ	wavelength
ν	Poisson's ratio
ρ	density
ρ_1	sphere density according to the test certificate
ρ_2	sample density
r_p	polymer density
s_Y	the temperature dependence of yield stress
τ	sphere movement time
j_{cl}	relative fraction of local order domains (nanoclusters)
χ	a relative fraction of elastically deformed polymer

PREFACE

Functional materials have played a significant part in humans' existence. They have a role in every aspect of modern life, including as health care, food, information technology, transportation, energy industries, etc. The speed of developments within the polymer sector is phenomenal and at same time crucial to meet demands of today's and future life. Specific applications for polymers range from adhesives, coatings, painting, foams and packaging to structural materials, composites, textiles, electronic and optical devices, biomaterials and many other uses in industries and daily life. Functional materials are the basis of natural and synthetic materials. They are macromolecules and, in nature, are the raw material for proteins and nucleic acids, which are essential for human bodies.

Cellulose, wool, natural rubber and synthetic rubber, and plastics are well-known examples of natural and synthetic types. Natural and synthetic polymers play a massive role in everyday life, and a life without functional materials really does not exist. These developments have made polymers such as polyesters contribute enormously to today's modern life. One of the most important applications is the furnishing sector (home, office, cars, aviation industry, etc.), which benefits hugely from the advances in technology. There are a number of requirements for a fabric to function in its chosen end use, for example, resistance to pilling and abrasion, as well as dimensional stability. Polyester is now an important part of the upholstery fabrics. The shortcomings attributed to the fiber in its early days have mostly been overcome. Now it plays a significant part in improving the life-span of a fabric, as well as its dimensional stability, which is due to its heat-setting properties.

About a half century has passed since synthetic leather, a composite material completely different from conventional ones, came to the market. Synthetic leather was originally developed for end-uses such as the upper of shoes. Gradually other uses like clothing steadily increased the production of synthetic leather and suede. Synthetic leathers and suede have a continuous ultrafine porous structure comprising a three-dimensional entangled nonwoven fabric and an elastic material principally made of polyurethane. Polymeric materials consisting of the synthetic leathers are polyamide and polyethylene terephthalate for the fiber and polyurethanes with various soft segments, such as aliphatic polyesters, polyethers and polycarbonates for the matrix.

The introduction of plastics is associated with the twentieth century but the first plastic material, celluloid, was made in 1865. During the 1970s, clothes of polyester became fashionable but by the 1980s synthetics lost the popularity in favor of natural materials. Although people were less enthusiastic about synthetic fabrics for everyday wear, Gore-Tex and other synthetics became popular for outdoor and workout clothing. At the same time as the use of synthetic materials in clothing declined, alternative uses were found. One great example is the use of polyester for making beverage bottles where it replaced glass with its shatterproof properties as a significant property.

In general it can be said that plastics enhance and even preserve life. Kevlar, for instance, when it is used in making canoes for recreation or when used to make a bulletproof vest. Polyester enhances life, such as when this highly nonreactive material is used to make replacement human blood vessels or even replacement skin for burn victims. With all the benefits attributed to plastics, they have their negative side. A genuine environmental problem exists due to the fact that the synthetic polymers do not break down easily compared with the natural polymers, hence the need not only to develop biodegradable plastics, but also to work on more effective means of recycling. A lot of research is needed to study the methods of degradation and stabilization of polymers in order to design polymers according to the end-use.

Among the most important and versatile of the hundreds of commercial plastics is polyethylene. Polyethylene is used in a wide variety of applications because it can be produced in many different forms. The first type to be commercially exploited was called low density polyethylene (LDPE). This polymer is characterised by a large degree of branching, forcing the molecules to pack together rather than loosely forming a low density material. LDPE is soft and pliable and has applications ranging from plastic bags, containers, textiles, and electrical insulation, to coatings for packaging materials.

Another form of polyethylene differing from LDPE in structure is high density polyethylene (HDPE). HDPE demonstrates little or no branching, resulting in the molecules to be tightly packed. HDPE is much more rigid than LDPE and is used in applications where rigidity is important. Major uses of HDPE are plastic tubing, bottles, and bottle caps. Other variations of polyethylene include high and ultra-high molecular mass ones. These types are used in applications where extremely tough and resilient materials are needed.

Natural polymers unlike the synthetic ones do possess very complex structure. Natural polymers such as cellulose, wool, and natural rubber are used in many products in large proportions. Cellulose derivatives are one of the most versatile groups of regenerated materials with various fields of application. Cellulose is found in nature in all forms of plant life, particularly in wood and cotton. The purest form of cellulose is obtained from the seed hairs of the cotton plant which contain up to 95% cellulose. The first cellulose derivatives came to the stage around 1845 when the nitration of starch and paper led to discovery of cellulose nitrate.

In 1865 for the first time a mouldable thermoplastic was made of cellulose nitrate and castor oil.

In 1865 the first acetylation of cellulose was carried out but the first acetylation process for use in industry was announced in 1894. In 1905 an acetylation process was introduced which yielded a cellulose acetate soluble in the cheap solvent, acetone. It was during the First World War when cellulose acetate dope found importance for weather proofing and stiffening the fabric of aircraft wings. There was a large surplus production capacity after the war which led to civilian end uses such as the production of cellulose acetate fibers by 1920s. Cellulose acetate became the main thermoplastic moulding material when the first modern injection moulding machines were designed. Among the cellulose derivatives, cellulose acetates are produced in the largest volume. Cellulose acetate can be made into fibers, transparent films and the less substituted derivatives are true thermoplastics. Cellulose acetates are mouldable and can be fabricated by the conventional processes. They have toughness, good appearance, and are capable of many color variations including white transparency.

New applications are being developed for polymers at a very fast rate all over the world at various research centers. Examples of these include electro active polymers, nano products, robotics, etc. Electro active polymers are special types of materials that can be used for example as artificial muscles and facial parts of robots or even in nano robots. These polymers change their shape when activated by electricity or even by chemicals. They are light weight but can bear a large force that is very useful when being utilized for artificial muscles. Electro active polymers together with nanotubes can produce very strong actuators. Currently research works are carried out to combine various types of electro active polymers with carbon nanotubes to make the optimal actuator. Carbon nanotubes are very strong, elastic, and the conduct electricity. When they are used as an actuator, in combination with an electro active polymer, the contractions of the artificial muscle can be controlled by electricity. Already works are under way to use electro active polymers in space. Various space agencies are investigating the possibility of using these polymers in space. This technology has a lot to offer for the future, and with the ever-increasing work on nanotechnology, electro active materials will play a very important part in modern life.

This book covers a broad range of functional materials and provides industry professionals and researchers in polymer science and technology with a single, comprehensive book summarizing all aspects involved in the functional materials production chain. The book focuses on industrially important materials, analytical techniques, and formulation methods, with chapters covering step-growth, radical, and co-polymerization, crosslinking and grafting, reaction engineering, advanced technology applications, including conjugated, dendritic, and nanomaterial polymers and emulsions, and characterization methods, which includes spectroscopy, light scattering, and microscopy.

The book introduces current state-of-the-art technology in functional materials with an emphasis on the rapidly growing technologies. It takes a unique approach by presenting specific materials and then progresses into a discussion of the ways in which these materials and processes are integrated into today's functioning manufacturing industry. It follows a more quantitative and design-oriented approach than other texts in the market, helping readers gain a better understanding of important concepts. Readers will also discover how material properties relate to the process variables in a given process as well as how to perform quantitative engineering analysis of manufacturing processes.

— Charles Wilkie, PhD, Georges Geuskens, PhD, and Victor Manuel de Matos Lobo, PhD

CHAPTER 1

A STUDY ON NATURAL HYBRID NANOCOMPOSITES

G. M. MAGOMEDOV, K. S. DIBIROVA, G. V. KOZLOV, and G. E. ZAIKOV

CONTENTS

1.1 Introduction 2
1.2 Experimental 2
1.3 Results and Discussion 3
1.4 Conclusions 9
Keywords 9
References 9

1.1 INTRODUCTION

It has been shown that at semicrystalline polymers with devitrificated amorphous phase consideration as natural hybrid nanocomposites their anomalous high reinforcement degree is realized at the expense of crystallites partial recrystallization (mechanical disordering), that means crystalline phase participation in these polymers elastic properties formation. It is obvious, that the proposed mechanism is inapplicable for the description of polymer nanocomposites with inorganic nanofillers.

As it is known [1], semicrystalline polymers, similar to widely applicable polyethylene and polypropylene, at temperatures of the order of room ones have devitrificated amorphous phase. This means that such phase elasticity modulus is small and makes up the value of the order of 10 MPa [2]. At the same time elasticity modulus of the semicrystalline polymers can reach values of ~ 1.0–1.4 GPa and is a comparable one with the corresponding parameter for amorphous glassy polymers. In case of the latters it has been shown [3, 4], that they can be considered as natural nanocomposites, in which local order domains (nanoclusters) serve as nanofiller and as loosely packed matrix of polymer within the framework of the cluster model of polymers amorphous state structure [5] is considered as matrix. In this case elasticity modulus of glassy loosely packed matrix makes up the value of order of 0.8 GPa and a corresponding parameter for polymer (e.g., polycarbonate or polyarylate) ~ 1.6 GPa. In other words, the reinforcement degree of loosely packed matrix by nanoclusters for amorphous glassy polymers is equal to ~ 2, whereas for the semicrystalline polymers it can exceed two orders. By analogy with amorphous [3, 4] and cross-linked [6] polymers semicrystalline polymers can be considered as natural hybrid nanocomposites, in which rubber-like matrix is reinforced by two kinds of nanofiller: nanoclusters (analog of disperse nanofiller with particles size of the order of ~ 1 nm [5]) and crystallites (analog of organoclay with platelets size of the order of ~ 30–50 nm [7]). The clarification of abnormally high reinforcement degree mechanism allows giving an answer to the question, would this mechanism be applicable to polymer nanocomposites, filled with inorganic nanofiller (e.g., organoclay). Therefore, the purpose of the present work is the study of reinforcement mechanism of rubber-like matrix of high-density polyethylene (HDPE) at its consideration as natural hybrid nanocomposite.

1.2 EXPERIMENTAL

The gas-phase HDPE of industrial production of mark HDPE-276, GOST 16338-85 with average weight molecular mass 1.4×10^5 and crystallinity degree 0.723, measured by the sample density, was used.

The testing specimens were prepared by method of casting under pressure on a casting machine Test Samples Molding Apparate RR/TS MP of firm Ray-Ran (Taiwan) at material cylinder temperature 473 K, compression mold temperature 333 K and pressure of blockage 8 MPa.

The impact tests have been performed by using a pendulum impact machine on samples without a notch according to GOST 4746-80, type II, within the testing temperatures range T=213–333 K. Pendulum impact machine was equipped with a piezoelectric load sensor, that allows to determine elasticity modulus E and yield stress σ_Y in impact tests according to the techniques [8] and [9], respectively.

Uniaxial tension mechanical tests have been performed on the samples in the shape of two-sided spade with sizes according to GOST 11262-80. The tests have been conducted on a universal testing apparatus Gotech Testing Machine CT-TCS 2000, production of German Federal Republic, within the testing temperatures range T=293–363 K and strain rate of 2×10^{-3} s^{-1}.

1.3 RESULTS AND DISCUSSION

In Fig. 1, the temperature dependences of elasticity modulus E for the studied HDPE have been adduced. As one can see, at comparable testing temperatures E value in case of quasistatic tests is about twice smaller, than in impact ones. Let us note, that this distinction is not due to tests type. Thus, HDPE with the same crystallinity degree at T=293 K E value can reach 1252 MPa [10]. Let us consider the physical grounds of this discrepancy. The value of fractal dimension d_f of polymer structure, which is its main characteristic, can be determined by several methods application. The first from them uses the following equation [11]:

$$d_f = (d-1)(1+\nu), \tag{1}$$

where d is dimension of Euclidean space, in which a fractal is considered (it is obvious, that in our case d=3), ν is Poisson's ratio, estimated according to the mechanical tests results with the aid of the equation [12]:

$$\frac{\sigma_Y}{E} = \frac{1-2\nu}{6(1+\nu)} \tag{2}$$

The second method assumes the value d_f calculation according to the equation [5]:

$$d_f = d - 6.44\times10^{-10}\left(\frac{\varphi_{cl}}{C_\infty S}\right)^{1/2}, \tag{3}$$

where φ_{cl} is relative fraction of local order domains (nanoclusters), C_{∞} is characteristic ratio, S is cross-sectional area of macromolecule.

For HDPE C_{∞}=7 [13], S=14.4 Å^2 [14] and φ_{cl} value can be calculated according to the following percolation relationship [5]:

$$\varphi_{cl} = 0.03(1-K)(T_m - T)^{0.55}, \tag{4}$$

where K is crystallinity degree, T_m and T are melting and testing temperatures, respectively. For HDPE T_m≈400K [15].

And at last, for semicrystalline polymers d_f value can be evaluated as follows [16]:

$$d_f = 2 + K \tag{5}$$

In Table 1, the comparison of d_f values, determined by the three indicated methods has been adduced (K change with temperature was estimated according to the data of paper [7]). As one can see, if in case of impact tests the calculation according to all three indicated methods gives coordinated results, then for quasistatic tests estimation according to the Eqs. (1) and (2) gives clearly understated d_f values, especially with appreciation of possible variation of this dimension for nonporous solids ($2 \le d_f \le 2.95$ [11]).

TABLE 1 The values of fractal dimension d_f of HDPE structure, calculated by different methods.

Tests type	T, K	d_f, the equation (1)	d_f, the equation (3)	d_f, the equation (5)
Quasistatic	293	2.302	2.800	2.723
	303	2.296	2.796	2.723
	313	2.272	2.802	2.713
	323	2.353	2.801	2.693
	333	2.248	2.799	2.673
	343	2.182	2.799	2.663
	353	2.170	2.800	2.643
	363	2.078	2.808	2.633

TABLE 1 *(Continued)*

Tests type	*T*, K	d_f, the equation (1)	d_f, the equation (3)	d_f, the equation (5)
Impact	213	2.764	2.734	2.723
	233	2.762	2.741	2.723
	253	2.700	2.727	2.723
	273	2.750	2.756	2.723
	293	2.680	2.729	2.723
	313	2.624	2.743	2.713
	333	2.646	2.766	2.763

Let us consider the causes of the indicated discrepancy in more details. At present it has been assumed [1], that in case of semicrystalline polymers with devitrificated amorphous phase deformation in elasticity region, i.e., at E value determination, the indicated amorphous phase is only deformed, that defines smaller values of both E and d_f. This conclusion is confirmed by disparity between d_f values, calculated on the basis of mechanical characteristics (the Eqs. (1) and (2)) and crystallinity degree (the Eq. (5)). And on the contrary, a good correspondence of d_f values, obtained by the three indicated methods (Table 1), assumes crystalline phase participance at HDPE deformation in elasticity region in case of impact tests (Fig. 1).

In Fig. 2, the temperature dependence of yield stress σ_Y has been adduced for HDPE in case of both types of tests. As one can see, the data for quasistatic and impact tests are described by the same curve. As it has been shown in work [5], σ_Y values are defined by contribution of both nanoclusters, i.e., amorphous phase, and crystallites. Combined consideration of the plots of Figs. 1 and 2 demonstrates, that the distinction of d_f values, determined according to the Eqs. (1), (2) and (5), is only due to the above indicated structural distinction of HDPE deformation in an elasticity region.

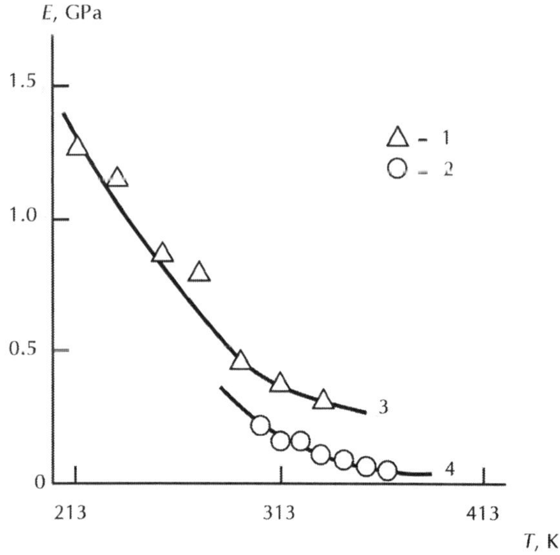

FIGURE 1 The dependences of elasticity modulus E on testing temperature T for HDPE, obtained in impact (1, 3) and quasistatic (2, 4) tests. 1, 2 – the experimental data; 3, 4 – calculation according to the Eq. (12).

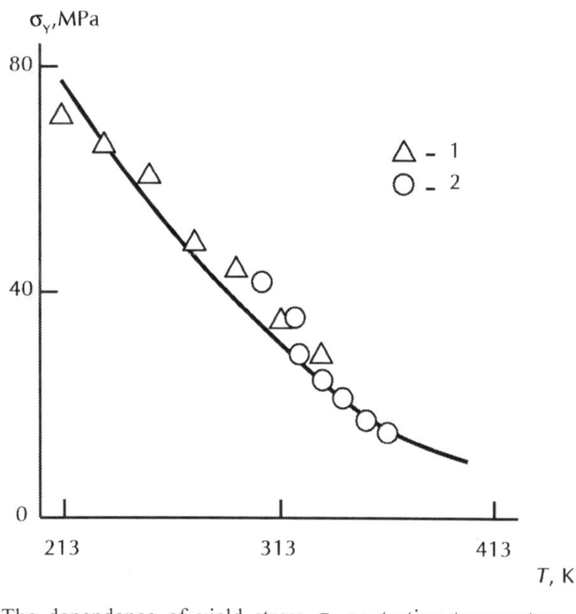

FIGURE 2 The dependence of yield stress σ_Y on testing temperature T for HDPE, obtained in impact (1) and quasistatic (2) tests.

The quantitative evaluation of crystalline regions contribution in HDPE elasticity can be performed within the framework of yield fractal conception [17], according to which the value of Poisson's ration in yield point ν_Y can be estimated as follows:

$$v_y = v\chi + 0.5(1-\chi), \tag{6}$$

where ν is Poisson's ratio in elastic strains region, determining according to the Eq. (2), χ is a relative fraction of elastically deformed polymer.

For amorphous glassy polymers it has been shown that χ value is equal to a relative fraction of loosely packed matrix $\varphi_{l.m.}$. In case of semicrystalline polymers in deformation process partial recrystallization (mechanical disordering) of a crystallites part can be realized, the relative fraction of which χ_{cr} is determined by the following equation [5]:

$$v_y = v\chi + 0.5(1-\chi) \tag{7}$$

If to consider HDPE as natural hybrid nanocomposite, then amorphous phase (the indicated nanocomposite matrix) elasticity modulus E_{am} can be determined within the framework of high-elasticity conception, using the known equation [2]:

$$G_{am} = kNT, \tag{8}$$

where G_{am} is a shear modulus of amorphous phase, k is Boltzmann constant, N is a number of active chains of polymer network.

As it is known [5], in amorphous phase two types of macromolecular entanglements are present: traditional macromolecular "binary hookings" and entanglements, formed by nanoclusters, networks density of which is equal to ν_e and ν_{cl}, respectively. ν_e value is determined within the framework of rubber high-elasticity conception [2]:

$$\nu_e = \frac{\rho_p N_A}{M_e}, \tag{9}$$

where ρ_p is polymer density, N_A is Avogadro number, M_e is molecular weight of polymer chain part between macromolecular "binary hookings," which is equal to 1390 [18] and ρ_p value can be accepted equal to 960 kg/m^3 [15].

In its turn, ν_{cl} value can be determined according to the following equation [5]:

$$G_{am} = kNT, \tag{10}$$

where l_0 is length of the main chain skeletal bond, for HDPE equal to 1.54 Å [13].

E_{am} and G_{am} are connected by a simple fractal formula [11]:

$$E_{am} = d_f G_{am} \tag{11}$$

Calculation according to the Eqs. (6) and (7) has shown that in case of HDPE quasistatic tests χ_{cr} value is close to zero and in case of impact tests χ_{cr}=0.400–0.146 within the range of T=231–333 K. Now reinforcement degree of HDPE, considered as hybrid nanocomposite, can be expressed as the ratio E/E_{am}. In Fig. 3, the dependence $E/E_{am}(\chi_{cr}^2)$ has been adduced (such form of dependence was chosen for its linearization). As one can see, this linear dependence, demonstrates E/E_{am} (or E) increase at χ_{cr} growth and is described analytically by the following relationship:

$$\frac{E}{E_{am}} = 590\chi_{cr}^2 \text{, MPa.} \tag{12}$$

The comparison of experimental E and calculated according to the Eq. (12) E^T elasticity modulus values for the studied HDPE has been adduced in Fig. 1. As one can see, the good correspondence between theory and experiment is obtained (the average discrepancy between E and E^T does not exceed 6%, i.e., comparable with an error of elasticity modulus experimental determination).

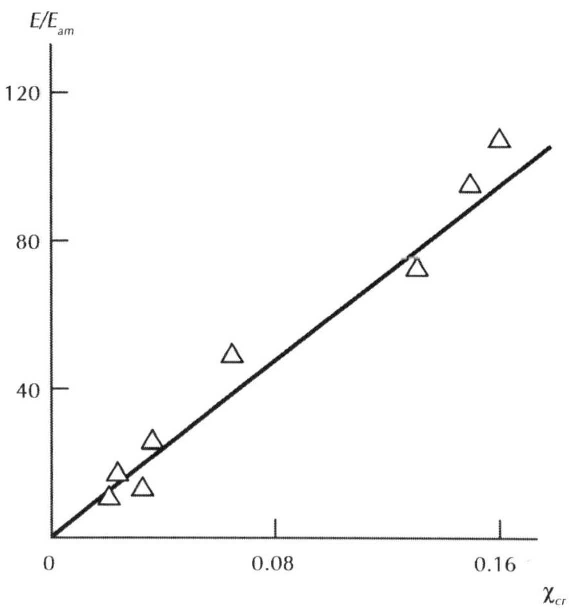

FIGURE 3 The dependence of reinforcement degree E/E_{am} on relative fraction of crystalline phase χ_{cr}, subjecting to partial recrystallization, for HDPE in impact tests

Thus, the performed estimations demonstrated, that high values of reinforcement degree E/E_{am} for semicrystalline polymers, considered as hybrid nanocomposites (in case of studied HDPE E/E_{am} value is varied within the limits of 10–110) were due to recrystallization process (mechanical disordering of crystallites) in elastic deformation process and, as consequence, to contribution of crystalline regions in polymers elastic properties formation. It is obvious, that this mechanism does not work in case of inorganic nanofiller (e.g., organoclay). Besides, a nanofiller (crystallites) is formed spontaneously in a polymer crystallization process, that automatically cancels the problem of its dispersion at large K of the order of 70 mass %, whereas to obtain exfoliated organoclay at contents larger than 3 mass % is difficult [19]. Thus, it is now clear that nanocomposites polymer/organoclay maximum reinforcement degree does not exceed 4 [20].

1.4 CONCLUSIONS

The performed analysis has shown that at the consideration of semicrystalline polymers with devitrificated amorphous phase as natural hybrid nanocomposites their abnormally high reinforcement degree is realized at the expense of crystallites partial recrystallization (mechanical disordering), that means crystalline phase participation in the formation of these polymers elastic properties. It is obvious, that the proposed mechanism is inapplicable for the description of reinforcement of polymer nanocomposites with inorganic nanofiller.

KEYWORDS

- **binary hookings**
- **high-density polyethylene**
- **nanoclusters**
- **semicrystalline polymers**

REFERENCES

1. Narisawa, I.; *Strength of Polymeric Materials*. Chemistry Publishing House (in rus.): Moscow, **1987,** p. 400.

2. Bartenev, G. M.; Frenkel, S. Ya. *Physics of Polymers*. Chemistry Publishing House (in rus.): Moscow, **1990,** p. 432.
3. Kozlov, G. V.; *Recent Patents on Chemical Engineering*, **2011,** *4(1)*, 53–77.
4. Kozlov, G. V.; Mikitaev, A. K.; *Polymers as Natural Nanocomposites: Unrealized Potential*. Lambert Academic Publishing: Saarbrücken, **2010,** p. 323.
5. Kozlov, G. V.; Ovcharenko, E. N.; Mikitaev, A. K.; *Structure of Polymer Amorphous State* (in rus.). RKhTU Publishing House: Moscow, **2009,** p. 392.
6. Magomedov, G. M.; Kozlov, G. V.; Zaikov, G. E.; *Structure and Properties of Cross-Linked Polymers*. A Smithers Group Company: Shawbury, **2011,** p. 492.
7. Tanabe, Y.; Strobl, G. R.; Fisher, E. W.; *Polymer*, **1986,** *27(8)*, 1147–1153.
8. Kozlov, G. V.; Shetov, R. A.; Mikitaev, A. K.; *Russian Polymer Science* (in rus.), **1987,** *29(5),* 1109–1110.
9. Kozlov, G. V.; Shetov, R. A.; Mikitaev, A. K.; *Russian Polymer Science* (in rus.), **1987,** *29(9),* 2012–2013.
10. Pegoretti, A.; Dorigato, A.; Penati, A. *EXPRESS Polymer Lett.;* **2007,** *1(3)*, 123–131.
11. Balankin, A. S.; *Synergetics of Deformable Body*. Ministry of Defence SSSR Publishing House: Moscow, **1991,** p. 404.
12. Kozlov, G. V.; Sanditov, D. S.; *Anharmonic Effects and Physical-Mechanical Properties of Polymers*. Nauka (Science) Publishing House (in rus.): Novosibirsk, **1994,** p. 264.
13. Aharoni, S. M.; *Macromolecules*, **1983,** *16(9),* 1722–1728.
14. Aharoni, S. M.; *Macromolecules*, **1985,** *18(12),* 2624–2630.
15. Kalinchev, E. L.; Sakovtseva, M. B.; *Properties and Processing of Thermoplastics* (in rus.). Chemistry Publishing House: Leningrad, **1983,** p. 288.
16. Aloev, V. Z.; Kozlov, G. V.; *Physics of Orientation Phenomena in Polymeric Materials* (in rus.). Polygraphservice and, T. Nal'chik, **2002,** p. 288.
17. Balankin, A. S.; Bugrimov, A. L.; *Russian Polymer Science* (in rus.), **1992,** *34(5),* 129–132.
18. Graessley, W. W.; Edwards, S. F.; *Polymer*, **1981,** *22(10),* 1329–1334.
19. Miktaev, A. K.; Kozlov, G. V.; Zaikov, G. E.; *Polymer Nanocomposites: Variety of Structural Forms and Applications*. Nova Science Publishers, Inc.: New York, **2008,** p. 319.
20. Liang, Z.-M.; Yin, J.; Wu, J.-H.; Qiu, Z.-X.; He, F.-F.; *Europ. Polymer J.* **2004,** *40(2),* 307–314.

CHAPTER 2

GREEN NANOFIBERS—PRODUCTION AND LIMITS

A. K. HAGHI and G. E. ZAIKOV

CONTENTS

2.1 Introduction 13
2.2 Experimental Part 15
2.3 Preparation of CHT-MWNTS Dispersions 16
2.4 Electrospinning Process 16
2.5 Measurements and Characterizations Method 17
2.6 Results 18
2.7 Silk Fibers 24
    2.7.1 Preparation of Regenerated SF Solution 26
    2.7.2 Preparation of the Spinning Solution 26
    2.7.3 Electrospinning 27
    2.7.4 Characterization 27
    2.7.5 Results and Discussion 28
2.8 Modeling 38
2.9 Concluding Remarks 40
Keywords 42
References 42
Appendix A 43
A1 Bast Fiber Biodamading 54
A2 Biodamaging of Artificial Fibers 57
A3 Wool Fiber Biodamaging 58
A4 Changes of Structure and Properties of Wool Fibers by Microorganisms 62

A5 Biodamages of Synthetic Fibers 69
A6 Changes in Structure and Properties of Polyamide Fibers Induced by Microorganisms 71
A7 Methods of Textile Material Protection Against Damaging by Microorganisms 75
Appendix B 79
Appendix C 81
C1 Measurement and Characterization 81
C2 Response Surface Methodology 82
C3 Artificial Neural Network 83
C4 The Analysis of Variance (Anova) 84
C5 Artificial Neural Network 87
C6 Effects of Significant Parameters on Response 88
C7 Optimizing The CA of Electrospun Fiber Mat 91
C8 Comparison Between RSM and ANN Model 91
Appendix D 94

2.1 INTRODUCTION

Over the recent decades, the fabrication of biodegradable nanofibers for many biomedical applications, such as tissue engineering, drug delivery, wound dressing, enzyme immobilization, and so forth, has been recorded. Nanofiber fabrics have unique characteristics, such as very large surface area, ease of functionalization for various purposes and superior mechanical properties. Electrospinning (Fig. 1) is an important technique that can be used for the production of polymer nanofibers with diameters from several micrometers down to tens of nanometers. In electrospinning, the charged jets of a polymer solution, collected on a target, are created by using an electrostatic force. Many parameters can influence the quality of fibers, including the solution properties (polymer concentration, solvent volatility and solution conductivity), the governing variables (flow rate, voltage and tip-to-collector distance), and the ambient parameters (humidity, solution temperature, air velocity in the electrospinning chamber).

In recent years, scientists have manifested increased interest in electrospinning of natural materials, such as collagen, fibrogen, gelatin, silk, chitin and chitosan, due to their high biocompatible and biodegradable properties. Chitin is the second most abundant natural polymer in the world and chitosan (poly (1-4)-2-amino-2-deoxy-β-D-glucose) is the deacetylated product of chitin[1-12].

Researchers are interested in this natural polymer because of its properties, including solid-state structure and chain conformations in dissolved state.

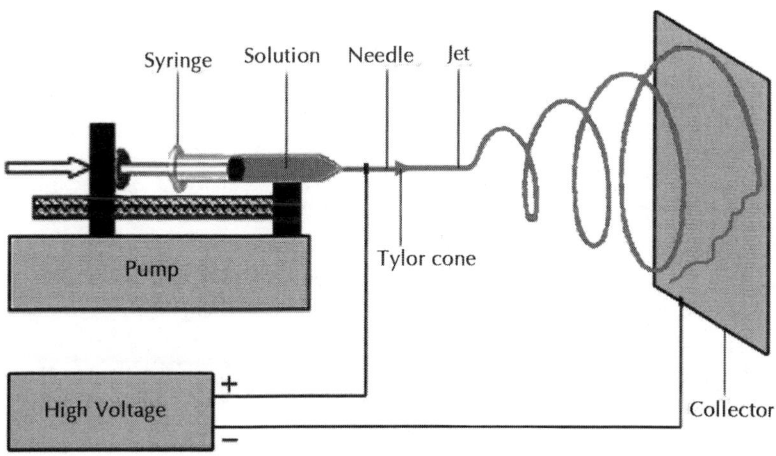

FIGURE 1 A typical image of Electrospinning process.

The physical characteristics of electrospun nanofibers such as fiber diameter depend on various parameters that are mainly divided into three categories: solution properties (solution viscosity, solution concentration, polymer molecular weight, and surface tension), processing conditions (applied voltage, volume flow rate, spinning distance, and needle diameter), and ambient conditions (temperature, humidity, and atmosphere pressure). Numerous applications require nanofibers with desired properties suggesting the importance of the process control. It does not come true unless having a comprehensive outlook of the process and quantitative study of the effects of governing parameters.

Besides physical characteristics, medical scientists showed a remarkable attention to biocompatibility and biodegradability of nanofibers made of biopolymers such as collagen, fibrogen, gelatin, silk, chitin and chitosan. Chitin is the second abundant natural polymer in the world and Chitosan (poly(1–4)-2-amino-2-deoxy-β-D-glucose) is the deacetylated product of chitin. CHT is well known for its biocompatible and biodegradable properties.

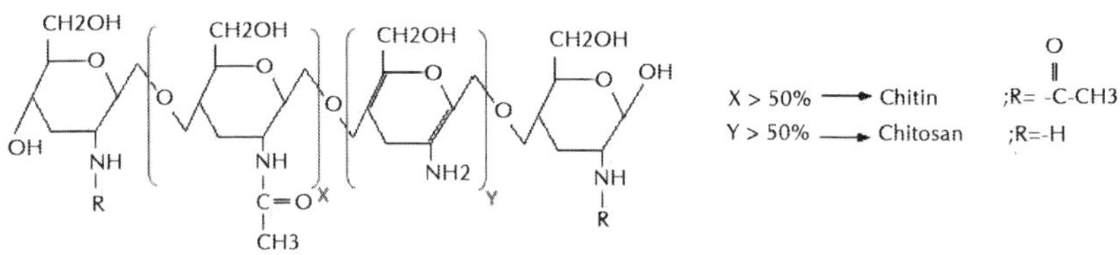

SCHEME 1 Chemical structures of Chitin and Chitosan biopolymers.

Chitosan (CHT) is insoluble in water, alkali, and most mineral acidic systems. However, though its solubility in inorganic acids is quite limited, CHT is in fact soluble in organic acids, such as dilute aqueous acetic, formic, and lactic acids. CHT also has free amino groups, which make it a positively charged polyelectrolyte. This property makes CHT solutions highly viscous and complicates its electrospinning. Furthermore, the formation of strong hydrogen bonds in a 3-D network prevents the movement of polymeric chains exposed to the electrical field.

Different strategies were used for bringing CHT in nanofiber form. The three top most abundant techniques include blending of favorite polymers for electrospinning process with CHT matrix, alkali treatment of CHT backbone to improve electro spinnability through reducing viscosity and employment of concentrated organic acid solution to produce nanofibers by decreasing of surface tension. Electrospinning of polyethylene oxide (PEO)/CHT and polyvinyl alcohol (PVA)/CHT blended nanofiber are two recent studies based on first strategy. In the second protocol, the molecular weight of CHT decreases through alkali treatment.

Solutions of the treated CHT in aqueous 70–90% acetic acid have been employed to produce nanofibers with appropriate quality and processing stability.

Using concentrated organic acids such as acetic acid and triflouroacetic acid (TFA) with and without dichloromethane (DCM) has been reported exclusively for producing neat CHT nanofibers. They similarly reported the decreasing of surface tension and at the same time enhancement of charge density of CHT solution without significant effect on viscosity. This new method suggests significant influence of the concentrated acid solution on the reducing of the applied field required for electrospinning.

The mechanical and electrical properties of neat CHT electrospun natural nanofiber mat can be improved by addition of the synthetic materials including carbon nanotubes (CNTs). CNTs are one of the key synthetic polymers that were discovered in 1991. CNTs either single walled nanotubes (SWNTs) or multiwalled nanotubes (MWNTs) combine the physical properties of diamond and graphite. They are extremely thermally conductive like diamond and appreciably electrically conductive like graphite. Moreover, the flexibility and exceptional specific surface area to mass ratio can be considered as significant properties of CNTs. The scientists are becoming more interested to CNTs for existence of exclusive properties such as superb conductivity and mechanical strength for various applications. To best of our knowledge, there has been no report on electrospinning of CHT/MWNTs blend, except those ones that use PVA to improve spinnability. Results showed uniform and porous morphology of the electrospun nanofibers. Despite adequate spinnability, total removing of PVA from nanofiber structure to form conductive substrate is not feasible. Moreover, thermal or alkali solution treatment of CHT/PVA/MWNTs nanofibers extremely influence on the structural morphology and mechanical stiffness. The CHT/CNT composite can be produced by the hydrogen bonds due to hydrophilic positively charged polycation of CHT due to amino groups and hydrophobic negatively charged of CNT due to carboxyl and hydroxyl groups.

2.2 EXPERIMENTAL PART

CHT with degree of deacetylation of 85% and molecular weight of 5×10^5 was supplied by Sigma-Aldrich. The MWNTs, supplied by Nutrino, have an average diameter of four nm and the purity of about 98%. All of the other solvents and chemicals were commercially available and used as received without further purification.

2.3 PREPARATION OF CHT-MWNTS DISPERSIONS

A Branson Sonifier 250 operated at 30W used to prepare the MWNTs dispersions in CHT /organic acid (90%w/w acetic acid, 70/30 TFA/DCM) solution based on different protocols. In first approach, 3 mg of as received MWNTs was dispersed into deionized water or DCM using solution sonicating for 10 min (current chapter, sample 1). Different amount of CHT was then added to MWNTs dispersion for preparation of a 8–12 w/w % solution and then sonicated for another 5 min. Figure 2 shows two different protocols used in this study.

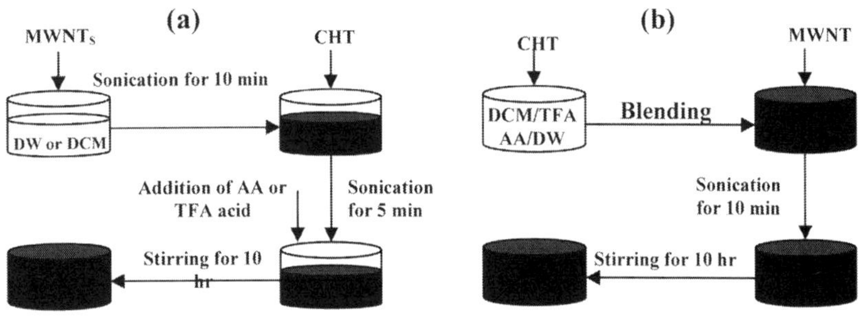

FIGURE 2 Two protocols used in this study for preparation of MWNTs/CHT dispersion (a) Current study (b) (AA/DW abbreviated for acetic acid diluted in de-mineralized water).

In next step, the organic acid solution added to obtain a CHT/MWNT solution with total volume of 5 mL and finally the dispersion was stirred for another 10 h. The sample 2 was prepared using the second technique. Same amount of MWNTs were dispersed in CHT solution, and the blend with total volume of 5 mL were sonicated for 10 min and dispersion was stirred for 10 h.

2.4 ELECTROSPINNING PROCESS

After the preparation of spinning solution, it transferred to a 5 mL syringe and became ready for spinning of nanofibers. The experiments were carried out on a horizontal electrospinning setup shown schematically in Fig. 1. The syringe containing CHT/MWNTs solution was placed on a syringe pump (New Era NE-100) used to dispense the solution at a controlled rate. A high voltage DC power supply (Gamma High Voltage ES-30) employed to generate the required electric field for electrospinning. The positive electrode and the grounding electrode of the high voltage supplier attaches respectively to the syringe needle and flat collector wrapped with aluminum foil where electrospun nanofibers accumulates via an alligator clip to form a nonwoven mat. The voltage and the tip-to-collector distance fixed respectively on 18–24 kV and 4–10 cm. In addition, the electrospinning carried out at room temperature and the aluminum foil removed from the collector.

2.5 MEASUREMENTS AND CHARACTERIZATIONS METHOD

A small piece of mat placed on the sample holder and gold sputter-coated (Bal-Tec). Thereafter, the micrograph of electrospun CHT/MWNTs nanofibers were obtained using scanning electron microscope (SEM, Phillips XL-30). Fourier transform infrared spectra (FT-IR) recorded using a Nicolet 560 spectrometer to investigate the interaction between CHT and MWNT in the range of 800–4000 cm^{-1} under a transmission mode. The size distribution of the dispersed solution was evaluated using dynamic light scattering technique (Zetasizer, Malvern Instruments). The conductivity of nanofiber samples was measured using a home-made four-probe electrical conductivity cell operating at constant humidity. The electrodes were circular pins with separation distance of 0.33 cm and fibers connect to pins by silver paint (SPI). Between the two outer electrodes, a constant DC current applies by Potentiostat/Galvanostat model 363 (Princeton Applied Research). The generated potential difference between the inner electrodes and the current flow between the outer electrodes recorded by digital multimeter 34401A (Agilent). Figure 3 illustrates the experimental setup for conductivity measurement. The conductivity (δ: S/cm) of the nanofiber with rectangular surface can then be calculated according to Eq. (1), which parameters call for length (L:cm), width (W:cm), thickness (t:cm), DC current applied (mA)and the potential drop across the two inner electrodes (mV). All measuring repeated at least five times for each set of samples.

$$\delta = \frac{I \times L}{V \times W \times t} \quad (1)$$

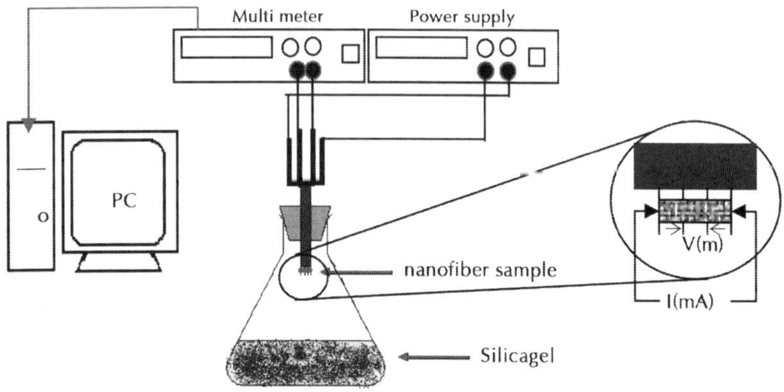

FIGURE 3 The experimental setup for four-probe electrical conductivity measurement of nanofiber thin film.

2.6 RESULTS

Utilization of MWNTs in biopolymer matrix initially requires their homogenous dispersion in a solvent or polymer matrix. Dynamic light scattering (DLS) is a sophisticated technique used for evaluation of particle size distribution. DLS provides many advantages for particle size analysis to measures a large population of particles in a very short time period, with no manipulation of the surrounding medium. DLS of MWNTs dispersions indicate that the hydrodynamic diameter of the nanotube bundles is between 150 and 400 nm after 10 min of sonication for sample 2 (Fig. 4).

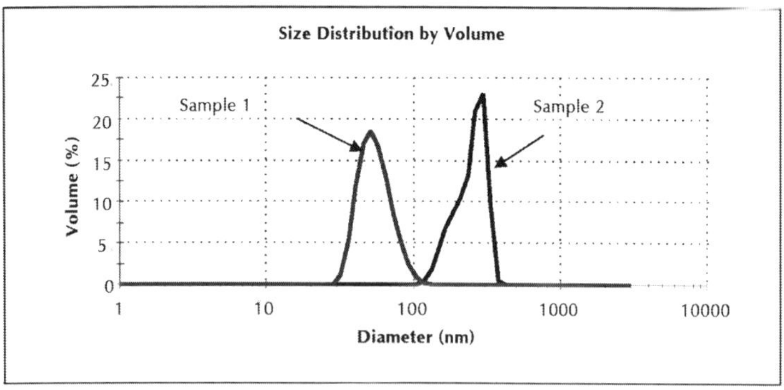

FIGURE 4 Hydrodynamic diameter distributions of MWNT bundles in CHT/acetic acid (1%) solution for different preparation technique

MWNTs bundle in sample 1(different approach but same sonication time compared to sample 2) shows a range of hydrodynamic diameter between 20–100 nm (Fig. 4). The lower range of hydrodynamic diameter for sample 1 correlates to more exfoliated and highly stable nanotubes strands in CHT solution. The higher stability of sample 1 compared to sample 2 over a long period of time was confirmed by solution stability test. The results presented in Fig. 5 indicate that procedure employed for preparation of sample 1 (current chapter) was an effective method for dispersing MWNTs in CHT/acetic acid solution. However, MWNTs bundles in sample 2 showed reagglomeration upon standing after sonication.

Despite the method reported in Ref. [35] neither sedimentation nor aggregation of the MWNTs bundles observed in first sample. Presumably, this behavior in sample 1 can be attributed to contribution of CHT biopolymer to forms an effective barrier against reagglomeration of MWNTs nanoparticles. In fact, using sonication energy, in first step without presence of solvent, make very tiny exfoliated but unstable particle in water as dispersant. Instantaneous addition of acetic acid as solvent and long mixing most likely helps the wrapping of MWNTs strands with CHT polymer chain[13-22].

Figure 6 shows the FT-IR spectra of neat CHT solution and CHT/MWNTs dispersions prepared using strategies explained in experimental part. The interaction between the functional group associated with MWNTs and CHT in dispersed form has been understood through recognition of functional groups. The enhanced peaks at ~1600 cm^{-1} can be attributed to (N-H) band and (C=O) band of amide functional group. However, the intensity of amide group for CHT/MWNTs dispersion increases presumably due to contribution of G band in MWNTs. More interestingly, in this region, the FT-IR spectra of MWNTs-CHT dispersion (sample 1) have been highly intensified compared to sample 2. It correlates to higher chemical interaction between acid functionalized C-C group of MWNTs and amide functional group in CHT.

This probably is the main reason of the higher stability and lower MWNTs dimension demonstrated in Figs. 4 and 5. Moreover, the intensity of protonated secondary amine absorbance at 2400 cm^{-1} for sample 1 prepared by new technique is negligible compared to sample 2 and neat CHT. Furthermore, the peak at 2123 cm^{-1} is a characteristic band of the primary amine salt, which is associated to the interaction between positively charged hydrogen of acetic acid and amino residues of CHT. In addition, the broad peaks at ~3410 cm^{-1} due to the stretching vibration of OH group superimposed on NH stretching bond and broaden due to hydrogen bonds of polysaccharides. The broadest peak of hydrogen bonds observed at 3137–3588 cm^{-1} for MWNTs/CHT dispersion prepared by new technique (sample1).

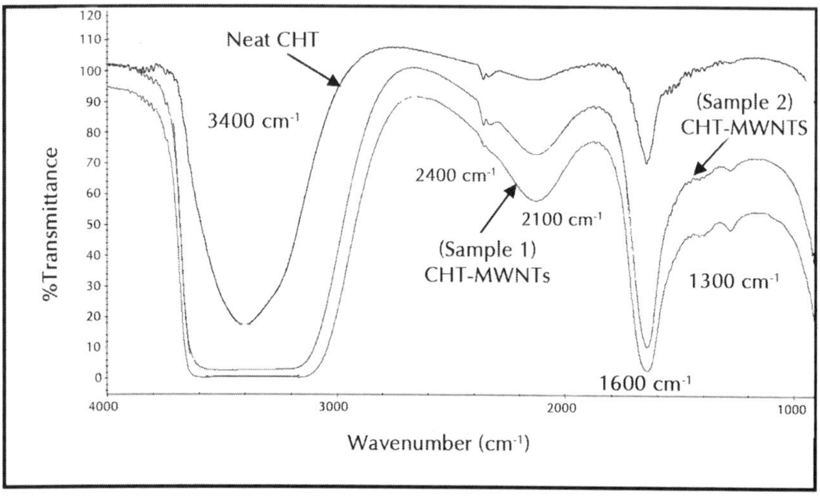

FIGURE 5 FT-IR spectra of CHT-MWCNT in 1% acetic acid with different techniques of dispersion

The different solvents including acetic acid 1–90%, pure formic acid, and TFA/DCM tested for preparation of spinning solution-using protocol explained for sample 1. Upon applying of the high voltage even more than 25 kV, no polymer jet forms using of acetic acid 1–30% and formic acid as the solvent for CHT/MWNT. However, experimental observation shows bead formation when the acetic acid (30–90%) used as the solvent. Therefore, one does not expect the formation of electrospun fiber of CHT/MWNTs using prescribed solvents (data not shown).

Figure 6 shows scanning electronic micrographs of the CHT/MWNTs electrospun nanofibers in different concentration of CHT in TFA/DCM (70:30) solvent. As presented in Fig. 6a, at low concentrations of CHT the beads deposited on the collector and thin fibers coexited among the beads. When the concentration of CHT increases as shown in Figs. 6a–c the bead density decreases. Figure 6c show the homogenous electrospun nanofibers with minimum beads, thin and interconnected fibers. More increasing of concentration of CHT lead to increasing of interconnected fibers at Figs. 6d–e. Figure 7 shows the effect of concentration on average diameter of MWNTs/CHT electrospun nanofibers. Our assessments indicate that the fiber diameter of CHT/MWNTs increases with the increasing of the CHT concentration. Hence, CHT/MWNTs (10 wt%) solution in TFA/DCM (70:30) considered as resulted in optimized concentration. An average diameter of 275 nm (Fig. 6c: diameter distribution, 148–385) investigated for these conditions.

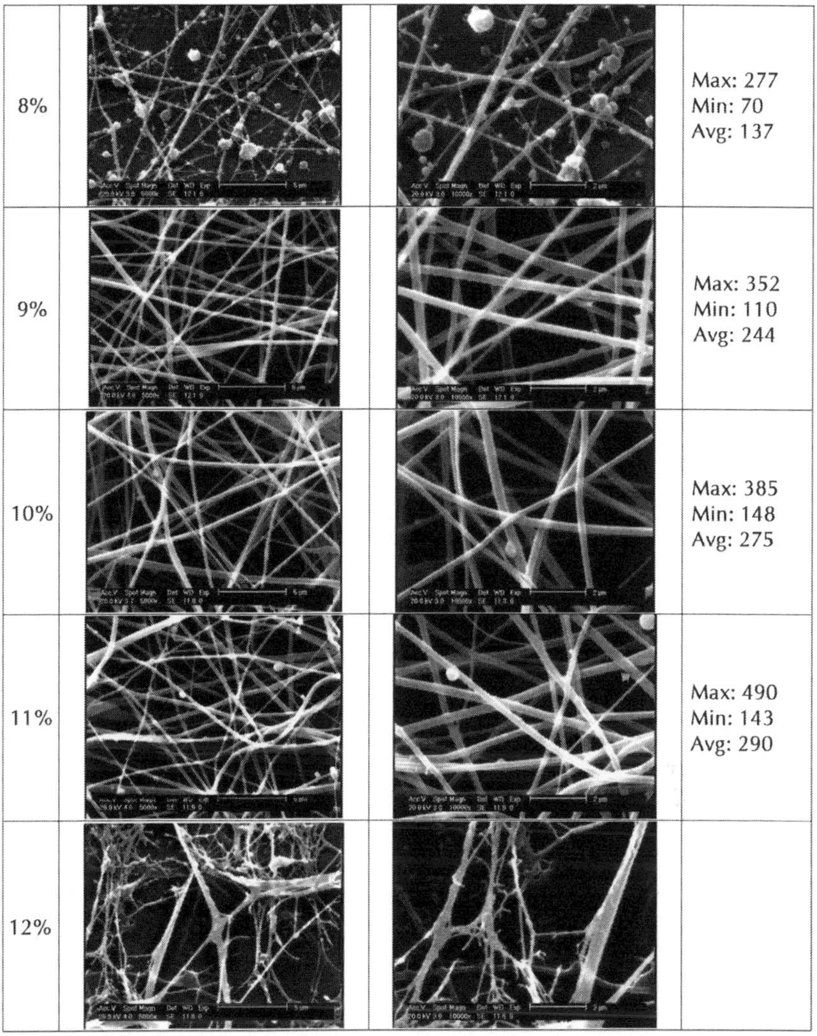

FIGURE 6 Scanning electron micrographs of electrospun nanofibers at different CHT concentration (wt %): (a) 8, (b) 9, (c) 10, (d) 11, (e) 12, 24 kV, 5 cm, TFA/DCM: 70/30, (0.06%wt MWNTs).

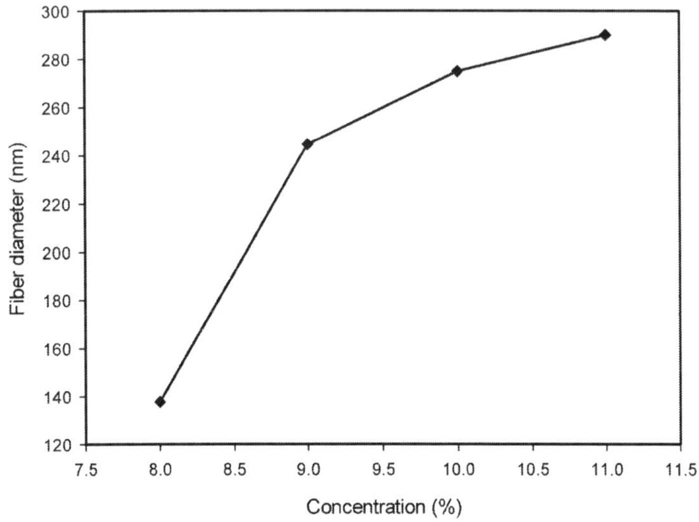

FIGURE 7 The effect of the CHT concentration in CHT/MWNT dispersion on nanofiber diameter.

Figure 7 shows the effect of the CHT concentration in CHT/MWNT dispersion on nanofiber diameter. Figure 8 shows the SEM image of CHT/MWNTs electrospun nanofibers produced in different voltage. In our experiments, 18 kV attains as threshold voltage, where fiber formation occurs.

For lower voltage, the beads and some little fiber deposited on collector (Fig. 8a). As shown in Figs. 8a–d, the beads decrease while voltage increasing from 18 kV to 24 kV. The collected nanofibers by applying 18 kV (9a) and 20 kV (9b) were not quite clear and uniform. The higher the applied voltage, the uniform nanofibers with narrow distribution starts to form. The average diameter of fibers, 22 kV (9c), and 24 kV (9d), respectively, were 204 (79–391), and 275 (148–385). The conductivity measurement given in Table 2 confirms our observation in first set of conductivity data. As can be seen from last row, the amount of electrical conductivity reaches to a maximum level of 9×10^{-5} S/cm at prescribed setup.

The distance between tips to collector is another parameter that controls the fiber diameter and morphology. Fig. 9 shows the change in morphologies of CHT/MWNTs electrospun nanofibers at different distance. When the distance is not long enough, the solvent could not find opportunity for separation. Hence, the interconnected thick nanofiber deposits on the collector (Fig. 9a). However, the adjusting of the distance on 5 cm (Fig. 9b) leads to homogenous nanofibers with negligible beads and interconnected areas. However, the beads increases by increasing of distance of the tip-to-collector as represented from Fig. 9b to Fig.

9f. Also, the results show that the diameter of electrospun fibers decreases by increasing of distance tip to collector in Figs. 9b–d, respectively, 275 (148–385), 170 (98–283), 132 (71–224). A remarkable defects and nonhomogeneity appears for those fibers prepared at a distance of 8 cm (Fig. 9e) and 10 cm (Fig. 10f). However, a 5 cm distance selected as proper amount for CHT/MWNTs electrospinning process.

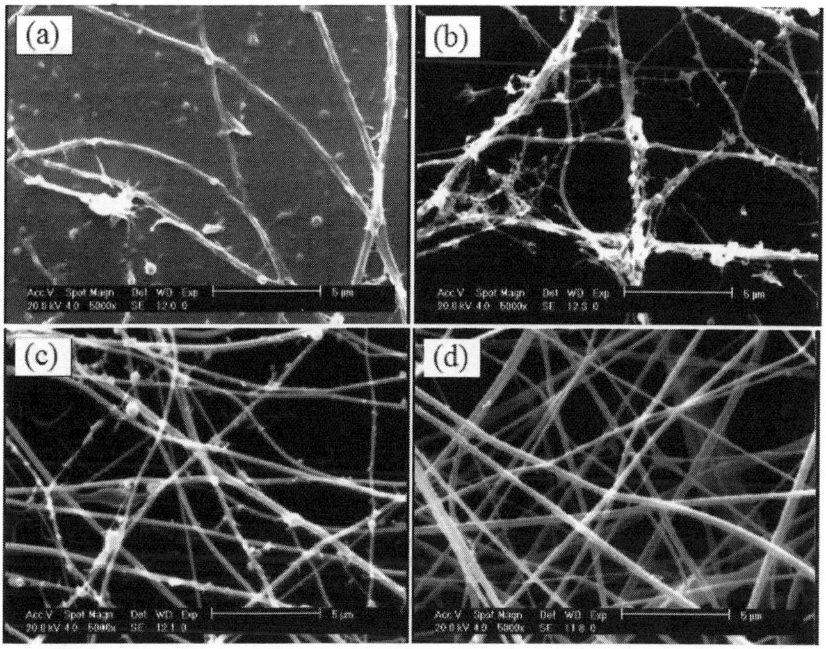

FIGURE 8 Scanning electronic micrographs of electrospun fibers at different voltage (kV): (a) 18, (b) 20, (c) 22, (d) 24, 5 cm, 10 wt%, TFA/DCM: 70/30. (0.06 wt% MWNTs).

The nonhomogeneity and huge bead densities plays as a barrier against electrical current and still a bead free and thin nanofiber mat shows higher conductivity compared to other samples. Experimental framework in this study was based on parameter adjusting for electrospinning of conductive CHT/MWNTs nanofiber. It can be expected that, the addition of nanotubes can boost conductivity and change morphological aspects, which is extremely important for biomedical applications.

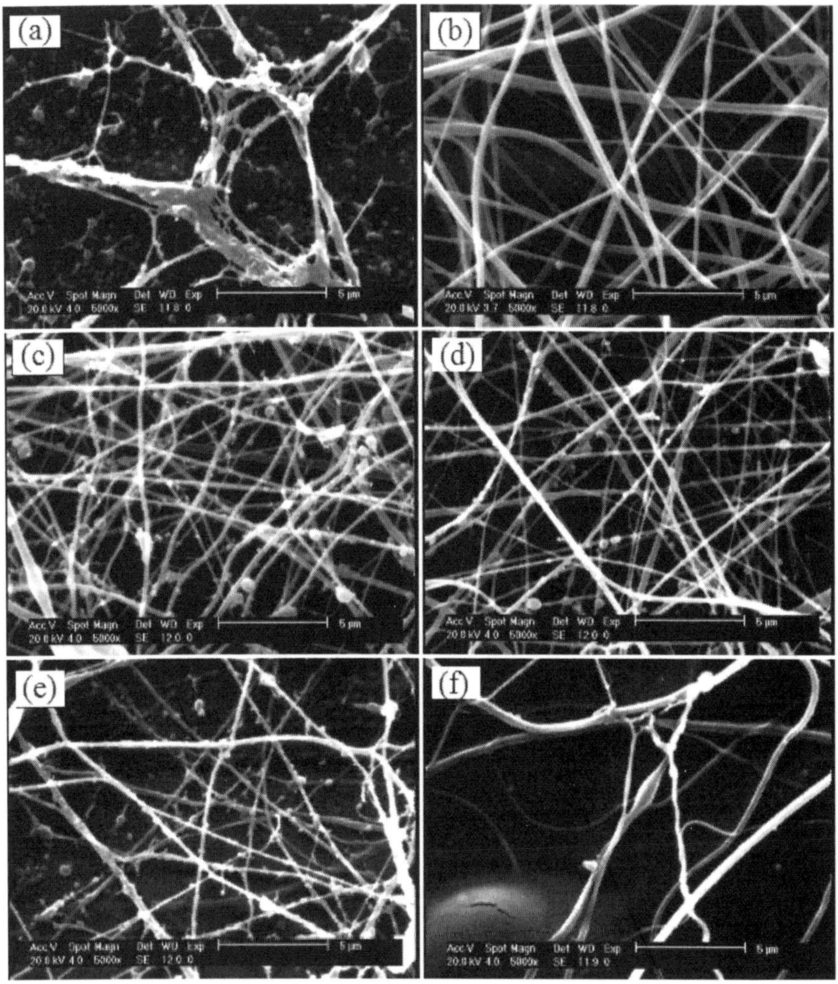

FIGURE 9 Scanning electronic micrographs of electrospun fibers of CHT/MWNT at different tip-to-collector distances (cm): (a) 4, (*b*) 5, (*c*) 6, (d) 7, (e) 8, (f) 10, 24 kV, 10 wt%, TFA/DCM: 70/30.

2.7 SILK FIBERS

In recent years, the electrospinning process has gained much attention because it is an effective method to manufacture ultrafine fibers or fibrous structures of many polymers with diameter in the range from several micrometers down to tens of nanometers. In the electrospinning process, a high voltage is used to create an electrically charged jet of a polymer solution or a molten polymer. This jet is collected on a target as a nonwoven fabric. The jet typically develops a bending instability and then solidifies to form fibers, which measures in the range of nanometers to 1 mm. Because these nanofibers have some useful properties such as high specific surface area and high porosity, they can be used as filters, wound dressings, tissue engineering scaffolds, and so forth.

Recently, the protein-based materials have been interested in biomedical and biotechnological fields. The silk polymer, a representative fibrous protein, has been investigated as one of promizing resources of biotechnology and biomedical materials due to its unique properties. Furthermore, high molecular weight synthetic polypeptides of precisely controlled amino acid composition and sequence have been fabricated using the recombinant tools of molecular biology. These genetic engineered polypeptides have attracted the researcher's attention as a new functional material for biotechnological applications including cellular adhesion promoters, biosensors, and suture materials.

Silks are generally defined as fibrous proteins that are spun into fibers by some Lepidoptera larvae such as silkworms, spiders, scorpions, mites and flies. Silkworm silk has been used as a luxury textile material since 3000 B.C., but it was not until recently that the scientific community realized the tremendous potential of silk as a structural material. While stiff and strong fibers (such as carbon, aramid, glass, etc.) are routinely manufactured nowadays, silk fibers offer a unique combination of strength and ductility which is unrivaled by any other natural or man-made fibers. Silk has excellent Properties such as lightweight (1.3 g/cm^3) and high tensile strength (up to 4.8 GPa as the strongest fiber known in nature). Silk is thermally stable up to 250 °C, allowing processing over a wide range of temperatures. The origins of this behavior are to be found in the organization of the silks at the molecular and supramolecular levels, which are able to impart a large load bearing capability together with a damage tolerant response. The analysis of the fracture micro mechanisms in silk may shed light on the relationship between its microstructure and the mechanical properties. The antiparallel β pleated structure of silk gives rise to its structural properties of combined strength and toughness[22-31].

B. Mori silk consists of two types of proteins, fibroin and sericin. Fibroin is the protein that forms the filaments of silkworm silk. Fibroin filaments made up of bundles of nanofibrils with a bundle diameter of around 100 nm. The nanofibrils are oriented parallel to the axis of the fiber, and are thought to interact strongly with each other. Fibroin contains 46% Glycine, 29% Alanine and 12% Serine. Fibroin is a giant molecule (4700 amino acids) comprizing a "crystalline" portion of about two-thirds and an "amorphous" region of about one-third. The crystalline portion comprises about 50 repeats of polypeptide of 59 amino acids whose sequence is known: Gly Ala-Gly Ala-Gly Scr-Gly Ala-Ala-Gly (Scr-Gly Ala-Gly

Ala-Gly)s-Tyr. This repeated unit forms, β-sheet and is responsible for the mechanical properties of the fiber.

Silk fibroin (SF) can be prepared in various forms such as gels, powders, fibers, and membranes. Number of researchers has investigated silk-based nanofibers as one of the candidate materials for biomedical applications, because it has several distinctive biological properties including good biocompatibility, good oxygen and water vapor permeability, biodegradability, and minimal inflammatory reaction. Several researchers have also studied processing parameters and morphology of electrospun silk nanofibers using Hexafluoroacetone, Hexafluoro-2-propanol and formic acid as solvents. In all these reports nanofibers with circular cross sections have been observed.

In this section, effects of electrospinning parameters are studied and nanofibers dimensions and morphology are reported. Morphology of fibers diameter of silk precursor was investigated varying concentration, temperature and applied voltage and observation of ribbon like silk nanofibers were reported. Furthermore, a more systematic understanding of the process conditions was studied and a quantitative basis for the relationships between average fiber diameter and electrospinning parameters was established using response surface methodology (RSM), which will provide some basis for the preparation of silk nanofibers with desired properties.

2.7.1 PREPARATION OF REGENERATED SF SOLUTION

Raw silk fibers (B.mori cocoons were obtained from domestic producer, Abrisham Guilan Co., IRAN) were degummed with 2 gr/L Na_2CO_3 solution and 10 gr/L anionic detergent at 100 °C for 1 h and then rinsed with warm distilled water. Degummed silk (SF) was dissolved in a ternary solvent system of $CaCl_2/CH_3CH_2OH/H_2O$ (1:2:8 in molar ratio) at 70 °C for 6 h. After dialysis with cellulose tubular membrane (Bialysis Tubing D9527 Sigma) in H_2O for 3 days, the SF solution was filtered and lyophilized to obtain the regenerated SF sponges.

2.7.2 PREPARATION OF THE SPINNING SOLUTION

SF solutions were prepared by dissolving the regenerated SF sponges in 98% formic acid for 30 min. Concentrations of SF solutions for electrospinning was in the range from 8% to 14% by weight.

2.7.3 ELECTROSPINNING

In the electrospinning process, a high electric potential (Gamma High voltage) was applied to a droplet of SF solution at the tip (0.35 mm inner diameter) of a syringe needle, The electrospun nanofibers were collected on a target plate which was placed at a distance of 10 cm from the syringe tip. The syringe tip and the target plate were enclosed in a chamber for adjusting and controlling the temperature. Schematic diagram of the electrospinning apparatus is shown in Fig. 10. The processing temperature was adjusted at 25, 50 and 75 °C. A high voltage in the range from 10 kV to 20 kV was applied to the droplet of SF solution.

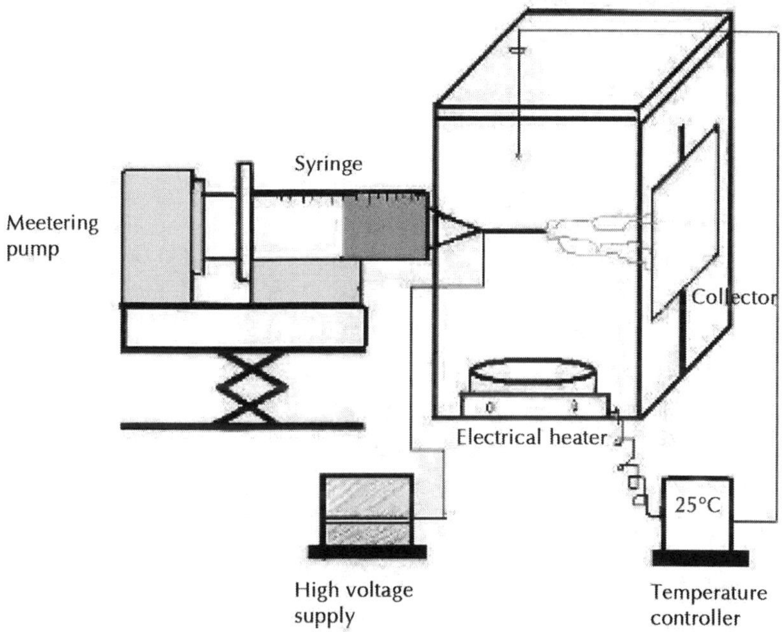

FIGURE 10 Schematic diagram of electrospinning apparatus.

2.7.4 CHARACTERIZATION

Optical microscope (Nikon Microphot-FXA) was used to investigate the macroscopic morphology of electrospun SF fibers. For better resolving power, morphology, surface texture and dimensions of the gold-sputtered electrospun nanofibers were determined using a Philips XL-30 scanning electron microscope. A measurement of about 100 random fibers was used to determine average fiber diameter and their distribution.

2.7.5 RESULTS AND DISCUSSION

2.7.5.1 EFFECT OF SILK CONCENTRATION

One of the most important quantities related with electrospun nanofibers is their diameter. Since nanofibers are resulted from evaporation of polymer jets, the fiber diameters will depend on the jet sizes and the solution concentration. It has been reported that during the traveling of a polymer jet from the syringe tip to the collector, the primary jet may be split into different sizes multiple jets, resulting in different fiber diameters. When no splitting is involved in electrospinning, one of the most important parameters influencing the fiber diameter is Concentration of regenerated silk solution. The jet with a low concentration breaks into droplets readily and a mixture of fibers, bead fibers and droplets as a result of low viscosity is generated. These fibers have an irregular morphology with large variation in size, on the other hand jet with high concentration don't break up but traveled to the grounded target and tend to facilitate the formation of fibers without beads and droplets. In this case, Fibers became more uniform with regular morphology.

At first, a series of experiments were carried out when the silk concentration was varied from 8 to 14% at the 15KV constant electric field and 25 °C constant temperature. Below the silk concentration of 8% as well as at low electric filed in the case of 8% solution, droplets were formed instead of fibers. Figure 11 shows morphology of the obtained fibers from 8% silk solution at 20 KV. The obtained fibers are not uniform. The average fiber diameter is 72 nm and a narrow distribution of fiber diameters is observed. It was found that continuous nanofibers were formed in the above silk concentration of 8% regardless of the applied electric field and electrospinning condition.

In the electrospinning of silk fibroin, when the silk concentration is more than 10%, thin and rod like fibers with diameters range from 60–450 nm were obtained. Figs. 12–14 show the SEM micrographs and diameter distribution of the resulted fibers.

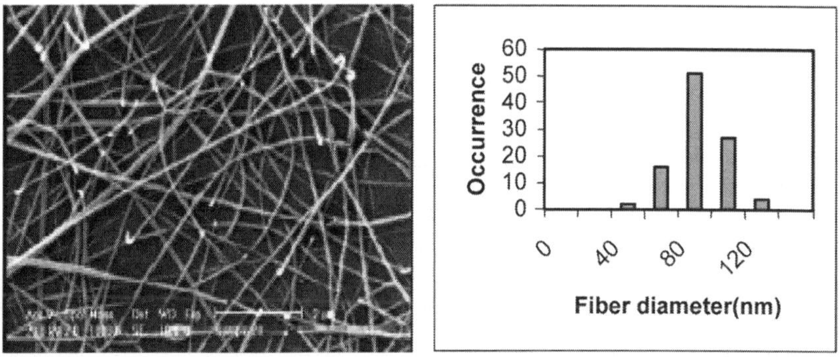

FIGURE 11 SEM micrograph and fiber distribution of 8 wt% of silk at 20 KV and 25 °C.

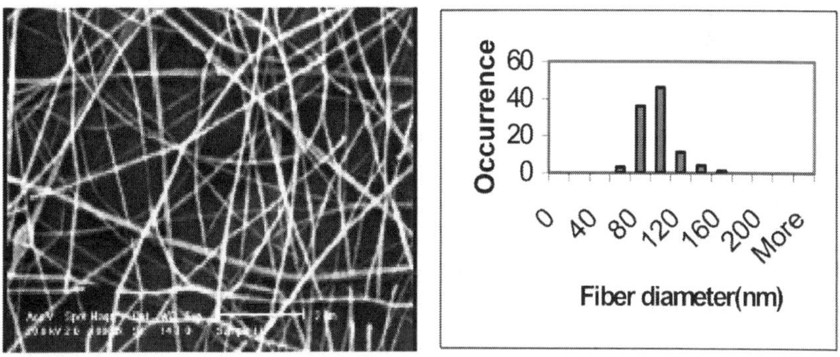

FIGURE 12 SEM micrograph and fiber distribution of 10 wt% of silk at 15 KV and 25 °C.

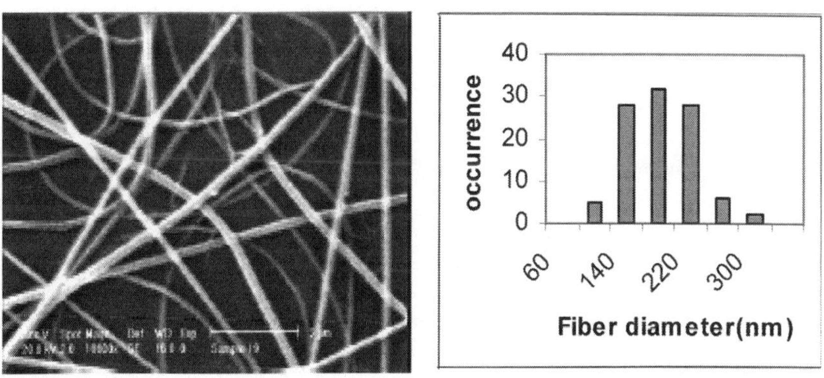

FIGURE 13 SEM micrograph and fiber distribution of 12 wt% of silk at 15 KV and 25 °C.

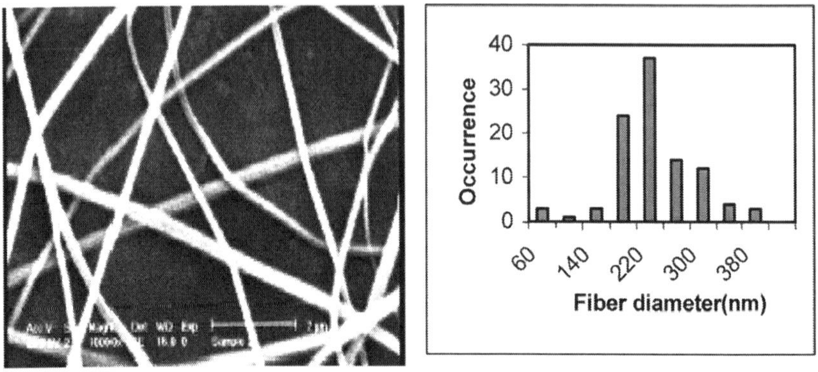

FIGURE 14 SEM micrograph and fiber distribution of 14 wt% of silk at 15 KV and 25 °C.

There is a significant increase in mean fiber diameter with the increasing of the silk concentration, which shows the important role of silk concentration in fiber formation during electrospinning process. Concentration of the polymer solution reflects the number of entanglements of polymer chains in the solution, thus solution viscosity. Experimental observations in electrospinning confirm that for forming fibers, a minimum polymer concentration is required. Below this critical concentration, application of electric field to a polymer solution results electrospraying and formation of droplets to the instability of the ejected jet. As the polymer concentration increased, a mixture of beads and fibers is formed. Further increase in concentration results in formation of continuous fibers as reported in this chapter. It seems that the critical concentration of the silk solution in formic acid for the formation of continuous silk fibers is 10%.

Experimental results in electrospinning showed that with increasing the temperature of electrospinning process, concentration of polymer solution has the same effect on fibers diameter at 25 °C. Figures 15–17 show the SEM micrographs and diameter distribution of the fibers at 50 °C.

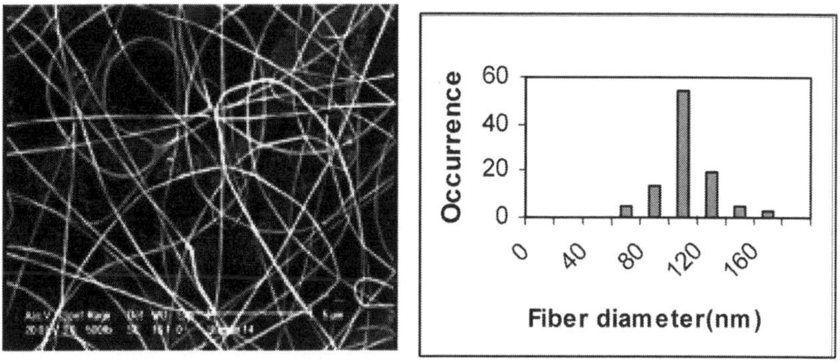

FIGURE 15 SEM micrograph and fiber distribution of 10 wt% of silk at 15 KV and 50 °C.

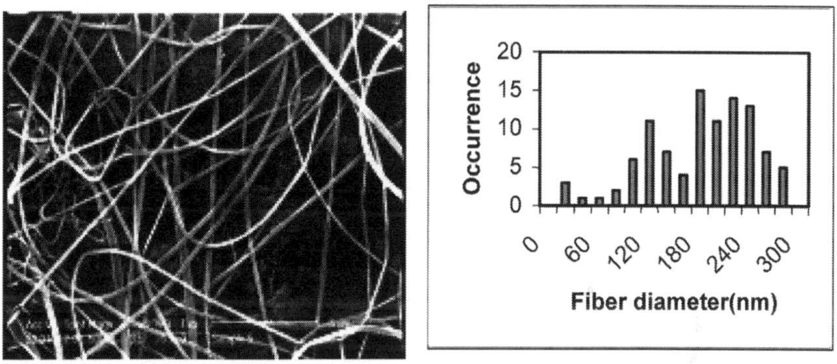

FIGURE 16 SEM micrograph and fiber distribution of 12 wt% of silk at 15 KV and 50 °C.

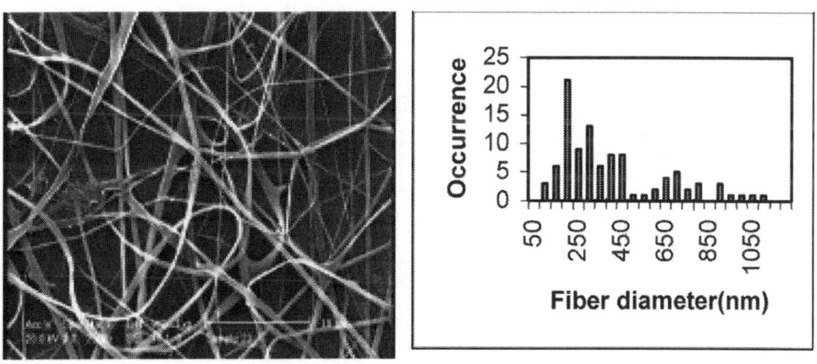

FIGURE 17 SEM micrograph and fiber distribution of 14 wt% of silk at 15 KV and 50 °C.

When the temperature of the electrospinning process increased, the circular cross section became elliptical and then flat, forming a ribbon with a cross-sectional perimeter nearly the same as the perimeter of the jet. Flat and ribbon like fibers have greater diameter than circular fibers. It seems that at high voltage the primary jet splits into different sizes multiple jets, resulting in different fiber diameters that nearly in a same range of diameter.

At 75 °C, the concentration and applied voltage have same effect as 50 °C. Figures 18–20 show the SEM micrographs and diameter distribution of Nano fibers at 75 °C.

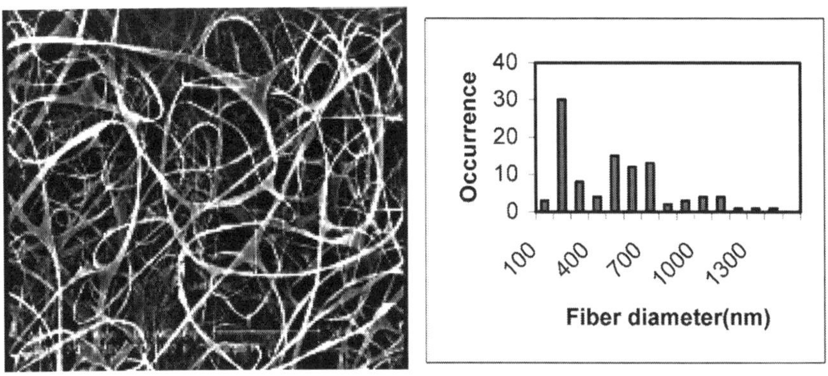

FIGURE 18 SEM micrograph and fiber distribution of 10 wt% of silk at 15 KV and 75 °C.

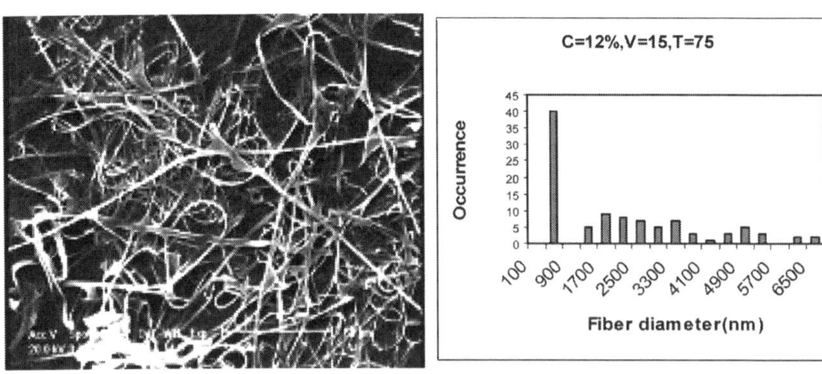

FIGURE 19 SEM micrograph and fiber distribution of 12 wt% of silk at 15 KV and 75 °C.

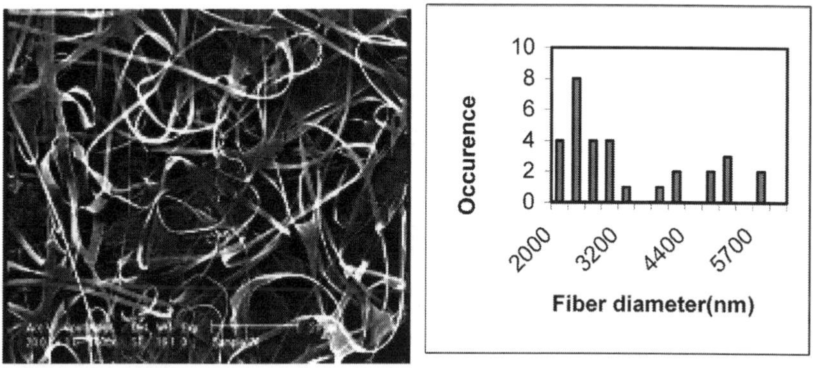

FIGURE 20 SEM micrograph and fiber distribution of 14 wt% of silk at 15 KV and 75 °C.

Figure 21 shows the relationship between mean fiber diameter and SF concentration at different electrospinning temperature. There is a significant increase in mean fiber diameter with increasing of the silk concentration, which shows the important role of silk concentration in fiber formation during electrospinning process. It is well known that the viscosity of polymer solutions is proportional to concentration and polymer molecular weight. For concentrated polymer solution, concentration of the polymer solution reflects the number of entanglements of polymer chains, thus have considerable effects on the solution viscosity. At fixed polymer molecular weight, the higher polymer concentration resulting higher solution viscosity. The jet from low viscosity liquids breaks up into droplets more readily and few fibers are formed, while at high viscosity, electrospinning is prohibit because of the instability flow causes by the high cohesiveness of the solution. Experimental observations in electrospinning confirm that for fiber formation to occur, a minimum polymer concentration is required. Below this critical concentration, application of electric field to a polymer solution results electro spraying and formation of droplets to the instability of the ejected jet. As the polymer concentration increased, a mixture of beads and fibers is formed. Further increase in concentration results in formation of continuous fibers as reported in this chapter. It seems that the critical concentration of the silk solution in formic acid for the formation of continuous silk fibers is 10% when the applied electric field was in the range of 10 to 20 kV.

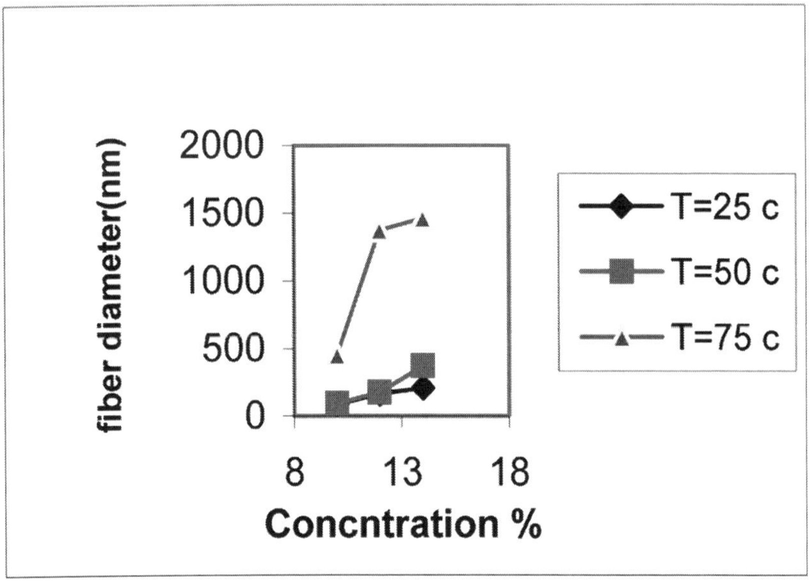

FIGURE 21 Mean fiber diameter of electrospun silk fibers at 25, 50 and 75 °C at15 kV.

2.7.5.2 EFFECT OF ELECTRIC FIELD

It was already reported that the effect of the applied electrospinning voltage is much lower than effect of the solution concentration on the diameter of electrospun fibers. In order to study the effect of the electric field, silk solution with the concentration of 10%, 12%, and 14% were electrospun at 10, 15, and 20 KV at 25 °C. The variation of the mean fiber diameter at different applied voltage for different concentration is shown in **fig. 22** This figure and SEM micrographs show that at all electric fields continuous and uniform fibers with different fiber diameter are formed. At a high solution concentration, Effect of applied voltage is nearly significant. It is suggested that, at this temperature, higher applied voltage causes multiple jets formation, which would provide decrees fiber diameter.

As the results of this finding it seems that electric field shows different effects on the nanofibers morphology. This effect depends on the polymer solution concentration and electrospinning conditions.

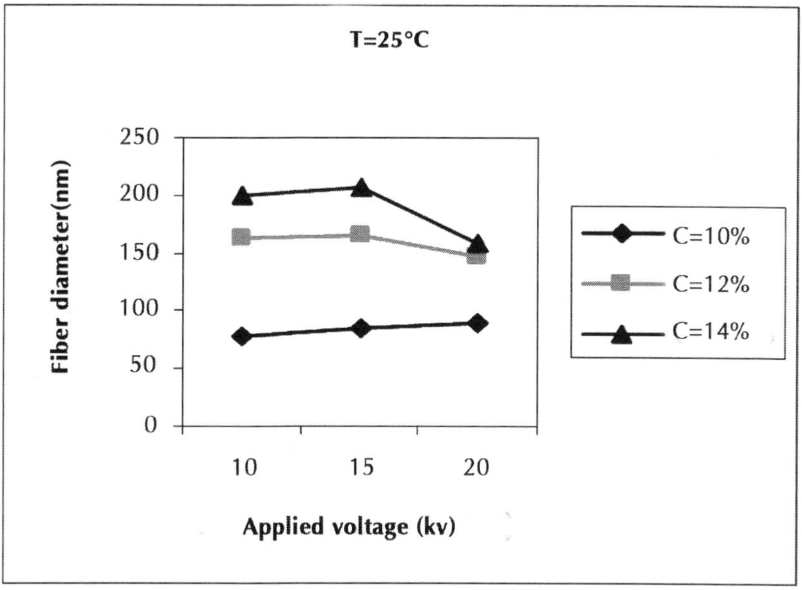

FIGURE 22 Mean fiber diameter of electrospun silk fibers at 10%, 15%, 20% and 25 °C.

2.7.5.3 EFFECT OF ELECTROSPINNING TEMPERATURE

One of the most important quantities related with electrospun nanofibers is their diameter. Since nanofibers are resulted from evaporation of polymer jets, the fiber diameters will depend on the jet sizes. The elongation of the jet and the evaporation of the solvent both change the shape and the charge per unit area carried by the jet. After the skin is formed, the solvent inside the jet escapes and the atmospheric pressure tends to collapse the tube like jet. The circular cross section becomes elliptical and then flat, forming a ribbon-like structure. In this chapter we believe that ribbon-like structure in the electrospinning of SF at higher temperature thought to be related with skin formation at the jets. With increasing the electrospinning temperature, solvent evaporation rate increases, which results in the formation of skin at the jet surface. Non- uniform lateral stresses around the fiber due to the uneven evaporation of solvent and/or striking the target make the nanofibers with circular cross-section to collapse into ribbon shape.

Bending of the electrospun ribbons were observed on the SEM micrographs as a result of the electrically driven bending instability or forces that occurred when the ribbon was stopped on the collector. Another problem that may be occurring in the electrospinning of SF at high temperature is the branching of jets. With increasing the temperature of electrospinning process, the balance between the surface tension and electrical forces can shift so that the shape of a jet becomes unstable. Such an unstable jet can reduce its local charge per unit surface area by ejecting a smaller jet from the surface of the primary jet or by splitting apart into two smaller jets. Branched jets, resulting from the ejection of the smaller jet on the surface of the primary jet were observed in electrospun fibers of SF. The axis of the cones from which the secondary jets originated were at an angle near 90° with respect to the axis of the primary jet. Figure 23 shows the SEM micrographs of flat, ribbon like and branched fibers.

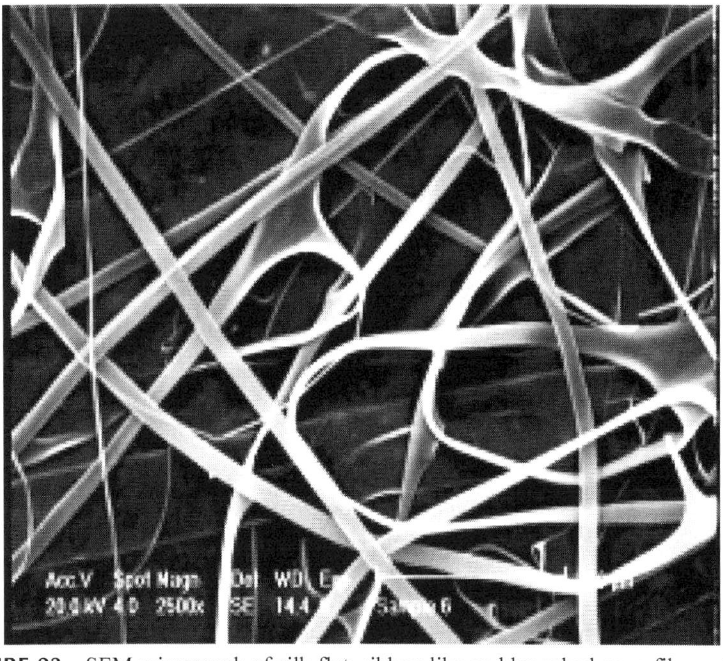

FIGURE 23 SEM micrograph of silk flat, ribbon like and branched nano fibers at high temperature.

In order to study the effect of electrospinning temperature on the morphology and texture of electrospun silk nanofibers, 12% silk solution was electrospun at various temperatures of 25, 50 and 75 °C. Results are shown in Fig. 24. Interestingly, the electrospinning of silk solution showed flat fiber morphology at 50 and 75 °C, whereas circular structure was observed at 25 °C. At 25 °C, the nanofibers with a rounded cross section and a smooth surface were collected on the target. Their diameter showed a size range of approximately 100 to 300 nm with 180 nm being the most frequently occurring. They are within the same range of reported size for electrospun silk nanofibers. With increasing the electrospinning temperature to 50 °C, The morphology of the fibers was slightly changed from circular cross section to ribbon like fibers. Fiber diameter was also increased to a range of approximately 20 to 320 nm with 180 nm the most occurring frequency. At 75 °C, The morphology of the fibers was completely changed to ribbon like structure. Furthermore, fibers dimensions were increased significantly to the range of 500 to 4100 nm with 1100 nm the most occurring frequency.

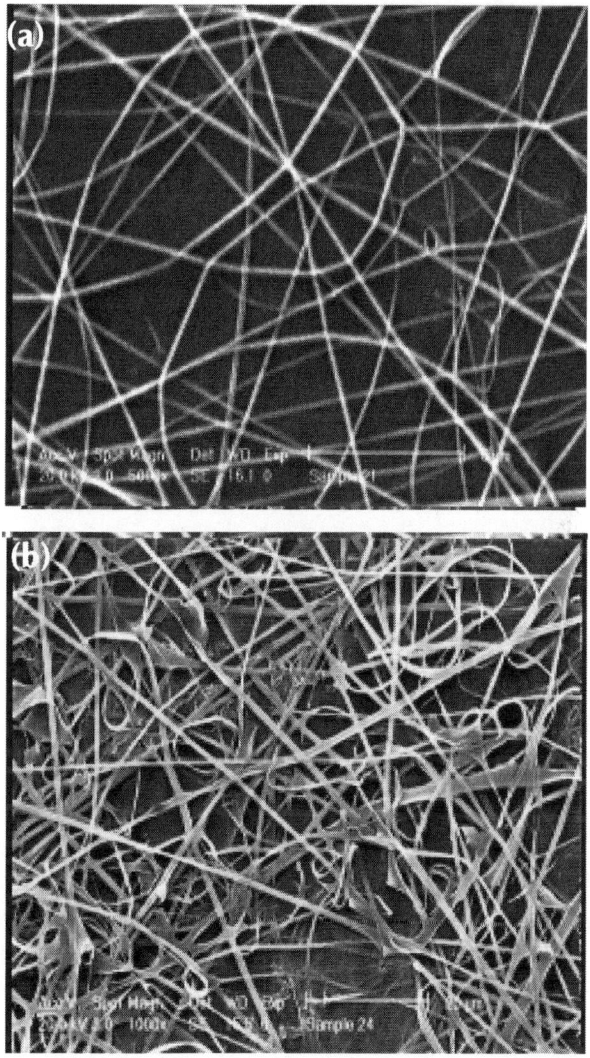

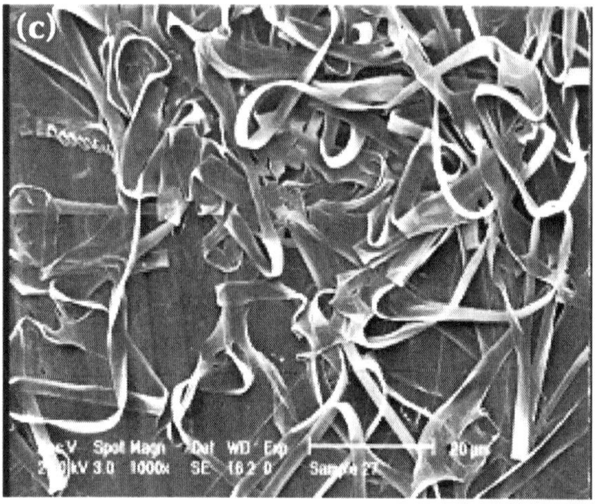

FIGURE 24 SEM micrograph of 12 wt% of silk at 20 KV and (a) 25 °C, (b) 50 °C, (c) 75 °C.

2.8 MODELING

Response surface methodology (RSM) is a collection of mathematical and statistical techniques for empirical model building (Appendix). By careful design of *experiments*, the objective is to optimize a *response* (output variable), which is influenced by several *independent variables* (input variables). An experiment is a series of tests, called *runs*, in which changes are made in the input variables in order to identify the reasons for changes in the output response.

In order to optimize and predict the morphology and average fiber diameter of electrospun silk, design of experiment was employed in the present chapter. Morphology of fibers and distribution of fiber diameter of silk precursor were investigated varying concentration, temperature and applied voltage. A more systematic understanding of these process conditions was obtained and a quantitative basis for the relationships between average fiber diameter and electrospinning parameters was established using response surface methodology (Appendix), which will provide some basis for the preparation of silk nanofibers.

A central composite design was employed to fit a second-order model for three variables. Silk concentration ($\times_1$), applied voltage ($\times_2$), and temperature ($\times_3$) were three independent variables (factors) considered in the preparation of silk nanofibers, while the fibers diameter were dependent variables (response). The actual and corresponding coded values of three factors ($\times_1$, X_2, and X_3) are given in Table 1. The following second-order model in X_1, X_2 and X_3 was fitted using the data in Table 1:

$$Y = \beta_0 + \beta_1 x_1 + \beta_2 x_2 + \beta_3 x_3 + \beta_1 x_1^2 + \beta_2 x_2^2 + \beta_3 x_3^2 + \beta_2 x_1 x_2 + \beta_3 x_1 x_3 + \beta_3 x_2 x_3 + \varepsilon$$

TABLE 1 Central composite design.
Coded values

X_i Independent variables		−1–0 1
X_1 Silk concentration (%)		10–12–14
X_2	applied voltage (KV)	10–15–20
X_3	temperature (°C)	25–50–75

The Minitab and Mathlab programs were used for analysis of this second-order model and for response surface plots (Minitab 11, Mathlab 7).

By Regression analysis, values for coefficients for parameters and P-values (a measure of the statistical significance) are calculated. When P-value is less than 0.05, the factor has significant impact on the average fiber diameter. If P-value is greater than 0.05, the factor has no significant impact on average fiber diameter. And R^2_{adj} (represents the proportion of the total variability that has been explained by the regression model) for regression models were obtained (Table 2) and main effect plots on fiber diameter (Fig. 25) were obtained and reported.

The fitted second-order equation for average fiber diameter is given by:

$$Y = 391 + 311 X_{1-} 164 X_2 + 57 X_{3-} 162 X_1^{2} + 69 X_{2-}^{2} + 391 X_{3-}^{2} - 159 X_1X_2 + 315 X_1X_{3-} 144 X_2X_3 \qquad (1)$$

where Y is the average fiber diameter.

TABLE 2 Regression Analysis for the three factors (concentration, applied voltage, temperature) and coefficients of the model in coded unit*.

Variables	Constant		*p*-value
	β_0	391.3	0.008
x_1	β_1	310.98	0.00
x_2	β_2	−164.0	0.015
x_3	β_3	57.03	0.00
x^2_1	β_{11}	161.8	0.143
x^2_2	β_{22}	68.8	0.516
x^2_3	β_{33}	390.9	0.002

TABLE 2 *(Continued)*

Variables	Constant		*p*-value
x_1x_2	β_{12}	–158.77	0.048
x_1x_3	β_{13}	314.59	0.001
x_2x_3	β_{23}	–144.41	0.069
F	*p* -value	R^2	R^2 (adj)
18.84	0.00	0.907	0.858

* Model: $Y=\beta o+\beta_1x_1+ \beta_2x_2+ \beta_3x_3+ \beta_{11}x^2_1+ \beta_{22}x^2_2+ \beta_{332}x^2_3 + \beta_{12}x_1x2+ \beta_{13}x_1x\ 3+ \beta_{13}x_2x3$, where "*y*" is average fiber diameter.

From the *p*-values listed in Table 2, it is obvious that *p*-value of term X_2 is greater than P-values for terms X_1 and X_3. And other P-values for terms related to applied voltage such as, X_2^2, X_1X_2, X_2X_3 are much greater than significance level of 0.05. That is to say, applied voltage has no much significant impact on average fiber diameter and the interactions between concentration and applied voltage, temperature and applied voltage are not significant, either. But *p*-values for term related to X_3 and X_1 are less than 0.05. Therefore, temperature and concentration have significant impact on average fiber diameter. Furthermore, R^2_{adj} is 0.858, That is to say, this model explains 86% of the variability in new data.

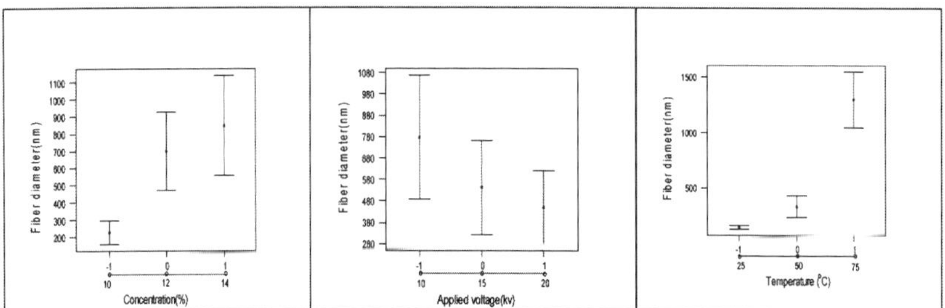

Relation between average fiber diameter and solution 's concentration	Relation between average fiber diameter and applied voltage	Relation between average fiber diameter and posses temperature

FIGURE 25 Effect of electrospinning parameters on silk nanofibers diameter.

2.9 CONCLUDING REMARKS

Several solvents, including acetic acid 1–90%, formic acid and TFA/DCM (70:30), were used for electrospinning of chitosan/carbon nanotube dispersion.

It has been observed that the TFA/DCM (70:30) solvent is the only solvent with a proper reliability for the electrospinnability of chitosan/carbon nanotube. This is a significant improvement in the electrospinning of chitosan/carbon nanotube dispersion. It was also observed that homogenous fibers with an average diameter of 455 nm (306–672) could be prepared with chitosan/carbon nanotube dispersion in TFA/DCM 70:30. In addition, the SEM images showed that the fiber diameter decreased by decreasing voltage and increasing the tip-to-collector distance.

The electrospinning of silk fibroin was processed and the average fiber diameters depend on the electrospinning condition. Morphology of fibers and distribution of diameter were investigated at various concentrations, applied voltages and temperature. The electrospinning temperature and the solution concentration have a significant effect on the morphology of the electrospun silk nanofibers. There effects were explained to be due to the change in the rate of skin formation and the evaporation rate of solvents. To determine the exact mechanism of the conversion of polymer into nanofibers require further theoretical and experimental work.

From the practical view the results of the present chapter can be condensed. Concentration of regenerated silk solution was the most dominant parameter to produce uniform and continuous fibers. The jet with a low concentration breaks into droplets readily and a mixture of fibers and droplets as a result of low viscosity is generated. On the other hand jets with high concentration do not break up but traveled to the target and tend to facilitate the formation of fibers without beads and droplets. In this case, fibers become more uniform with regular morphology. In the electrospinning of silk fibroin, when the silk concentration is more than 10%, thin and rod like fibers with diameters range from 60–450 nm were obtained. Furthermore, in the electrospinning of silk fibroin, when the process temperature is more than 25 °C, flat, ribbon like and branched fibers with diameters range from 60–7000 nm were obtained.

Two-way analysis of variance was carried out at the significant level of 0.05 to study the impact of concentration, applied voltages and temperature on average fiber diameter. It was concluded that concentration of solution and electrospinning temperature were the most significant factors impacting the diameter of fibers. Applied voltage had no significant impact on average fiber diameter. The average fiber diameter increased with polymer concentration and electrospinning temperature according to proposed relationship under the experimental conditions studied in this chapter [1–31].

KEYWORDS

- **amorphous region**
- **chitin**
- **chitosan**
- **collagen**
- **crystalline portion**
- **fibroin**
- **gelatin**
- **silk**

REFERENCES

1. Miwa, M.; Nakajima, A.; Fujishima, A.; Hashimoto, K.; Watanabe, T.; *Langmuir,* **2000,** *16,* 5754.
2. Ö ner, D.; McCarthy, T. J. *Langmuir,* **2000,** *16,* 7777.
3. Abdelsalam, M. E.; Bartlett, P. N.; Kelf, T.; Baumberg, J.; *Langmuir,* **2005,** *21,* 1753,
4. A.; Nakajima, K.; Hashimoto, T.; Watanabe, K.; Takai, G.; Yamauchi, A.; Fujishima, *Langmuir,* **2000,** *16,* 7044.
5. Zhong, W.; Liu, S.; Chen, X.; Wang, Y.; Yang, W.; *Macromolecules,* **2006,** *39,* 3224.
6. Shams Nateri, A.; Hasanzadeh, M.; *J. Comput. Theor. Nanosci.;* **2009,** *6,* 1542.
7. Kilic, A.; Oruc, F.; Demir, A.; *Text. Res. J.* **2008,** *78,* 532.
8. Reneker, D. H.; Chun, I.; *Nanotechnology* **1996,** *7,* 216.
9. Shin, Y. M.; Hohman, M. M.; Brenner, M. P.; Rutledge, G. C.; *Polymer,* **2001,** *42,* 9955.
10. Reneker, D. H.; Yarin, A. L.; Fong, H.; Koombhongse, S.; *J. Appl. Phys.;* **2000,** *87,* 4531.
11. Zhang, S.; Shim, W. S.; Kim, J.; *Mater. Design,* **2009,** *30,* 3659.
12. Yördem, O. S.; Papila, M.; Menceloğlu, Y. Z.; *Mater. Design,* **2008,** *29,* 34.
13. Chronakis, I. S.; *J. Mater. Process. Tech.;* **2005,** *167,* 283.
14. Dotti, F.; Varesano, A.; Montarsolo, A.; Aluigi, A.; Tonin, C.; Mazzuchetti, G.; *J. Ind. Text.* **2007,** *37,* 151.
15. Lu, Y.; Jiang, H.; Tu, K.; Wang, L.; *Acta Biomater.* **2009,** *5,* 1562.
16. Lu, H.; Chen, W.; Xing, Y.; Ying, D.; Jiang, B.; *J. Bioact. Compat. Pol.;* **2009,** *24,* 158.
17. Nisbet, D. R.; Forsythe, J. S.; Shen, W.; Finkelstein D. I.; Horne, M. K.; *J. Biomater. Appl.* **2009,** *24,* 7.
18. Ma, Z.; Kotaki, M.; Inai, R.; Ramakrishna, S.; *Tissue Eng.* **2005,** *11,* 101.
19. Hong, K. H.; *Polym. Eng. Sci.*; **2007,** *47,* 43.
20. Zhang, W.; Pintauro, P. N.; *ChemSusChem.*; **2011,** *4,* 1753.

21. Lee, S.; Obendorf, S. K.; *Text. Res. J.;* **2007,** *77,* 696.
22. Myers, R. H.; Montgomery D. C.; Anderson-cook, C. M.; *Response surface methodology: process and product optimization using designed experiments,* 3rd ed.; John Wiley and Sons, USA, **2009**.
23. Gu, S. Y.; Ren J.; Vancso, G. J.; *Eur. Polym. J.* **2005,** *41,* 2559.
24. Dev, V.; Venugopal, R. G.; Senthilkumar, J. R.; Gupta, M. D.; Ramakrishna, S.; *J. Appl. Polym. Sci.;* **2009,** *113,* 3397.
25. Galushkin, A. L.; *Neural networks Theory*, Springer, Moscow Institute of Physics and Technology **2007**.
26. Ma, M.; Mao, Y.; Gupta, M.; Gleason K. K.; Rutledge, G. C.; *Macromolecules,* **2005,** *38,* 9742.
27. Haghi, A. K.; Akbari, M.; *Phys. Status. Solidi. A.;* **2007,** *204,* 1830.
28. Ziabari, M.; Mottaghitalab, V.; Haghi, A. K.; in *Nanofibers: Fabrication, Performance, and Applications,* Chang, W. N.; Nova Science Publishers, USA, **2009**.
29. Ramakrishna, S.; Fujihara, K.; Teo, W. E.; Lim, T. C.; Ma, Z. *An Introduction to Electrospinning and Nanofibers*, World Scientific Publishing, Singapore, **2005**.
30. Zhang, S.; Shim, W. S.; Kim, J.; *Mater. Design,* **2009,** *30,* **3659**.
31. Kasiri, M. B.; Aleboyeh H.; Aleboyeh, A.; *Environ. Sci. Technol.;* **2008,** *42,* 7970.

APPENDIX A

Materials produced from fibers and threads are classified as textile: fabrics, nonwoven materials, fur fabric, carpets and rugs, and so forth.

Textile fibers are the main raw material of the textile industry. According to their origin, these fibers are divided into natural and chemical ones.

Natural fibers are plant (cotton, bast fibers), animal (wool, silk) and mineral (asbestos) origin ones.

Chemical fibers are produced from modified natural or synthetic high-molecular substances and are classified as artificial ones obtained by chemical processing of natural raw material, commonly cellulose (viscose, acetate), and synthetic ones obtained from synthetic polymers (nylon-6, polyester, acryl, PVC fibers, etc.).

Textile materials are damaged by microorganisms, insects, rodents and other biodamaging agents. Fibers and fabrics resistance to biodamages primarily depends upon chemical nature of the fibers from which they are made. Most frequently we have to put up with microbiological damages of textile materials based on natural fibers – cotton, linen, and so forth, used by saprophyte microflora. Chemical fibers and fabrics, especially synthetic ones, are higher biologically resistant, but microorganisms-biodegraders can also adapt to them.

Textile material degradation by microorganisms depends on their wear rate, kind and origin, organic composition, temperature and humidity conditions, degree of aeration, and so forth.

With increased humidity and temperature, and restricted air exchange microorganisms damage fibers and fabrics at different stages of their manufacture and application, starting from the primary processing of fibers including spinning, weaving, finishing and storage, transportation and operation of textile materials and articles from them. Fiber and fabric biodamage intensity sharply increases when contacting with the soil and water, specifically in the regions with warm and humid climate.

Textile materials are damaged by bacteria and microscopic fungi. Bacterial degradation of textile materials is more intensive than the fungal one. The damaging bacterium genus are: *Cytophaga, Micrococcus, Bacterium*, *Bacillus, Cellulobacillus, Pseudomonas, Sarcina.* Among fungi damaging textile materials in the air and in the soil the following are detected: *Aspergillus, Penicillium, Alternaria, Cladosporium, Fusarium, Trichoderma*, and so forth.

Annual losses due to microbiological damaging of fabrics reach hundreds of millions of dollars.

Fiber and fabric biodamaging by microorganisms is usually accompanied by the mass loss and mechanical strength of the material as a result, for example, of fiber degradation by microorganism metabolites: enzymes, organic acids, and so forth.

CURIOUS FACTS

The immediate future of the textile industry belongs to biotechnology. Even today suggestions on the synthesis of various polysaccharides using microbiological methods are present. These methods may be applied to synthesis of fiber-forming monomers and polymers.

Scientists have demonstrated possibilities of microbiological synthesis of some monomers to produce dicarboxylic acids, caprolactam, and so forth. Some kinds of fiber-forming polymers, polyethers, in particular, can also be obtained by microbiological synthesis.

Some kinds of fiber-forming polypeptides have already been obtained by the microbiological synthesis. In some cases, concentration of these products may reach 40% of the biomass weight and they can be used as a perspective raw material for synthetic fibers. Studies in this direction are widely performed in many countries all over the world.

The impact of microorganisms on textile materials that causes their degradation is performed, at least, by two main ways (direct and indirect):

- fungi and bacteria use textile materials as the nutrient source (assimilation);
- textile materials are damaged by microorganism metabolism (degradation).

Biodamages of textile materials induced by microorganisms and their metabolites manifest in coloring (occurrence of spots on textile materials or their coatings), defects (formation of bubbles on colored surfaces of textile materials), bond breaks in fibrous materials, penetration deep inside (penetration of microorganisms into the cavity of the natural fiber), deterioration of mechanical properties (e.g., strength at break reduction), mass loss, a change of chemical properties (cellulose degradation by microorganisms), liberation of volatile substances and changes of other properties.

It is known that when microorganisms completely consume one part of the substrate they then are able to liberate enzymes degrading other components of the culture medium. It is found that, degrading fiber components each group of microorganisms due to their physiological features decomposes some definite part of the fiber, damages it differently and in a different degree. It is found that along with the enzymes, textile materials are also degraded by organic acids produced by microorganisms: lactic, gluconic, acetic, succinic, fumaric, malic, citric, oxalic, and so forth. It is also found that enzymes and organic acids liberated by microorganisms continue degrading textile materials even after microorganisms die. As noted, the content of cellulose, proteins, pectins and alcohol-soluble waxs increases with the fiber damage degree in it; pH increases, and concentration of water-soluble substances increases that, probably, is explained by increased accumulation of metabolites and consumption of nutrients by microorganisms for their vital activity.

The typical feature of textile material damaging by microorganisms is occurrence of honeydew, red violet or olive spots with respect to a pigment produced by microorganisms and fabric color. As microorganism's pigment interacts with the fabric dye, spots of different hue and tints not removed by laundering or by hydrogen peroxide oxidation. They may sometimes be removed by hot treatment in blankly solution. Spot occurrence on textile materials is usually accompanied by a strong musty odor.

CURIOUS FACTS

Textile materials are also furnished by biotechnological methods based on the use of enzymes performing various physicochemical processes. Biotechnologies are mostly full for preparation operations (cloth softening, boil-off and bleaching of cotton fabrics, wool washing) and bleaching effects of jeans and other fabrics, biopolishing, and making articles softer.

Enzymes are used to improve sorption properties of cellulose fibers, to increase specific area and volume of fibers, to remove pectin "companions" of cotton and linen cellulose. Enzymes also hydrolyze ether bonds on the surface of polyether fibers.

Cotton fabric dyeing technologies in the presence of enzymes improving coloristic parameters of prepared textile articles under softer conditions, which reduce pollutants in the sewage. To complete treatment of the fabric surface, that is, removal of surface fiber fibrilla, enzymes are applied. The enzymatic processing also decreases and even eliminates the adverse effect of long wool fiber puncturing. The application of enzymes to treatment of dyed tissues is industrially proven. This processing causes irregular bleaching that adheres the articles fashionable "worn" style.

Temperature and humidity are conditions promoting biodamaging of fibrous materials. The comparison of requirements to biological resistance of textile materials shall be based on the features of every type of fibers, among which mineral fiber are most biologically resistant.

Different microorganism damaging rate for fabrics is due to their different structure. Thinner fabrics with the lower surface density and higher through porosity are subject to the greatest biodamaging, because these properties of the material provide large contact area for microorganisms and allow their easy penetration deep into it. The thread bioresistance increases with the yarning rate.

Cotton fiber is a valuable raw material for the textile industry. Its technological value is due to a complex of properties that should be retained during harvesting, storage, primary and further processing to provide high quality of products. One of the factors providing retention of the primary fiber properties is its resistance to bacteria and fungi impacts. This is tightly associated with the chemical and physical features of cotton fiber structures.

Ripe cotton fiber represents a unit extended plant cell shaped as a flattened tube with a corkscrew waviness. The upper fiber end is cone-shaped and dead-ended. The lower end attached to the seed is a torn open channel.

The cotton fiber structure is formed during its maturation, when cellulose is biosynthesized and its macromolecules are regularly disposed.

The basic elements of cotton fiber morphological structure are known to be the cuticle, primary wall, convoluted layer, secondary wall, and tertiary wall with the central channel. The fiber surface is covered by a thin layer of wax substances – the primary wall (cuticle). This layer is a protective one and possesses rather high chemical resistance.

There is a primary wall under the cuticle consisting of a cellulose framework and fatty-wax-pectin substances. The upper layer of the primary wall is less densely packed as compared with the inner one, due to fiber surface expansion during its growth. Cellulose fibrils in the primary wall are not regularly oriented.

The primary layer covers the convoluted layer having structure different from the secondary wall layers. It is more resistant to dissolution, as compared with the main cellulose mass.

The secondary cotton fabric wall is more homogeneous and contains the greatest amount of cellulose. It consists of densely packed, concurrently oriented cellulose fibrils composed in thin layers. The fibril layers are spiral-shaped twisted around the fiber axis. There are just few micropores in this layer.

The tertiary wall is an area adjacent to the fiber channel. Some authors think that the tertiary wall contains many pores and consists of weakly ordered cellulose fibrils and plenty of protein admixtures, protoplasm, and pectin substances.

The fiber channel is filled with protoplasm residues, which are proteins, and contains various mineral salts and a complex of microelements. For the mature fiber, channel cross-section is 4–8% of total cross-section.

Cotton fiber is formed during maturation. This includes not only cellulose biosynthesis, but also ordering of the cellulose macromolecules shaped as chains and formed by repetitive units consisting two β-D-glucose residues bound by glucosidic bonds.

Among all plant fibers, cotton contains the maximum amount of cellulose (95–96%).

The morphological structural unit of cellulose is a cluster of macromolecules – a fibril 1.0–1.5 μm long and 8–15 nm thick, rather than an individual molecule. Cellulose fibers consist of fibril clusters uniform oriented along the fiber or at some angle to it.

It is known that cellulose consists not only of crystalline areas – micelles where molecule chains are concurrently oriented and bound by intermolecular forces, but also of amorphous areas.

Amorphous areas in the cellulose fiber are responsible for the finest capillaries formation, that is, a "submicroscopic" space inside the cellulose structure is formed. The presence of the submicroscopic system of capillaries in the cellulose fibers is of paramount significance, because it is the channel of chemical reactions by which water-soluble reagents penetrate deep in the cellulose structure. More active hydroxyl groups interacting with various substances are also disposed here.

It is known that hydrogen bonds are present between hydroxyl groups of cellulose molecules in the crystalline areas. Hydroxyl groups of the amorphous area may occur free of weakly bound and, as a consequence, they are accessible for sorption. These hydroxyl groups represent active sorption centers able to attract water.

Among all plant fibers, cotton has the highest quantity of cellulose (95–96%). Along with cellulose, the fibers contain some fatty, wax, coloring mineral substances (4–5%). Cellulose concomitant substances are disposed between macromolecule clusters and fibrils. Raw cotton contains mineral substances (K, Na, Ca, Mg) that promote mold growth and also contains microelements (Fe, Cu, Zn) stimulating growth of microorganisms. Moreover, it contains sulfates, phosphorus, glucose, glycidols and nitrogenous substances, which also stimulate growth of microbes. Differences in their concentrations are one of the reasons for different aggressiveness of microorganisms in relation to the cotton fiber.

The presence of cellulose, pectin, nitrogen containing and other organic substances in the cotton fiber, as well as its hygroscopicity makes it a good culture medium for abundant microflora.

Cotton is infected by microorganisms during harvesting, transportation and storage. When machine harvested, the raw cotton is clogged by various admixtures. It obtains multiple fractures of leaves and cottonseed hulls with humidity higher than of the fibers. Such admixtures create a humid macrozone around, where microorganisms intensively propagate. Fiber humidity above 9% is the favorable condition for cotton fiber degradation by microorganisms.

It is found that cotton fiber damage rate directly in hulls may reach 42–59%; hence, the fiber damage rate depends on a number of factors, for example, cultivation conditions, harvesting period, type of selection, and so forth.

Cotton fiber maturity is characterized by filling with cellulose. As the fiber becomes more mature, its strength, elasticity and coloring value increase.

Low quality cotton having higher humidity is damaged by microorganisms to a greater extent. The fifths fibers contain microorganisms 3–5 times more than the first quality fibers. When cotton hulls open, the quantity of microorganisms sharply increases in them, because along with dust wind brings fungus spores and bacteria to the fibers.

Cotton is most seriously damaged during storage: in compartments with high humidity up to 24% of cotton is damaged. Cotton storage in compact bales covered by tarpaulin is of high danger, especially after rains. For instance, after one and half months of such storage the fiber is damaged by 50% or higher.

Cotton microflora remains active under conditions of the spinning industry. As a result, the initial damage degree of cotton significantly increases.

CURIOUS FACTS

On some textile enterprises of Ivanovskaya Oblast, sickness cases of spinning-preparatory workshops were observed. When processing biologically contaminated cotton, plenty of dust particles with microorganisms present on them are liberated to the air. This can affect the health status of the employees.

Under natural conditions, cotton products are widely used in contact with the soil (fabrics for tents) receiving damage both from the inside and the outside. The main role here is played by cellulose degrading bacteria and fungi.

It was considered over a number of years that the main role in cotton fiber damage is played by cellulose degrading microorganisms. Not denying participation of cellulose degrading bacteria and fungi in damaging of the cotton fiber, it is noted that a group of bacteria with yellow pigmented mucoid colonies representing epiphytic microflora always present on the cotton plant dominate in the process of the fiber degradation. Nonsporeforming epiphytic bacteria inhabiting in the cotton plants penetrate from their seeds into the fiber channel and begin developing there. Using chemical substances of the channel, these microorganisms then permeate into the submicroscopic space of the tertiary wall primarily consuming pectins of the walls and proteins of the channels.

Enzymes and metabolites produced by microorganisms induce hydrolysis of cellulose macromolecules, increasing damage of internal areas of the fiber. Thus, the fiber delivered to processing factories may already be significantly damaged by microorganisms that inevitably affects production of raw yarn, fabric, and so forth.

One hundred and thirty five strains of fungi of different genus capable of damaging cotton fibers are currently determined. It is found that the population of phytopathogenic fungi is much lower than that of cellulose degraders: *Chaetomium globosum, Aspergillus flavus, Aspergillus niger, Rhizopus nigricans, Trichothecium roseum.* These species significantly deteriorate the raw cotton condition, sharply reduce spinning properties of the fiber, in particular.

It is also found that the following species of fungi are usually present on cotton fibers: *Mucor* (consumes water-soluble substances), *Aspergillus* and *Penicillium* (consumes insoluble compounds), *Chaetomium, Trichoderma*, and so forth. (degrade cellulose). This points to the fact that some species of mold fungi induce the real fiber decomposition that shall be distinguished from simple surface growth of microorganisms. For example, *Mucor* fungi incapable of inducing cellulose degradation may actively vegetate on the yarn finish. Along with fungi, bacteria are always present on the raw cotton, most represented by *Bacillus* and *Pseudomonas* genus species.

Figures A1 and A2 present surface micrographs of the first and fifth quality grade primary cotton fibers.

Figure A3, a micrograph, shows the cotton fiber surface after the impact of spontaneous microflora during 7 days. It is observed that bacterial cells are accumulated in places of fiber damage, at clearly noticeable cracks. Figure 4 shows the first quality cotton fiber surface after impact of *Aspergillus niger* culture during 14 days. On the fiber surface mycelium is observed. Figures A5 and A6 show photos of cotton fiber surfaces infected by *Bac. subtilis* (14-day exposure): bacterial cells on first quality fibers are separated by the surface, forming no conglomerates, whereas on the fifth quality fibers conglomerates are observed that indicates their activity (Fig. A7).

Academician A.A. Imshenetsky has demonstrated that aerobic cellulose bacteria are able to propagate under increased humidity, whereas fungi propagate at lower humidity. Textile products are destroyed by fungi at their humidity about 10%, whereas bacteria destroy them at humidity level of, at least, 20%. As a consequence, the main attention at cotton processing to yarn should be paid to struggle against fungi, and at wet spinning and finishing fabrics and knitwear not only fungi, but mostly bacteria should be struggled against.

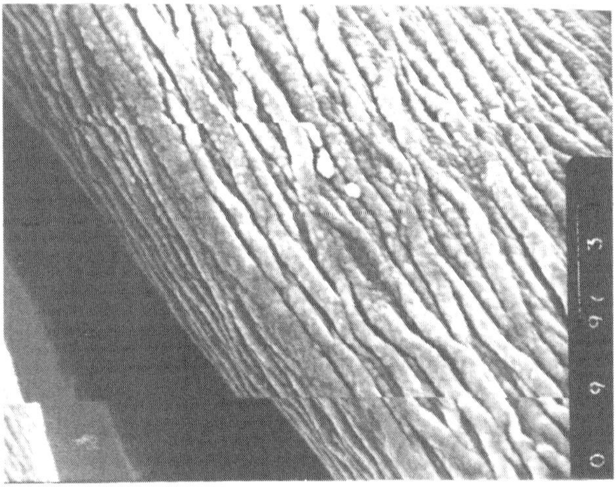

FIGURE A1 The surface of initial first quality cotton specimen (x 4500).

FIGURE A2 The surface of initial fifth quality cotton specimen (x 4500).

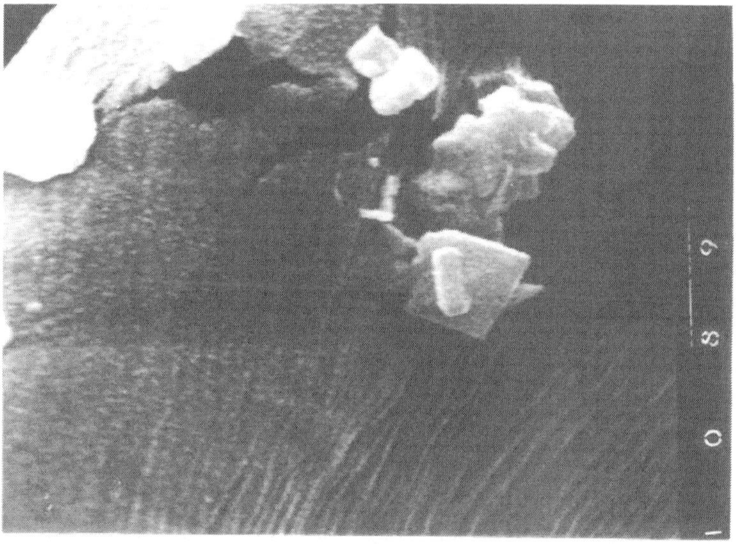

FIGURE A3 The first quality cotton, 7-day exposure (spontaneous microflora) (x 10000).

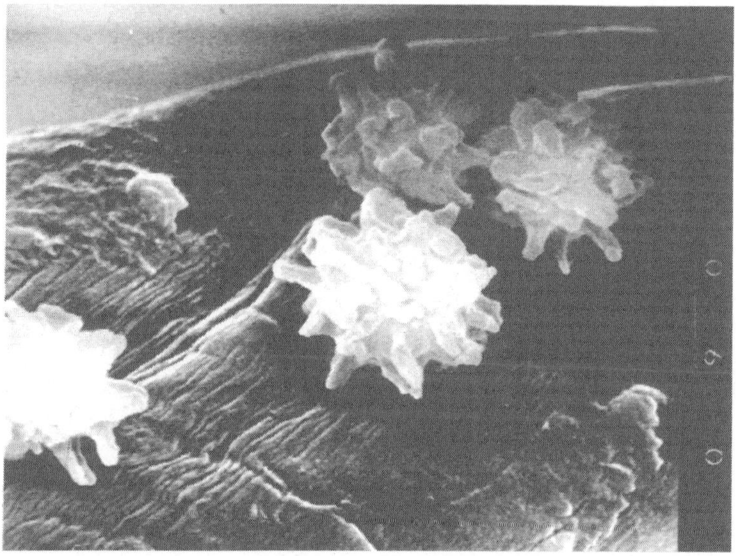

FIGURE A4 The first quality cotton, *Asp. niger* contaminated, 14-day exposure (x 3000).

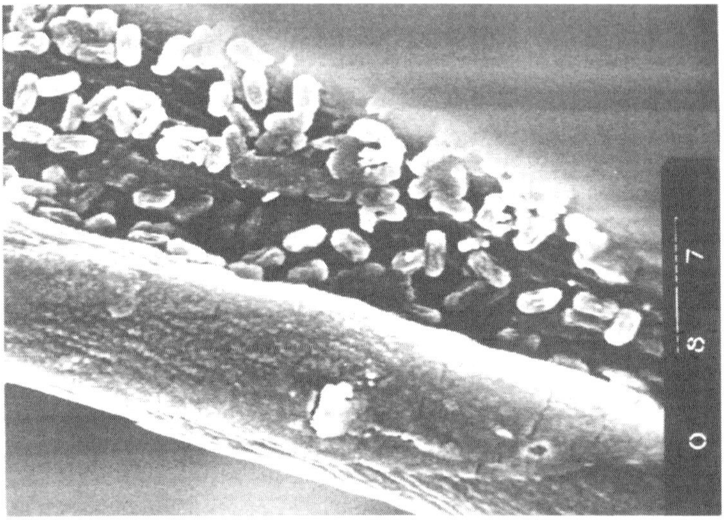

FIGURE A5 The first quality cotton, *Bacillus subtilis* contaminated, 14-day exposure (x 4500).

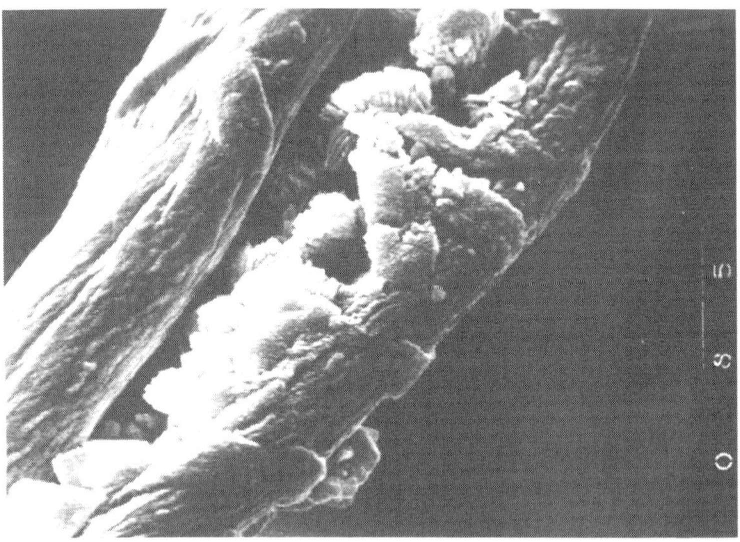

FIGURE A6 The fifth quality cotton, *Bacillus subtilis* contaminated, 14-day exposure (x 4500).

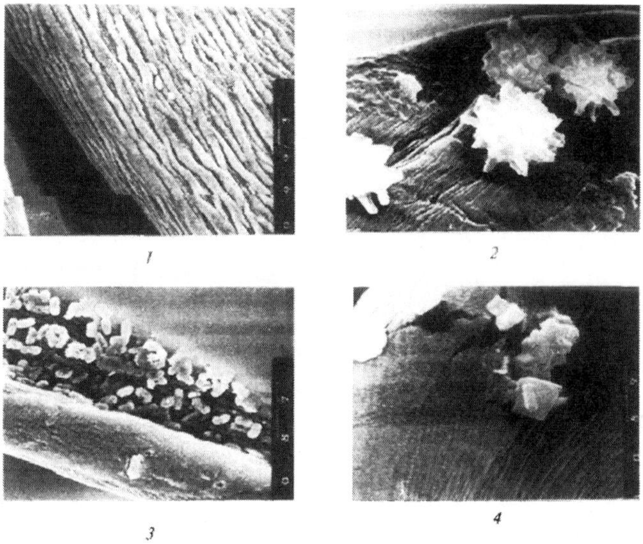

FIGURE A7 Cotton fiber micrographs: *1* – initial fiber (× 4500); *2–4:* fibers damages by different microorganisms: *2* – *Aspergillus niger* (× 3000); *3* – *Bacillus subtilis* (× 4500); *4* – *Pseudomonas fluorescens* (× 10000)

Cotton damage leads to:

- significant decrease of strength of the fibers and articles from them;
- disturbance of technological process (the smallest particles of sticky mucus excreted by some species of bacteria and fungi become the reason for sticking executive parts of machines);
- abruptness increase;
- waste volume increase.

Damaging of cotton fibers, fabrics and textile products by microorganisms is primarily accompanied by occurrence of colored yellow, orange, red, violet, and so forth, spots and then by putrefactive odor, and, finally, the product loses strength and degrades.

The effect of microorganisms results in noticeable changes in chemical composition and physical structure of cotton fibers.

As found by electron microscopy, cotton fiber degradation by enzymes is most intensive in the zones of lower fibril structure density.

In the cotton fiber damaged by microorganisms, cellulose concentration decreases by 7.5%, pectin substances – by 60.7%, hemicellulose – by 20%, and noncellulose polysaccharides content also decreases. Cellulose biostability increases with its crystallinity degree and macromolecule orientation, as well as with hydroxyl group replacement by other functional groups. Microscopic fungi and bacteria are able to degrade cellulose and as a result glucose is accumulated in the medium, used as a source of nutrition by microorganisms. However, some part of cellulose is not destroyed and completely preserves its primary structure.

Cellulose of undamaged cotton fiber has 76.5% of well-ordered area, 7.8% of weakly ordered area, and 15.7% of disordered area. Microbiological degradation reduces the part of disordered area to 12.7%, whereas the part of well-ordered area increases to 80.4%. The ratio of weakly ordered area changes insignificantly. This goes to prove that the order degree of cotton cellulose increases due to destruction of disordered areas.

A definite type of fiber degradation corresponds to each stage of the cotton fiber damage. The initial degree of damage is manifested in streakiness, when the fiber surface obtains cracks of different length and width due to its wall break.

Swellings are formed resulting abundant accumulation of microorganisms and their metabolites in a definite part of the fiber. They may be accompanied by fiber wall break induced by biomass pressure. In this case, microorganisms and their metabolites splay out that causes blobs formation from the fiber and breaks in the yarn, as well as irregular fineness and strength.

The external microflora induces the wall damage. The highest degradation stage is fiber decomposition and breakdown into separate fibrils. Hence, perfect fiber structure is absent in this case.

In all cases of damage, a high amount of fungal mycelium may be present on the fiber surface, which hyphae penetrate through the fiber or wrap about it thus preventing spinning and coloring of textile materials.

Enzymatic activity of fungi is manifested in strictly defined places of cellulose microfibrils, and the strength loss rate depends on both external climate conditions and contamination conditions. Cotton fabrics inoculated by the microscopic fungus *Aspergillus niger* under laboratory conditions at a temperature +29°C lose 66% of the initial strength 2–3 weeks after contamination, whereas inoculation by *Chaetomium globosum* induces 98.7% loss of strength, that is, completely destroys the material.

The same fabric exposed to soil at +29°C during 6 days loses 92% of the initial strength.

And cotton fabric exposed to seawater for 65 days loses up to 90% of strength.

A1 BAST FIBER BIODAMADING

Fibers produced from stalks, leaves or fruit covers of plants are called bast fibers. Hemp stalks give strong, coarse fibers – the hemp used for packing cloth and ropes. Coarse technical fibers: jute, ambary, ramie, and so forth, are produced from stalks of cognominal plants. Among all bast fibers, linen ones are most widely used.

The linen complex fiber, from which yarn and fabrics are manufactured, represents a batch of agglutinated filaments (plant cells) stretched and arrow-headed. The linen filament represents a plant cell with thick walls, narrow channel and knee-shaped nodes called shifts. Shifts are traces of fractures or bends of the fiber occurred during growth, and especially during mechanical treatment. Fiber ends are arrow-shaped, and the channel is closed. The cross-section represents an irregular polygon with five or six edges and a channel in the center. Coarser fibers have oval cross-section with wider and slightly flattened channel.

Complex fibers consist of filament batches (15–30 pieces in a batch) linked by middle lamellae. Middle lamellae consist of various substances: pectins, lignin, hemicellulose, and so forth.

Bast fibers contain a bit lower amount of cellulose (about 70%) than cotton ones. Moreover, they contain such components as lignin (10%), wax and trace amounts of antibiotics, some of which increase biostability of the fiber. The presence of lignins induces coarsening (lignifications) of plant cells that promotes the loss of softness, flexibility, elasticity, and increased friability of fibers.

The main method for fiber separation from the flax is microbiological one, in which vital activity of pectin degrading microorganisms degrade pectins linking bast batches to the stalk tissues. After that the fiber can be easily detached by mechanical processing.

Microorganisms affect straw either at its spreading directly at the farm that lasts 20–30 days or at its retting at a flax-processing plant where retting lasts 2–4 days.

In the case of spreading and retting of spread straw by atmospheric fallouts and dew under anaerobic conditions, the main role is played by microscopic fungi. According to data by foreign investigators, the following fungi are the most widespread at straw spreading: *Pullularia* (spires in the stalk bark); *Cladosporium* (forms a velvet taint of olive to dark green color); *Alternaria* (grows through the bark by a flexible colorless chain and unambiguously plays an important role at dew spreading).

The studies indicate that *Cladosporium* fungus is the most active degrader of flax straw pectins.

When retting linen at flax-processing plants, conditions different from spreading are created for microflora. Here flax is submerged to the liquid with low oxygen content due to its displacement from straws by the liquid and consumption by aerobic bacteria, which propagate on easily accessible nutrients extracted from the straw.

These conditions are favorable for multiplication of anaerobic, pectin degrading clostridia related to the group of soil spore bacteria, which includes just few species. Most of them are thermophiles and, therefore, the process takes 2–4 days in the warmed up water; however, at lower temperature (+15…20°C) it takes 10–15 days.

CURIOUS FACTS

In Russia and Check Republic, spreading is the most popular way of processing flax. In Poland, Romania and Hungary, the flax is processed at flax-processing plants by retting, and I Netherlands – by retting and partly by spreading.

The linen fiber obtained by different methods (spread or retted straw) has different spinning properties. The spread straw is now considered to be the best, where the main role in degradation of stalk pectins is played by mold fungi. In production of retted fiber, this role is played by pectin degrading bacteria, some strains of which being able to form an enzyme (cellulose) that degrades cellulose itself. Such impact may be one of the damaging factors in the processes of linen retting. Thus, biostability of the flax depends on the method of fiber production.

The studies show that all kinds of biological treatment increase the quantity of various microbial damages of the fiber. Meanwhile, the spread fiber had lower total number of microscopic damages compared with any other industrial method.

There are other methods for flax production, steaming, for example, that give steamed fiber. It has been found that steamed flax is the most biostable fiber. Possible reasons for so high biostability are high structure ordering of this fiber and high content of modified lignin in it. Moreover, during retting and spreading the fiber is enriched with microorganisms able to degrade cellulose under favorable conditions, whereas steaming sterilizes the fiber.

When exposed to microorganisms, pectins content in the linen fiber decreases by 38%, whereas cellulose content – by 1.2% only. The quantity of wax and ash content of the fiber exposed to microorganisms do not virtually change.

The ordered area share in the linen cellulose is 83.6%, the weakly ordered area – 5.1%, and disordered area – 15.7%. During microbiological degradation the share of disordered areas in the linen cellulose decreases to 7.8%, and the share of ordered areas increases to 86.9%. The share of weakly ordered areas varies insignificantly.

Microbiological damages of linen, jute and other bast fibers and fabrics are manifested by separate staining (occurrence of splotches of color or fiber darkening) and putrefactive odor. On damaged bast fibers, microscopic cross fractures and chips, and microholes and scabs in the fiber walls are observed.

The studies of relative biostability of bast fibers demonstrate that Manilla hemp and jute are most stable, whereas linen and cannabis fibers have the lowest stability.

Natural biostability of bast fibers is generally low and in high humidity and temperature conditions, when exposed to microorganisms, physicochemical and strength indices of both fibers and articles from them rapidly deteriorate. Generally, bast fibers are considered to have virtually the same biostability, as cotton fibers do.

Biostability of cellulose fivers is highly affected by further treatment with finishing solutions (sizing and finishing) containing starch, powder, resins and other substances which confer wearing capacity, wrinkle resistance, fire endurance, and so forth, to textile materials. Many of these substances represent a good culture medium for microorganisms. Therefore, at the stage of yarn and fabric sizing and finishing, the main attention is paid to strict compliance with sanitary and technological measures, which are to prevent fabric infection, by microorganisms and further biodamaging.

A2 BIODAMAGING OF ARTIFICIAL FIBERS

Artificial fibers and fabrics are produced by chemical treatment of natural cellulose obtained from spruce, pine tree and fir. Artificial fibers based on cellulose are viscose, acetate, and so forth, ones. These fibers obtained from natural raw material have higher amorphous structure as compared with high-molecular natural material and, therefore, have lower stability, higher moisture and swelling capacity.

By chemical structure and microbiological stability viscose fibers are similar to common cotton fibers. Biostability of these fibers is low: many cellulosolytic microorganisms are capable of degrading them. Under laboratory conditions, some species of mold fungi shortly (within a month) induces complete degradation of viscose fibers, whereas wool fibers under the same conditions preserve up to 50% of initial stability. For viscose fabrics, the loss of stability induced by soil microorganisms during 12–14 days gives 54 to 76%. These parameters of artificial fibers and fabrics are somewhat higher than for cotton.

Acetate fibers are produced from acetyl cellulose – the product of cellulose etherification by acetic anhydride. Their properties significantly differ from those of viscose fibers and more resemble artificial fibers. For instance, they possess lower moisture retaining property, lesser swelling and loss of strength under wet condition. They are more stable to damaging effect of cellulosolytic enzymes of bacteria and microscopic fungi, because contrary to common cellulose fibers possessing side hydroxyl groups in macromolecules, acetate fiber macromolecules have side acetate groups hindering interaction of macromolecules with enzymes.

Among artificial textile materials of the new generation, textile fibers from bamboo, primarily obtained by Japanese, are highlighted. Bamboo possesses the reference antimicrobial properties due to the presence of "bambocane" substance in the fiber. Bamboo fivers possess extremely porous structure that makes them much more hygroscopic that cotton. Clothes from bamboo fibers struggle against sweat secretion – moisture is immediately absorbed and evaporated by fabric due to presence of pores, and high antimicrobial properties of bamboo prevent perspiration odor.

Various modified viscose fibers, micromodal and modal, for example, produced from beech were not studies for biostability. Information on biostability of artificial fibers produced from lactic casein, soybean protein, maize, peanut and corn is absent.

A3 WOOL FIBER BIODAMAGING

By wool the animal hair is called, widely used in textile and light industry. The structure and chemical composition of the wool fiber significantly differ it from other types of fibers and shows great variety and heterogeneity of properties. Sheep, camel, goat and rabbit wool is used as the raw material.

After thorough cleaning, the wool fiber can be considered virtually consisting of a single protein – keratin. The wool contains the following elements (in %): carbon – 50; hydrogen – 6–7; nitrogen – 15–21; oxygen 21–24; sulfur – 2–5, and other elements.

The chemical feature of wool is high content of various amino acids. It is known that wool is a copolymer of, at least, 17 amino acids, whereas the most of synthetic fibers represent copolymers of two monomers.

Different content of amino acids in wool fibers promotes the features of their chemical properties. Of the great importance is the quantity of cystine containing virtually all sulfur, which is extremely important for the wool fiber properties. The higher sulfur content in the wool is, the better its processing properties are, the higher resistance to chemical and other impacts is and the higher physic-mechanical properties are.

Wool fiber layers, in turn, differ by the sulfur content: it is higher in the cortical layer that in the core.

Among all textile fibers, wool has the most complex structure.

The fine merino wool fiber consists of two layers: external flaky layer or cuticle and internal cortical layer – the cortex. Coarser fibers have the third layer – the core.

The cuticle consists of flattened cells overlapping one another (the flakes) and tightly linked to one another and the cortical layer inside.

Cuticular cells have a membrane, the so-called epicuticle, right around. It is found that epicuticle gives about 2% of the fiber mass. Cuticle cells limited by walls quite tightly adjoin one another, but, nevertheless, there is a thin layer of intercellular protein substance between them, which mass is 3–4% of the fiber mass.

The cortical layer, the cortex, is located under the cuticle and forms the main mass of the fiber and, consequently, defines basic physicomechanical and many other properties of the wool. Cortex is composed of spindle-shaped cells connive to one another. Protein substance is also located between the cells.

The cortical layer cells are composed of densely located cylindrical, thread-like macrofibrils of about 0.05–0.2 μm in diameter. Macrofibrils of the cortical layer are composed of microfibrils with the average diameter of 7–7.5 nm.

Microfibrils, sometimes call the secondary agents, are composed of primary aggregates – protofibrils. Protofibril represents two or three twisted α-spiral chains.

It is suggested [27, 28] that α-spirals are twisted due to periodic repetition of amino acid residues in the chain, hence, side radicals of the same spiral are disposed in the inner space of another α-spiral providing strong interaction, including for the account of hydrophobic bonds, because each seventh residue has a hydrophobic radical.

According to the data by English investigator J.D. Leeder, the wool fiber can be considered as a collection of flaky and cortical cells bound by a cell membrane complex (CMC) which thus forms a uniform continuous phase in the keratin substance of the fiber. This intercellular cement can easily be chemically and microbiologically degraded, that is, a δ-layer about 15 nm thick (CMC or intercellular cement) is located between cells filling in all gaps.

The studies show that the composition of intercellular material between flaky cells may differ from that of the material between cortical cells. In the cuticle-cuticle, cuticle-cortex and cortex-cortex complexes the intercellular "cement" has different chemical compositions.

Although the cell membrane complex gives only 6% of the wool fiber mass, there are proofs that it causes the main effect on many properties of the fiber and fabric. For instance, a suggestion was made that cmC components may affect such mechanical properties, as wear resistance and torsion fatigue, as well as such chemical properties, as resistance to acids, proteolytic enzymes and chemical finishing agents.

The core layer is present in the fibers of coarser wool with the core cell content up to 15%. Disposition and shape of the core layer cells significantly vary with respect to the fiber type. This layer can be continuous (along the whole fiber) or may be separated in sections. The cell carcass of the core layer is composed of protein similar to microfibril cortex protein.

By its chemical composition, wool is a protein substance. The main substance forming wool is keratin – a complex protein containing much sulfur in contrast with other proteins. Keratin is produced during amino acid biosynthesis in the hair bag epidermis in the hide. Keratin structure represents a complex of high-molecular chain batches interacting both laterally and transversally. Along with keratin, wool contains lower amounts of other substances.

Wool keratin reactivity is defined by its primary, secondary and tertiary structures, that is, the structure of the main polypeptide chains, the nature of side radicals and the presence of cross bonds.

Among all amino acids, only cystine forms cross bonds; their presence considerably defines wool insolubility in many reagents. Cystine bond decomposition simplifies wool damaging by sunlight, oxidants and other agents. Cystine contains almost all sulfur present in the wool fibers. Sulfur is very important for the wool quality, because it improves chemical properties, strength and elasticity of fibers.

Along with general regularities in the structure of high-molecular compounds, fibers differ from one another by chemical composition, monomer structure, polymerization degree, orientation, intermolecular bond strength and type, and so forth, that defines different physicomechanical and chemical properties of the fibers.

The main chemical component of wool – keratin, is a nutrition for microorganisms.

Microorganisms may not directly consume proteins. Therefore, they are only consumed by microbes having proteolytic enzymes – exoproteases that are excreted by cells to the environment.

Wool damage may start already before sheep shearing, that is, in the fleece, where favorable nutritive (sebaceous matters, wax, epithelium), temperature, aeration and humidity conditions are formed.

Contrary to microorganisms damaging plant fibers, the wool microflora is versatile, generally represented by species typical of the soil and degrading plant residues.

Initiated in the fleece, wool fiber damages are intensified during its storage, processing and transportation under unfavorable conditions.

Specific epiphytic microflora typical of this particular fiber is always present on its surface. Representatives of this microflora excrete proteolyric enzymes (mostly pepsin), which induce hydrolytic keratin decay by polypeptide bonds to separate amino acids.

Wool is degraded in several stages: first, microorganisms destroy the flaky layer and then penetrate into the cortical layer of the fiber, although the cortical layer itself is not destroyed, because intercellular substance located between the cells is the culture medium. As a result, the fiber structure is disturbed: flakes and cells are not bound yet, the fiber cracks and decays.

The mechanism of wool fiber hydrolysis by microorganisms suggested by American scientist E. Race represents a sequence of transformations: proteins – peptones – polypeptides – water + ammonia + carboxylic acids.

The most active bacteria: *Alkaligenes bookeri, Pseudomonas aeroginosa, Proteus vulgaris, Bacillus agri, B. mycoides, B. mesentericus, B. megatherium, B. subtilis, and microscopic fungi: Aspergillus, Alternaria, Cephalothecium, De-*

matium, Fusarium, Oospora, Penicillium, Trichoderma, were extracted from the wool fiber surface.

However, the dominant role in the wool degradation is played by bacteria. Fungi are less active in degrading wool. Consuming fat and dermal excretion, fungi create conditions for further vital activity of bacteria-degraders. The role of microscopic fungi may also be reduced to splitting the ends of fibers resulting mechanical efforts of growing hyphae. Such splitting allows bacteria to penetrate into the fiber. Fungi weakly use wool as the source of carbon.

In 1960 s, the data on the effect of fat and dirt present on the surface of unclean fibers on the wool biodamaging were published. It is found that unclean wool is damaged much faster than clean one. The presence of fats on unclean wool promotes fungal microflora development.

The activity of microbiological processes developing on the wool depends on mechanical damages of the fiber and preliminary processing of the wool.

It is found that microorganism penetration may happen through fiber cuts or microcracks in the flaky layer. Cracks may be of different origins – mechanical, chemical, and so forth. It is also found that wool subject to intensive mechanical or chemical treatment is easier degraded by microorganisms than untreated one.

For instance, high activity of microorganisms during wool bleaching by hydrogen peroxide in the presence of alkaline agents and on wool washed in the alkaline medium was observed. When wool is treated in a weak acid medium, the activity of microorganisms is abruptly suppressed. This also takes place on the wool colored by chrome and metal-containing dyes. The middle activity of microorganisms is observed on the wool colored by acid dyes.

When impacted by microorganisms, structural changes in the wool are observed: flaky layer damages, its complete exfoliation, and lamination of the cortical layer.

Wool fiber damages can be reduced to several generalized types provided by their structural features:

- channeling and overgrowth – accumulation of bacteria or fungal hyphae and their metabolites on the fiber surface;
- flaky layer damage, local and spread;
- cortical layer lamination to spindle-shaped cells;
- spindle-shaped cell destruction.

Along with the fiber structure damage, some bacteria and fungi decrease its quality by making wool dirty blue or green that may not be removed by water or detergents. Splotches of color also occur on wool, for example, due to the impact of *Pseudomonas aeruginosa bacteria; in this case, color depends on medium pH: green splotches occurred in a weakly alkaline medium, and in weak acid medium they are red. Green splotches may also be caused by development of* Dermatophilus *congolensis* fungi. Black color of wool is provided by *Pyronellaea glomerata* fungi.

Thus, wool damage reduces its strength, increases waste quantity at combing and imparts undesirable blue, green or dirty color and putrefactive odor.

However, wool is degraded by microorganisms slower than plant fibers.

A4 CHANGES OF STRUCTURE AND PROPERTIES OF WOOL FIBERS BY MICROORGANISMS

To evaluate bacterial contamination of wool fibers, it is suggested to use an index suggested by A.I. Sapozhnikova, which characterizes discoloration rate of resazurin solution, a weak organic dye and currently hydrogen acceptor. It is also indicator of both presence and activity of reductase enzyme.

The method is based on resazurin ability to lose color in the presence of reductase, which is microorganisms' metabolite, due to redox reaction proceeding. This enables judging about quantity of active microorganisms present in the studied objects by solution discoloration degree.

Discoloration of the dye solution was evaluated both visually and spectrophotometrically by optical density value.

Table 1 shows results of visual observations of color transitions and optical density measurements of incubation solutions after posing the reductase test.

As follows from the data obtained, coloration of water extracts smoothly changed from blue-purple for control sterile physiological solution (D = 0.889) to purple for initial wool samples (D_{thin} = 0.821 and D_{coarse} = 0.779), crimson (D_{thin} = 0.657 and D_{coarse} = 0.651) and light crimson at high bacterial contamination (D_{thin} = 0.548 and D_{coarse} = 0.449 and 0.328) depending on bacterial content of the fibers.

TABLE A1 Visual coloration and optical density of incubation solutions with wool fibers at the wavelength λ = 600 nm and different stages of spontaneous microflora development.

Time of microorganism development, days	Thin merino wool		Coarse caracul wool	
	Optical density	Color (visual assessment)	Optical density	Color (visual assessment)
Control, physiological saline	0.889	Blue purple	0.889	Blue purple
0 (init.)	0.821	Purple	0.779	Purple
7	0.712	Purple	0.657	Crimson
14	0.651	Crimson	0.449	Light crimson
28	0.548	Light crimson	0.328	Light crimson

This dependence can be used for evaluation of bacterial contamination degree for wool samples applying color standard scale.

It is found that the impact of microorganisms usually reduces fiber strength, especially for coarse caracul wool: after 28 days of impact strength decreased by 57–65%. The average rate of strength reduction is about 2% per day (Fig. A8).

It is found that after 28 days of exposure, the highest reduction of breaking load of wool fibers is induced by *Bac. subtilis* bacteria.

Figure A9 clearly shows that 14 days after exposure to microorganisms the surface of coarse wool fiber is almost completely covered by bacterial cells. Meanwhile, note also (Fig. A10) that cuticular cells themselves are not damaged, but their bonding is disturbed that grants access to cortical cells for microorganisms. Figure A11 shows the wool fiber decay to separate fibrils caused by microorganisms.

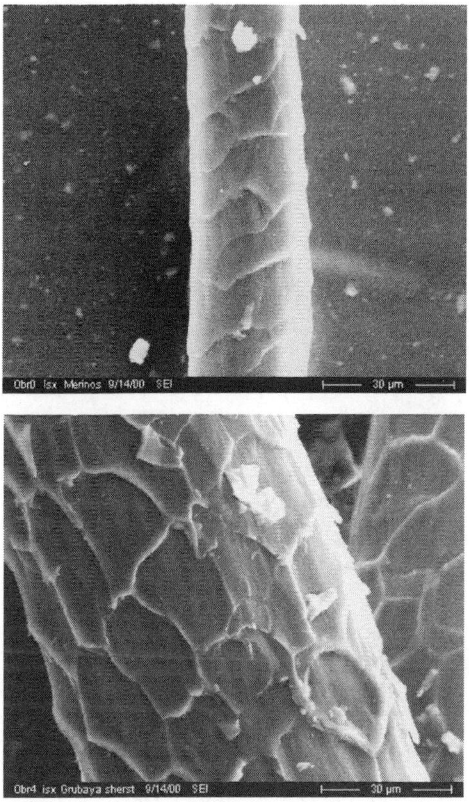

FIGURE A8 Micrographs of initial wool fibers (×1000): a – thin merino wool; *b* – coarse caracul wool

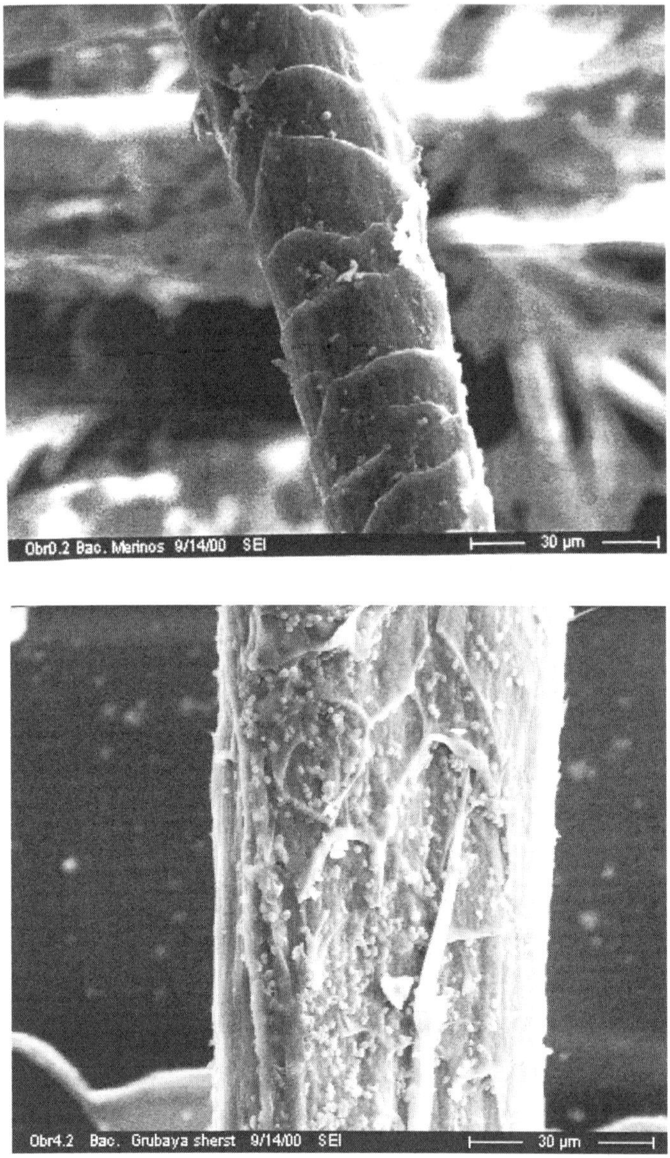

FIGURE A9 Micrographs of thin merino wool (a) and coarse caracul wool (*b*) after 14 days of exposure to *Bac. subtilis* (×1000)

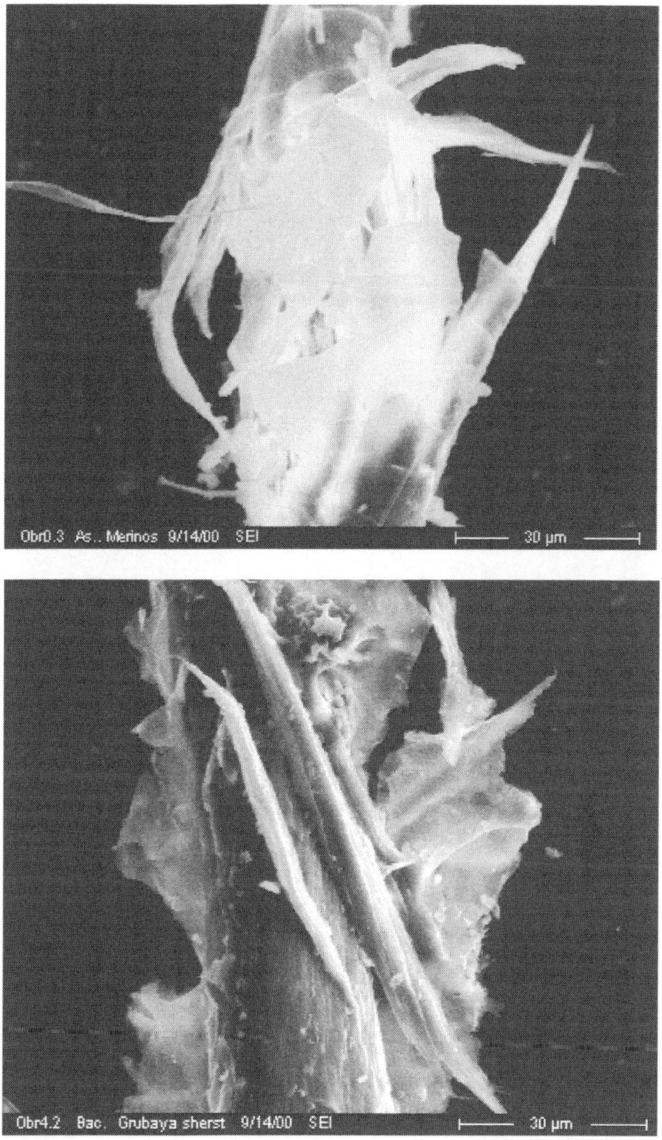

FIGURE A10 Micrographs of flaky layer destruction of thin (a) and coarse (*b*) wool fibers after 14 days of exposure to spontaneous microflora (×1000)

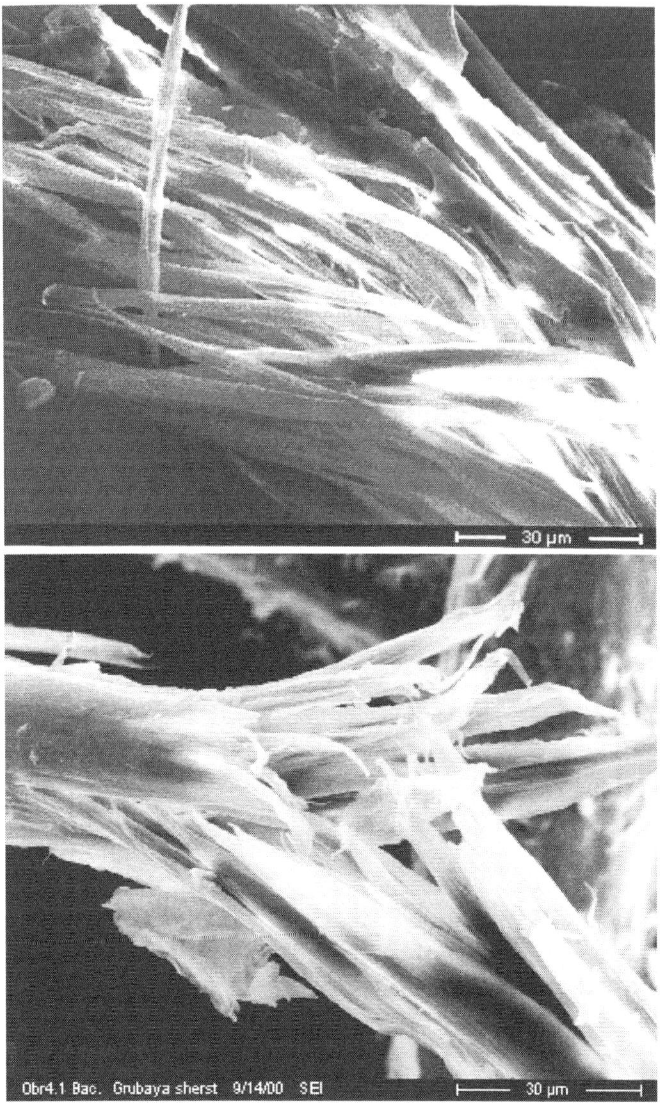

FIGURE A11 Micrographs of thin (a) and coarse (*b*) wool fiber fibrillation after 28 days of exposure to microorganisms (×1000)

Wool fiber biodegradation changes the important quality indices, such as whiteness and yellowness. This process can be characterized as "yellowing" of wool fibers.

Table A2 shows data obtained on yellowness of thin merino and coarse caracul wool fibers exposed to spontaneous microflora, *Bac. subtilis* bacteria and *Asp. niger* fungi during 7, 14 and 28 days.

TABLE A2 Yellowness of thin merino and coarse caracul wool fibers after different times of exposure to different microorganisms.

Time of exposure to microorganisms, days Type of fibers	Yellowness, %		
	Impacting microorganisms		
	Bac. subtilis	*Asp. niger*	Spontaneous microflora
Thin, 0	27.7	27.7	27.7
Thin, 7	36.4	37.3	29.1
Thin, 14	45.4	42.5	34.8
Thin, 28	53.9	48.1	39.9
Coarse 0	39.3	39.3	39.3
Coarse 7	46.8	44.2	41.2
Coarse 14	51.8	46.4	42.1
Coarse 28	57.1	49.5	43.3

In terms of detection of biodegradation mechanism and changes of material properties, determination of relations between changes in properties and structure of fibers affected by microorganisms is of the highest importance. Yellowness increase of wool fibers affected by microorganisms testifies occurrence of additional coloring centers.

For the purpose of elucidating the mechanism of microorganisms' action on the wool fibers that leads to significant changes of properties and structure of the material, the changes of amino acid composition of wool keratin proteins being the nutrition source of microorganisms damaging the material are studied.

Exposure to microorganisms during 28 days leads to wool fiber degradation and, consequently, to noticeable mass reduction of all amino acids in the fiber composition. To the greatest extent, these changes are observed for coarse wool, where total quantity of amino acids is reduced by 10–12 rel. %, and for merino wool slightly lower reduction (4.7 rel. %) is observed.

It should be noted that at comparatively low reduction of the average amount of amino acids in the system (not more than 12 rel. %) all types of wool fibers demonstrate a significant reduction of the quantity of some amino acids, such as serine, cystine, methionine, and so forth (up to 25–33 rel. %).

Analysis of the data obtained testifies that in all types of wool fibers (but to different extent) reduction of quantity of amino acids with disulfide bonds (cystine, methionine) and ones related to polar (hydrophilic) amino acids, including serine, glycine, threonine and tyrosine, is observed. These very amino acids provide hydrogen bonds imparting stability to keratin structure.

In the primary structure of keratin, serine is the N-end group, and Tyrosine is C-end group. In this connection, reduction of the quantity of these amino acids testifies degradation of the primary structure of the protein.

Changes observed in the amino acid composition of wool fiber proteins exposed to spontaneous microflora may testify that microorganisms degrade peptide and disulfide bonds, which provide stability of the primary structure of proteins, and break hydrogen bonds, which play the main role in stabilization of spatial structure of proteins (secondary, tertiary and quaternary).

Very important data were obtained in the study of the wool fiber structure by IR-spectroscopy method. It is found that when microorganisms affect wool fibers, their surface layers demonstrate increasing quantity of hydroxyl groups that indicates accumulation of functional COO-groups, nitrogen concentration in keratin molecule decreases, and protein chain configuration partly changes – β-configuration (stretched chains) transits to α-configuration (a spiral). This transition depends on α- and β-forms ratio in the initial fiber and is more significant for thin fibers, which mostly have β-configuration of chains in the initial fibers.

Thus, it is found that microorganisms generally affect the cell membrane complex and degrade amino acids, such as cystine, methionine, serine, glycine, threonine and tyrosine. Microorganisms reduce breaking load and causes "yellowing" of the wool fibers (Fig. A12).

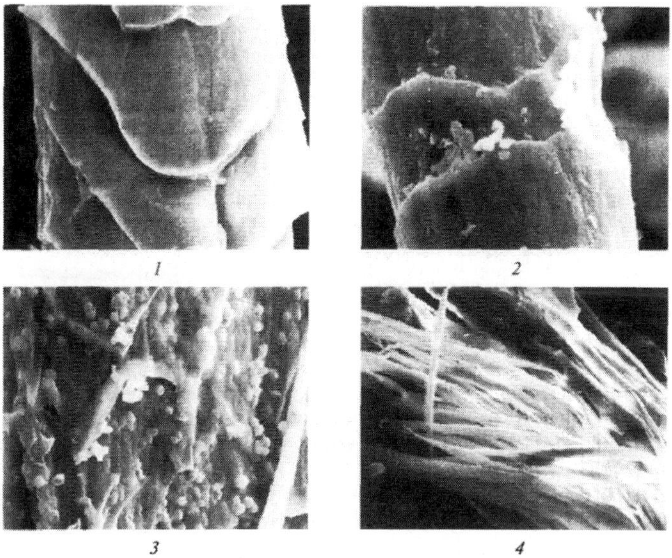

FIGURE A12 Micrographs of wool fibers: 1 – initial undamaged fiber (× 3000); 2, 3 – bacterial cells on the fiber surface (× 3000); 4 – fiber fibrillation after exposure to microorganisms during 4 weeks (× 1000)

A5 BIODAMAGES OF SYNTHETIC FIBERS

Synthetic fibers are principally different from natural and artificial ones by structure and, being an alien substrate for microorganisms, are harder damaged by them. Since occurrence of synthetic fabrics in 1950 s, it is suggested that they are "everlasting" and are not used by microorganisms. However, it has been found with time that, firstly, microorganisms although slower, but yet are capable of colonizing synthetic fabrics and using their carbon in the course of development (i.e. causing biodamage), and secondly, there are both more and less microorganism resistant fabrics among synthetic ones.

Among microorganisms damaging synthetic fibers, Trichoderma genus fungi are identified, at the initial stages developing due to lubricants and finishing agents without fiber damage and then wrap them with mycelium, loosen threads and, hence, reduce fabric strength.

When studying fabrics from nitrone, lavsan, caprone, it has been found that soil fungi and bacteria cause roughly the same effect on characteristics of these fabrics increasing the fiber swelling degree by 20–25%, reducing strength by 10–15% and elongation at break by 15–20%.

Synthetic fibers represent potential source of energy and nutrition for microorganisms. The ability of microorganisms to attach to surfaces of insoluble solids, then using them as the nutritive substrate, is well known. Living cells of microorganisms have complex structure, just on the surface of bacterial cells complexes of proteins, lipids and polysaccharides were found; it contains hydrophilic and hydrophobic areas, various functional groups and mosaic electric charge (at total negative charge of the cells).

The first stage of microorganism interaction with synthetic fibers can be rightfully considered in terms of the adhesion theory with provision for the features of structure and properties of microorganisms as a biological system.

The entire process of microorganism impact of the fiber can conditionally be divided into several stages: attachment to the fiber, growth and multiplication on it and consumption of it, as the nutrition and energy source.

Enzymes excreted by bacteria act just in the vicinity of bacterial membrane. Been adsorbed onto the fiber, living cells attach to the surface and adapt to new living conditions. The ability to be adsorbed onto the surface of synthetic fibers is caused by:

- the features of chemical structure of the fibers. For instance, fibers adsorbing microorganisms are polyamide and polyvinyl alcohol ones; the fiber not adsorbing microorganisms is, for example, ftorin;
- physical structure of the fiber. For example, fibers with smaller linear density, with a lubricant on the surface absorb greater amount of microorganisms;
- the presence of electric charge on the surface, its value and sign. Positively charged chemical fibers adsorb virtually all bacteria, fibers having no electric charge adsorb the majority of bacteria, and negatively charged fibers do not adsorb bacteria.

Supermolecular structure also stipulates the possibility for microorganisms and their metabolites to diffuse inside the internal areas of the fiber. Microorganism assimilation of the fiber starts from the surface, and further degradation processes and their rate are determined by microphysical state of the fiber. Microorganism metabolite penetration into inner areas of the fiber and deep layers of a crystalline material is only possible in the presence of capillaries.

Chemical fiber damaging and degradation starting from the surface are, in many instances, promoted by defects like cracks, chips or hollows which may occur in the course of fiber production and finishing.

Along with physical inhomogeneity, chemical inhomogeneity may promote biodegradation of synthetic fibers. Chemical inhomogeneity occurs during polymer synthesis and its thermal treatments, manifesting itself in different content of monomers and various end groups. The possibility for microorganism metabolites to penetrate inside the structure of synthetic fibers depends on the quantity and accessibility of functional end groups in the polymer, which are abundant in oligomers.

The ability to synthetic fibers to swell also makes penetration of biological agents inside low ordered areas of fibers and weakens intermolecular interactions, off-orientation of macromolecules, and degradation in the amorphous and crystalline zones. Structural changes result in reduction of strength properties of fibers.

Theoretical statements that synthetic fibers with the lower ordered structure and higher content of oligomers possess lower stability to microorganism impact than fibers with highly organized structure and lower content of low-molecular compounds.

Thus, the most rapid occurrence and biodegradation of synthetic fibers are promoted by low ordering and low orientation of macromolecules in the fibers, their low density, low crystallinity and the presence of defects in macro and microstructure of the fibers, pores and cavities in their internal zones.

Carbochain polymer based fibers are higher resistant to microbiological damages. These polymers are: polyolefins, polyvinyl chloride, polyvinyl fluoride, polyacrylonitrile, polyvinyl alcohol. Fibers based on heterochain polymers: polyamide, polyether, polyurethane, and so forth, are less bioresistant.

Comparative soil tests for biostability of artificial and synthetic fibers demonstrate that viscose fiber is completely destroyed on the 17th day of tests; bacterium and fungus colonies occur on lavsan on the 20th day; caprone is overgrown by fungus mycelium on the 30th day. Chlorin and ftorlon have the highest biostability. The initial signs of their biodamage are only observed 3 months after the test initiation.

The studies of nitron, lavsan and capron fabric biostability have found that soil fungi and bacteria cause nearly equal influence on parameters of these fabrics, increasing swelling degree of the fibers by 20–25%, reducing strength by 10–15% and elongation at break by 15–20%. Meanwhile, nitron demonstrated higher biostability, as compared with lavsan and capron.

A6 CHANGES IN STRUCTURE AND PROPERTIES OF POLYAMIDE FIBERS INDUCED BY MICROORGANISMS

In contrast with natural fibers, chemical fibers have no permanent and particular microflora. Therefore, the most widespread species of microorganisms possessing increased adaptability are the main biodegraders of these materials.

Occurrence and progression of biological degradation of polyamide fibers is, in many instances, is induced by their properties and properties of affecting microorganisms, and their species composition. Generally, the species of microorganisms degrading polyamide and other chemical fibers are determined by their operation conditions, which form microflora, and its adaptive abilities.

Polyamide fibers are most frequently used in mixtures with natural fibers. Natural fibers contain specific microflora on the surface and inside. Therefore, capron fibers mixed with cotton, wool or linen are affected by their microflora. It is found that capron fiber degradation by microorganisms obtained from wool is characterized as deep fiber decay; microorganisms extracted from natural silk cause streakiness of capron fibers; microorganisms extracted from cotton cause fading and decomposition; microorganisms extracted from linen cause fading, streakiness and decomposition.

The microorganism interaction with polyamide fibers is most fully studied in the works by scientists. For the purpose of detecting bacteria-degraders of polyamide fibers, microorganisms were extracted from fibers damaged in the medium of active sewage silt, soil, microflora of natural fibers and test-bacteria complex selected as degraders of polyamide materials. Capron fibers were inoculated by extracted cultures of microorganisms and types of damages were reproduced.

Polyamide fiber materials were natural nutritive and energy source for these microorganisms. Therefore, bacterial strains extracted from damaged fibers were different from the initial strains. It is proven that the existence on a new substrate has stipulated changes of intensity and direction of physiological-biochemical processes of bacterial cells, the change of their morphological and culture properties.

Extraction, cultivation and use of such adaptive strains are of both scientific and practical interest. Using bacterial strains adaptive to capron, polyamide production waste, warn products, toxic substances may be used. This allows obtaining of secondary raw materials and solving the problem of environmental protection.

It is proved experimentally that polycaproamide fibers possess high adsorbability, which value depends on the properties of impacting bacteria. For instance, gram-positive, especially sporeforming bacteria *Bacillus subtilis, Bac. mesentericus* (from 84.5 to 99.3% of living cells), are most highly adsorbed, and adsorption of gram-positive bacteria varies significantly.

The extensive research of test-bacterium complex impact was performed. These bacteria were chosen as degraders of polyamide fibers, along with microflora of active sewage silts, linen and jute microorganisms as degraders of a complex capron thread. It is found that the test-bacterium complex injected by the author, after 7 months of exposure, increases biodegradation index to 1.27 that testifies about intensive degradation of the fiber microstructure. The highest degradation of complex polycaproamide (PCA) thread is caused by from active sewage silt microorganisms.

High activity of silt microorganisms and test-bacteria is explained by the fact that they include bacterium strains *Bacillus subtilis* and *Bac. mesentericus*, which according to the data by a number of authors may induce full degradation of caprolactam to amino acids using it, as the source of carbon and nitrogen.

Thus, adaptive forms of microorganisms induce the highest degradation of polyamide fibers.

Polyamide fibers are characterized by physical structure inhomogeneity, which occurs during processing and is associated with differences in crystallinity and orientation of macromolecules determining fiber accessibility for microorganisms and their metabolites penetration. The surface layer is damaged during orientational stretching and, consequently, has lower molecular alignment. That is why the surface layer is most intensively changed by microorganisms. The study of polyamide fiber macrostructure after exposure to microorganisms shows that streakiness and cover damage are the main damages of these fibers.

The studies of supermolecular structure of capron fiber surface show that after exposure to microorganisms the capron fiber cover becomes loose and uneven [39, 43]. Surface supermolecular structure degradation increases with the microbial impact: fibrils and their yarns become split and misaligned both laterally and transversely, multiple defects in the form of pores and cavities, and cracks of various depths are formed.

Chemical inhomogeneity of polyamide fibers also promotes changes in the fiber structure, when exposed to microorganisms. It is found that the polyamide fiber degradation increases with low-molecular compound (LMC) content in them; meanwhile, at the same content of low-molecular compounds, thermally treated fibers were higher biostable, as compared with untreated specimens.

Along with the morphological characters, which characterize biodamaging of the fibers, functional features, such as strength decrease and increase of deformation properties of the fiber, were detected. The greatest strength decrease (by 46.4%) was observed for thermally untreated capron fiber containing 3.4% LMC, and the smallest decrease (by 5%) was observed for the fibers with 3.2% LMC, thermally treated at the optimum time of 5 s.

IR-spectroscopy method was applied to detect polycaproamide fiber damage by microorganisms [39]. It is found that carboxyl and amide groups are accumulated during their biodegradation.

The change of various property indices of polyamide fibers also results from macro, micro and chemical changes.

Of interest are studies of the microorganism effect on polyamide fabric quality. Fabrics (both bleached and colored) from capron monofilament were exposed to a set of test-cultures: *Bac.subtilis, Ps.fluorecsens, Ps.herbicola, Bac.mesentericus*. After 3–9 month exposure, yellowness and dark spots occurred, coloring intensity decreased, and an odor appeared. The optical microscopy studies indicated that all fibers exposed to microorganisms, had damages typical of synthetic fibers – overgrowing, streakiness, bubbles, wall damages. The increasing quantity of biodamages with time results in tensile strength reduction: by 6–8% for capron fibers after 9 months, at inconsiderable change of relative elongation.

It all goes to show that development of microorganisms on polyamide fibrous materials results in changes of fibers morphology, their molecular and supermolecular structure and, as a consequence, reduction of strength properties, color change, and odor.

To clear up the mechanism of polycaproamide fiber degradation by polarographic investigation, a possibility of ε-amino caproic acid accumulation by *Bacillus subtilis k1* culture during degradation of polycaproamide fibers 0.3 and 0.7 tex fineness was studied (fig.A13).

As a source of carbon, ε-amino caproic acid (10 mg/l) or PCA (0.5 g/l) was injected into the mineral medium.

It is found that at PCA fibrous materials exposure to *Bacillus subtilis k1*strain, the maximum quantity of ε-amino caproic acid is liberated on the fifth day: 32 mg/l, 66 mg/l.

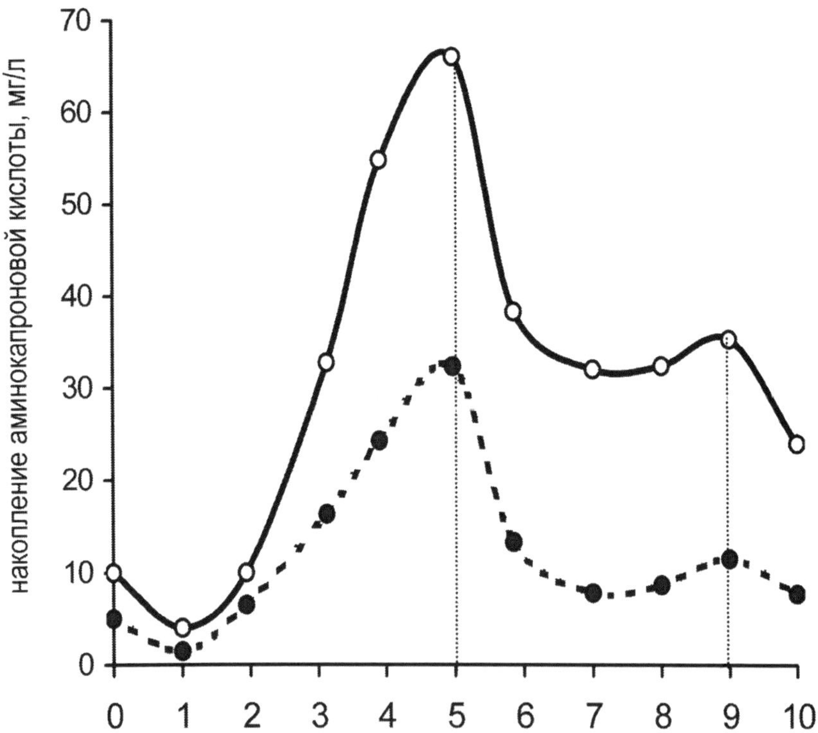

FIGURE A13 The change of ε-amino caproic acid (ACA) concentration during *Bacillus subtilis k1* development on PCA material fibers: – 0.7 tex; – 0.3 tex.

It is known that if the medium includes several substrates metabolized by a particular strain of microorganisms, the substrate providing the maximum culture propagation rate is consumed first. As this substrate is going to exhaust, bacteria subsequently consume other substrates, which provide lower rates of cell multiplication.

In the mineral medium containing chemically pure ε-amino caproic acid (with the initial concentration of 10 mg/L) as the source of carbon, its concentration decreased gradually, and 5 days after it was not detected in the solution.

Thus, basing on the data obtained, one may conclude that, firstly, PCA fibers of 0.3 tex fineness are more accessible for microorganisms; secondly, *Bacillus subtilis kl* strain can be used to use polycaproamide fibrous materials; thirdly, polarographic analysis has proven the mechanism of PCA fibrous material degradation with ε-amino caproic acid liberation. As a consequence, suggested *Bacillus subtilis kl* strain VKM No.V-1676D degrades PCA fibrous materials at the both macro and microstructure levels, with ε-amino caproic acid formation.

A7 METHODS OF TEXTILE MATERIAL PROTECTION AGAINST DAMAGING BY MICROORGANISMS

Imparting antimicrobial properties to textile materials pursues two main aims: protection of the objects contacting with textile materials against actions of microorganisms and pathogenic microflora.

In the first case, we speak about imparting biostability to materials and, as a consequence, about passive protection. The second case concerns creation of conditions for preventive attack of a textile material on pathogenic bacteria and fungi to prevent their impact on the protected object.

The basic method of increasing biostability of textile materials is application of antimicrobial agents (biocides). The requirements to the "ideal" biocide are the following:

- The efficacy against the most widespread microorganisms at minimal concentration and maximal action time;
- Non-toxicity of applied concentrations for people;
- The absence of color and odor;
- Low price and ease of application;
- Retaining of physicomechanical, hygienic and other properties of the product;
- Compatibility with other finishing agents and textile auxiliaries;
- light stability and weather ability.

At any time, nearly every class of chemical compounds was applied to impart textile materials antibacterial or antifungal activity. Today, application of nanotechnologies, specifically injection of silver and iodine nanoparticles, to impart textile antimicrobial properties is of the greatest prospect.

At all times, copper, silver, tin, mercury, and so forth, salts were used to protect fibrous materials against biodamages. Among these biocides, the most widespread are copper salts due to their low cost and comparatively toxicity. The use of zinc salts is limited by their low biocide action, whereas mercury, tin and arsenic salts are highly toxic for the man. However, there are organomercury preparations applied to synthetic and natural fibrous materials used as linings and shoe plates, widely advertised for antibacterial and antifungal finishing.

The data are cited that impregnation of textile materials by a mixture of neomycin with tartaric, propionic, stearic, phthalic and some other acids imparts them the bacteriostatic effect. Acids were dissolved in water, methyl alcohol or butyl alcohol and were sprayed on the material.

The methods of imparting textile materials biostability can be divided into the following groups:

- impregnation by biocides, chemical and physical modification of fibers and threads, which then form a textile material;
- cloth impregnation by antimicrobial agent solutions of emulsions, its chemical modification;
- injection of antimicrobial agents into the binder (at nonwoven material manufacture by the chemical method);
- imparting antimicrobial properties to textile materials during their coloring and finishing;
- application of disinfectants during chemical cleaning or laundry of textile products.

However, impregnation of fibers and cloth does not provide firm attachment of reagents. As a result, the antimicrobial action of such materials is nondurable. The most effective methods of imparting biocide properties to textile materials are those providing chemical bond formation, that is, chemical modification methods. Chemical modification methods for fibrous materials represent processing that leads to clathrate formation, for example, injection of biological active agents into spinning melts or solutions.

At the stage of capron polymerization, an antibacterial organotin compound (tributyltin oxide or hydroxide) is added that retains the antibacterial effect after multiple laundries. Methods of imparting antimicrobial properties to textile materials by injection of nitrofuran compounds into spinning melts with further fixing them at molding in the fine structure of fibers similar to clathrates were designed.

There are data on imparting antimicrobial properties to synthetic materials during oiling. Prior to drafting, fibers are treated by compounds based on oxyquinoline derivatives, by aromatic amines or nitrofuran derivatives. Such fibers possess durable antimicrobial effect.

Nanotechnologies are actively intruded in the light industry, allowing obtaining of materials with antimicrobial properties. The following directions using nanotechnologies, which are now investigated, should be outlined, including creation of new textile materials on the account of:

primarily, the use of textile nonfibers and threads in the materials;

secondly, the use of nanodispersions and nanoemulsions for textile finishing.

It may be said that nanotechnologies allow a significant decrease of expenses at the main production stage, where consumption of raw materials and semiproducts is considerable. For nanoparticles imparting antimicrobial properties, silver, copper, palladium, and so forth, particles are widely used. Silver is the natural antimicrobial agent, which properties are intensified by the nanoscale of particles (the surface area sharply increases), so that such textile is able to kill multiple microorganisms and viruses. In this form, silver also reduces necessity in fabric cleaning, eliminates sweat odor as a result of microorganism development on the human body during wearing.

Properties of materials designed with the use of silver nanoparticles, which prevent multiplication of various microorganisms, may be useful in medicine, for example. The examples are surgical retention sutures, bandages, plasters, surgical boots, medical masks, whites, skullcaps, towels, and so forth. Customers know well sports clothes, prophylactic socks with antimicrobial properties.

Along with chemical fibers and threads, natural ones are treated by nanoparticles. For example, silver and palladium nanoparticles (5–20 nm in diameter) were synthesized in citric acid, which prevented their agglutination, and then natural fibers were dipped in the solution with these negatively charged particles. Nanoparticles imparted antibacterial properties and even ability to purify air from pollutants and allergens to clothes and underwear.

When these products appeared at the world market, disputes about ecological properties and the influence of these technologies on the human organism have arisen. There are no accurate data yet how these developments may affect the human organism. However, it should be noted that some specialists do not recommend everyday use of antibacterial socks, because these antibacterial properties affect the natural skin microflora.

Nanomaterials are primarily hazardous due to their microscopic size. Firstly, owing to small size they are chemically more active because of a great total area of the nanosubstance. As a result, low toxic substance may become extremely toxic. Secondly, chemical properties of the nanosubstance may significantly change due to manifestations of quantum effects that, finally, may make a safe substance extremely hazardous. Thirdly, due to small size, nanoparticles freely permeate through cellular membranes damaging bioplasts and disturbing the cell operation.

Physical modification of fibers or threads is the direct change of their composition (without new chemical formations and transformations), structure (supermolecular and textile), properties, production technology and processing. Modernization of the structure and increase of the fiber crystallinity degree induces biostability increase. However, in contrast with chemical modification, physical modification does not impart antimicrobial properties to the fibers, but may increase biostability.

By no means always textile materials produced completely from antimicrobial fibers are required. Even a small fracture of highly active antimicrobial fiber (e.g., 1/3 or even 1/4) is able to provide sufficient biostability to the entire material. The studies show that antimicrobial fibers were found not only protected themselves against microorganism damage, but also capable of shielding plant fibers from their impact.

Manufacture of antimicrobial nonwoven materials by injection of active microcapsule ingredients into it is of interest. Microcapsules can contain solid particles of microdrops of antimicrobial substances liberated under particular conditions (e.g., by friction, pressure, dissolution of capsule coatings or their biodegradation).

Biostability of fibrous materials may be significantly affected by the dye selection. Dyes possessing antimicrobial activity on the fiber are known: salicylic acid derivatives capable of bonding copper, triphenylmethane, acridic, thiazonic, and so forth, ones. For instance, chromium-containing dyes possess antibacterial action, but resistance to mold fungi is not imparted.

It is known that synthetic fibers dyed by dispersed pigments are more intensively degraded by microorganisms. It is suggested that these pigments make the fiber surface more accessible for bacteria and fungi.

Single bath coloring and bioprotective finishing of textile materials are also applied. A combination of these processes is not only of theoretical interest, but is also perspective in terms of technology and economy.

Processing of textile materials by silicones also imparts antimicrobial properties to these clothes. Some authors state that textile material sizing by water repellents imparts them sufficient antimicrobial activity. Water repellency of materials may reduce the adverse impact of microorganisms, because the quantity of adsorbed moisture is reduced. However, Hydrophobic finishing itself may not fully eliminate the adverse effect of microorganisms. Therefore, antimicrobial properties imparted to some textile materials during silicone finishing may be related to application of metal salts as catalysts, such as copper, chromium and aluminum.

Disinfectants, for example, at laundry, may be applied by the customer himself. The method of sanitizing substance application for carpets, which is spraying or dispensing of a disinfectant on the surface of floor covers during operation is known. The acceptable disinfection level may be obtained during laundry of textile products by such detergents, which may create residual fungal and bacteriostatic activity.

APPENDIX B

Figure B1 shows the scanning electronic micrographs of carbon chitosan/nanotube electrospun fibers, at different chitosan concentrations, in a TFA/DCM (70:30) solvent. As presented in Fig. B1a, at low concentrations of chitosan, the beads deposited on the collector and the thin fibers coexisted. As the concentration of chitosan increased (Figs. B1a–c), the beads decreased significantly. Figure B1c shows homogenous electrospun fibers with minimum beads, thin fibers and interconnected fibers. The increase of chitosan concentration leads to an increase of the interconnected fibers, as shown in Figures B1d-e. The average diameter of chitosan/carbon nanotube fibers increased when increasing the concentration of chitosan (Figs. B1 a–e). Hence, a chitosan/carbon nanotube solution of TFA/DCM (70:30) with 10 wt% of chitosan assured optimized conditions for electrospinning of this solution. When the voltage was low, the beads were deposited on the collector (Fig. B2a). As shown in Fig. B2a–d, the number of beads decreased with increasing voltage from 18 to 24 kV. In our study, the average diameter of fibers prepared by 18 kV was measured as 307 nm. As the applied voltage increased, the average fiber diameters also increased. The average diameter of fibers for 20 kV (2b), 22 kV (2c), and 24 kV (2d), was 308 (194–792), 448 (267–656) and 455 (306–672), respectively.

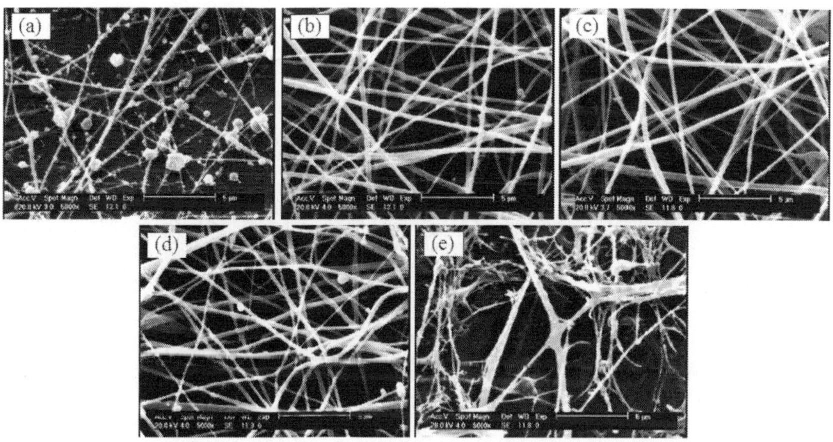

FIGURE B1 Scanning electron micrographs of electrospun fibers at different chitosan concentrations (wt%): (a) 8, (b) 9, (c) 10, (d) 11, (e) 12, 24 kV, 5 cm, TFA/DCM: 70/30.

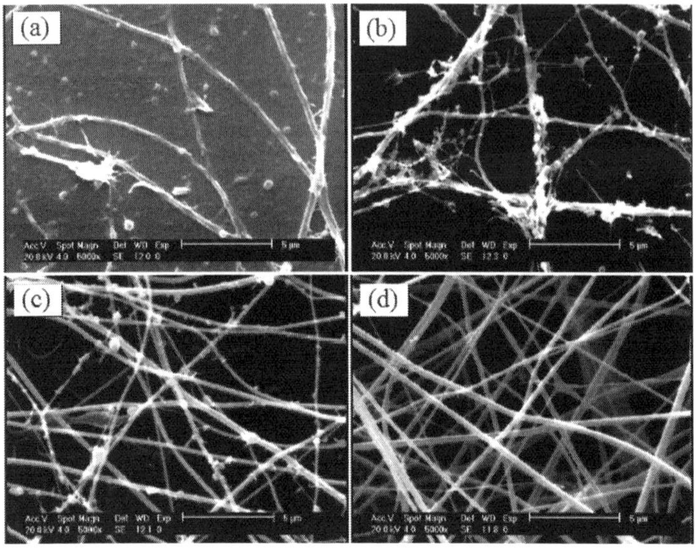

FIGURE B2 Scanning electron micrographs of electrospun fibers at different voltages (kV): (a) 18, (b) 20, (c) 22, (d) 24, 5 cm, 10 wt%, TFA/DCM: 70/30.

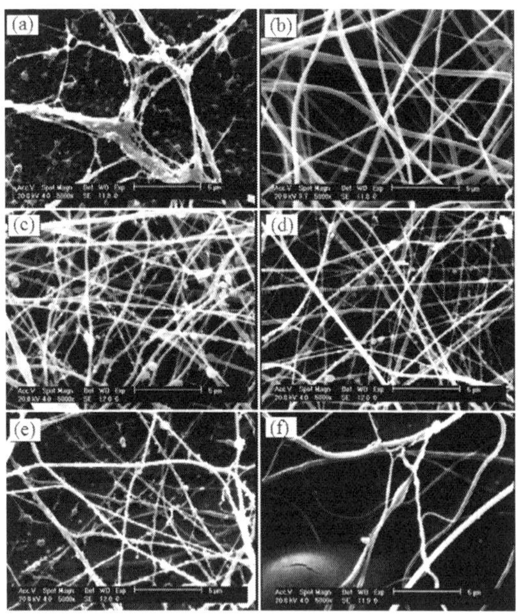

FIGURE B3 Scanning electron micrographs of electrospun fibers of chitosan/carbon nanotubes at different tip-to-collector distances (cm): (a) 4, (b) 5, (c) 6, (d) 7, (e) 8, (f) 10, 24 kV, 10 wt%, TFA/DCM: 70/30.

The morphologies of chitosan/carbon nanotube electrospun fibers at different tip-to-collector distances are presented in Fig. B3. When the tip-to-collector distance was low, a little interconnected fiber (with high fiber diameter) was deposited on the collector (as shown in Fig. B3a). At a 5 cm tip-to-collector distance (Fig. 3b), more homogenous fibers with negligible beads were obtained. However, the beads increased with increasing the tip-to-collector distance (Figs. B3b–f).

Also, our study has demonstrated that the diameter of electrospun fibers decreased by increasing tip-to-collector distance (as shown in Figs. B3b–d; 455 (306–672), 134 (87–163), 107 (71–196)). The fibers prepared within a distance of 8 cm (Fig. B3e) and 10 cm (Fig. B3f) presented defects and nonhomogenous diameter. However, a 5-cm tip-to-collector distance appears to be reliable for electrospinning.

APPENDIX C

In this appendix, effects of four electrospinning parameters, including solution concentration (wt.%), applied voltage (kV), tip to collector distance (cm), and volume flow rate (mL/h), on contact angle (CA) of polyacrylonitrile (PAN) nanofiber mat are presented. To optimize and predict the CA of electrospun fiber mat, RSM and artificial neural network (ANN) are employed and a quantitative relationship between processing variables and CA of electrospun fibers is established.

C1 MEASUREMENT AND CHARACTERIZATION

The morphology of the gold-sputtered electrospun fibers were observed by scanning electron microscope (SEM, Philips XL-30). The average fiber diameter and distribution was determined from selected SEM image by measuring at least 50 random fibers. The wettability of electrospun fiber mat was determined by CA measurement. The CA measurements were carried out using specially arranged microscope equipped with camera and PCTV vision software as shown in Figure 2. The droplet used was distilled water and was 1 μl in volume. The CA experiments were carried out at room temperature and were repeated five times. All contact angles measured within 20 s of placement of the water droplet on the electrospun fiber mat. A typical SEM image of electrospun fiber mat, its corresponding diameter distribution and CA image are shown in Figs. C1 and C2.

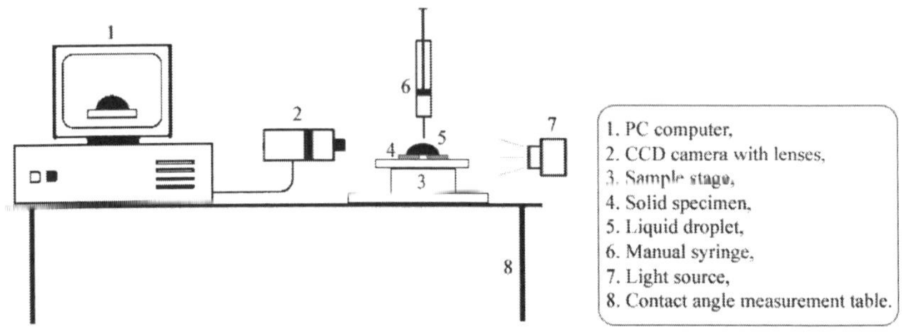

FIGURE C1 Schematic of CA measurement set up.

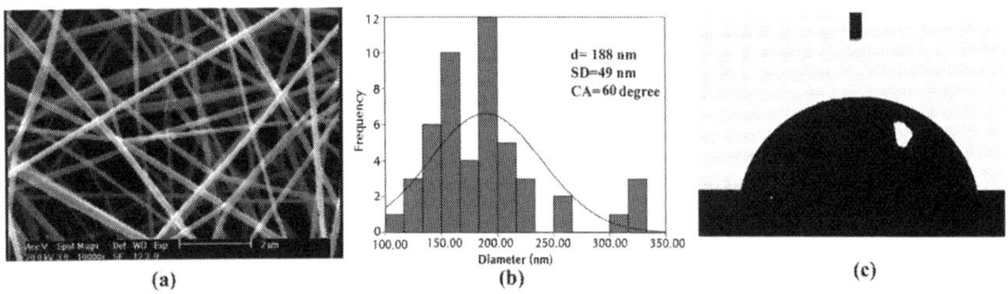

FIGURE C2 A typical (a) SEM image, (*b*) fiber diameter distribution, and (*c*) CA of electrospun fiber mat.

C2 RESPONSE SURFACE METHODOLOGY

Response surface methodology (RSM) is a combination of mathematical and statistical techniques used to evaluate the relationship between a set of controllable experimental factors and observed results. This optimization process is used in situations where several input variables influence some output variables (responses) of the system.

In the present study, central composite design (CCD) was employed to establish relationships between four electrospinning parameters and the CA of electrospun fiber mat. The experiment was performed for at least three levels of each factor to fit a quadratic model. Based on preliminary experiments, polymer solution concentration (wt.%), applied voltage (kV), tip to collector distance (cm), and volume flow rate (mL/h) were determined as critical factors with significance effect on CA of electrospun fiber mat. These factors were four independent variables and chosen equally spaced, while the CA of electrospun fiber mat was dependent variable. The values of −1, 0, and 1 are coded variables corresponding to low, intermediate and high levels of each factor respectively. The experimental parameters and their levels for four independent variables are shown in Table C1. The regression analysis of the experimental data was carried out to obtain an empirical model between processing variables. The contour surface plots were obtained using Design-Expert® software.

TABLE C1 Design of experiment (factors and levels).

Factor	Variable	Unit	Factor level		
			–1	0	1
X_1	Solution concentration	(wt.%)	10	12	14
X_2	Applied voltage	(kV)	14	18	22
X_3	Tip to collector distance	(cm)	10	15	20
X_4	Volume flow rate	(mL/h)	2	2.5	3

The quadratic model, Eq. (1) including the linear terms, was fitted to the data.

$$Y = \beta_0 + \sum_{i=1}^{4} \beta_i x_i + \sum_{i=1}^{4} \beta_{ii} x_i^2 + \sum_{i=1}^{3} \sum_{j=2}^{4} \beta_{ij} x_i x_j \quad \text{(C1)}$$

where, Y is the predicted response, x_i and x_j are the independent variables, β_0 is a constant, β_i is the linear coefficient, β_{ii} is the squared coefficient, and β_{ij} is the second-order interaction coefficients.

The quality of the fitted polynomial model was expressed by the determination coefficient (R^2) and its statistical significance was performed with the Fisher's statistical test for analysis of variance (ANOVA).

C3 ARTIFICIAL NEURAL NETWORK

Artificial neural network (ANN) is an information processing technique, which is inspired by biological nervous system, composed of simple unit (neurons) operating in parallel. A typical ANN consists of three or more layers, comprizing an input layer, one or more hidden layers and an output layer. Every neuron has connections with every neuron in both the previous and the following layer. The connections between neurons consist of weights and biases. The weights between the neurons play an important role during the training process. Each neuron in hidden layer and output layer has a transfer function to produce an estimate as target. The interconnection weights are adjusted, based on a comparison of the network output (predicted data) and the actual output (target), to minimize the error between the network output and the target.

In this study, feed forward ANN with one hidden layer composed of four neurons was selected. The ANN was trained using back-propagation algorithm. The same experimental data used for each RSM designs were also used as the input variables of the ANN. There are four neurons in the input layer corresponding to four electrospinning parameters and one neuron in the output layer corresponding to CA of electrospun fiber mat. Figure C3 illustrates the topology of ANN used in this investigation.

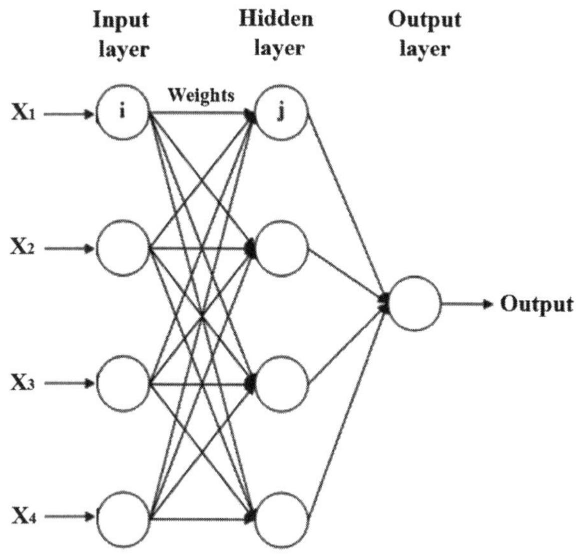

FIGURE C3 The topology of artificial neural network used in this study.

C4 THE ANALYSIS OF VARIANCE (ANOVA)

All 30 experimental runs of CCD were performed according to Table C2. A significance level of 5% was selected; that is, statistical conclusions may be assessed with 95% confidence. In this significance level, the factor has significant effect on CA if the p-value is less than 0.05. On the other hand, when p-value is greater than 0.05, it is concluded the factor has no significant effect on CA.

The results of ANOVA for the CA of electrospun fibers are shown in Table C3. Eq. (2) is the calculated regression equation.

TABLE C2 The actual design of experiments and response.

No.	Electrospinning parameters				Response
	X_1 Concen-tration	X_2 Voltage	X_3 Distance	X_4 Flow rate	CA (°)
1	10	14	10	2	44±6
2	10	22	10	2	54±7
3	10	14	20	2	61±6
4	10	22	20	2	65±4
5	10	14	10	3	38±5
6	10	22	10	3	49±4
7	10	14	20	3	51±5
8	10	22	20	3	56±5
9	10	18	15	2.5	48±3
10	12	14	15	2.5	30±3
11	12	22	15	2.5	35±5
12	12	18	10	2.5	22±3
13	12	18	20	2.5	30±4
14	12	18	15	2	33±4
15	12	18	15	3	25±3
16	12	18	15	2.5	26±4
17	12	18	15	2.5	29±3
18	12	18	15	2.5	28±5
19	12	18	15	2.5	25±4
20	12	18	15	2.5	24±3
21	12	18	15	2.5	21±3
22	14	14	10	2	31±4
23	14	22	10	2	35±5
24	14	14	20	2	33±6
25	14	22	20	2	37±4
26	14	14	10	3	19±3
27	14	22	10	3	28±3
28	14	14	20	3	39±5
29	14	22	20	3	36±4
30	14	18	15	2.5	20±3

$$CA = 25.80 - 9.89X_1 + 2.17X_2 + 4.33X_3 - 2.33X_4$$
$$-1.63X_1X_2 - 1.63X_1X_3 + 1.63X_1X_4 - 0.88X_2X_3 - 0.63X_2X_4 + 0.37X_3X_4 \quad \text{(C2)}$$
$$+7.90X_1^2 + 6.40X_2^2 - 0.096X_3^2 + 2.90X_4^2$$

TABLE C3 Analysis of variance for the CA of electrospun fiber mat.

Source	SS	DF	MS	F-value	Probe > F	Remarks
Model	4175.07	14	298.22	32.70	<0.0001	Significant
X_{1-}Concentration	1760.22	1	1760.22	193.01	<0.0001	Significant
X_{2-}Voltage	84.50	1	84.50	9.27	0.0082	Significant
X_{3-}Distance	338.00	1	338.00	37.06	<0.0001	Significant
X_{4-}Flow rate	98.00	1	98.00	10.75	0.0051	Significant
X_1X_2	42.25	1	42.25	4.63	0.0481	Significant
X_1X_3	42.25	1	42.25	4.63	0.0481	Significant
X_1X_4	42.25	1	42.25	4.63	0.0481	Significant
X_2X_3	12.25	1	12.25	1.34	0.2646	
X_2X_4	6.25	1	6.25	0.69	0.4207	Significant
X_3X_4	2.25	1	2.25	0.25	0.6266	
X_1^2	161.84	1	161.84	17.75	0.0008	Significant
X_2^2	106.24	1	106.24	11.65	0.0039	Significant
X_3^2	0.024	1	0.024	0.0026	0.9597	
X_4^2	21.84	1	21.84	2.40	0.1426	
Residual	136.80	15	9.12			
Lack of Fit	95.30	10	9.53	1.15	0.4668	

From the *p*-values presented in Table C3, it can be concluded that the p-values of terms x_3^2, x_4^2, x_2x_3, x_2x_4 and x_3x_4 is greater than the significance level of 0.05, therefore they have no significant effect on the CA of electrospun fiber mat. Since the above terms had no significant effect on CA of electrospun fiber mat, these terms were removed. The fitted equations in coded unit are given in Eq. (C3).

$$\begin{aligned} CA = 26.07 - 9.89X_1 + 2.17X_2 + 4.33X_3 - 2.33X_4 \\ -1.63X_1X_2 - 1.63X_1X_3 + 1.63X_1X_4 \\ +9.08X_1^2 + 7.58X_2^2 \end{aligned} \tag{C3}$$

Now, all the p-values are less than the significance level of 0.05. Analysis of variance showed that the RSM model was significant ($p<0.0001$), which indicated that the model has a good agreement with experimental data. The determination coefficient (R^2) obtained from regression equation was 0.958.

C5 ARTIFICIAL NEURAL NETWORK

In this section, the best prediction, based on minimum error, was obtained by ANN with one hidden layer. The suitable number of neurons in the hidden layer was determined by changing the number of neurons

The good prediction and minimum error value were obtained with four neurons in the hidden layer. The weights and bias of ANN for CA of electrospun fiber mat are given in Table C4. The R^2 and mean absolute percentage error were 0.965 and 5.94% respectively, which indicates that the model was shows good fitting with experimental data.

TABLE C4 Weights and bias obtained in training ANN.

Hidden layer	Weights	IW_{11}	IW_{12}	IW_{13}	IW_{14}
		1.0610	1.1064	21.4500	3.0700
		IW_{21}	IW_{22}	IW_{23}	IW_{24}
		−0.3346	2.0508	0.2210	−0.2224
		IW_{31}	IW_{32}	IW_{33}	IW_{34}
		−0.6369	−1.1086	−41.5559	0.0030
		IW_{41}	IW_{42}	IW_{43}	IW_{44}
		−0.5038	−0.0354	0.0521	0.9560
	Bias	b_{11}	b_{21}	b_{31}	b_{41}
		−2.5521	−2.0885	−0.0949	1.5478

TABLE C4 *(Continued)*

Output layer	Weights	LW_{11}
		0.5658
		LW_{21}
		0.2580
		LW_{31}
		−0.2759
		LW_{41}
		−0.6657
	Bias	b
		0.7104

C6 EFFECTS OF SIGNIFICANT PARAMETERS ON RESPONSE

The morphology and structure of electrospun fiber mat, such as the nanoscale fibers and interfibrillar distance, increases the surface roughness as well as the fraction of contact area of droplet with the air trapped between fibers. It is proved that the CA decrease with increasing the fiber diameter, therefore the thinner fibers, due to their high surface roughness, have higher CA than the thicker fibers. Hence, we used this fact for comparing CA of electrospun fiber mat. The interaction contour plot for CA of electrospun PAN fiber mat are shown in Figure C4.

As mentioned in the literature, a minimum solution concentration is required to obtain uniform fibers from electrospinning. Below this concentration, polymer chain entanglements are insufficient and a mixture of beads and fibers is obtained. On the other hand, the higher solution concentration would have more polymer chain entanglements and less chain mobility. This causes the hard jet extension and disruption during electrospinning process and producing thicker fibers. Figure C4(a) show the effect of solution concentration and applied voltage at middle level of distance (15 cm) and flow rate (2.5 mL/h) on CA of electrospun fiber mat. It is obvious that at any given voltage, the CA of electrospun fiber mat decrease with increasing the solution concentration.

Figure C4(b) shows the response contour plot of interaction between solution concentration and spinning distance at fixed voltage (18 kV) and flow rate (2.5 mL/h). Increasing the spinning distance causes the CA of electrospun fiber mat to increase. Because of the longer spinning distance could give more time for the solvent to evaporate, increasing the spinning distance will decrease the nanofiber diameter and increase the CA of electrospun fiber mat. As demonstrated in Figure C4(b), low solution concentration cause the increase in CA of electrospun fiber mat at large spinning distance.

The response contour plot in Figure C4(c) represented the CA of electrospun fiber mat at different solution concentration and volume flow rate. Ideally, the volume flow rate must be compatible with the amount of solution removed from the tip of the needle. At low volume flow rates, solvent would have sufficient time to evaporate and thinner fibers were produced, but at high volume flow rate, excess amount of solution fed to the tip of needle and thicker fibers were resulted. Therefore the CA of electrospun fiber mat will be decreased.

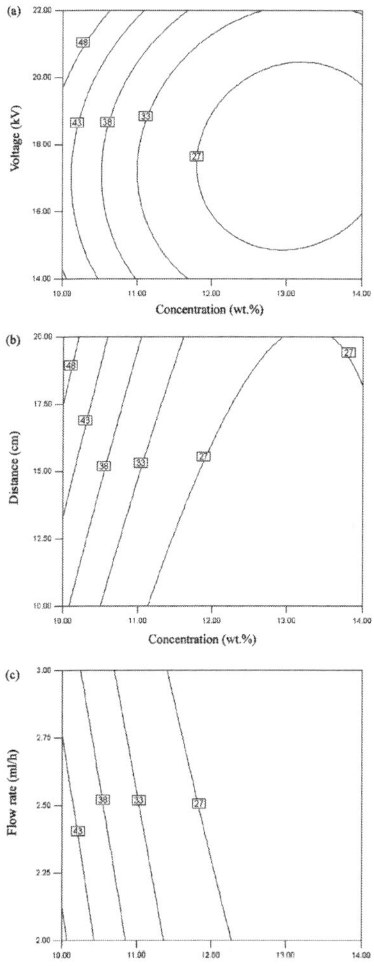

FIGURE C4 Contour plots for contact angle of electrospun fiber mat showing the effect of: (a) solution concentration and applied voltage, (b) solution concentration and spinning distance, (c) solution concentration and volume flow rate.

As shown by Eq. (4), the relative importance (RI) of the various input variables on the output variable can be determined using ANN weight matrix.

$$RI_j = \frac{\sum_{m=1}^{N_h}((|IW_{jm}|/\sum_{k=1}^{N_i}|IW_{km}|)\times|LW_{mn}|)}{\sum_{k=1}^{N_i}\left\{\sum_{m=1}^{N_h}((|IW_{km}|/\sum_{k=1}^{N_i}|IW_{km}|)\times|LW_{mn}|)\right\}}\times 100 \qquad \text{(C4)}$$

where RI_j is the relative importance of the jth input variable on the output variable, N_i and N_h are the number of input variables and neurons in hidden layer, respectively (N_i =4, N_h =4 in this study), IW and LW are the connection weights, and subscript "n" refer to output response (n=1).

The relative importance of electrospinning parameters on the value of CA calculated by Eq. (C4) and is shown in Fig. C5. It can be seen that, all of the input variables have considerable effects on the CA of electrospun fiber mat. Nevertheless, the solution concentration with relative importance of 49.69% is found to be most important factor affecting the CA of electrospun nanofibers. These results are in close agreement with those obtained with RSM.

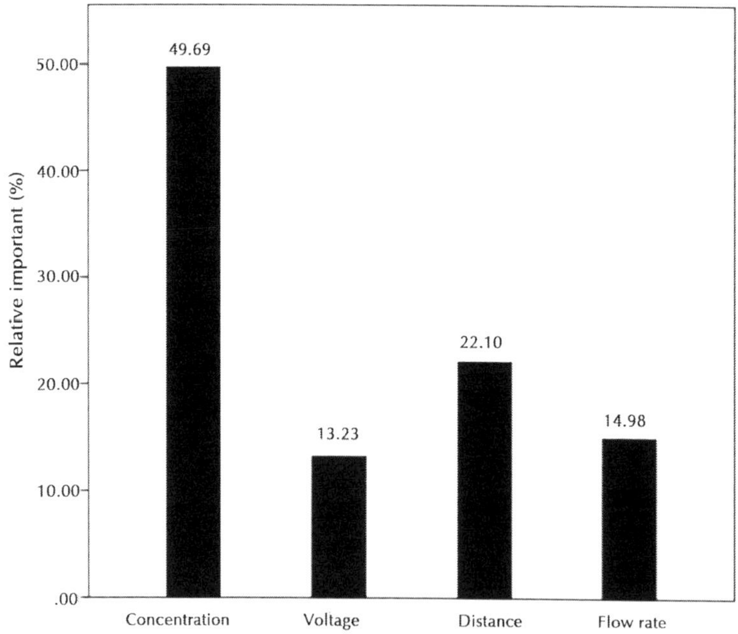

FIGURE C5 Relative importance of electrospinning parameters on the CA of electrospun fiber mat.

C7 OPTIMIZING THE CA OF ELECTROSPUN FIBER MAT

The optimal values of the electrospinning parameters were established from the quadratic form of the RSM. Independent variables (solution concentration, applied voltage, spinning distance, and volume flow rate) were set in range and dependent variable (CA) was fixed at minimum. The optimal conditions in the tested range for minimum CA of electrospun fiber mat are shown in Table C5. This optimum condition was a predicted value, thus to confirm the predictive ability of the RSM model for response, a further electrospinning and CA measurement was carried out according to the optimized conditions and the agreement between predicted and measured responses was verified. Figure C6 shows the SEM, average fiber diameter distribution and corresponding CA image of electrospun fiber mat prepared at optimized conditions.

TABLE C5 Optimum values of the process parameters for minimum CA of electrospun fiber mat.

Solution concentration (wt.%)	Applied voltage (kV)	Spinning distance (cm)	Volume flow rate (mL/h)	Predicted CA (°)	Observed CA (°)
13.2	16.5	10.6	2.5	20	21

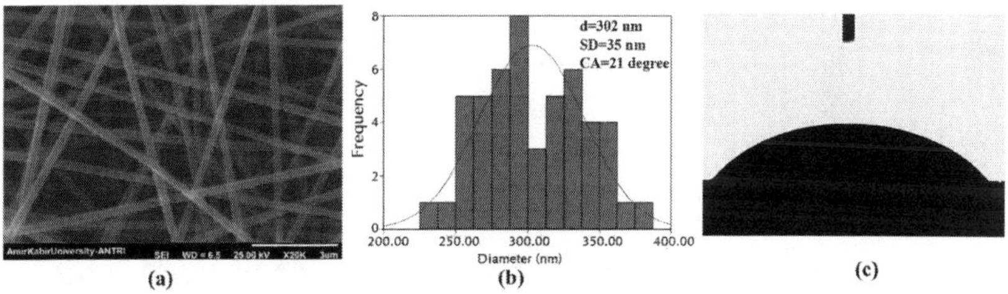

FIGURE C6 (a) SEM image, (b) fiber diameter distribution, and (c) CA of electrospun fiber mat prepared at optimized conditions.

C8 COMPARISON BETWEEN RSM AND ANN MODEL

Table C6 gives the experimental and predicted values for the CA of electrospun fiber mat obtained from RSM as well as ANN model. It is demonstrated that both models performed well and a good determination coefficient was obtained for both RSM and ANN. However, the ANN model shows higher determination coefficient (R^2=0.965) than the RSM model (R^2=0.958). Moreover, the absolute percentage error in the ANN prediction of CA was found to be around 5.94%, while for the RSM model, it was around 7.83%. Therefore, it can be suggested that the ANN model shows more accurately result than the RSM model. The plot of actual and predicted CA of electrospun fiber mat for RSM and ANN is shown in Figure C7.

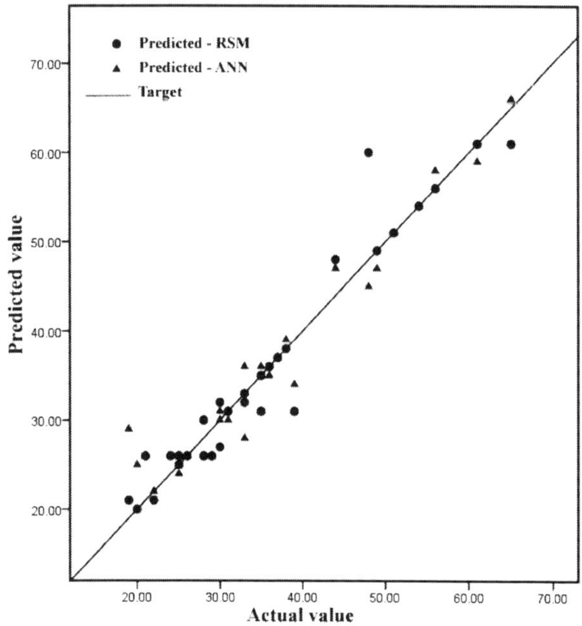

FIGURE C7 Comparison between the actual and predicted contact angle of electrospun nanofiber for RSM and ANN model.

TABLE C6 Experimental and predicted values by RSM and ANN models.

No.	Experimental	Predicted		Absolute error (%)	
		RSM	ANN	RSM	ANN
1	44	47	48	6.41	9.97
2	54	54	54	0.78	0.46
3	61	59	61	3.70	0.42
4	65	66	61	2.06	6.06
5	38	39	38	2.37	0.54
6	49	47	49	5.10	0.68

TABLE C6 *(Continued)*

No.	Experimental	Predicted		Absolute error (%)	
		RSM	ANN	RSM	ANN
7	51	51	51	0.35	0.45
8	56	58	56	4.32	0.17
9	48	45	60	6.17	24.37
10	30	31	27	4.93	9.35
11	35	36	31	2.34	11.15
12	22	22	21	1.18	4.15
13	30	30	32	1.33	6.04
14	33	28	33	13.94	0.60
15	25	24	25	5.04	0.87
16	26	26	26	0.27	1.33
17	29	26	26	10.10	9.16
18	28	26	26	6.89	5.91
19	25	26	26	4.28	5.38
20	24	26	26	8.63	9.77
21	21	26	26	24.14	25.45
22	31	30	31	2.26	0.57
23	35	31	35	10.34	0.66
24	33	36	32	8.18	2.18
25	37	37	37	0.59	0.34
26	19	29	21	52.11	10.23
27	28	30	30	7.07	8.20
28	39	34	31	12.05	21.30
29	36	35	36	1.72	0.04
30	20	25	20	26.30	2.27
R^2		0.958	0.965		
Mean absolute error (%)				7.83	5.94

APPENDIX D

Variables that potentially can alter the electrospinning process (Fig. D1) are large. Hence, investigating all of them in the framework of one single research would almost be impossible. However, some of these parameters can be held constant during experimentation. For instance, performing the experiments in a controlled environmental condition, which is concerned in this study, the ambient parameters (i.e. temperature, air pressure, and humidity) are kept unchanged. Solution viscosity is affected by polymer molecular weight, solution concentration, and temperature. For a particular polymer (constant molecular weight) at a fixed temperature, solution concentration would be the only factor influencing the viscosity. In this circumstance, the effect of viscosity could be determined by the solution concentration. Therefore, there would be no need for viscosity to be considered as a separate parameter.

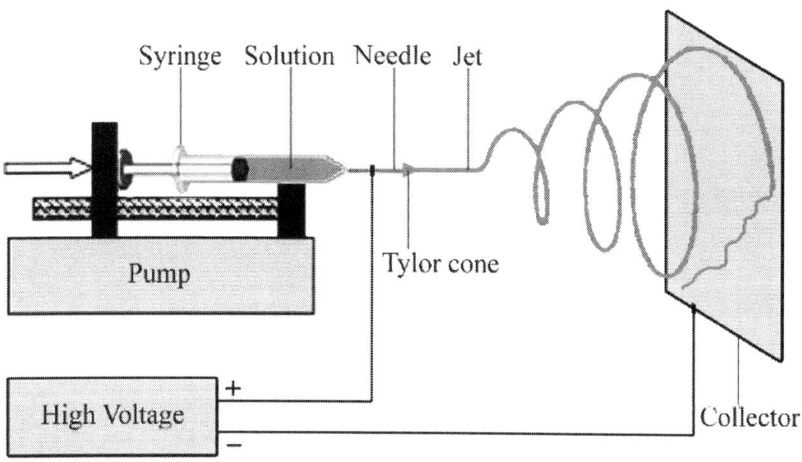

FIGURE D1 A typical image of Electrospinning process.

In this regard, solution concentration ©, spinning distance (*d*), applied voltage (*V*), and volume flow rate (*Q*) were selected to be the most influential parameters. The next step is to choose the ranges over which these factors are varied. Process knowledge, which is a combination of practical experience and theoretical understanding, is required to fulfill this step. The aim is here to find an appropriate range for each parameter where dry, bead-free, stable, and continuous fibers without breaking up to droplets are obtained. This goal could be achieved by conducting a set of preliminary experiments while having the previous works in mind along with using the reported relationships.

The relationship between intrinsic viscosity ($[\eta]$) and molecular weight (M) is given by the well-known Mark-Houwink-Sakurada equation as follows:

$$[\eta] = KM^a \quad \text{(D1)}$$

where K and a are constants for a particular polymer-solvent pair at a given temperature. Polymer chain entanglements in a solution can be expressed in terms of Berry number (B), which is a dimensionless parameter and defined as the product of intrinsic viscosity and polymer concentration ($B=[\eta]C$). For each molecular weight, there is a lower critical concentration at which the polymer solution cannot be electrospun.

As for determining the appropriate range of applied voltage, referring to previous works, it was observed that the changes of voltage lay between 5 kV to 25 kV depending on experimental conditions; voltages above 25 kV were rarely used. Afterwards, a series of experiments were carried out to obtain the desired voltage domain. At V<10 kV, the voltage was too low to spin fibers and 10 kV$\leq V<$15 kV resulted in formation of fibers and droplets; in addition, electrospinning was impeded at high concentrations. In this regard, 15 kV$\leq V \leq$25 kV was selected to be the desired domain for applied voltage.

The use of 5 cm – 20 cm for spinning distance was reported in the literature. Short distances are suitable for highly evaporative solvents whereas it results in wet coagulated fibers for nonvolatile solvents due to insufficient evaporation time. Afterwards, this was proved by experimental observations and 10 cm$\leq d \leq$20 cm was considered as the effective range for spinning distance.

Few researchers have addressed the effect of volume flow rate. Therefore in this case, the attention was focused on experimental observations. At $Q<$0.2 mL/h, in most cases especially at high polymer concentrations, the fiber formation was hindered due to insufficient supply of solution to the tip of the syringe needle. Whereas, excessive feed of solution at $Q>$0.4 mL/h incurred formation of droplets along with fibers. As a result, 0.2 $mL/h \leq Q \leq$0.4 mL/h was chosen as the favorable range of flow rate in this study.

Consider a process in which several factors affect a response of the system. In this case, a conventional strategy of experimentation, which is extensively used in practice, is the *one-factor-at-a-time* approach. The major disadvantage of this approach is its failure to consider any possible interaction between the factors, say the failure of one factor to produce the same effect on the response at different levels of another factor. For instance, suppose that two factors A and B affect a response. At one level of A, increasing B causes the response to increase, while at the other level of A, the effect of B totally reverses and the response decreases with increasing B. As interactions exist between electrospinning parameters, this approach may not be an appropriate choice for the case of the present chapter. The correct strategy to deal with several factors is to use a full factorial design. In this method, factors are all varied together; therefore all possible combinations of the levels of the factors are investigated. This approach is very efficient, makes the most use of the experimental data and takes into account the interactions between factors.

It is trivial that in order to draw a line at least two points and for a quadratic curve at least three points are required. Hence, three levels were selected for each parameter in this study so that it would be possible to use quadratic models. These levels were chosen equally spaced. A full factorial experimental design with four factors (solution concentration, spinning distance, applied voltage, and flow rate) each at three levels (3^4 design) were employed resulting in 81 treatment combinations. This design is shown in Fig. D 2.

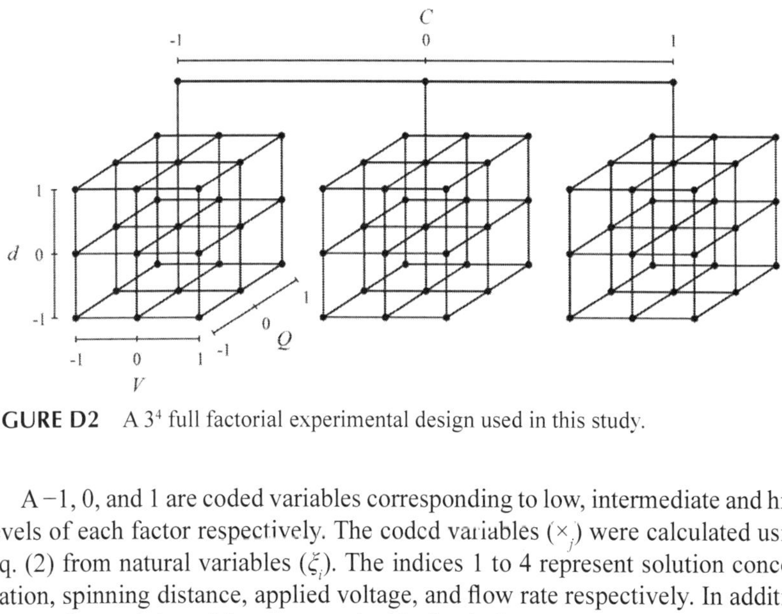

FIGURE D2 A 3^4 full factorial experimental design used in this study.

A −1, 0, and 1 are coded variables corresponding to low, intermediate and high levels of each factor respectively. The coded variables (x_j) were calculated using Eq. (2) from natural variables (ξ_i). The indices 1 to 4 represent solution concentration, spinning distance, applied voltage, and flow rate respectively. In addition to experimental data, 15 treatments inside the design space were selected as test data and used for evaluation of the models. The natural and coded variables for experimental data (numbers 1–81) as well as test data (numbers 82–96) are listed in Table A8 in Appendix.

$$x_j = \frac{\xi_j - [\xi_{hj} + \xi_{lj}]/2}{[\xi_{hj} - \xi_{lj}]/2} \tag{D2}$$

The mechanism of some scientific phenomena has been well understood and models depicting the physical behavior of the system have been drawn in the form of mathematical relationships. However, there are numerous processes at the moment, which have not been, sufficiently been understood to permit the theoretical approach. RSM is a combination of mathematical and statistical techniques useful for empirical modeling and analysis of such systems. The application of RSM is in situations where several input variables are potentially influence some performance measure or quality characteristic of the process – often called responses. The relationship between the response (y) and k input variables (ξ_1, ξ_2, …, ξ_k) could be expressed in terms of mathematical notations as follows:

$$y = f(\xi_1, \xi_2, \ldots, \xi_k) \tag{D3}$$

where the true response function f is unknown. It is often convenient to use coded variables (x_1, x_2,.., x_k) instead of natural (input) variables. The response function will then be:

$$y = f(x_1, x_2, \ldots, x_k) \tag{D4}$$

Since the form of true response function f is unknown, it must be approximated. Therefore, the successful use of RSM is critically dependent upon the choice of appropriate function to approximate f. Low-order polynomials are widely used as approximating functions. First order (linear) models are unable to capture the interaction between parameters, which is a form of curvature in the true response function. Second order (quadratic) models will be likely to perform well in these circumstances. In general, the quadratic model is in the form of:

$$y = \beta_0 + \sum_{j=1}^{k} \beta_j x_j + \sum_{j=1}^{k} \beta_j x_j^2 + \sum_{i<j} \sum_{j=2}^{k} \beta_j x_i x_j + \varepsilon \tag{D5}$$

where ε is the error term in the model. The use of polynomials of higher order is also possible but infrequent. The βs are a set of unknown coefficients needed to be estimated. In order to do that, the first step is to make some observations on the system being studied. The model in Eq. (5) may now be written in matrix notations as:

$$y = X\beta + \varepsilon \tag{D6}$$

where $\boldsymbol{y}$ is the vector of observations, X is the matrix of levels of the variables, β is the vector of unknown coefficients, and **ε** is the vector of random errors. Afterwards, method of least squares, which minimizes the sum of squares of errors, is employed to find the estimators of the coefficients ($\hat{\beta}$) through:

$$\hat{\beta} = (X'X)^{-1}X'y \tag{D7}$$

The fitted model will then be written as:

$$\hat{y} = X\hat{\beta} \tag{D8}$$

Finally, response surfaces or contour plots are depicted to help visualize the relationship between the response and the variables and see the influence of the parameters. As you might notice, there is a close connection between RSM and linear regression analysis.

After the unknown coefficients (βs) were estimated by least squares method, the quadratic models for the mean fiber diameter (MFD) and standard deviation of fiber diameter (StdFD) in terms of coded variables are written as:

$$\begin{aligned} \text{MFD} = {} & 282.031 + 34.953x_1 + 5.622x_2 - 2.113x_3 + 9.013x_4 \\ & -11.613x_1^2 - 4.304x_2^2 - 15.500x_3^2 \\ & -0.414x_4^2 + 12.517x_1x_2 + 4.020x_1x_3 - 0.162x_1x_4 + 20.643x_2x_3 + 0.741x_2x_4 + 0.877x_3x_4 \end{aligned} \tag{D9}$$

$$\begin{aligned} \text{StdFD} = {} & 36.1574 + 4.5788x_1 - 1.5536x_2 + 6.4012x_3 + 1.1531x_4 \\ & -2.2937x_1^2 - 0.1115x_2^2 - 1.1891x_3^2 + 3.0980x_4^2 \\ & -0.2088x_1x_2 + 1.0010x_1x_3 + 2.7978x_1x_4 + 0.1649x_2x_3 - 2.4876x_2x_4 + 1.5182x_3x_4 \end{aligned} \tag{D10}$$

In the next step, a couple of very important hypothesis-testing procedures were carried out to measure the usefulness of the models presented here. First, the test for significance of the model was performed to determine whether there is a subset of variables, which contributes significantly in representing the response variations. The appropriate hypotheses are:

$$\begin{aligned} & H_0 : \beta_1 = \beta_2 = \cdots = \beta_k \\ & H_1 : \beta_j \neq 0 \quad \text{for at least one } j \end{aligned} \tag{D11}$$

The F statistics (the result of dividing the factor mean square by the error mean square) of this test along with the p-values (a measure of statistical significance, the smallest level of significance for which the null hypothesis is rejected) for both models are shown in Table D1.

TABLE D1 Summary of the results from statistical analysis of the models.

	F	p-value	R^2	R^2_{adj}	R^2_{pred}
MFD	106.02	0.000	95.74%	94.84%	93.48%
StdFD	42.05	0.000	89.92%	87.78%	84.83%

The p-values of the models are very small (almost zero), therefore it could be concluded that the null hypothesis is rejected in both cases suggesting that there are some significant terms in each model. There are also included in Table D1, the values of R^2, R^2_{adj}, and R^2_{pred}. R^2 is a measure for the amount of response variation which is explained by variables and will always increase when a new term is added to the model regardless of whether the inclusion of the additional term is statistically significant or not. R^2_{adj} is the adjusted form of R^2 for the number of terms in the model; therefore it will increase only if the new terms improve the model and decreases if unnecessary terms are added. R^2_{pred} implies how well the model predicts the response for new observations, whereas R^2 and R^2_{adj} indicate how well the model fits the experimental data. The R^2 values demonstrate that 95.74% of MFD and 89.92% of StdFD are explained by the variables. The R^2_{adj} values are 94.84% and 87.78% for MFD and StdFD respectively, which account for the number of terms in the models. Both R^2 and R^2_{adj} values indicate that the models fit the data very well. The slight difference between the values of R^2 and R^2_{adj} suggests that there might be some insignificant terms in the models. Since the R^2_{pred} values are so close to the values of R^2 and R^2_{adj}, models does not appear to be overfit and have very good predictive ability.

The second testing hypothesis is evaluation of individual coefficients, which would be useful for determination of variables in the models. The hypotheses for testing of the significance of any individual coefficient are:

$$H_0 : \beta_j = 0$$
$$H_1 : \beta_j \neq 0 \qquad \text{(D12)}$$

The model might be more efficient with inclusion or perhaps exclusion of one or more variables. Therefore the value of each term in the model is evaluated using this test, and then eliminating the statistically insignificant terms, more efficient models could be obtained. The results of this test for the models of MFD and StdFD are summarized in Table AD2 and Table D3, respectively. T statistic in these tables is a measure of the difference between an observed statistic and its hypothesized population value in units of standard error.

TABLE D2 The test on individual coefficients for the model of mean fiber diameter (MFD).

Term (coded)	Coeff.	T	p-value
Constant	282.031	102.565	0.000
C	34.953	31.136	0.000
d	5.622	5.008	0.000

TABLE D2 *(Continued)*

Term (coded)	Coeff.	T	*p*-value
V	–2.113	–1.882	0.064
Q	9.013	8.028	0.000
C^2	–11.613	–5.973	0.000
d^2	–4.304	–2.214	0.030
V^2	–15.500	–7.972	0.000
Q^2	–0.414	–0.213	0.832
Cd	12.517	9.104	0.000
CV	4.020	2.924	0.005
CQ	–0.162	–0.118	0.906
dV	20.643	15.015	0.000
dQ	0.741	0.539	0.592
VQ	0.877	0.638	0.526

TABLE D3 The test on individual coefficients for the model of standard deviation of fiber diameter (StdFD).

Term (coded)	Coef	T	*p*-value
Constant	36.1574	39.381	0.000
C	4.5788	12.216	0.000
D	–1.5536	–4.145	0.000
V	6.4012	17.078	0.000
Q	1.1531	3.076	0.003
C^2	–2.2937	–3.533	0.001
d^2	–0.1115	–0.172	0.864
V^2	–1.1891	–1.832	0.072
Q^2	3.0980	4.772	0.000
Cd	–0.2088	–0.455	0.651

TABLE D3 *(Continued)*

Term (coded)	Coef	T	*p*-value
CV	1.0010	2.180	0.033
CQ	2.7978	6.095	0.000
dV	0.1649	0.359	0.721
dQ	–2.4876	–5.419	0.000
VQ	1.5182	3.307	0.002

As depicted, the terms related to Q^2, CQ, dQ, and VQ in the model of MFD and related to d^2, CD, and dV in the model of StdFD have very high p-values, therefore they do not contribute significantly in representing the variation of the corresponding response. Eliminating these terms will enhance the efficiency of the models. The new models are then given by recalculating the unknown coefficients in terms of coded variables in Eqs. (D13) and (14), and in terms of natural (uncoded) variables in Eqs. (D15), and (D16).

$$\begin{aligned} MFD = {} & 281.755 + 34.953x_1 + 5.622x_2 - 2.113x_3 + 9.013x_4 \\ & -11.613x_1^2 - 4.304x_2^2 - 15.500x_3^2 \\ & +12.517x_1x_2 + 4.020x_1x_3 + 20.643x_2x_3 \end{aligned} \tag{D13}$$

$$\begin{aligned} StdFD = {} & 36.083 + 4.579x_1 - 1.554x_2 + 6.401x_3 + 1.153x_4 \\ & -2.294x_1^2 - 1.189x_3^2 + 3.098x_4^2 \\ & +1.001x_1x_3 + 2.798x_1x_4 - 2.488x_2x_4 + 1.518x_3x_4 \end{aligned} \tag{D14}$$

$$\begin{aligned} MFD = {} & 10.3345 + 48.7288C - 22.7420d + 7.9713V + 90.1250Q \\ & -2.9033C^2 - 0.1722d^2 - 0.6120V^2 \\ & +1.2517Cd + 0.4020CV + 0.8257dV \end{aligned} \tag{D15}$$

$$\begin{aligned} StdFD = {} & -1.8823 + 7.5590C + 1.1818d + 1.2709V - 300.3410Q \\ & -0.5734C^2 - 0.0476V^2 + 309.7999Q^2 \\ & +0.1001CV + 13.9892CQ - 4.9752dQ + 3.0364VQ \end{aligned} \tag{D16}$$

The results of the test for significance as well as R^2, R^2_{adj}, and R^2_{pred} for the new models are given in Table D4. It is obvious that the p-values for the new models are close to zero indicating the existence of some significant terms in each model. Comparing the results of this table with Table D1, the F statistic increased for the new models, indicating the improvement of the models after eliminating the insignificant terms. Despite the slight decrease in R^2, the values of R^2_{adj}, and R^2_{pred} increased substantially for the new models. As it was mentioned earlier in the chapter, R^2 will always increase with the number of terms in the model. Therefore, the smaller R^2 values were expected for the new models, due to the fewer terms. However, this does not necessarily suggest that the pervious models were more efficient. Looking at the tables, R^2_{adj}, which provides a more useful tool for comparing the explanatory power of models with different number of terms, increased after eliminating the unnecessary variables. Hence, the new models have the ability to better explain the experimental data. Due to higher R^2_{pred}, the new models also have higher prediction ability. In other words, eliminating the insignificant terms results in simpler models, which not only present the experimental data in superior form, but also are more powerful in predicting new conditions.

TABLE D4 Summary of the results from statistical analysis of the models after eliminating the insignificant terms.

	F	*p*-value	R^2	R^2_{adj}	R^2_{pred}
MFD	155.56	0.000	95.69%	95.08%	94.18%
StdFD	55.61	0.000	89.86%	88.25%	86.02%

The test for individual coefficients was performed again for the new models. The results of this test are summarized in Table A5 and Table A6. This time, as it was anticipated, no terms had higher *p*-value than expected, which need to be eliminated. Here is another advantage of removing unimportant terms. The values of *T* statistic increased for the terms already in the models implying that their effects on the response became stronger.

TABLE D5 The test on individual coefficients for the model of mean fiber diameter (MFD) after eliminating the insignificant terms.

Term (coded)	Coeff.	T	*p*-value
Constant	281.755	118.973	0.000
C	34.953	31.884	0.000
d	5.622	5.128	0.000
V	–2.113	–1.927	0.058
Q	9.013	8.221	0.000
C^2	–11.613	–6.116	0.000
d^2	–4.304	–2.267	0.026

TABLE D5 *(Continued)*

Term (coded)	Coeff.	T	*p*-value
V^2	–15.500	–8.163	0.000
Cd	12.517	9.323	0.000
CV	4.020	2.994	0.004
dV	20.643	15.375	0.000

TABLE D6 The test on individual coefficients for the model of standard deviation of fiber diameter (StdFD) after eliminating the insignificant terms.

Term (coded)	Coef	T	*p*-value
Constant	36.083	45.438	0.000
C	4.579	12.456	0.000
d	–1.554	–4.226	0.000
V	6.401	17.413	0.000
Q	1.153	3.137	0.003
C^2	–2.294	–3.602	0.001
V^2	–1.189	–1.868	0.066
Q^2	3.098	4.866	0.000
CV	1.001	2.223	0.029
CQ	2.798	6.214	0.000
dQ	–2.488	–5.525	0.000
VQ	1.518	3.372	0.001

After developing the relationship between parameters, the test data were used to investigate the prediction ability of the models. Root mean square errors (RMSE) between the calculated responses (C_i) and real responses (R_i) were determined using Eq. (D17) for experimental data as well as test data for the sake of evaluation of both MFD and StdFD models.

$$\text{RMSE} = \sqrt{\frac{\sum_{i=1}^{n}(C_i - R_i)^2}{n}} \tag{D17}$$

CHAPTER 3

POLYMERIC NANOSYSTEMS

A. K. HAGHI and G. E. ZAIKOV

CONTENTS

3.1 Introduction.. 107
3.2 Properties of Nanotherapeutics 112
3.3 Metal-Based Nanoparticles..113
3.4 Lipid-Based Nanoparticles 113
3.5 Polymer-Based Nanoparticles..113
3.6 Biological Nanoparticles..114
3.7 Nanocarriers in Drug Delivery 114
3.7.1 Phagocytosis Pathway..114
3.7.2 Non-Phagocytic Pathway..115
3.8 Experimental.. 120
3.8.1 Magnetic Sorbent Synthesis... 120
3.8.2 Protein Coatings Formation 120
3.8.3 Study of Proteins Adsorption on MNPs.......................... 121
3.8.4 Coating Stability Analysis and Analysis of Selectivity of Free Radical Process... 123
3.8.5 Enzyme Activity Estimation 123
3.9 Results and discussion 124
3.10 Transfer of Heat in a Nanosystem 131
3.10.1 Heat.. 131
3.10.2 Convection Heat Transfer 133
3.10.3 Conduction Heat Transfer 134
3.10.4 Radiation Heat Transfer 135
3.10.5 Combined Heat Transfer Coefficient............................. 137
3.10.6 Porosity and Pore Size Distribution in Nanofabric........ 137

3.10.7 Thermal Equilibrium and Comfort 138
3.10.8 Moisture in Nanofibers 139
3.11 Conclusion 139
Keywords 141
References 141
Additional Readings 145

3.1 INTRODUCTION

The appearance of "nanoscience" and "nanotechnology" stimulated the burst of terms with "nano-" prefix. Historically the term "nanotechnology" appeared before and it was connected with the appearance of possibilities to determine measurable values up to 10^{-9} of known parameters: 10^{-9} m-nm (nanometer), 10^{-9} s-ns (nanosecond), 10^{-9} degree (nanodegree, shift condition). Nanotechnology and molecular nanotechnology comprise the set of technologies connected with transport of atoms and other chemical particles (ions, molecules) at distances contributing the interactions between them with the formation of nanostructures with different nature.

When scanning tunnel microscope was invented there was an opportunity to influence atoms of a substance thus stimulating the work in the field of probe technology, which resulted in substantiation and practical application of nanotechnological methods in 1994. With the help of this technique it is possible to handle single atoms and collect molecules or aggregates of molecules, construct various structures from atoms on a certain substrate (base). Naturally, such a possibility cannot be implemented without preliminary computer designing of so-called "nanostructures architecture." Nanostructures architecture assumes a certain given location of atoms and molecules in space that can be designed on computer and afterwards transferred into technological program of nanotechnological facility.

The term "chemical assembly" appeared together with the development of chemistry and physics of surface after the birth of "electron spectroscopy for chemical analysis."

Chemical assembly comprises the interaction of chemical particles with the surface and "grafting" of functional groups to the surface or interface boundary "gas–solid."

It should be noted that in nanoproduct obtained fullerene C_{60} predominates by its content, represents an enclosed cluster of 60 carbon atoms. This cluster has a stable and symmetrical structure. Furthermore in 1991 nonotubes were discovered. Afterwards the investigations in the field of nanoparticles and nanosystems started spreading all over the world.

The development of these trends predetermined the appearance and development of so-called "nanoscience." In the same way as "nanotechnology" and "nanoscience" is determined as a combination of scientific knowledge from various disciplines, such as physics, chemistry, biology, mathematics, programming, etc., adapted to nanostructures and nanosystems. Nanoscience comprises fundamental and applied scientific knowledge. Therefore, it is possible to speak about nanochemistry, nanometallurgy, nanoelectronics, nanomachine-building, science of nanomaterials and similar disciplines. If we think of the world from the point that the nature is unified and different disciplines in the science were created by people for the convenience of perception and understanding of the world around, the appearing areas of nanoscience closely connected with nanotechnology represent a vast aggregation of disciplines the list of which will still be incomplete analogously to [4]. Let us be restricted to definitions connected with science of nanomaterials.

The notion "science of nanomaterials" assumes scientific knowledge for obtaining, composition, properties and possibilities to apply nanostructures, nanosystems and nanomaterials. A simplified definition of this term can be as follows: material science dealing with materials comprising particles and phases with nanometer dimensions. To determine the existence area for nanostructures and nanosystems it is advisable to find out the difference of these formations from analogous material objects.

From the analysis of literature the following can be summarized: the existence area of nanosystems and nanoparticles with any structure is between the particles of molecular and atomic level determined in pictometers and aggregates of molecules or per molecular formations over micron units. Here it should be mentioned that in polymer chemistry particles with nanometer dimensions belong to the class of per molecular structures, such as globules and fibrils by one of parameters, for example, by diameter or thickness. In chemistry of complex compounds clusters with nanometer dimensions are also known.

The notion "cluster" assumes energy-wise compensated nucleus with a shell, the surface energy of which is rather small, as a result under given conditions the cluster represents a stable formation.

In chemical literature a cluster is equated with a complex compound containing a nucleus and a shell. Usually a nucleus consists of metal atoms combined with metallic bond, and a shell – of ligands. Manganese carbonyls $[(Co)_5MnMn(Co)_5]$ and cobalt carbonyls $[Co_6(Co)_{18}]$, nickel pentadienyls $[Ni_6(C_5H_6)_6]$ belong to elementary clusters.

In recent years the notion "cluster" has got an extended meaning. At the same time the nucleus can contain not only metals or not even contain metals. In some clusters, for instance, carbon ones there is no nucleus at all. In this case their shape can be characterized as a sphere (icosahedron, to be precise) – fullerenes, or as a cylinder – fullerene tubules. Surely a certain force field is formed by atoms on internal walls inside such particles. It can be assumed that electrostatic, electromagnetic and gravitation fields conditioned by corresponding properties of atoms contained in particle shells can be formed inside tubules and fullerenes. If analyze papers recently published, it should be noted that a considerable exceeding of surface size over the volume and, consequently, a relative growth of the surface energy in comparison with the growth of volume and potential energy is the main feature of clusters. If particle dimensions (diameters of "tubes" and "spheres") change from 1 up to several hundred nanometers, they would be called nanoparticles. In some papers the area of nanoclusters existence is within 1–10 nm.

Based on classical definitions, in given paper metal nanoparticles and nanocrystals are referred to as nanoclusters. Apparently, the difference of nanoparticles from other particles (smaller or larger) is determined by their specific characteristics. The search of nanoworld distinctions from atomic-molecular, micro and macroworld can lead to finding analogies and coincidences in colloid chemistry, chemistry of polymers and coordination compounds. First of all it should be noted that nanoparticles usually represent a small collective aggregation of atoms being within the action of adjacent atoms, thus conditioning the shape of nanoparticles. A nanoparticle shape can vary depending upon the nature of adjacent atoms and character of formation medium. Obviously, the properties of separate atoms and molecules (of small size) are determined by their energy and geometry characteristics, the determinative role being played by electron properties. In particular, electron interactions determine the geometry of molecules and atomic structures of small size and mobility of these chemical particles in media, as well as their activity or reactivity.

When the number of atoms in chemical particle exceeds 30, a certain stabilization of its shape being also conditioned by collective influence of atoms constituting the particle is observed. At the same time, the activity of such a particle remains high but the processes with its participation have a directional character. The character of interactions with the surroundings of such structures is determined by their formation mechanism.

During polymerization or copolymerization the influence of macromolecule growth parameters changes with the increase of the number of elementary acts of its growth. After 7–10 acts the shape or geometry of nanoparticles formed becomes the main determinative factor providing the further growth of macromolecule (chain development). A nanoparticle shape is usually determined not only by its structural elements but also by its interactions with surrounding chemical particles.

From the aforesaid it can be concluded that the possibility of self-organization of nanoparticles with the formation of corresponding nanosystems and nanomaterials is the main distinction of nanoworld from pico, micro, and macroworld. Recently, much attention has been paid to synergetics or the branch of science dealing with self-organization processes since these processes, in many cases, proceed with small energy consumption and, consequently, are more ecologically clear in comparison with existing technological processes.

In turn, nanoparticle dimensions are determined by its formation conditions. When the energy consumed for macroparticle destruction or dispersion over the surface increases, the dimensions of nanomaterials are more likely to decrease. The notion "nanomaterial" is not strictly defined. Several researchers consider nanomaterials to be aggregations of nanocrystals, nanotubes or fullerenes. At the same time there is a lot of information available that nanomaterials can represent materials containing various nanostructures. The most attention researchers pay to metallic nanocrystals. Special attention is paid to metallic nanowires and nanofibers with different compositions.

Here are some names of nanostructures:

1) fullerenes, 2) gigantic fullerenes, 3) fullerenes filled with metal ions, 4) fullerenes containing metallic nucleus and carbon (or mineral) shell, 5) one-layer nanotubes, 6) multilayer nanotubes, 7) fullerene tubules, 8) "scrolls", 9) conic nanotubes, 10) metal-containing tubules, 11) "onions", 12) "Russian dolls", 13) bamboo-like tubules, 14) "beads", 15) welded nanotubes, 16) bunches of nanotubes, 17) nanowires, 18) nanofibers, 19) nanoropes, 20) nanosemispheres (nanocups), 21) nanobands and similar nanostructures, as well as various derivatives from enlisted structures. It is quite possible that a set of such structures and notions will be enriched.

In most cases nanoparticles obtained are bodies of rotation or contain parts of bodies of rotation. In natural environment there are minerals containing fullerenes or representing thread-like formations comprising nanometer pores or structures. In the first case we are talking about schungite that is available in quartz rock in unique deposit in Prionezhje. Similar mineral can also be found in the river Lena basin but it consists of micro and macrodimensional cones, spheroids and complex fibers. In the second case we are talking about kerite from pegmatite on Volyn (Ukraine) that consists of polycrystalline fibers, spheres and spirals mostly of micron dimensions, or fibrous vetcillite from the state of Utah (USA); globular anthraxolite and asphaltite.

Diameters of some internal channels are up to 20–50 nm. Such channels can be of interest as nanoreactors for the synthesis of organic, carbon and polymeric substances with relatively low energy consumption. In case of directed location of internal channels in such matrixes and their inner-combinations the spatial structures of certain purpose can be created. Terminology in the field of nanosystems existence is still being developed, but it is already clear that nanoscience obtains qualitatively new knowledge that can find wide application in various areas of human practice thus significantly decreasing the danger of people's activities for themselves and environment.

The system classification by dimensional factor is known, based on which we consider the following:

- microobjects and microparticles 10^{-6}–10^{-3} m in size;
- nanoobjects and nanoparticles 10^{-9}–10^{-6} m in size;
- picoobjects and picoparticles 10^{-12}–10^{-9} m in size.

Assuming that nanoparticle vibration energies correlate with their dimensions and comparing this energy with the corresponding region of electromagnetic waves, we can assert that energy action of nanostructures is within the energy region of chemical reactions. System self-organization refers to synergetics. Quite often, scientists considered that nanotechnology is based on self-organization of metastable systems. As assumed, self-organization can proceed by dissipative (synergetic) and continual (conservative) mechanisms. At the same time, the system can be arranged due to the formation of new stable ("strengthening") phases or due to the growth provision of the existing basic phase. This phenomenon underlies the arising nanochemistry. Below is one of the possible definitions of nanochemistry.

Nanochemistry is a science investigating nanostructures and nanosystems in metastable ("transition") states and processes flowing with them in near-"transition" state or in "transition" state with low activation energies.

To carry out the processes based on the notions of nanochemistry, the directed energy action on the system is required, with the help of chemical particle field as well, for the transition from the prepared near-"transition" state into the process product state (in our case – into nanostructures or nanocomposites). The perspective area of nanochemistry is the chemistry in nanoreactors. Nanoreactors can be compared with specific nanostructures representing limited space regions in which chemical particles orientate creating "transition state" prior to the formation of the desired nanoproduct. Nanoreactors have a definite activity, which predetermines the creation of the corresponding product. When nanosized particles are formed in nanoreactors, their shape and dimensions can be the reflection of shape and dimensions of the nanoreactor.

In the recent years, a lot of scientific information in the field of nanotechnology and science of nanomaterials appeared. Scientists defined the interval from 1 to 1000 nm as the area of nanostructure existence, the main feature of which is to regulate the system self-organization processes. However, scientists limited the upper threshold at 100 nm. At the same time it was not well substantiated and taken from. Now many nanostructures varying in shapes and sizes are known. These nanostructures have sizes that fit into the interval, and are active in the processes of self-organization, and also demonstrate specific properties.

Problems of nanostructure activity and the influence of nanostructure super small quantities on the active media structural changes are explained.

The molecular nanotechnology ideology is analyzed [19]. In accordance with the development tendencies in self-organizing systems under the influence of nanosized excitations the reasons for the generation of self-organization in the range 10^{-6}–10^{-9} m should be determined.

Based on the law of energy conservation the energy of nanoparticle field and electromagnetic waves in the range 1–1000 nm can transfer, thus corresponding to the range of energy change from soft X-ray to near IR radiation. This is the range of energies of chemical reactions and self-organization (structuring) of systems connected with them.

Apparently the wavelengths of nanoparticle oscillations near the equilibrium state are close or correspond to their sizes. Then based on the concepts of ideologists of nanotechnology in material science the definition of nanotechnology can be as follows:

Nanotechnology is a combination of knowledge in the ways and means of conducting processes based on the phenomenon of nano-sized system self-organization and utilization of internal capabilities of the systems that results in decreasing the energy consumption required for obtaining the targeted product while preserving the ecological cleanness of the process.

Nanotechnology expands itself as one of the major area of science [1–25]. An important and exciting aspect of nanomedicine is the use of nanoparticle for target specific drug delivery systems and it allows a new innovative therapeutic approaches. Due to their small size, these drug delivery systems are promising tools in therapeutic approaches such as selective or targeted drug delivery towards a specific tissue or organ. It also enhances drug transport across biological barriers and intracellular drug delivery. Nanotherapeutic agents work successfully against traditional drugs, which exhibits very less bioavailability and are often associated with gastro-intestinal side effects. Nanotherapeutics improves the delivery of drugs that cannot normally be taken orally and it also improves the safety and efficacy of low molecular weight drugs. It also improves the stability and absorption of proteins that normally cannot be taken orally.

3.2 PROPERTIES OF NANOTHERAPEUTICS

A good therapeutic agent or nanoparticle should have ability to: (1) cross one or various biological membranes (e.g., mucosa, epithelium, endothelium) before (2) diffusing through the plasma membrane to (3) finally gain access to the appropriate organelle where the biological target is located.

Different methods of drug delivery are listed below:

1. Oral Drug Delivery
2. Injection Based Drug Delivery
3. Transdermal Drug Delivery
4. Bone Marrow Infusion
5. Control Release Systems
6. Targeted Drug Delivery:
    a. Therapeutical Monoclonal
    b. Antibodies
    c. Liposomes
    d. Microparticles
    e. Modified Blood Cells
    f. Nanoparticles
7. Implant Drug Delivery System

3.3 METAL-BASED NANOPARTICLES

Metallic nanoparticles are either spherical metal or semiconductor particles with nanometer-sized diameters. Similar to nanoparticles "nanocrystal" (NC) or "nanocrystallite" term is used for nanoparticles made up of semiconductors. Generally semiconductor materials from the groups II-VI (CdS, CdSe, CdTe), III-V (InP, InAs) and IV-VI (PbS, PbSe, PbTe) are of particular interest. Metal Based nanoparticle are used for nanoscale transistors, biological sensors, and next generation photovoltaics [26–49].

3.4 LIPID-BASED NANOPARTICLES

Lipid based nanoparticles are widely used for drug targeting and drug delivery. Solid lipid nanoparticles (SLN) introduced in 1991 represent an alternative carrier system for drugs [4]. Nanoparticles made from solid lipids are attracting major attention as novel colloidal drug carrier for intravenous applications as they have been proposed as an alternative particulate carrier system. SLN are submicron colloidal carriers ranging from 50 to 1000 nm, which are composed of physiological lipid [23–52].

Lipid based system are many used because:

1. Lipids enhance oral bioavailability and reduce plasma profile variability.
2. Better characterization of lipoid excipients.
3. An improved ability to address the key issues of technology transfers and manufactures scale-up.

3.5 POLYMER-BASED NANOPARTICLES

Polymeric nanoparticles are nanoparticles, which are prepared from polymers. Polymeric nanoparticles forms (1) the micronization of a material into nanoparticles and (2) the stabilization of the resultant nanoparticles. As for the micronization, one can start with either small monomers or a bulk polymer. The drug is dissolved, entrapped, encapsulated or attached to a nanoparticles and one can obtain different nanoparticles, nanospheres or nanocapsules according to methods of preparation. Gums, Gelatin Sodium alginate Albumin are used for polymer based drug delivery. Polymeric nanoparticles are prepared by Cellulosics, Poly(2-hydroxy ethyl methacrylate), Poly(N-vinyl pyrrolidone), Poly(vinyl alcohol), Poly(methyl methacrylate), Poly(acrylic acid), Polyacrylamide, Poly(ethylene-covinyl acetate) like polymeric materials. Polymer used in drug delivery must have following qualities like it should be chemically inert, nontoxic and free of leachable impurities [39–60].

3.6 BIOLOGICAL NANOPARTICLES

For proper interaction with molecular targets, a biological or molecular coating or layer is required which act as a bioinorganic interface, and it should be attached to the nanoparticle for proper interaction. These biological coating may be antibodies, biopolymers like collagen. Layers of small molecules that make the nanoparticles biocompatible are also used as bioinorganic interface. For detection and analysis these nanoparicles should be optically active and they should either fluoresces or change their color in different intensities of light.

3.7 NANOCARRIERS IN DRUG DELIVERY

Nanoparticles and Liposomes get absorbed by blood streams for its mode of action. For particular function theses nanoparticles must have specific structure and composition. These particles rapidly get cleared from blood stream by macrophages of reticuloendothelial system (RES) [13–28].

Nanoparticles can be delivered to targets by phagocytosis pathway and nonphagocytosis pathway.

3.7.1 PHAGOCYTOSIS PATHWAY

Phagocytosis occurs by profession phagocytes and by nonprofessional phagocytes. A professional phagocyte includes macrophages, monocytes, neutrophils and dendritic cell, while nonprofessional phagocytes are fibroblast, epithelial and endothelial cell.

Phagocytosis takes place by:

1. Recognition of the opsonized particles in the blood stream.
2. Adhesion of particles to macrophages
3. Ingestion of the particle

Opsonization of the nanoparticles is a major step and it takes place before phagocytosis. In this process nanoparticles get tagged by a protein known as *opsonins*. Due to this tagging nanoparticles make a visible complex for macrophages, this whole process takes place in blood streams. These activated particles then get attached to macrophages and this interaction is similar to receptor-ligand interaction [61–80].

As actin is depolymerized from the phagosome, the newly denuded vacuole membrane becomes accessible to early endosomes. Through a series of fusion and fission events, the vacuolar membrane and its contents will mature, fusing with late endosomes and ultimately lysosomes to form a phagolysosome (Fig. 1). The rate of these events depends on the surface properties of the ingested particle, typically from half to several hours. The phagolysosomes become acidified due to the vacuolar proton pump ATPase located in the membrane and acquire many enzymes, including esterases and cathepsins [81–95].

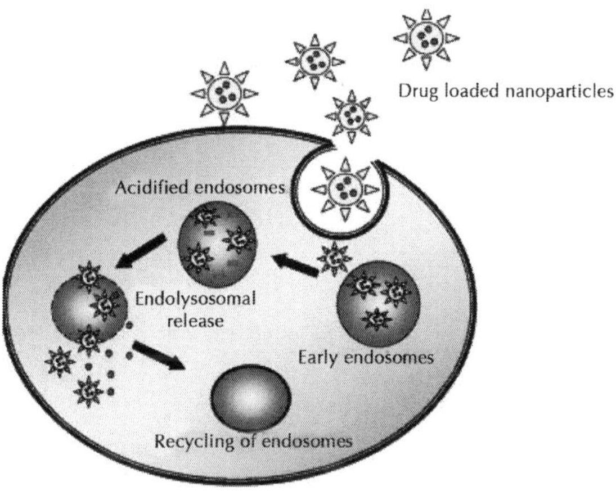

FIGURE 1 Nanoparticles enter the cell through receptor-mediated endocytosis and get localized in the endolysosomal compartment.

3.7.2 NON-PHAGOCYTIC PATHWAY

This is also known as pinocytosis or cell drinking method. It is basically taking of fluids and solutes. Very small nanoparticles get entered in well and absorbed by pinocytosis as phagocytosis is restricted to size of particle. The process of Pinocytosis occurs in specialized cell. It may be clathrin mediated endocytosis, caveolae-mediated endocytosis, macropinocytosis and caveolae-independent endocytosis

3.7.2.1 CLATHRIN-MEDIATED ENDOCYTOSIS

Endocytosis via clathrin-coated pits, or clathrin-mediated endocytosis (CME), occurs constitutively in all mammalian cells, and fulfills crucial physiological roles, including nutrient uptake and intracellular communication. For most cell types, CME serves as the main mechanism of internalization for macromolecules and plasma membrane constituents. CME via specific receptor-ligand interaction is the best described mechanism, to the extent that it was previously referred to as "receptor-mediated endocytosis" (RME). However, it is now clear that alternative nonspecific endocytosis via clathrin-coated pits also exists (as well as receptor-mediated but clathrin-independent endocytosis). Notably, the CME, either receptor-dependent or independent, causes the endocytosed material to end up in degradative lysosomes. This has an important impact in the drug delivery field since the drug-loaded nanocarriers may be tailored in order to become metabolized into the lysosomes, thus releasing their drug content intracellularly as a consequence of lysosomal biodegradation [96–103].

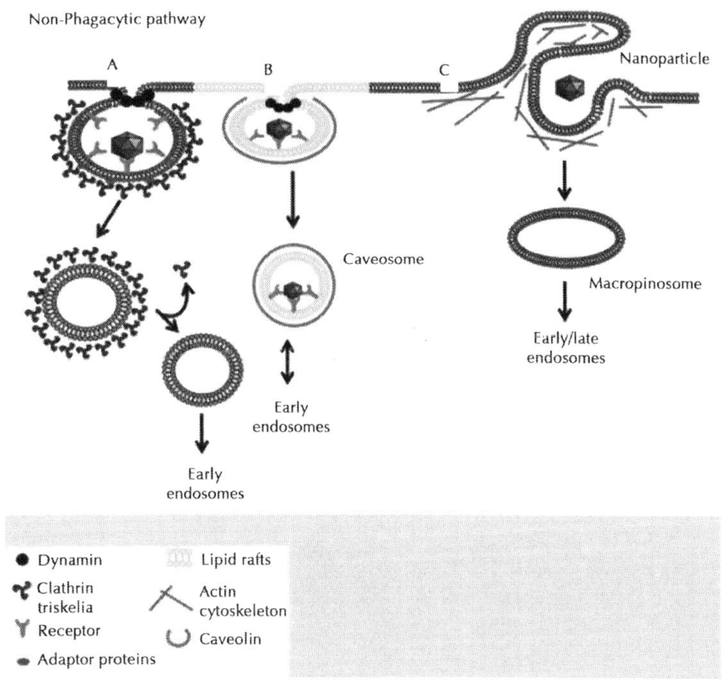

FIGURE 2 Diagrammamtic presentation of (a) Clathrin-mediated endocytosis (b) Caveolae-mediated endocytosis (© Macropinocytosis).

3.7.2.2 CAVEOLAE-MEDIATED ENDOCYTOSIS

Although CME is the predominant endocytosis mechanism in most cells, alternative pathways have been more recently identified, caveolae-mediated endocytosis (CvME) being the major one. Caveolae are characteristic flask-shaped membrane invaginations, having a size generally reported in the lower end of the 50–100 nm range, typically 50–80 nm. They are lined by caveolin, a dimeric protein, and enriched with cholesterol and sphingolipids (Fig. 2). Caveolae are particularly abundant in endothelial cells, where they can constitute 10–20% of the cell surface, but also smooth muscle cells and fibroblasts. CvMEs are involved in endocytosis and trancytosis of various proteins; they also constitute a port of entry for viruses (typically the SV40 virus) and receive increasing attention for drug delivery applications using nanocarriers.

Unlike CME, CvME is a highly regulated process involving complex signaling, which may be driven by the cargo itself. After binding to the cell surface, particles move along the plasma membrane to caveolae invaginations, where they may be maintained through receptor-ligand interactions. Fission of the caveolae from the membrane, mediated by the GTPase dynamin, then generates the cytosolic caveolar vesicle, which does not contain any enzymatic cocktail. Even this pathway is employed by many pathogens to escape degradation by lysosomal enzymes. The use of nanocarriers exploiting CvME may therefore be advantageous to by-pass the lysosomal degradation pathway when the carried drug(e.g., peptides, proteins, nucleic acids, etc.) is highly sensitive to enzymes. On the whole, the uptake kinetics of CvME is known to occur at a much slower rate than that of CME. Ligands known to be internalized by CvME include folic acid, albumin and cholesterol.

3.7.2.3 MACROPINOCYTOSIS

Macropinocytosis is another type of clathrin-independent endocytosis pathway, occurring in many cells, including macrophages. It occurs via formation of actin-driven membrane protusions, similarly to phagocytosis. However, in this case, the protusions do not zipper up along the ligand-coated particle; instead, they collapse onto and fuse with the plasma membrane. This generates large endocytic vesicles, called macropinosomes, which sample the extracellular milieu and have a size generally bigger than 1 lm (and sometimes as large as 5 lm). The intracellular fate of macropinosomes varies depending on the cell type, but in most cases, they acidify and shrink. They may eventually fuse with lysosomal compartments or recycle their content to the surface (Fig. 2). Macropinosomes have not been reported to contain any specific coating, nor do they concentrate receptors.

This endocytic pathway does not seem to display any selectivity, but is involved, among others, in the uptake of drug nanocarriers [80–103].

3.7.2.4 MAGNETIC NANOPARTICLES (MNPS)

Magnetic nanoparticles (MNPs) have many applications in different areas of biology and medicine. MNPs are used for hyperthermia, magnetic resonance imaging, immunoassay, purification of biologic fluids, cell and molecular separation, tissue engineering. The design of magnetically targeted nanosystems (MNSs) for a smart delivery of drugs to target cells is a promising direction of nanobiotechnology. They traditionally consist on one or more magnetic cores and biological or synthetic molecules, which serve as a basis for polyfunctional coatings on MNPs surface. The coatings of MNSs should meet several important requirements. They should be biocompatible, protect magnetic cores from influence of biological liquids, prevent MNSs agglomeration in dispersion, provide MNSs localization in biological targets and homogeneity of MNSs sizes. The coatings must be fixed on MNPs surface and contain therapeutic products (drugs or genes) and biovectors for recognition by biological systems. The model, which is often used when MNSs are developed, is presented in Fig. 3.

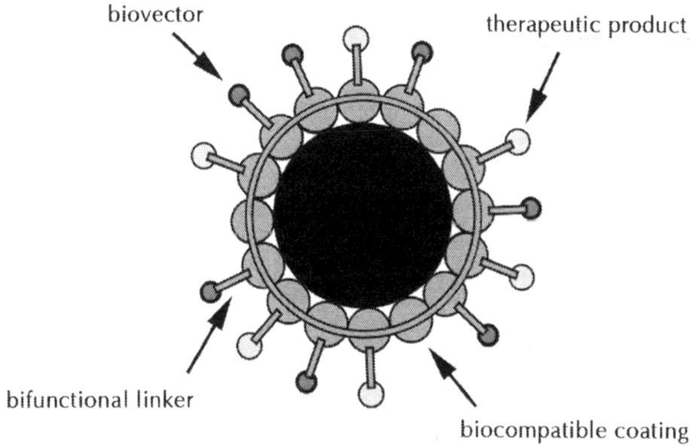

FIGURE 3 The classical scheme of magnetically targeted nanosystem for a smart delivery of therapeutic products.

Proteins are promising materials for creation of coatings on MNPs for biology and medicine. When proteins are used as components of coatings it is of the first importance that they keep their functional activity. Protein binding on MNPs surface is a difficult scientific task. Traditionally bifunctional linkers (glutaraldehyde, carbodiimide) are used for protein cross-linking on the surface of MNPs and modification of coatings by therapeutic products and biovectors. Researchers modified MNPs surface with aminosilanes and performed protein molecules attachment using glutaraldehyde. In the issue bovine serum albumin (BSA) was adsorbed on MNPs surface in the presence of carbodiimide. These works revealed several disadvantages of this way of protein fixing which make it unpromising. Some of them are clusters formation as a result of linking of protein molecules adsorbed on different MNPs, desorption of proteins from MNSs surface as a result of incomplete linking, uncontrollable linking of proteins in solution (Fig. 4). The creation of stable protein coatings with retention of native properties of molecules still is an important biomedical problem.

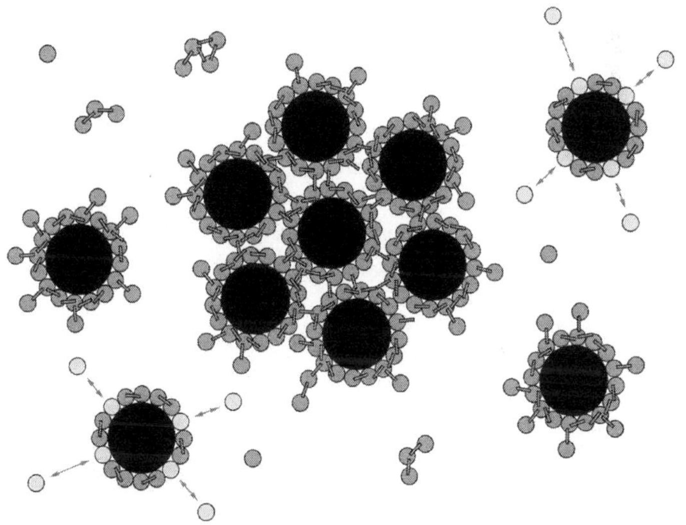

FIGURE 4 Nonselective linking of proteins on MNPs surface by bifunctional linkers leading to clusters formation and desorption of proteins from nanoparticles surface.

It is known that proteins can be chemically modified in the presence of free radicals with formation of cross-links. The goals of the work were to create stable protein coating on the surface of individual MNPs using a fundamentally novel approach based on the ability of proteins to form interchain covalent bonds under the action of free radicals and estimate activity of proteins in the coating.

3.8 EXPERIMENTAL

3.8.1 MAGNETIC SORBENT SYNTHESIS

Nanoparticles of magnetite Fe_3O_4 were synthesized by coprecipitation of ferrous and ferric salts in water solution at 4 °C and in the alkaline medium:

$$Fe^{2+} + 2Fe^{3+} + 8OH- \rightarrow Fe_3O_4\downarrow + 4H_2O$$

1.4 g of $FeSO_4 \cdot 7H_2O$ and 2.4 g of $FeCl_3 \cdot 6H_2O$ were dissolved in 50 ml of distilled water so that molar ratio of Fe^{2+}/Fe^{3+} was equal to 1:2. After filtration of the solution 10 mL of 25 mass% NH_4OH was added to it on a magnetic stirrer. 2.4 g of PEG 2 kDa) was added previously in order to reduce the growth of nanoparticles during the reaction. After the precipitate was formed the solution (with 150 mL of water) was placed on a magnet. Magnetic particles precipitated on it and supernatant liquid was deleted. The procedure of particles washing was repeated for 15 times until neutral pH was obtained. MNPs were stabilized by double electric layer with the use of disperser. To create the double electric layer 30 ml of 0.1 M phosphate-citric buffer solution (0.05 M NaCl) with pH value of 4 was introduced. MNPs concentration in hydrosol was equal to 37 mg/mL.

3.8.2 PROTEIN COATINGS FORMATION

Bovine serum albumin and thrombin with activity of 92 units per 1 mg were used for protein coating formation. Several types of reaction mixtures were created: "A1-MNP-0", "A2-MNP-0", "A1-MNP-1", "A2-MNP-1", "A2-MNP-1-acid", "T1-MNP-0", "T1-MNP-0" and "T1–0-0." All of them contained:

(1) 2.80 mL of protein solution ("A1" or "A2" means that there is BSA solution with concentration of 1 mg/mL or 2 mg/mL in 0.05 M phosphate buffer with pH 6.5 (0.15 M NaCl) in the reaction mixture; "T1" means that there is thrombin solution with concentration of 1 mg/mL in 0.15M NaCl with pH 7.3);

(2) 0.35 mL of 0.1 M phosphate-citric buffer solution (0.05 M NaCl) or MNPs hydrosol ("MNP" in the name of reaction mixture means that it contains MNPs);

(3) 0.05 mL of distilled water or 3 mass% H_2O_2 solution ("0" or "1" in the reaction mixture names correspondingly).

Hydrogen peroxide interacts with ferrous ion on MNPs surface with formation of hydroxyl-radicals by Fenton reaction:

$$Fe^{2+} + H_2O_2 \rightarrow Fe^{3+} + OH^{\cdot} + OH^{-}$$

"A2-MNP-1-acid" is a reaction mixture, containing 10 μL of ascorbic acid with concentration of 152 mg/mL. Ascorbic acid is known to form free radicals in reaction with H_2O_2 and generate free radicals in solution but not only on MNPs surface.

The sizes of MNPs, proteins and MNPs in adsorption layer were analyzed using dynamic light scattering (Zetasizer Nano S "Malvern", England) with detection angle of 173° at temperature 25 °C.

3.8.3 STUDY OF PROTEINS ADSORPTION ON MNPS

The study of proteins adsorption on MNPs was performed using ESR-spectroscopy of spin labels. The stable nitroxide radical used as spin label is presented in Fig. 5. Spin labels technique allows studying adsorption of macromolecules on nano-sized magnetic particles in dispersion without complicated separation processes of solution components. The principle of quantitative evaluation of adsorption is the following. Influence of local fields of MNPs on spectra of radicals in solution depends on the distance between MNPs and radicals. If this distance is lower than 40 nm for magnetite nanoparticles with the average size of 17 nm ESR spectra lines of the radicals broaden strongly and their intensity decreases to zero. The decreasing of the spectrum intensity is proportional to the part of radicals, which are located inside the layer of 40 nm in thickness around MNP. The same happens with spin labels covalently bound to protein macromolecules. An intensity of spin labels spectra decreases as a result of adsorption of macromolecules on MNPs (Fig. 6). We have shown that spin labels technique can be used for the study of adsorption value, adsorption kinetics, calculation of average number of molecules in adsorption layer and adsorption layer thickness, concurrent adsorption of macromolecules.

1

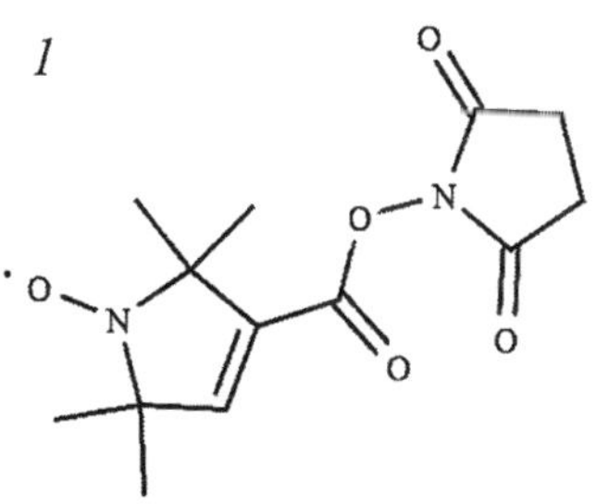

2

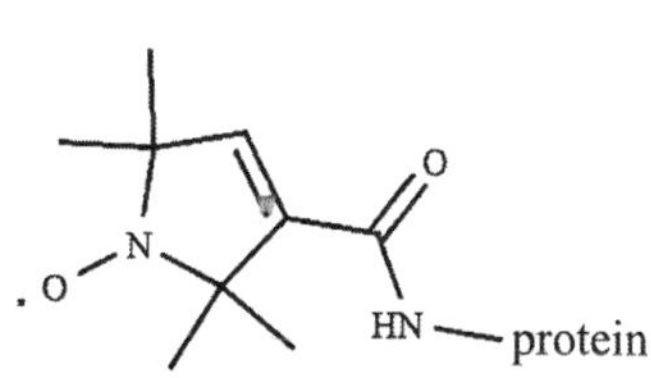

FIGURE 5 The stable nitroxide radical used for labeling of macromolecules containing aminogroups (*1*) and spin label attached to protein macromolecule (*2*).

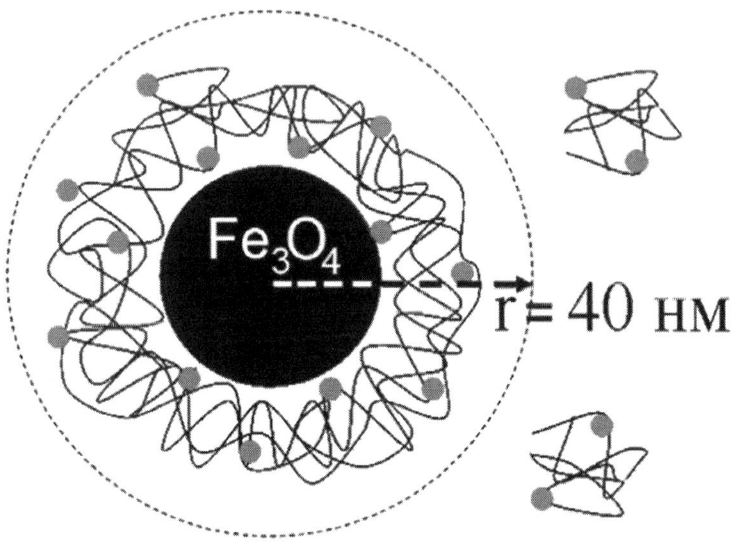

FIGURE 6 Magnetic nanoparticles and spin-labeled macromolecules in solution.

The reaction between the radical and protein macromolecules was conducted at room temperature. 25 μL of radical solution in 96% ethanol with concentration of 2.57 mg/mL was added to 1 mL of protein solution. The solution was incubated for 6 h and dialyzed. The portion of adsorbed protein was calculated from intensity of the low-field line of nitroxide radical triplet I_{+1}.

The method of ferromagnetic resonance was also used to study adsorption layer formation.

The spectra of the radicals and magnetic nanoparticles were recorded at room temperature using spectrometer at a microwave power of 5 mW, modulation frequency 100 kHz and amplitude 1 G. The first derivative of the resonance absorption curve was detected. The samples were placed into the cavity of the spectrometer in a quartz flat cell. Magnesium oxide powder containing Mn^{2+} ions was used as an external standard in ESR experiments. Average amount of spin labels on protein macromolecules reached 1 per 4–5 albumin macromolecules and 1 per 2–3 thrombin macromolecules. Rotational correlation times of labels were evaluated as well as a fraction of labels with slow motion.

3.8.4 COATING STABILITY ANALYSIS AND ANALYSIS OF SELECTIVITY OF FREE RADICAL PROCESS

In our previous works it was shown that fibrinogen (FG) adsorbed on MNPs surface forms thick coating and micron-sized structures. Also FG demonstrates an ability to replace BSA previously adsorbed on MNPs surface. This was proved by complex study of systems containing MNPs, spin-labeled BSA and FG with spin labels technique and ferromagnetic resonance. The property of FG to replace BSA from MNPs surface was used in this work for estimating BSA coating stability. 0.25 mL of FG solution with concentration of 4 mg/mL in 0.05 M phosphate buffer with pH 6.5 was added to 1 mL of the samples "A1-MNP-0", "A2-MNP-0", "A1-MNP-1", "A2-MNP-1." The clusters formation was observed by dynamic light scattering.

The samples "A2-MNP-0", "A2-MNP-1", "T1-MNP-0", "T1-MNP-1" were centrifugated at 120,000 g during 1 h. On these conditions MNPs precipitate, but macromolecules physically adsorbed on MNPs remain in supernatant liquid. The precipitates containing MNPs and protein fixed on MNPs surface were dissolved in buffer solution with subsequent evaluation of the amount of protein by Bradford colorimetric method. Spectrophotometer was used as well.

Free radical modification of proteins in supernatant liquids of "A2-MNP-0", "A2-MNP-1" and the additional sample "A2-MNP-1-acid" were analyzed by IR-spectroscopy using FTIR-spectrometer with DTGS-detector with 2 cm^{-1} resolution. Comparison of "A2-MNP-0", "A2-MNP-1" and "A2-MNP-1-acid" helps to reveal the selectivity of free radical process in "A2-MNP-1."

3.8.5 ENZYME ACTIVITY ESTIMATION

Estimation of enzyme activity of protein fixed on MNPs surface was performed on the example of thrombin. This protein is a key enzyme of blood clotting system, which catalyzes the process of conversion of fibrinogen to fibrin. Thrombin may lose its activity as a result of free radical modification and the rate of the enzyme reaction may decrease. So estimation of enzyme activity of thrombin cross-linked on MNPs surface during free radical modification was performed by comparison of the rates of conversion of fibrinogen to fibrin under the influence of thrombin contained in reaction mixtures. 0.15 mL of the samples "T1-MNP-0", "T1-MNP-1" and "T1–0-0" was added to 1.4 mL of FG solution with concentration of 4 mg/mL. Kinetics of fibrin formation was studied by Rayleigh light scattering on spectrometer with multibit 64-channel correlator.

3.9 RESULTS AND DISCUSSION

ESR spectra of spin labels covalently bound to BSA and thrombin macromolecules (Fig. 7) allow obtaining information about their microenvironment. The spectrum of spin labels bound to BSA is a superposition of narrow and wide lines characterized by rotational correlation times of 10^{-9} s and 2×10^{-8} s respectively. This is an evidence of existence of two main regions of spin labels localization on BSA macromolecules. The portion of labels with slow motion is about 70%. So a considerable part of labels are situated in internal areas of macromolecules with high microviscosity. The labels covalently bound to thrombin macromolecules are characterized by one rotational correlation time of 0.26 ns. These labels are situated in areas with equal microviscosity.

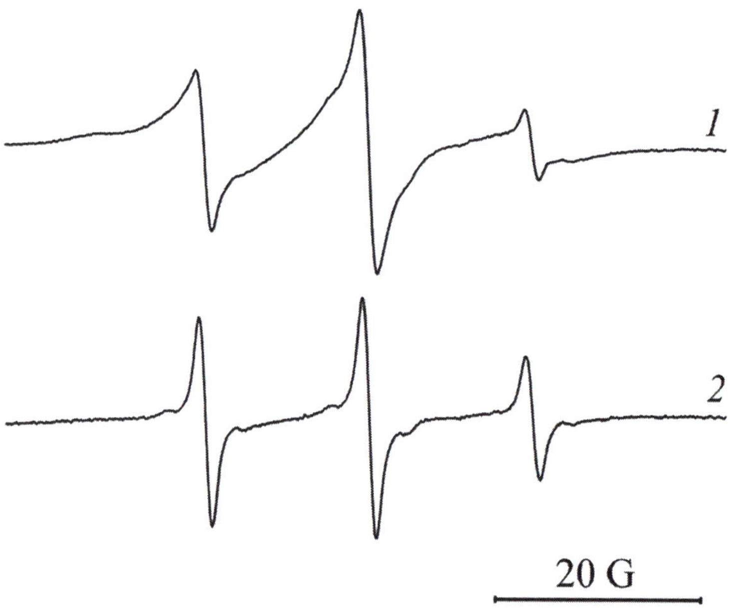

FIGURE 7 ESR spectra of spin labels on BSA (*1*) and thrombin (*2*) at 25 °C.

The signal intensity of spin-labeled macromolecules decreased after introduction of magnetic nanoparticles into the solution that testifies to the protein adsorption on MNPs (Fig. 8). Spectra of the samples "A1-MNP-0" and "T1-MNP-0" consist of nitroxide radical triplet, the third line of sextet of Mn^{2+} (the external standard) and ferromagnetic resonance spectrum of MNPs. Rotational correlation time of spin labels does not change after MNPs addition. The dependences of spectra lines intensity for spin-labeled BSA and thrombin in the presence of MNPs on incubation time are shown in Table 1. Signal intensity of spin-labeled BSA changes insignificantly. These changes correspond to adsorption of approximately 12% of BSA after the sample incubation for 100 min. The study of adsorption kinetics allows establishing that adsorption equilibrium in "T1-MNP-0" takes place when the incubation time equals to 80 min and approximately 41% of thrombin is adsorbed. The value of adsorption *A* may be estimated using the data on the portion of macromolecules adsorbed and specific surface area calculated from MNPs density (5200 mg/m^3), concentration and size. Hence BSA adsorption equals to 0.35 mg/m^2 after 100 min incubation. The dependence of thrombin adsorption value on incubation time is shown in Fig. 9. Thrombin adsorption equals to 1.20 mg/m^2 after 80 min incubation.

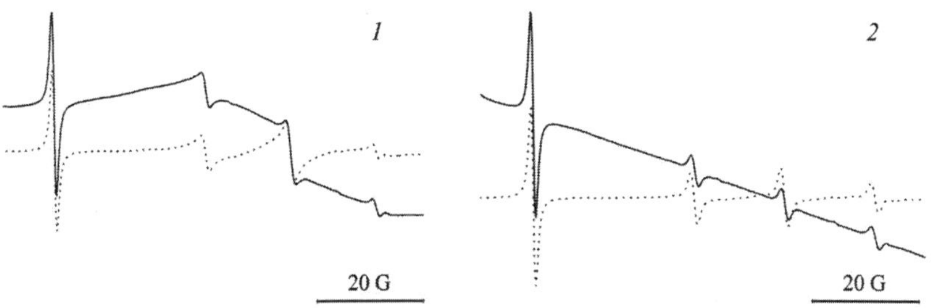

FIGURE 8 ESR spectra of spin labels on BSA (*1*) and thrombin (2) macromolecules before (dotted line) and 75 min after (solid line) MNPs addition to protein solution at 25 °C. External standard – MgO powder containing Mn^{2+}.

TABLE 1 The dependence of relative intensity of low-field line of triplet I_{+1} of nitroxide radical covalently bound to BSA and thrombin macromolecules, and the portion *N* of the protein adsorbed on incubation time *t* of the samples "A1-MNP-0" and "T1-MNP-0."

	Spin-labeled BSA		Spin-labeled thrombin	
***t*, min.**	**I_{+1}, rel. units**	***N*,%**	**I_{+1}, rel. units**	***N*,%**
0	0.230 ± 0.012	0 ± 5	0.25 ± 0.01	0 ± 4
15	—	—	0.17 ± 0.01	32 ± 4
35	0.205 ± 0.012	9 ± 5	0.16 ± 0.01	36 ± 4
75	0.207 ± 0.012	10 ± 5	0.15 ± 0.01	40 ± 4
95	—	—	0.15 ± 0.01	40 ± 4
120	0.200 ± 0.012	13 ± 5	0.14 ± 0.01	44 ± 4

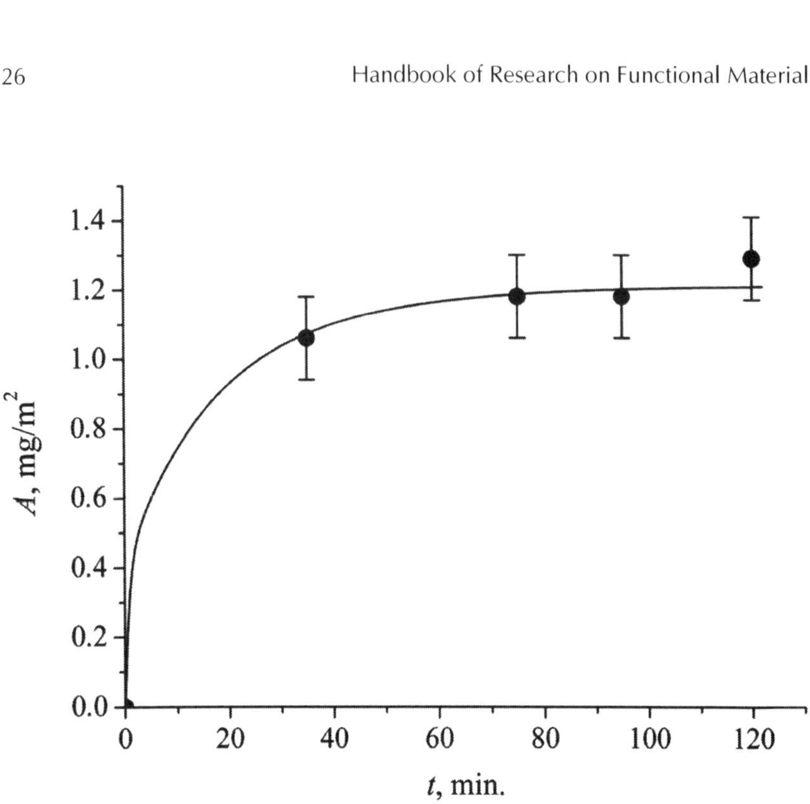

FIGURE 9 Kinetics of thrombin adsorption on magnetite nanoparticles at 25 °C. Concentration of thrombin in the sample is 0.9 mg/mL, MNPs – 4.0 mg/mL.

The FMR spectra of the samples "A1-MNP-0", "T1-MNP-0" and MNPs are characterized by different position in magnetic field (Fig. 10). The center of the spectrum of MNPs is 3254 G, while the center of "A1-MNP-0" and "T1-MNP-0" spectra is 3253 G and 3449 G, respectively. Resonance conditions for magnetic nanoparticles in magnetic field of spectrometer include a parameter of the shift of FMR spectrum $|M_1| = \frac{3}{2}|H_1|$, where H_1 is a local field created by MNPs in linear aggregates, which form in spectrometer field. $H_1 = 2\sum_{1}^{\infty}\frac{2\mu}{(nD)^3}$, where D is a distance between MNPs in linear aggregates, μ is MNPs magnetic moment, n is a number of MNPs in aggregate. Coating formation and the thickness of adsorption layer influence on the distance between nanoparticles decrease dipole interactions and particles ability to aggregate. As a result the center of FMR spectrum moves to higher fields. This phenomenon of FMR spectrum center shift we observed in the system "A1-MNP-0" after FG addition. The spectrum of MNPs with thick coating becomes similar to FMR spectra of isolated MNPs. So the similar center positions of FMR spectra of MNPs without coating and MNPs in BSA coating point to a very thin coating and low adsorption of protein in this case. In contrast according to FMR center position the thrombin coating on MNPs is thicker than albumin coating. This result is consistent with the data obtained by ESR spectroscopy.

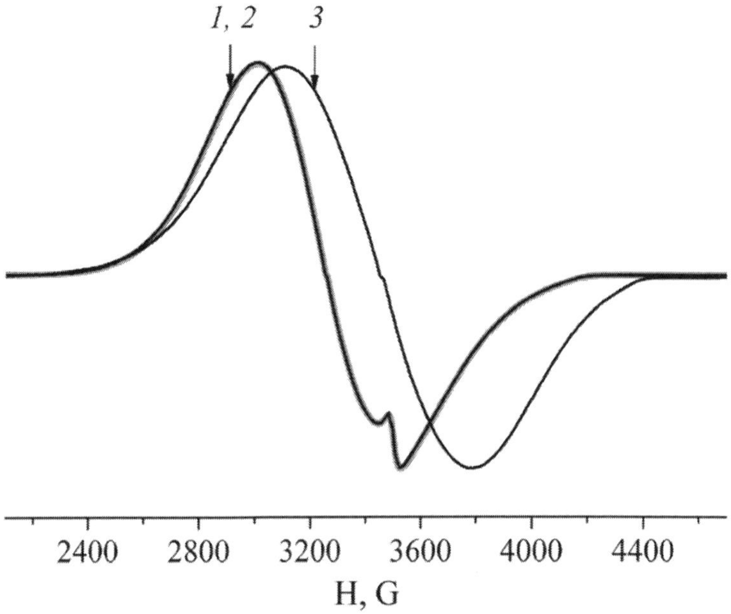

FIGURE 10 FMR spectra of MNPs (*1*), MNPs in the mixture with BSA (the sample "A1-MNP-0") after incubation time of 120 min (*2*) and MNPs in the mixture with thrombin (the sample "T1-MNP-0") after incubation time of 120 min (*3*).

FG ability to replace BSA in adsorption layer on MNPs surface is demonstrated in Fig. 11. Initially there is bimodal volume distribution of particles over sizes in the sample "A2-MNP-0" that can be explained by existence of free (unabsorbed) BSA and MNPs in BSA coating. After FG addition the distribution changes. Micron-sized clusters form in the sample that proves FG adsorption on MNPs [18]. In the case of "A2-MNP-1" volume distribution is also bimodal. The peak of MNPs in BSA coating is characterized by particle size of maximal contribution to the distribution of ~23 nm. This size is identical to MNPs in BSA coating in the sample "A2-MNP-0." It proves that H_2O_2 addition does not lead to uncontrollable linking of protein macromolecules in solution or cluster formation. Since MNPs size is 17 nm, the thickness of adsorption layer on MNPs is approximately 3 nm.

After FG addition to "A2-MNP-1" micron-sized clusters do not form. So adsorption BSA layer formed in the presence of $H_2O_{2\,k}$eeps stability. This stability can be explained by formation of covalent bonds between protein macromolecules [13] in adsorption layer as a result of free radicals generation on MNPs surface. Stability of BSA coating on MNPs was demonstrated for the samples "A1-MNP-1" and "A2-MNP-1" incubated for more than 100 min before FG addition. Clusters are shown to appear if the incubation time is insufficient.

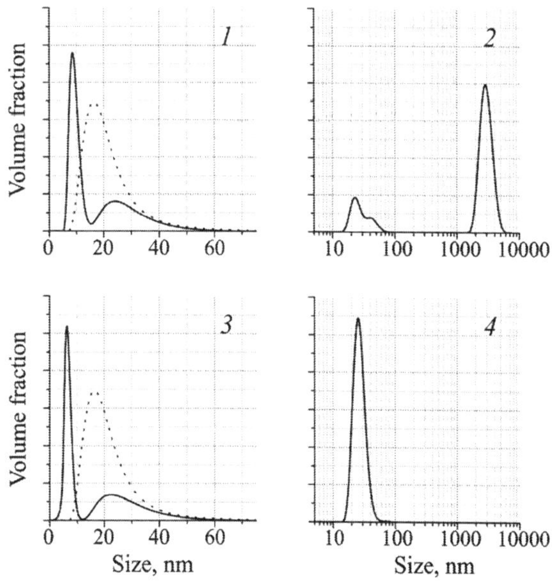

FIGURE 11 Volume distributions of particles in sizes in systems without (*1*, *2*) and with (*3*, *4*) H_2O_2 ("A2-MNP-0", "A2-MNP-1") incubated for 2 h before (*1*, *3*) and 20 min after (*2*, *4*) FG addition. Dotted line is the volume distribution of nanoparticles in sizes in dispersion.

The precipitates obtained by ultracentrifugation of "A2-MNP-0", "A2-MNP-1", "T1-MNP-0" and "T1-MNP-1" were dissolved in buffer solution. The amount of protein in precipitates was evaluated by colorimetric method (Table 2). The results showed that precipitates of systems with H_2O_2 contained more protein than the same systems without H_2O_2. Therefore in the samples containing H_2O_2 the significant part of protein molecules does not leave MNPs surface when centrifuged while in the samples "A2-MNP-0" and "T1-MNP-0" the most of protein molecules leaves the surface. This indicates the stability of adsorption layer formed in the presence of free radical generation initiator and proves cross-links formation.

TABLE 2 The amount of protein in precipitates after centrifugation of the samples "A2-MNP-0", "A2-MNP-1", "T1-MNP-0" and "T1-MNP-1" of 3.2 mL in volume.

Sample name	Amount of protein in precipitates, mg
"A2-MNP-0"	0.05
"A2-MNP-1"	0.45
"T1-MNP-0"	0.15
"T1-MNP-1"	1.05

Analysis of content of supernatant liquids obtained after ultracentrifugation of reaction systems containing MNPs and BSA that differed by H_2O_2 and ascorbic acid presence ("A2-MNP-0", "A2-MNP-1" and "A2-MNP-1-acid") allows evaluating the scale of free radical processes in the presence of H_2O_2. As it was above-mentioned in the presence of ascorbic acid free radicals generate not only on MNPs surface but also in solution. So both molecules on the surface and free molecules in solution can undergo free radical modification in this case. From Fig. 12, we can see that the IR-spectrum of "A2-MNP-1-acid" differs from the spectra of "A2-MNP-0" and "A2-MNP-1", while the spectra of "A2-MNP-0" and "A2-MNP-1" almost have no differences. The IR-spectra differ in the region of 1200–800 cm^{-1}. The changes in this area are explained by free radical oxidation of amino acid residues of methionine, tryptophane, histidine, cysteine, phenylalanine. These residues are sulfur-containing and cyclic ones which are the most sensitive to free radical oxidation. The absence of differences in "A2-MNP-0" and "A2-MNP-1" proves that cross-linking of protein molecules in the presence of H_2O_2 is selective and takes place only on MNPs surfaces.

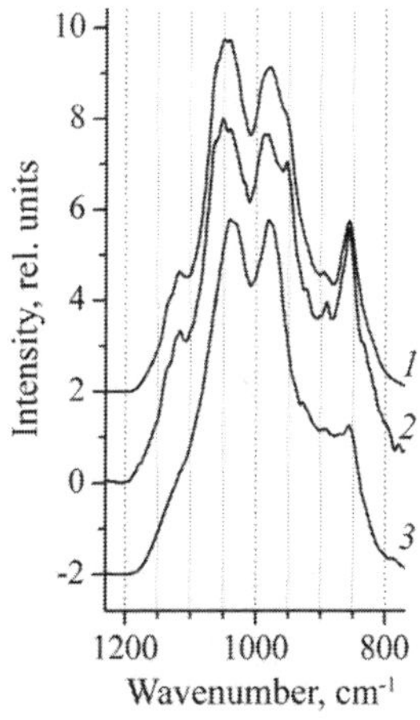

FIGURE 12 IR-spectra of supernatant solutions obtained after centrifugation of the samples "A2-MNP-0" (*1*), "A2-MNP-1" (*2*) and "A2-MNP-1-acid" (*3*).

When proteins are used as components of coating on MNPs for biology and medicine their functional activity retaining is very important. Proteins fixed on MNPs can lose their activity as a result of adsorption on MNPs or free radical modification, which is cross-linking, and oxidation but it was shown that they do not lose it. Estimation of enzyme activity of thrombin cross-linked on MNPs surface was performed by comparison of the rates of conversion of fibrinogen to fibrin under the influence of thrombin contained in reaction mixtures "T1-MNP-0", "T1-MNP-1" and "T1-0-0." The curves for the samples containing thrombin and MNPs, which differ by the presence of H_2O_2, had no fundamental differences that illustrate preservation of enzyme activity of thrombin during free radical cross-linking on MNPs surface. Fibrin gel was formed during ~15 min in both cases. Rayleigh light scattering intensity was low when "T1-0-0" was used and small fibrin particles were formed in this case. The reason of this phenomenon is autolysis (self-digestion) of thrombin. Enzyme activity of thrombin, one of serine proteinases, decreases spontaneously in solution. So the proteins can keep their activity longer when adsorbed on MNPs. This way, the method of free radical cross-linking of proteins seems promising for enzyme immobilization (Fig. 13).

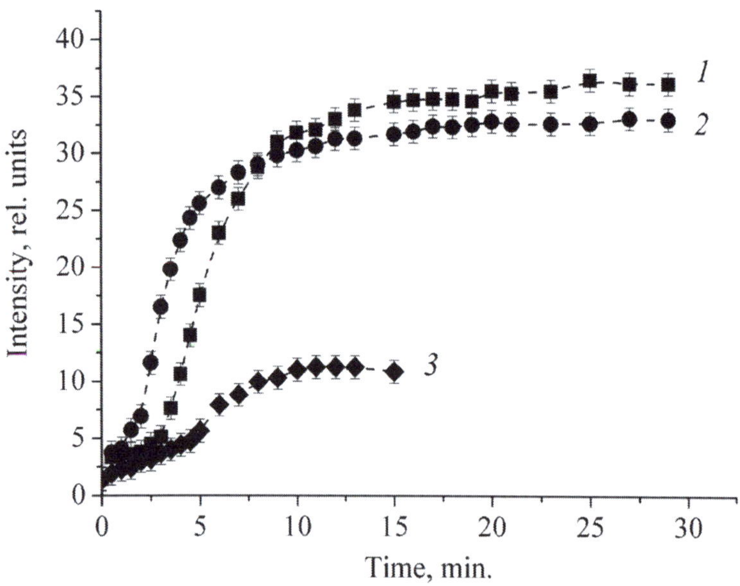

FIGURE 13 Kinetics curves of growth of Rayleigh light scattering intensity in the process of fibrin gel formation in the presence of "T1-MNP-0" (*1*), "T1-MNP-1" (*2*) and "T1-0-0" (*3*).

3.10 TRANSFER OF HEAT IN A NANOSYSTEM

Heat and mass transfer in wet nanostructure are coupled in a complicated way. The structure of the solid matrix varies widely in shape. There is, in general, a distribution of void sizes, and the nanostructures may also be locally irregular. Energy transport in such a medium occurs by conduction in all of the phases. Mass transport occurs within voids of the medium. In an unsaturated state these voids are partially filled with a liquid, whereas the rest of the voids contain some gas. It is a common misapprehension that nonhygroscopic fibers (i.e., those of low intrinsic for moisture vapor) will automatically produce a hydrophobic fabric. The major significance of the fine geometry of a nanostructure in contributing to resistance to water penetration can be stated in the following manner:

For instance the requirements of a water repellent fabric are (*a*) that the nanofibers shall be spaced uniformly and as far apart as possible and (*b*) that they should be held so as to prevent their ends drawing together. In the meantime, wetting takes place more readily on surfaces of high fiber density and in a fabric where there are regions of high fiber density such as yarns, the peripheries of the yarns will be the first areas to wet out and when the peripheries are wetted, water can pass unhindered through the nanofabric.

For thermal analysis of wet nanofabrics, the liquid is water and the gas is air. Evaporation or condensation occurs at the interface between the water and air so that the air is mixed with water vapor. A flow of the mixture of air and vapor may be caused by external forces, for instance, by an imposed pressure difference. The vapor will also move relative to the gas by diffusion from regions where the partial pressure of the vapor is higher to those where it is lower.

Again, heat transfer by conduction, convection, and radiation and moisture transfer by vapor diffusion are the most important mechanisms in very cool or warm environments from the skin.

Meanwhile, nanotextile manufacturing involves a crucial energy-intensive drying stage at the end of the process to remove moisture left from dye setting. Determining drying characteristics for nanotextiles, such as temperature levels, transition times, total drying times and evaporation rates, etc. is vitally important so as to optimize the drying stage. In general, drying means to make free or relatively free from a liquid. We define it more narrowly as the vaporization and removal of water from textiles.

3.10.1 HEAT

When a wet nanofabric is subjected to thermal drying two processes occur simultaneously, namely:

(a) Transfer of heat to raise the wet nanofabric temperature and to evaporate the moisture content.

(b) Transfer of mass in the form of internal moisture to the surface of the nanofabric and its subsequent evaporation.

The rate at which drying is accomplished is governed by the rate at which these two processes proceed. Heat is a form of energy that can across the boundary of a system. Heat can, therefore, be defined as "the form of energy that is transferred between a nanosystem and its surroundings as a result of a temperature difference." There can only be a transfer of energy across the boundary in the form of heat if there is a temperature difference between the system and its surroundings. Conversely, if the nanosystem and surroundings are at the same temperature there is no heat transfer across the boundary.

Strictly speaking, the term "heat" is a name given to the particular form of energy crossing the boundary. However, heat is more usually referred to in thermodynamics through the term "heat transfer", which is consistent with the ability of heat to raise or lower the energy within a nanosystem.

There are three modes of heat transfer:

(1) Convection;

(2) Conduction; and

(3) Radiation.

All these three modes are different. Convection relies on movement of a fluid. Conduction relies on transfer of energy between molecules within a solid or fluid. Radiation is a form of electromagnetic energy transmission and is independent of any substance between the emitter and receiver of such energy. However, all three modes of heat transfer rely on a temperature difference for the transfer of energy to take place.

The greater the temperature difference the more rapidly will the heat be transferred. Conversely, the lower the temperature difference, the slower will be the rate at which heat is transferred. When discussing the modes of heat transfer it is the rate of heat transfer Q that defines the characteristics rather than the quantity of heat.

As it was mentioned earlier, there are three modes of heat transfer, convection, conduction and radiation. Although two, or even all three, modes of heat transfer may be combined in any particular thermodynamic situation, the three are quite different and will be introduced separately.

The coupled heat and liquid moisture transport of nano-porous material has wide industrial applications in textile engineering and functional design of apparel products. Heat transfer mechanisms in nano-porous textiles include conduction by the solid material of fibers, conduction by intervening air, radiation, and convection. Meanwhile, liquid and moisture transfer mechanisms include vapor diffusion in the void space and moisture sorption by the fiber, evaporation, and capillary effects. Water vapor moves through textiles as a result of water vapor concentration differences. Nanofibers absorb water vapor due to their internal chemical compositions and structures. The flow of liquid moisture through the textiles is caused by fiber-liquid molecular attraction at the surface of fiber materials, which is determined mainly by surface tension and effective capillary pore distribution and pathways. Evaporation and/or condensation take place, depending on the temperature and moisture distributions. The heat transfer process is coupled with the moisture transfer processes with phase changes such as moisture sorption and evaporation.

Mass transfer in the drying of a wet nanofabric will depend on two mechanisms: movement of moisture within the fabric which will be a function of the internal physical nature of the solid and its moisture content; and the movement of water vapor from the material surface as a result of water vapor from the material surface as a result of external conditions of temperature, air humidity and flow, area of exposed surface and supernatant pressure.

3.10.2 CONVECTION HEAT TRANSFER

A very common method of removing water from textiles is convective drying. Convection is a mode of heat transfer that takes place as a result of motion within a fluid. If the fluid, starts at a constant temperature and the surface is suddenly increased in temperature to above that of the fluid, there will be convective heat transfer from the surface to the fluid as a result of the temperature difference. Under these conditions the temperature difference causing the heat transfer can be defined as:

$$\Delta T = (\text{surface temperature}) - (\text{mean fluid temperature})$$

Using this definition of the temperature difference, the rate of heat transfer due to convection can be evaluated using Newton's law of cooling:

$$Q = h_c A \Delta T \quad (1)$$

where A is the heat transfer surface area and is the coefficient of heat transfer from the surface to the fluid, referred to as the "convective heat transfer coefficient."

The units of the convective heat transfer coefficient can be determined from the units of other variables:

$$Q = h_c A \Delta T$$

$$W = (h_c) m^2 K$$

So the units of are.

The relationship given in Eq. (1) is also true for the situation where a surface is being heated due to the fluid having higher temperature than the surface. However, in this case the direction of heat transfer is from the fluid to the surface and the temperature difference will now be

(Mean fluid temperature) – (Surface temperature)

The relative temperatures of the surface and fluid determine the direction of heat transfer and the rate at which heat transfer take place.

As given in Eq. (1), the rate of heat transfer is not only determined by the temperature difference but also by the convective heat transfer coefficient. This is not a constant but varies quite widely depending on the properties of the fluid and the behavior of the flow. The value of must depend on the thermal capacity of the fluid particle considered, that is, for the particle. The higher the density and of the fluid the better the convective heat transfer.

Two common heat transfer fluids are air and water, due to their widespread availability. Water is approximately 800 times denser than air and also has a higher value of. If the argument given above is valid then water has a higher thermal capacity than air and should have a better convective heat transfer performance.

3.10.3 CONDUCTION HEAT TRANSFER

If a fluid could be kept stationary there would be no convection taking place. However, it would still be possible to transfer heat by means of conduction. Conduction depends on the transfer of energy from one molecule to another within the heat transfer medium and, in this sense, thermal conduction is analogous to electrical conduction.

Conduction can occur within both solids and fluids. The rate of heat transfer depends on a physical property of the particular solid of fluid, termed its thermal conductivity k, and the temperature gradient across the medium. The thermal conductivity is defined as the measure of the rate of heat transfer across a unit width of material, for a unit cross-sectional area and for a unit difference in temperature.

From the definition of thermal conductivity k it can be shown that the rate of heat transfer is given by the relationship:

$$Q = \frac{kA\Delta T}{x}$$

ΔT is the temperature difference $T_1 - T_2$, defined by the temperature on the either side of the porous surface. The units of thermal conductivity can be determined from the units of the other variables:

$$Q = kA\Delta T / x$$
$$W = (k)m^2 K / m$$

The unit of k is $W / m^2 K / m$.

3.10.4 RADIATION HEAT TRANSFER

The third mode of heat transfer, radiation, does not depend on any medium for its transmission. In fact, it takes place most freely when there is a perfect vacuum between the emitter and the receiver of such energy. This is proved daily by the transfer of energy from the sun to the earth across the intervening space.

Radiation is a form of electromagnetic energy transmission and takes place between all matters providing that it is at a temperature above absolute zero. Infra-red radiation form just part of the overall electromagnetic spectrum. Radiation is energy emitted by the electrons vibrating in the molecules at the surface of a body. The amount of energy that can be transferred depends on the absolute temperature of the body and the radiant properties of the nanosurface.

A body that has a surface that will absorb all the radiant energy it receives is an ideal radiator, termed a "black body". Such a body will not only absorb radiation at a maximum level but will also emit radiation at a maximum level. However, in practice, bodies do not have the surface characteristics of a black body and will always absorb, or emit, radiant energy at a lower level than a black body.

It is possible to define how much of the radiant energy will be absorbed, or emitted, by a particular surface by the use of a correction factor, known as the "emissivity" and given the symbol ε. The emissivity of a surface is the measure of the actual amount of radiant energy that can be absorbed, compared to a black body. Similarly, the emissivity defines the radiant energy emitted from a surface compared to a black body. A black body would, therefore, by definition, have an emissivity ε of 1. It should be noted that the value of emissivity is influenced more by the nature of texture of clothes, than its color. The practice of wearing white clothes in preference to dark clothes in order to keep cool on a hot summer's day is not necessarily valid. The amount of radiant energy absorbed is more a function of the texture of the clothes rather than the color.

Since World War II, there have been major developments in the use of micro-waves for heating applications. After this time it was realized that microwaves had the potential to provide rapid, energy-efficient heating of materials. These main applications of microwave heating today include food processing, wood drying, plastic and rubber treating as well as curing and preheating of ceramics. Broadly speaking, microwave radiation is the term associated with any electromagnetic radiation in the microwave frequency range of 300 MHz-300 Ghz. Domestic and industrial microwave ovens generally operate at a frequency of 2.45 Ghz corresponding to a wavelength of 12.2 cm. However, not all materials can be heated rapidly by microwaves. Materials may be classified into three groups, that is, conductors' insulators and absorbers. Materials that absorb microwave radiation are called dielectrics, thus, microwave heating is also referred to as dielectric heating. Dielectrics have two important properties:

(1) They have very few charge carriers. When an external electric field is applied there is very little change carried through the material matrix.
(2) The molecules or atoms comprising the dielectric exhibit a dipole movement distance. An example of this is the stereochemistry of covalent bonds in a water molecule, giving the water molecule a dipole movement. Water is the typical case of nonsymmetric molecule. Dipoles may be a natural feature of the dielectric or they may be induced. Distortion of the electron cloud around nonpolar molecules or atoms through the presence of an external electric field can induce a temporary dipole movement. This movement generates friction inside the dielectric and the energy is dissipated subsequently as heat.

The interaction of dielectric materials with electromagnetic radiation in the microwave range results in energy absorbance. The ability of a material to absorb energy while in a microwave cavity is related to the loss tangent of the material.

This depends on the relaxation times of the molecules in the material, which, in turn, depends on the nature of the functional groups and the volume of the molecule. Generally, the dielectric properties of a material are related to temperature, moisture content, density and material geometry.

An important characteristic of microwave heating is the phenomenon of "hot spot" formation, whereby regions of very high temperature form due to nonuniform heating. This thermal instability arises because of the nonlinear dependence of the electromagnetic and thermal properties of material on temperature. The formation of standing waves within the microwave cavity results in some regions being exposed to higher energy than others.

Microwave energy is extremely efficient in the selective heating of materials as no energy is wasted in "bulk heating" the sample. This is a clear advantage that microwave heating has over conventional methods. Microwave heating processes are currently undergoing investigation for application in a number of fields where the advantages of microwave energy may lead to significant savings in energy consumption, process time and environmental remediation.

Compared with conventional heating techniques, microwave heating has the following additional advantages:

- higher heating rates;
- no direct contact between the heating source and the heated material;
- selective heating may be achieved;
- greater control of the heating or drying process.

3.10.5 COMBINED HEAT TRANSFER COEFFICIENT

For most practical situations, heat transfer relies on two, or even all three modes occurring together. For such situations, it is inconvenient to analyze each mode separately. Therefore, it is useful to derive an overall heat transfer coefficient that will combine the effect of each mode within a general situation. The heat transfer in moist fabrics takes place through three modes, conduction, radiation, and the process of distillation. With a dry fabric, only conduction and radiation are present.

3.10.6 POROSITY AND PORE SIZE DISTRIBUTION IN NANOFABRIC

The amount of porosity, that is, the volume fraction of voids within the nanofabric, determines the capacity of a nanofabric to hold water; the greater the porosity, the more water the fabric can hold. Porosity is obtained by dividing the total volume of water extruded from fabric sample by the volume of the sample:

Porosity = volume of water/volume of fabric

= (volume of water per gram sample)(density of sample)

It should be noted that most of water is stored between the yarns rather than within them. In the other words, all the water can be accommodated by the pores within the yarns, and it seems likely that the water is chiefly located there. It should be noted that pores of different sizes are distributed within a nanofabric. By a porous medium we mean a material contained a solid matrix with an interconnected void. The interconnectedness of the pores allows the flow of fluid through the fabric. In the simple situation ("single phase flow") the pores is saturated by a single fluid. In "two-phase flow" a liquid and a gas share the pore space.

The usual way of driving the laws governing the macroscopic variables are to begin with standard equations obeyed by the fluid and to obtain the macroscopic equations by averaging over volumes or areas contained many pores.

In defining porosity we may assume that all the pore space is connected. If in fact we have to deal with a fabric in which some of the pore space is disconnected from the reminder, then we have to introduce an "effective porosity", defined as the ratio of the connected pore to total volume.

A further complication arises in forced convection in fabric which is a porous medium. There may be significant thermal dispersion, i.e., heat transfer due to hydrodynamic mixing of the fluid at the pore scale. In addition to the molecular diffusion of heat, there is mixing due to the nature of the fabric.

3.10.7 THERMAL EQUILIBRIUM AND COMFORT

Some of the issues of clothing comfort that are most readily involve the mechanisms by which clothing materials influence heat and moisture transfer from skin to the environment. Heat flow by conduction, convection, and radiation and moisture transfer by vapor diffusion are the most important mechanisms in very cool or warm environments from the skin

It has been recognized that the moisture transport process in clothing under a humidity transient is one of the most important factors influencing the dynamic comfort of a wearer in practical wear situations. However, the moisture transport process is hardly a single process since it is coupled with the heat-transfer process under dynamic conditions. Some materials will posses properties promoting rapid capillary and diffusion movement of moisture to the surface and the controlling factor will be the rate at which surface evaporation can be secured. In the initial stages of drying materials of high moisture content, also it is important to obtain the highest possible rate of nanosurface evaporation. This surface evaporation is essentially the diffusion of vapor from the surface of the fabric to the surrounding atmosphere through a relatively stationary film of air in contact with its surface. This air film, in addition to presenting a resistance to the vapor flow, is itself a heat insulates. The thickness of this film rapidly decreases with increase in the velocity of the air in contact with it while never actually disappearing. The inner film of air in contact with the wet fabric remains saturated with vapor so long as the fabric surface has free moisture present. This result in a vapor pressure gradient through the film from the wetted solid surface to the outer air and, with large air movements, the rate of moisture diffusion through the air film will be considerable. The rate of diffusion, and hence evaporation of the moisture will be directly proportional to the exposed area of the fabric, inversely proportional to the film thickness and directly proportional to the inner film surface and the partial pressure of the water vapor in the surrounding air. It is of importance to note at this point that, since the layer of air film in contact with the wetted fabric undergoing drying remains saturated at the temperature of the area of contact, the temperature of the fabric surface while still possessing free moisture will lie very close to wet-bulb temperature of the air.

3.10.8 MOISTURE IN NANOFIBERS

The amount of moisture that a fiber can take up varies markedly, at low relative humidifies, below 0.35, water is adsorbed mono-molecularly by many natural fibers. From thermodynamic reasoning, we expect the movement of water through a single fiber to occur at a rate that depends on the chemical potential gradient. Meanwhile, moisture has a profound effect on the physical properties of many fibers. Hygroscopic fibers will swell as moisture is absorbed and shrink as it is driven off. Very wet fabrics lose the moisture trapped between the threads first, and only when the threads themselves dry out will shrinkage begin. The change in volume on shrinkage is normally assumed to be linear with moisture content. With hydrophilic materials moisture is found to reduce stiffness and increase creep, probably as a result of plasticization. Variations in moisture content can enhance creep. To describe movement of moisture at equilibrium relative humidity below unity, the idea of absorptive diffusion can be applied. Only those molecules with kinetic energies greater than the activation energy of the moisture fiber bonds can migrate from one site to another. The driving force for absorptive diffusion is considered to be the spreading pressure, which acts over molecular surfaces in two-dimensional geometry and is similar to the vapor pressure, which acts over three-dimensional spaces.

3.11 CONCLUSION

Now-a-days *n*-number of drugs are available in market, so nanotherapeutics is particularly very important to minimize health hazards caused by drugs. The most important aspect of nanotherapeutics is, it should be noninvasive and target oriented, it should have very less or no side effects. In-spite of using nonmaterial one can create nanodevices for better functioning. As most of the peptides and proteins have short half-lives so frequent dosing or injections are required, which is not applicable for many situations by using nanotherapeutics one can solve these problems. Thus, nanotherapeutics is expanding day by day in drug delivery system.

The novel method of fixation of proteins on MNPs proposed in the work was successfully realized on the example of albumin and thrombin. The blood plasma proteins are characterized by a high biocompatibility and allow decreasing toxicity of nanoparticles administered into organism. The method is based on the ability of proteins to form interchain covalent bonds under the action of free radicals. The reaction mixture for stable coatings obtaining should consist on protein solution, nanoparticles containing metals of variable valence (for example, Fe, Cu, Cr) and water-soluble initiator of free radicals generation. In this work albumin and thrombin were used for coating being formed on magnetite nanoparticles. Hydrogen peroxide served as initiator. By the set of physical (ESR-spectroscopy, ferromagnetic resonance, dynamic and Rayleigh light scattering, IR-spectroscopy) and biochemical methods it was proved that the coatings obtained are stable and formed on individual nanoparticles because free radical processes are localized strictly in the adsorption layer. The free radical linking of thrombin on the surface of nanoparticles has been shown to almost completely keep native properties of the protein molecules. Since the method provides enzyme activity and formation of thin stable protein layers on individual nanopaticles it can be successfully used for various biomedical goals concerning a smart delivery of therapeutic products and biologically active substances (including enzymes). It reveals principally novel technologies of one-step creation of biocompatible magnetically targeted nanosystems with multiprotein polyfunctional coatings, which meet all the requirements and contain both biovectors and therapeutic products (Fig. 14).

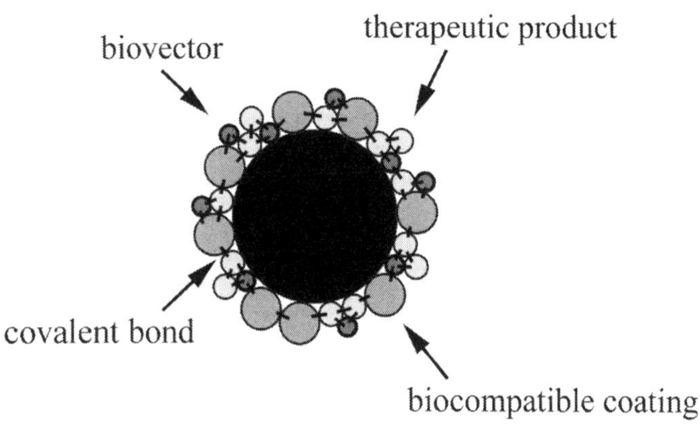

FIGURE 14 The scheme of magnetically targeted nanosystem for a smart delivery of therapeutic products based on the free radical protein cross-linking.

For Heat flow analysis of wet porous nanosystems, the liquid is water and the gas is air. Evaporation or condensation occurs at the interface between the water and air so that the air is mixed with water vapor. A flow of the mixture of air and vapor may be caused by external forces, for instance, by an imposed pressure difference. The vapor will also move relative to the gas by diffusion from regions where the partial pressure of the vapor is higher to those where it is lower.

Heat flow in porous nanosystems is the study of energy movement in the form of heat which occurs in many types of processes. The transfer of heat in porous nanostructure fabrics occurs from the high to the low temperature regions. Therefore, a temperature gradient has to exist between the two regions for heat transfer to happen. It can be done by conduction (within one porous solid or between two porous solids in contact), by convection (between two fluids or a fluid and a porous solid in direct contact with the fluid), by radiation (transmission by electromagnetic waves through space) or by combination of the above three methods.

KEYWORDS

- **black body**
- **chemical assembly**
- **cluster**
- **convective heat transfer**
- **electron spectroscopy for chemical analysis**
- **fibrinogen**
- **nanostructures architecture**
- **science of nanomaterials**

REFERENCES

1. Wu, Y.; Clark, R. L.; *Journal of Colloid and Interface Science*, **2007**, *310, 2,* 529.
2. Jaworek, A.; Sobczyk, A. T.; *Journal of Electrostatics*, **2008**, *66(3–4),* 197.
3. Koombhongse, S.; Liu, W.; Reneker, D. H.; *Journal of Polymer Science Part B: Polymer Physics*, **2001**, *39(21)*, 2598.
4. Yu, Q. Z.; Li, Y.; Wang, M.; Chen, H. Z.; *Chinese Chemical Letters*, **2008**, *19(2),* 223.
5. Luo, C. J.; Nangrejo, M.; Edirizinghe, M.; *Polymer*, **2010**, *51(7),* 1654.
6. Zheng, J.; He, A.; Li, J.; Xu, J.; Han, C. C.; *Polymer*, **2006**, *47(20),* 7095.
7. Leong, M. F.; Chian, K. S.; Mhaisalkar, P. S.; Ong, W. F.; Ratner, B. D.; *Journal of Biomedical Materials Research. Part, A.;* **2009**, *89(4),* 1040.
8. Servoli, E.; Ruffo, G. A.; Migliaresi, C. *Polymer*, **2010**, *51(11),* 2337.
9. Dayal, P.; Kyu, T.;*Journal of Applied Physics*, **2006**, *100(4),* 043512.
10. Megelski, S.; Stephens, J. S.; Chase, D. B.; Rabolt, J. F.; *Macromolecules*, **2002**, *35(22),* 8456.
11. Rangkupan, R.; Reneker, D. H.; *Journal of Metals, Materials and Minerals*, **2003**, *12(2),* 81.
12. Dalton, P. D.; Klinkhammer, K.; Salber, J.; Klee, D.; Möller, M.; *Biomacromolecules*, **2006**, *7(3)*, 686.

13. Shin, J. W.; Shin, H. J.; Heo, S. J.; Lee, Y. J.; Hwang, Y. M.; Kim, D. H.; Kim, J. H.; J. W.; Shin in *Proceedings of the 3rd Kuala Lumpur International Conference on Biomedical Engineering (IFMBE)*, Kuala Lumpur, Malaysia, **2006**, p. 692.
14. Bonani, W.; Maniglio, D.; Motta, A.; Tan, W.; Migliaresi, C.; *Journal of Biomedical Materials Research Part B: Applied Biomaterials*, **2011**, *96(2),* 276.
15. Lin, T.; Wang, H.; Wang, X. *Advanced Materials*, **2005**, *17(22),* 2699.
16. Gupta, P.; Wilkes, G. L.; *Polymer*, **2003**, *44(20),* 6353.
17. Jiang, H.; Hu, Y.; Li, Y.; Zhao, P.; Zhu, K.; Chen, W. *Journal of Controlled Release*, **2005**, *108,* 2–3, 237.
18. Díaz, J. E.; Fernández-Nieves, A.; Barrero, A.; Márquez, M.; Loscertales, I. G.; *Journal of Physics: Conference Series*, **2008**, *127, 1,* 012008.63 *Advanced Electrospinning Setups and Special Fibre and Mesh Morphologies.*
19. Stankus, J. J.; Guan, J.; Fujimoto, K.; Wagner, W. R.; *Biomaterials*, **2006**, *27(5),* 735.
20. Brauker, J. H.; Carr-Brendel, V. E.; Martinson, L. A.; Crudele, J.; Johnston, W. D.; Johnson, R. C.; *Journal of Biomedical Materials Research*, **1995**, *29(12),* 1517.
21. Lu, B.; Wang, Y.; Liu, Y.; Duan, H.; Zhou, J.; Zhang, Z.; Wang, Y.; Li, X.; Wang, W.; Lan, W.; Xie, E. *Small*, **2010**, *6(15),* 1612.
22. Fang, D.; Hsiao, B. S.; Chu, B. *Polymer Preprints*, **2003**, *44(2),* 59.
23. Zhou, F-L.; Gong, R-H.; Porat, I. *Journal of Applied Polymer Science*, **2010**, *115, 5,* 2591.
24. Theron, S. A.; Yarin, A. L.; Zussman, E.; Kroll, E. *Polymer*, **2005**, *46(9),* 2889.
25. Varesano, A.; Carletto, R. A.; Mazzuchetti, G.; *Journal of Materials Processing Technology*, **2009**, *209(11),* 5178.
26. Kim, G.; Cho, Y-S.; Kim, W. D.; *European Polymer Journal*, **2006**, *42(9),* 2031.
27. Yamashita, Y.; Ko, F.; Miyake, H.; Higashiyama, A.; *Sen'i Gakkaishi*, **2008**, *64(1),* 24.61 *Advanced Electrospinning Setups and Special Fibre and Mesh Morphologies*
28. Zhou, F-L.; Gong, R-H.; Porat, I. *Polymer Engineering and Science*, **2009**, *49(12),* 2475.
29. Kumar, A.; Wei, M.; Barry, C.; Chen, J.; Mead, J. *Macromolecular Materials and Engineering*, **2010**, *295, 8,* 701.
30. Dosunmu, O. O.; Chase, G. G.; Kataphinan, W.; Reneker, D. H.; *Nanotechnology*, **2006**, *17(4),* 1123.
31. Badrossamay, M. R.; McIlwee, H. A.; Goss, J. A.; Parker, K. K.; *Nano Letters*, **2010**, *10(6),* 2257.
32. Srivastava, Y.; Marquez, M.; Thorsen, T.; *Journal of Applied Polymer Science*, **2007**, *106(5),* 3171.
33. Jirsak, O.; Sanetrnik, F.; Lukas, D.; Kotek, V.; Martinova, L.; Chaloupek, J.; Inventors, Technicka Universita V Liberci, assignee; US Patent 7585437B2, **2009**.
34. Lukas, D.; Sarkar, A.; Pokorny, P. *Journal of Applied Physics*, **2008**, *103(8),* 084309.
35. Yarin, A.; Zussman, E. *Polymer*, **2004**, *45(9),* 2977.
36. Thoppey, N. M.; Bochinski, J. R.; Clarke, L. I.; Gorga, R. E.; *Polymer*, **2010**, *51(21),* 4928.
37. Tang, S.; Zeng, Y.; Wang, X. *Polymer Engineering and Science*, **2010**, *50(11),* 2252.
38. Cengiz, F.; Dao, T. A.; Jirsak, O. *Polymer Engineering and Science*, **2010**, *50(5),* 936.
39. Jirsak, O.; Syzel, P.; Sanetrnik, F.; Hruza, J.; Chaloupek, J. *Journal of Nanomaterials*, **2010**, 2010, 1.

40. Wang, X.; Niu, H.; Lin, T.; Wang, X.; *Polymer Engineering and Science*, **2009**, *49(8),* 1582.
41. Liu, Y.; He, J-H.; Yu, J-Y. *Journal of Physics: Conference Series*, **2008**, *96,* 012001.
42. Liu, Y.; Dong, L.; Fan, J.; Wang, R.; Yu, J-Y. *Journal of Applied Polymer Science*, **2011**, *120, 1,* 592.62 *Electrospinning for Advanced Biomedical Applications and Therapies*
43. Salem, D. R.; in *Nanofibers and Nanotechnology in Textiles*, Brown, P. J.; Stevens, K. (eds).; CRC Press, Boca Raton, FL, USA, **2007**, p.1.
44. Kelly, A. J.; *Journal of Aerosol Science*, **1994**, *25(6),* 1159.
45. Sun, D.; Chang, C.; Li, S.; Lin, L. *Nano Letters*, **2006**, ***6(4),*** 839.
46. Chang, C.; Limkrailassiri, K.; Lin, L. *Applied Physics Letters*, **2008**, *93(12),* 123111.
47. Levit, N.; Tepper, G. *The Journal of Supercritical Fluids*, **2004**, *31(3),* 329.
48. Larrondo, L.; Manley, R. S. J.; *Journal of Polymer Science. Part B.; Polymer Physics*, **1981**, *19(6),* 909.
49. Liu, Y.; Wang, X.; Li, H.; Yan, H.; Yang, W. *Society of Plastics Engineers Plastics Research Online*, **2010**, 10.1002/spepro.003055.
50. Dalton, P. D.; Grafahrend, D.; Klinkhammer, K.; Klee, D.; Möller, M.*Polymer*, **2007**, *48(23),* 6823.
51. Hou, Q.; Grijpma, D. W.; Feijen, J. *Biomaterials*, **2003**, *24(11),* 1937.
52. Nam, J.; Huang, Y.; Agarwal, S.; Lannutti, J. *Tissue Engineering*, **2007**, *13(9),* 2249.
53. Kim, T. G.; Chung, H. J.; Park, T. G.; *Acta Biomaterialia*, **2008**, ***4(6),*** 1611.
54. Baker, B. M.; Gee, A. O.; Metter, R. B.; Nathan, A. S.; Marklein, R. A.; Burdick, J. A.; Mauck, R. L.; *Biomaterials*, **2008**, *29(15),* 2348.
55. Whited, B. M.; Whitney, J. R.; Hofmann, M. C.; Xu, Y.; Rylander, M. N.; *Biomaterials*, **2011**, *32(9),* 2294.
56. Gentsch, R.; Boyzen, B.; Lankenau, A.; Börner, H. G.; *Macromolecular Rapid Communications*, **2010**, *31(1),* 59.
57. Soliman, S.; Pagliari, S.; Rinaldi, A.; Forte, G.; Fiaccavento, R.; Pagliari, F.; Franzese, O.; Minieri, M.; Di Nardo, P.; Licoccia, S.; Traversa, E. *Acta Biomaterialia*, **2010**, ***6(4),*** 1227.
58. Santos, M. I.; K.; Tuzlakoglu, S.; Fuchs, Gomes, M. E.; K.; Peters, Unger, R. E.; E.; Piskin, Reis, R. L.; Kirkpatrick, C. J.; *Biomaterials*, **2008**, *29(32),* 4306.
59. Tuzlakoglu, K.; Bolgen, N.;Salgado, A. J.; Gomes, M. E.; Piskin, E.; Reis, R. L.; *Journal of Materials Science: Materials in Medicine*, **2005**, *16(12),* 1099.65 *Advanced Electrospinning Setups and Special Fibre and Mesh Morphologies*
60. Thorvaldsson, A.; Stenhamre, H.; Gatenholm, P.; Walkenström, P.; *Biomacromolecules*, **2008**, ***9(3),*** 1044.
61. Smit, E.; Büttner, U.; Sanderson, R. D.; *Polymer*, **2005**, *46(8),* 2419.
62. Khil, M-S.; Bhattarai, S. R.; Kim, H-Y.; Kim S-Z.; Lee, K-H.; *Journal of Biomedical Materials Research. Part, B.; Applied Biomaterials*, **2005**, *72(1),* 117.
63. Teo, W. E.; Ramakrishna, S. *Nanotechnology*, **2006**, *17(14),* R89.64 *Electrospinning for Advanced Biomedical Applications and Therapies*
64. Simonet, M.; Schneider, O. D.; Neuenschwander, P.; Stark, W. J.; *Polymer Engineering and Science Engineering*, **2007**, *47(12),* 2020.
65. Leong, M. F.; Rasheed, M. Z.; Lim, T. C.; Chian, K. S.; *Journal of Biomedical Materials Research Part, A.;* **2009**, *91(1),* 231.

66. Schneider, O. D.; Weber, F.; Brunner, T. J.; Loher, S.; Ehrbar, M.; Schmidlin, P. R.; Stark, W. J.; *Acta Biomaterialia*, **2009**, *5(5)*, 1775.
67. Nam, Y. S.; Yoon, J. J.; Park, T. G.; *Journal of Biomedical Materials Research*, **2000**, *53(1)*, 1.
68. Lee, Y. H.; Lee, J. H.; An, I-G.; Kim, C.; Lee, D. S.; Lee, Y. K.; Nam, J-D.; *Biomaterials*, **2005**, *26(16)*, 3165.
69. Curtis, S. G.; Gadegaard, A. N.; Dalby, M. J.; Riehle, M. O.; Wilkinson, D. W.; Aitchison, C. G. *IEEE Transactions in Nanobioscience*, **2004**, *3(1)*, 61.
70. Wang, Y.; Bella, E.; Lee, S. D.; Migliaresi, C. C.; Pelcastre, L.; Schwartz, Z.; Boyan, B. D.; Motta, A. *Biomaterials*, **2010**, *31(17)*, 4672.
71. Kim, G.; Son, J.; Park, S.; Kim, W. *Macromolecular Rapid Communications*, **2008**, *29(19)*, 1577.
72. Ahn, S. H.; Koh, Y. H.; Kim, G. H.; *Journal of Micromechanics and Microengineering*, **2010**, *20(6)*, 065015.
73. Yoon, H.; Kim, G. *Journal of Pharmaceutical Sciences*, **2011**, *100(2)*, 424.
74. Tayalia, P.; Mendonca, C. R.; Baldacchini, T.; Mooney, D. J.; E Mazur, *Advanced Materials*, **2008**, *20(23)*, 4494.
75. Sachlos, E.; Czernuszka, J. T.; *European Cells and Materials*, **2003**, *5*, 29.
76. Mota, C.; Puppi, D.; Dinucci, D.; Errico, C.; Bártolo, P.; Chiellini, F. *Materials*, **2011**, *4(3)*, 527.
77. Wang, L.; Pai, C. -L.; Boyce, M. C.; Rutledge, G. C.; *Applied Physics Letters*, **2009**, *94(15)*, 151916.
78. Casper, C. L.; Stephens, J. S.; Tassi, N. G.; Chase, D. B.; Rabolt, J. F.; *Macromolecules*, **2004**, *37(2)*, 573.
79. Rosso, F.; Giordano, A.; Barbarisi, M.; Barbarisi, A. *Journal of Cellular Physiology*, **2004**, *199, 2*, 174.
80. Boyan, B. D.; Lossdörfer, S.; Wang, L.; Zhao, G.; Lohmann, C. H.; Cochran, D. L.; Z Schwartz, *European Cells and Materials*, **2003**, *6*, 22.
81. Chung, T-W.; Liu, D-Z.; Wang, S-Y.; Wang, S-S.; *Biomaterials*, **2003**, *24(25)*, 4655.
82. Moroni, L.; Licht, R.; de Boer, J.; de Wijn, J. R.; van Blitterswijk, C. A. *Biomaterials*, **2006**, *27(28)*, 4911.
83. Truong, Y. B.; Glattauer, V.; Lang, G.; Hands, K.; Kyratzis, I. L.; Werkmeister, J. A.; J. Ramshaw, A. M.; *Biomedical Materials*, **2010**, *5(2)*, 25005.
84. Lee, K. Y.; Yuk, S. H.; *Progress in Polymer Science*, **2007**, *32(7)*, 669.
85. Tao, S. L.; Desai, T. A.; *Advanced Drug Delivery Reviews*, **2003**, *55(3)*, 315.66 *Electrospinning for Advanced Biomedical Applications and Therapies*
86. Moroni, L.; de Wijn, J. R.; van Blitterswijk, C. A. *Journal of Biomaterials Science. Polymer Edition*, **2008**, *19(5)*, 543.
87. Luginbuehl, V.; Meinel, L.; Merkle, H. P.; B Gander, *European Journal of Pharmaceutics and Biopharmaceutics*, **2004**, *58(2)*, 197.
88. Willerth, S. M.; Sakiyama-Elbert, S. E.; *Advanced Drug Delivery Reviews*, **2007**, *59,* 4–5, 325.
89. Biondi, M.; Ungaro, F.; Quaglia, F.; Netti, P. A.; *Advanced Drug Delivery Reviews*, **2008**, *60(2)*, 229.
90. Munir, M. M.; Suryamas, A. B.; Iskandar, F.; Okuyama, K. *Polymer*, **2009**, *50(20)*, 4935.

91. Qi, Z.; Yu, H.; Chen, Y.; Zhu, M. *Materials Letters*, **2009**, *63,* 3–4, 415.
92. Luna-Bárcenas, G.; Kanakia, S. K.; Sanchez, I. C.; Johnston, K. P.; *Polymer*, **1995**, *36(16),* 3173.67 *Advanced Electrospinning Setups and Special Fibre and Mesh Morphologies*
93. Liu, J.; Shen, Z.; Lee, S-H.; Marquez, M.; McHugh, M. A. *The Journal of Supercritical Fluids*, **2010**, *53,* 1–3, 142.
94. McCann, J. T.; Marquez, M.; Xia, Y. *Journal of the American Chemical Society*, **2006**, *128, 5,* 1436.
95. You, Y.; Youk, J. H.; Lee, S. W.; Min, B-M.; Lee, S. J.; Park, W. H.; *Materials Letters*, **2006**, *60(6),* 757.
96. Lyoo, W. S.; Youk, J. H.; Lee, S. W.; Park, W. H.; *Materials Letters*, **2005**, *59(28),* 3558.
97. Chakraborty, S.; Liao, I-C.; Adler, A.; Leong, K. W.; *Advanced Drug Delivery Reviews*, **2009**, *61(12),* 1043.
98. Srikar, R.; Yarin, A. L.; Megaridis, C. M.; Bazilevsky, A. V.; Kelley, E. *Langmuir*, **2008**, *24(3),* 965.
99. Gandhi, M.; Srikar, R.; Yarin, A. L.; Megaridis, C. M.; Gemeinhart, R. A.; *Molecular Pharmaceutics*, **2009**, *6(2)***,** 641.
100. Wang, M.; Yu, J. H.; Kaplan, D. L.; Rutledge, G. C.; *Macromolecules*, **2006**, *39(3),* 1102.
101. Gupta, A. K.; Gupta, M.; *Biomaterials* **2005***, 26,* 3995
102. Vatta, L. L.; Sanderson, D. R.; Koch, K. R.; *Pure Appl. Chem.;* **2006**, *78,* 1793.
103. Laurent, S.; Forge, D.; Port, M.; Roch, A.; Robic, C.; Elst, L. V.; Muller, R. N.; *Chem. Rev.;* **2008**, *108,* 2064.

ADDITIONAL READINGS

Deitzel, J.; Kleinmeyer, J. D.; Hirvonen, J. K.; Beck Tan, N. C.; *Polymer*, **2001**, *42(19),* 8163.

Hong, J. K.; Madihally, S. V.; *Tissue Engineering*, **2011**, *17(2),* 125.

Huang, Z-M.; Zhang, Y-Z.; Kotaki, M.; Ramakrishna, S. *Composites Science and Technology*, **2003**, *63(15),* 2223.

Katta, P.; Alessandro, M.; Ramsier, R. D.; Chase, G. G.; *Nano Letters*, **2004**, *4, 11,* 2215.

Li, D.; Wang, Y.; Xia, Y. *Advanced Materials*, **2004**, *16(4),* 361.

Li, D.; Wang, Y.; Xia, Y. *Nano Letters*, **2003**, *3(8),* 1167.

Teo, W. E.; Kotaki, M.; Mo, X. M. Ramakrishna, S. *Nanotechnology*, **2005**, *16(6),* 918.

Theron, A.; Zussman, E.; Yarin, A. L.; *Nanotechnology*, **2001**, *12(3),* 384.

Thorvaldsson, A.; Engström, J.; Gatenholm, P.; Walkenström, P. *Journal of Applied Polymer Science*, **2010**, *118(1),* 511.

Volpato, F. Z.; S. Ramos, L. F.; Motta, A.; Migliaresi, C. *Journal of Bioactive and Compatible Polymers*, **2011**, *26(1),* 35.

CHAPTER 4

CARBON NANOTUBES AND RELATED STRUCTURES

A. K. HAGHI and G. E. ZAIKOV

CONTENTS

4.1 Introduction 148
4.1.1 Graphene 150
4.1.2 Carbon Nanotubes 151
4.1.3 Fullerenes 153
4.1.4 Classification of Nanotubes 154
4.2 Effective Parameters 159
4.2.1 Capillary Effects 159
4.2.2 Specific Electrical Resistance of Carbon Nanotubes 160
4.3 Slippage of the Fluid Particles Near the Wall 171
4.3.1 The Density of the Liquid Layer Near a Wall of Carbon Nanotube 173
4.3.2 The Effective Viscosity of the Liquid in a Nanotube 176
4.3.3 Energy Release Due to the Collapse of the Nanotube 180
4.4 Fluid Flow In Nanotubes 185
4.4.1 Some of the Ideas and Approaches for Modeling in Nanohydromecanics 194
4.4.2 Results 200
4.4.3 The Flow of Fluid With an Empty Interlayer 201
Appendix A—Modeling and Simulation Techniques 204
A1 Molecular Scale Methods 204
A2 Microscale Methods 209
A2.1 Brownian Dynamics 210
A2.2 Dissipative Particle Dynamics 211

A2.3 Lattice Boltzmann 211
A2.4 Time-Dependent Ginzburg–Landau Method 212
A2.5 Dynamic DFT Method 212
A3 Mesoscale and Macroscale Methods 213
A4 Micromechanics 214
A4.1 Basic Concepts 214
A4.2 Halpin–Tsai Model 216
A4.3 Mori–Tanaka Model 217
A4.4 Equivalent-Continuum and Self-Similar Approaches 217
A4.5 Finite Element Method 219
A5 Multi Scale Modeling of Mechanical Properties 220
Appendix B 231
Appendix C 240
C1 Model and Basic Relations 241
Keywords 245
References 245

4.1 INTRODUCTION

The appearance of "nanoscience" and "nanotechnology" stimulated the burst of terms with "nano-" prefix. Historically the term "nanotechnology" appeared before and it was connected with the appearance of possibilities to determine measurable values up to 10^{-9} of known parameters: 10^{-9} m-nm (nanometer), 10^{-9} s-ns (nanosecond), 10^{-9} degree (nanodegree, shift condition). Nanotechnology and molecular nanotechnology comprise the set of technologies connected with transport of atoms and other chemical particles (ions, molecules) at distances contributing the interactions between them with the formation of nanostructures with different nature. Researchers showed the possibility to develop technologies on nanometer level, Eric Drexler is considered to be the founder and ideologist of nanotechnology. When scanning tunnel microscope was invented there was an opportunity to influence atoms of a substance thus stimulating the work in the field of probe technology, which resulted in substantiation and practical application of nanotechnological methods in 1994. With the help of this technique it is possible to handle single atoms and collect molecules or aggregates of molecules, construct various structures from atoms on a certain substrate (base). Naturally, such a possibility cannot be implemented without preliminary computer designing of so-called "nanostructures architecture." Nanostructures architecture assumes a certain given location of atoms and molecules in space that can be designed on computer and afterwards transferred into technological program of nanotechnological facility.

The notion "science of nanomaterials" assumes scientific knowledge for obtaining, composition, properties and possibilities to apply nanostructures, nanosystems and nanomaterials. A simplified definition of this term can be as follows: material science dealing with materials comprizing particles and phases with nanometer dimensions. To determine the existence area for nanostructures and nanosystems it is advisable to find out the difference of these formations from analogous material objects.

From the analysis of literature the following can be summarized: the existence area of nanosystems and nanoparticles with any structure is between the particles of molecular and atomic level determined in picometers and aggregates of molecules or permolecular formations over micron units. Herc, it should be mentioned that in polymer chemistry particles with nanometer dimensions belong to the class of permolecular structures, such as globules and fibrils by one of parameters, for example, by diameter or thickness. In chemistry of complex compounds clusters with nanometer dimensions are also known.

In 1991, Japanese researcher Idzhima was studying the sediments formed at the cathode during the spray of graphite in an electric arc. His attention was attracted by the unusual structure of the sediment consisting of microscopic fibers and filaments. Measurements made with an electron microscope showed that the diameter of these filaments does not exceed a few nanometers and a length of one to several microns.

Having managed to cut a thin tube along the longitudinal axis, the researchers found that it consists of one or more layers, each representing a hexagonal grid of graphite, which is based on hexagon with vertices located at the corners of the carbon atoms. In all cases, the distance between the layers is equal to 0.34 nm that is the same as that between the layers in crystalline graphite.

Typically, the upper ends of tubes are closed by multilayer hemispherical caps, each layer is composed of hexagons and pentagons, reminiscent of the structure of half a fullerene molecule.

The extended structure consisting of rolled hexagonal grids with carbon atoms at the nodes are called nanotubes.

Lattice structure of diamond and graphite are shown in Fig. 1 Graphite crystals are built of planes parallel to each other, in which carbon atoms are arranged at the corners of regular hexagons. Each intermediate plane is shifted somewhat toward the neighboring planes, as shown in the Fig. 1.

The elementary cell of the diamond crystal represents a tetrahedron, with carbon atoms in its center and four vertices. Atoms located at the vertices of a tetrahedron form a center of the new tetrahedron, and thus, are also surrounded by four atoms each, etc. All the carbon atoms in the crystal lattice are located at equal distance (0.154 nm) from each other.

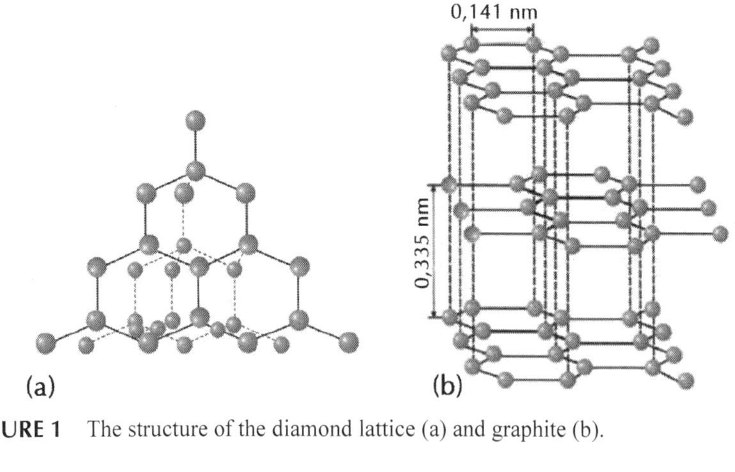

FIGURE 1 The structure of the diamond lattice (a) and graphite (b).

Nanotubes are rolled into a cylinder (hollow tube) graphite plane, which is lined with regular hexagons with carbon atoms at the vertices of a diameter of several nanometers. Nanotubes can consist of one layer of atoms – single-wall nanotubes SWNT and represent a number of "nested" one into another layer pipes – multiwalled nanotubes – MWNT.

Nanostructures can be built not only from individual atoms or single molecules, but the molecular blocks. Such blocks or elements to create nanostructures are graphene, carbon nanotubes and fullerenes.

4.1.1 GRAPHENE

Graphene is a single flat sheet, consisting of carbon atoms linked together and forming a grid, each cell is like a bee's honeycombs (Fig. 2). The distance between adjacent carbon atoms in graphene is about 0.14 nm.

Graphite, from which slates of usual pencils are made, is a pile of graphene sheets (Fig. 3). Graphenes in graphite is very poorly connected and can slide relative to each other. So, if you conduct the graphite on paper, then after separating graphene from sheet the graphite remains on paper. This explains why graphite can write.

FIGURE 2 Schematic illustration of the graphene. Light balls – the carbon atoms, and the rods between them – the connections that hold the atoms in the graphene sheet.

FIGURE 3 Schematic illustration of the three sheets of graphene, which are one above the other in graphite

4.1.2 CARBON NANOTUBES

Many perspective directions in nanotechnology are associated with carbon nanotubes.

Carbon nanotubes are a carcass structure or a giant molecule consisting only of carbon atoms.

Carbon nanotube is easy to imagine, if we imagine that we fold up one of the molecular layers of graphite – graphene (Fig. 4).

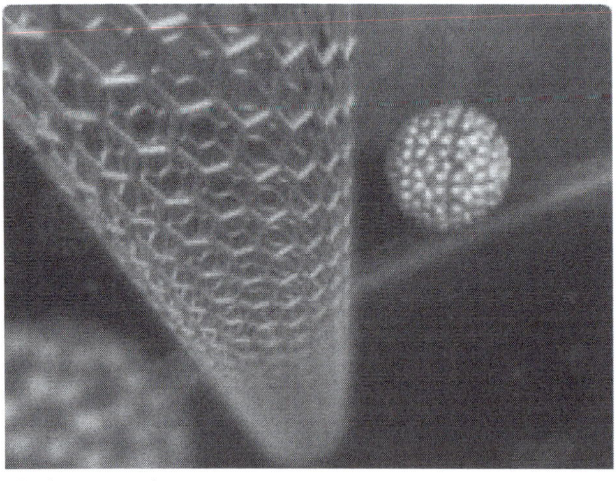

FIGURE 4 Carbon nanotubes.

The way of folding nanotubes – the angle between the directions of nanotube axis relative to the axis of symmetry of graphene (the folding angle) – largely determines its properties.

Of course, no one produces nanotubes, folding it from a graphite sheet. Nanotubes formed themselves, for example, on the surface of carbon electrodes during arc discharge between them. At discharge, the carbon atoms evaporate from the surface, and connect with each other to form nanotubes of all kinds – single, multilayered and with different angles of twist (Fig. 5).

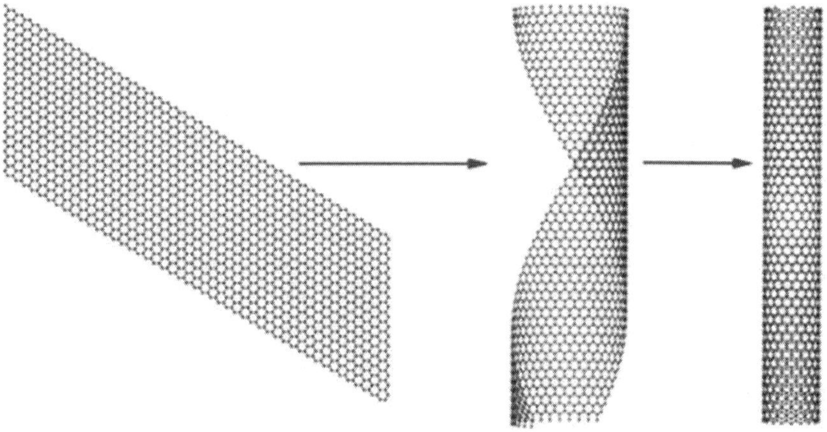

FIGURE 5 One way of imaginary making nanotube (right) from the molecular layer of graphite (left).

The diameter of nanotubes is usually about 1 nm and their length is a thousand times more, amounting to about 40 microns. They grow on the cathode in perpendicular direction to surface of the butt. The so-called self-assembly of carbon nanotubes from carbon atoms occurs. Depending on the angle of folding of the nanotube they can have conductivity as high as that of metals, and they can have properties of semiconductors.

Carbon nanotubes are stronger than graphite, although made of the same carbon atoms, because carbon atoms in graphite are located in the sheets. And everyone knows that sheet of paper folded into a tube is much more difficult to bend and break than a regular sheet. That's why carbon nanotubes are strong. Nanotubes can be used as a very strong microscopic rods and filaments, as Young's modulus of single-walled nanotube reaches values of the order of 1–5 TPa, which is much more than steel! Therefore, the thread made of nanotubes, the thickness of a human hair is capable to hold down hundreds of kilos of cargo.

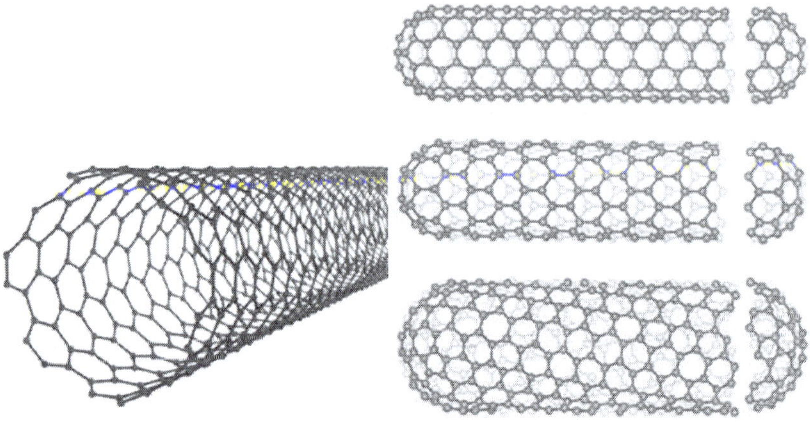

FIGURE 6 Left – schematic representation of a single-layer carbon nanotubes, on the right (top to bottom) – two-ply, straight and spiral nanotubes.

It is true that at present the maximum length of nanotubes is usually about a hundred microns – which is certainly too small for everyday use. However, the length of the nanotubes obtained in the laboratory is gradually increasing – now scientists have come close to the millimeter border. So there is every reason to hope that in the near future, scientists will learn how to grow a nanotube length in centimeters and even meters.

4.1.3 FULLERENES

The carbon atoms, evaporated from a heated graphite surface, connecting with each other, can form not only nanotube, but also other molecules, which are closed convex polyhedra, for example, in the form of a sphere or ellipsoid. In these molecules, the carbon atoms are located at the vertices of regular hexagons and pentagons, which make up the surface of a sphere or ellipsoid.

The molecules of the symmetrical and the most studied fullerene consisting of 60 carbon atoms (C_{60}), form a polyhedron consisting of 20 hexagons and 12 pentagons and resembles a soccer ball (Fig. 7). The diameter of the fullerene C_{60} is about 1 nm.

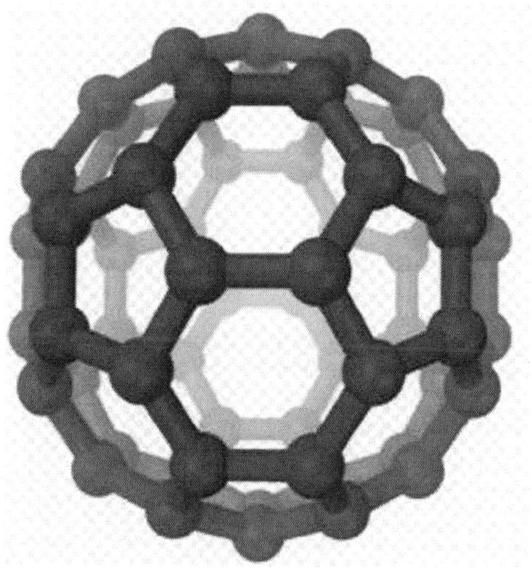

FIGURE 7 Schematic representation of the fullerene C_{60}.

4.1.4 CLASSIFICATION OF NANOTUBES

The main classification of nanotubes is conducted by the number of constituent layers.

Single-walled nanotubes – the simplest form of nanotubes. Most of them have a diameter of about 1 nm in length, which can be many thousands of times more. The structure of the nanotubes can be represented as a "wrap" a hexagonal network of graphite (graphene), which is based on hexagon with vertices located at the corners of the carbon atoms in a seamless cylinder. The upper ends of the tubes are closed by hemispherical caps, each layer is composed of hexa – and pentagons, reminiscent of the structure of half of a fullerene molecule. The distance d between adjacent carbon atoms in the nanotube is approximately equal to d = 0,15 nm (Fig. 8).

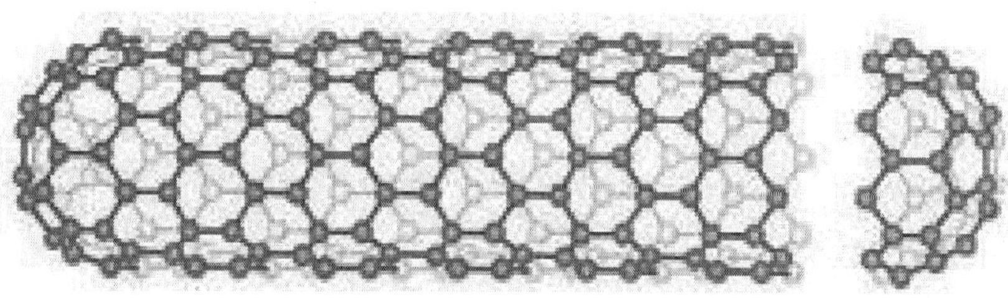

FIGURE 8 Graphical representation of single-walled nanotube.

Multi-walled nanotubes consist of several layers of graphene stacked in the shape of the tube. The distance between the layers is equal to 0.34 nm, that is the same as that between the layers in crystalline graphite (Fig. 9).

Carbon nanotubes could find applications in numerous areas:

- additives in polymers;
- catalysts (autoelectronic emission for cathode ray lighting elements, planar panel of displays, gas discharge tubes in telecom networks);
- absorption and screening of electromagnetic waves;
- transformation of energy;
- anodes in lithium batteries;
- keeping of hydrogen;
- composites (filler or coating);
- nanosondes;
- sensors;
- strengthening of composites;
- supercapacitors.

FIGURE 9 Graphic representation of a multiwalled nanotube.

More than a decade, carbon nanotubes, despite their impressive performance characteristics have been used, in most cases, for scientific research.

To date, the most developed production of nanotubes has Asia, the production capacity that is 2–3 times higher than in North America and Europe combined. Is dominated by Japan, which is a leader in the production of MWNT. Manufacturing North America, mainly focused on the SWNT. Growing at an accelerated rate production in China and South Korea. In the coming years, China will surpass the level of production of the U.S. and Japan, and by 2013, a major supplier of all types of nanotubes, according to experts, could be South Korea.

4.1.5 CHIRALITY

Chirality is a set of two integer positive indices (n, m), which determines how the graphite planes folds and how many elementary cells of graphite at the same time fold to obtain the nanotube.

From the value of parameters (n, m) are distinguished as direct (achiral) high-symmetry carbon nanotubes.

- o armchair $n = m$
- o zigzag $m = 0$ or $n = 0$
- o helical (chiral) nanotube

In Fig. 10a is shown a schematic image of the atomic structure of graphite plane – graphene, and shown how a nanotube can be obtained from it. The nanotube is fold up with the vector connecting two atoms on a graphite sheet. The cylinder is obtained by folding this sheet so that were combined the beginning and end of the vector. That is, to obtain a carbon nanotube from a graphene sheet, it should turn so that the lattice vector $\overline{R}$ has a circumference of the nanotube in Fig. 10b. This vector can be expressed in terms of the basis vectors of the elementary cell graphene sheet $\vec{R} = n\vec{r_1} + m\vec{r_2}$. Vector $\overline{R}$, which is often referred to simply by a pair of indices (n, m), called the chiral vector. It is assumed that $n > m$. Each pair of numbers (n, m) represents the possible structure of the nanotube.

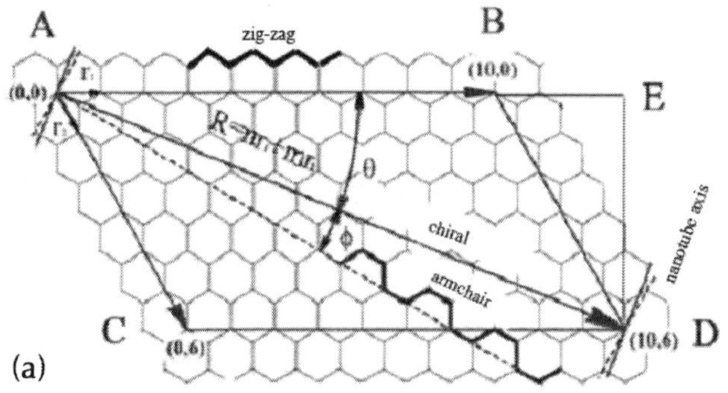

FIGURE 10 *(Continued)*

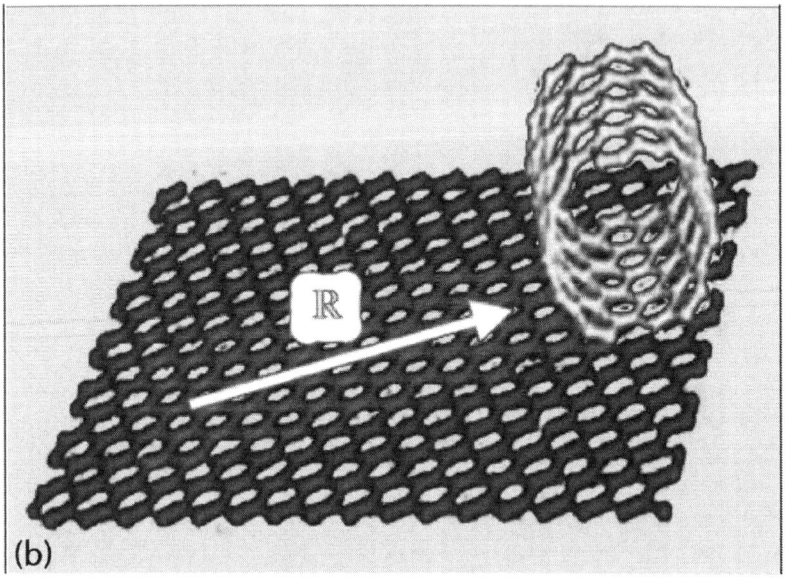

FIGURE 10 Schematic representation of the atomic structure of graphite plane.

In other words the chirality of the nanotubes (n, m) indicates the coordinates of the hexagon, which as a result of folding the plane has to be coincide with a hexagon, located at the beginning of coordinates (Fig. 11).

Many of the properties of nanotubes (e.g., zonal structure or space group of symmetry) strongly depend on the value of the chiral vector. Chirality indicates what property has a nanotube – a semiconductor or metallicheskm. For example, a nanotube (10,10) in the elementary cell contains 40 atoms and is the type of metal, whereas the nanotube (10, 9) has already in 1084 and is a semiconductor (Fig. 12).

If the difference $n-m$ is divisible by 3, then these CNTs have metallic properties. Semimetals are all achiral tubes such as "chair". In other cases, the CNTs show semiconducting properties. Just type chair CNTs $(n = m)$ are strictly metal.

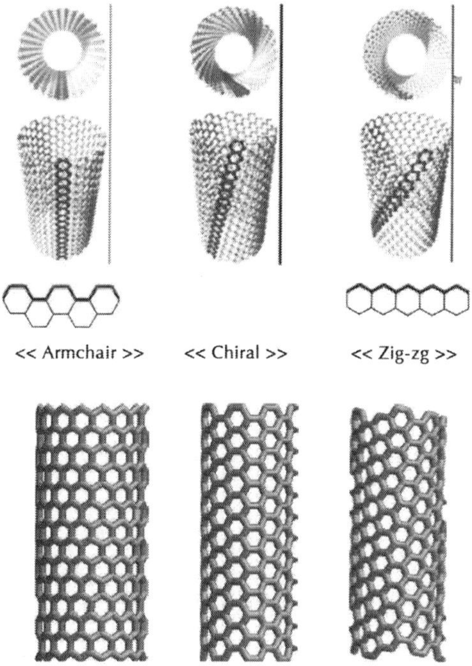

FIGURE 11 Single-walled carbon nanotubes of different chirality (in the direction of convolution). Left to right: the zigzag (16,0), armchair (8,8) and chiral (10,6) carbon nanotubes.

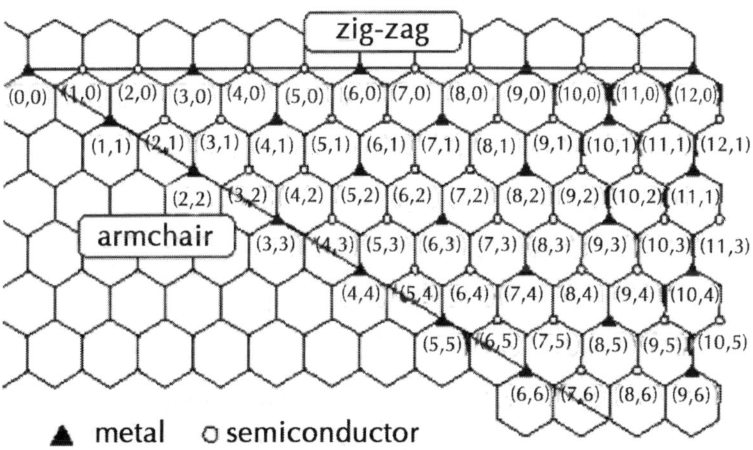

FIGURE 12 The scheme of indices (n, m) of lattice vector $\overline{R}$ tubes having semiconductor and metallic properties.

4.2 EFFECTIVE PARAMETERS

In addition to size effects that occur in micro and nano, we should note the following factors that determine the processes in low-dimensional systems:

- surface roughness (resistance to flow effects, interactions with the particles, etc.);
- dissolved gases (formation of bubbles, sticking to the surface, etc.);
- chemical surface properties (chemical reactions, etc.);
- hydrophobic – hydrophilic of surface;
- contaminants;
- heating due to uncontrollable processes;
- electrical properties of the surface (double layer, the change of surface and volume charge, charge transfer, etc.).

Viscous forces in the fluid can lead to large dispersion flow along the axis of motion. They have a significant impact, both on the scale of individual molecules, and the scale of microflows – near the borders of the liquid-solid (beyond a few molecular layers), during the motion on a complex and heterogeneous borders.

Influence of the effect of boundary regions on the particles and fluxes have been observed experimentally in the range of molecular thicknesses up to hundreds of nanometers. If the surface has a superhydrophobic properties, this range can extend to the micron thickness. Molecular theory can predict the effect of hydrophobic surfaces in the system only up to tens of nanometers.

Fluids, the flow of liquid or gas, have properties that vary continuously under the action of external forces. In the presence of fluid shear forces are small in magnitude, leads large changes in the relative position of the element of fluid. In contrast, changes in the relative positions of atoms in solids remain small under the action of any small external force. Termination of action of the external forces on the fluid does not necessarily lead to the restoration of its initial form.

4.2.1 CAPILLARY EFFECTS

To observe the capillary effects, you must open the nanotube, that is, remove the upper part – lids. Fortunately, this operation is quite simple.

The first study of capillary phenomena have shown that there is a relationship between the magnitude of surface tension and the possibility of its being drawn into the channel of the nanotube. It was found that the liquid penetrates into the channel of the nanotube, if its surface tension is not higher than 200 mN/m. Therefore, for the entry of any substance into the nanotube using solvents having a low surface tension. For example concentrated nitric acid with surface tension of 43 mN/m is used to inject certain metals into the channel of a nanotube. Then annealing is conducted at 4000 °C for 4 h in an atmosphere of hydrogen, which leads to the recovery of the metal. Thus, the obtained nanotubes containing nickel, cobalt and iron.

Along with the metals carbon nanotubes can be filled with gaseous substances, such as hydrogen in molecular form. This ability is of great practical importance, since opening the ability to safely store hydrogen, which can be used as a clean fuel in internal combustion engines.

4.2.2 SPECIFIC ELECTRICAL RESISTANCE OF CARBON NANOTUBES

Due to small size of carbon nanotubes only in 1996 they succeeded to directly measure their electrical resistivity ρ. The results of direct measurements showed that the resistivity of the nanotubes could be varied within wide limits to 0.8 ohm/cm. The minimum value is lower than that of graphite. Most of the nanotubes have metallic conductivity, and the smaller shows properties of a semiconductor with a bandgap of 0.1 to 0.3 eV.

The resistance of single-walled nanotube is independent of its length, because of this it is convenient to use for the connection of logic elements in microelectronic devices. The permissible current density in carbon nanotubes is much greater than in metallic wires of the same cross section, and one hundred times better achievement for superconductors.

The results of the study of emission properties of the material, where the nanotubes were oriented perpendicular to the substrate, have been very interesting for practical use. Attained values of the emission current density of the order of 0.5 mA/mm^2. The value obtained is in good agreement with the Fowler-Nordheim expression.

The most effective and common way to control microflow substances is *electrokinetic* and *hydraulic*. At the same time the most technologically advanced and automated considered electrokinetic.

Charges transfer in mixtures occurs as a result of the directed motion of charge carriers – ions. There are different mechanisms of such transfer, but usually are *convection, migration and diffusion.*

Convection is called mass transfer the macroscopic flow. *Migration* – the movement of charged particles by electrostatic fields. The velocity of the ions depends on field strength. In microfluidics a special role is played *electrokinetic processes* that can be divided into four types: *electro-osmosis, electrophoresis, streaming potential and sedimentation potential.* These processes can be qualitatively described as follows:

(a) *electro-osmosis* – the movement of the fluid volume in response to the applied electric field in the channel of the electrical double layers on its wetted surfaces.
(b) *Electrophoresis* – the forced motion of charged particles or molecules, in mixture with the acting electric field.
(c) *Streamy potential* – the electric potential, which is distributed through a channel with charged walls, in the case when the fluid moves under the action of pressure forces. Joule electric current associated with the effect of charge transfer is flowing stream.
(d) *The potential of sedimentation* – an electric potential is created when charged particles are in motion relative to a constant fluid. The driving force for this effect – usually gravity.

In general, for the microchannel cross-section S amount of introduced probe (when entering electrokinetic method) depends on the applied voltage U, time t during which the received power, and mobility of the sample components μ:

$$Q = \frac{\mu S U t}{L} \cdot c$$

where;
c – probe concentration in the mixture,
L – the channel length.

Amount of injected substance is determined by the electrophoretic and total electroosmotic mobilities μ.

In the hydrodynamic mode of entry by the pressure difference in the channel or capillary of circular cross section, the volume of injected probe V_c:

$$V_c = \frac{4}{128} \cdot \frac{\Delta p \pi d t}{\eta L}$$

where;
Δp – pressure differential,
d – diameter of the channel,
η – viscosity.

In the simulation of processes in micron-sized systems the following basic principles are fundamental:

1. hypothesis of *laminar* flow (sometimes is taken for granted when it comes to microfluidics);
2. continuum hypothesis (detection limits of applicability);
3. laws of formation of the velocity profile, mass transfer, the distribution of electric and thermal fields;

4. boundary conditions associated with the geometry of structural elements (walls of channels, mixers zone flows, etc.).

Since we consider the physical and chemical transport processes of matter and energy, mathematical models, most of them have the form of systems of differential equations of second order partial derivatives. Methods for solving such equations are analytical (Fourier and its modifications, such as the method of Greenberg, Galerkin, in some cases, the method of d'Alembert and the Green's functions, the Laplace operator method, etc.) or numerical (explicit or, more effectively, implicit finite difference schemes) – traditional. The development involves, basically, numerical methods and follows the path of saving computing resources, and increasing the speed of modern computers.

Laminar flow – a condition in which the particle velocity in the liquid flow is not a random function of time. The small size of the microchannels (typical dimensions of 5 to 300 microns) and low surface roughness create good conditions for the establishment of laminar flow. Traditionally, the image of the nature of the flow gives the dimensionless characteristic numbers: the Reynolds number and Darcy's friction factor.

In the motion of fluids in channels the turbulent regime is rarely achieved. At the same time, the movement of gases is usually turbulent.

Although the liquid – are quantized in the length scale of intermolecular distances (about 0.3 nm to 3 nm in liquids and for gases), they are assumed to be continuous in most cases, microfluidics. Continuum hypothesis (continuity, continuum) suggests that the macroscopic properties of fluids consisting of molecules, the same as if the fluid was completely continuous (structurally homogeneous). Physical characteristics: mass, momentum and energy associated with the volume of fluid containing a sufficiently large number of molecules must be taken as the sum of all the relevant characteristics of the molecules.

Continuum hypothesis leads to the concept of fluid particles. In contrast to the ideal of a point particle in ordinary mechanics, in fluid mechanics, particle in the fluid has a finite size.

At the atomic scale we would see large fluctuations due to the molecular structure of fluids, but if we the increase the sample size, we reach a level where it is possible to obtain stable measurements. This volume of probe must contain a sufficiently large number of molecules to obtain reliable reproducible signal with small statistical fluctuations. For example, if we determine the required volume as a cube with sides of 10 nm, this volume contains some of the molecules and determines the level of fluctuations of the order of 0.5%.

The most important position in need of verification is to analyze the admissibility of mass transfer on the basis of the continuum model that can be used instead of the concentration dependence of the statistical analysis of the ensemble of individual particles. The position of the continuum model is considered as a necessary condition for microfluidics.

The applicability of the hypothesis is based on comparison of free path length of a particle λ in a liquid with a characteristic geometric size d. The ratio of these lengths – the Knudsen number: $Kn = \lambda / d$. Based on estimates of the Knudsen number defined two important statements:

a) $Kn < 10^{-3}$ – justified hypothesis of a continuous medium,

b) $Kn < 10^{-1}$ – allowed the use of adhesion of particles to the solid walls of the channel.

Wording of the last condition can also be varied: both in form $U = 0$ and in a more complex form, associated with shear stresses. The calculation of λ can be carried out as $\lambda \approx \sqrt[3]{\overline{V} / Na}$, where;

$\overline{V}$ – molar volume,

Na – Avogadro's number.

Under certain geometrical approximations of the particles of substance free path length can be calculated as $\lambda \approx 1/\left(\sqrt{2}\pi r_S^2 Na\right)$, if used instead r_S Stokes radius, as a consequence of the spherical approximation of the particle. On the other hand, for a rigid model of the molecule r_S should be replaced by the characteristic size of the particles R_g – the radius of inertia, calculated as $R_g = n_i \cdot \delta_l / \sqrt{6}$. Here δ_l – the length of a fragment of the chain (link), n_i – the number of links.

Of course, the continuum hypothesis is not acceptable when the system under consideration is close to the molecular scale. This happens in nanoliquid, such as liquid transport through nano-pores in cell membranes or artificially made nano-channels.

In contrast to the continuum hypothesis, the essence of modeling the molecular dynamics method is as follows. We consider a large ensemble of particles, which simulate atoms or molecules, i.e., all atoms are material points. It is believed that the particles interact with each other and, moreover, may be subject to external influence. Interatomic forces are represented in the form of the classical potential force (the gradient of the potential energy of the system).

The interaction between atoms is described by means of van der Waals forces (intermolecular forces), mathematically expressed by the Lennard-Jones potential:

$$V(r) = \frac{Ae^{-\sigma r}}{r} - \frac{C_6}{r^6}$$

where;

A and C_6 – some coefficients depending on the structure of the atom or molecule,

σ— the smallest possible distance between the molecules.

In the case of two isolated molecules at a distance of r_0 the interaction force is zero, that is, the repulsive forces balance attractive forces. When $r > r_0$ the resultant force is the force of gravity, which increases in magnitude, reaching a maximum at $r = r_m$ and then decreases. When $r < r_0$ – a repulsive force. Molecule in the field of these forces has potential energy $V(r)$, which is connected with the force of $f(r)$ by the differential equation

$$dV = -f(r)dr$$

At the point $r = r_0$, $f(r) = 0$, $V(r)$ reaches an extremum (minimum).

The chart of such a potential is shown in Fig. 13. The upper (positive) half-axis r corresponds to the repulsion of the molecules, the lower (negative) half-plane – their attraction. We can say simply: at short distances the molecules mainly repel each, on the long – draw each other. Based on this hypothesis, and now an obvious fact, the van der Waals received his equation of state for real gases.

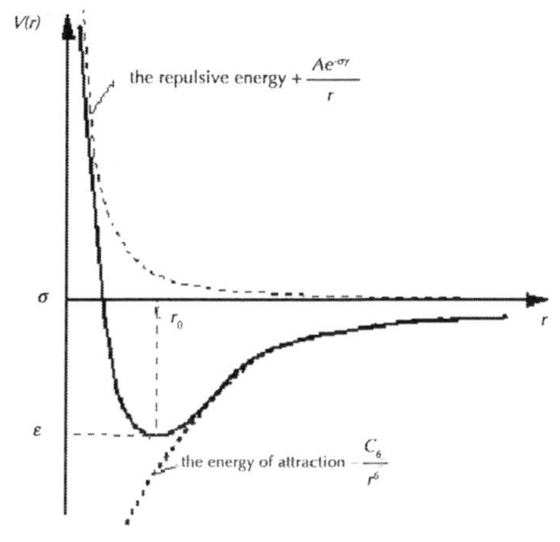

FIGURE 13 The chart of the potential energy of intermolecular interaction.

The exponential summand in the expression for the potential describing the repulsion of the molecules at small distances, often approximated as

$$\frac{Ae^{-\sigma r}}{r} \approx \frac{C_{12}}{r^{12}}$$

In this case we obtain the Lennard-Jones potential:

$$V(r) = \frac{C_{12}}{r^{12}} - \frac{C_6}{r^6} \qquad (1)$$

The interaction between carbon atoms is described by the potential

$$V_{CC}(r) = K(r-b)^2,$$

where,
K is – constant tension (compression) connection,
$b = 1{,}4A$ – the equilibrium length of connection,
r – current length of the connection.

The interaction between the carbon atom and hydrogen molecule is described by the Lennard-Jones

$$V(r) = 4\varepsilon\left[\left(\frac{\sigma}{r}\right)^{12} - \left(\frac{\sigma}{r}\right)^{6}\right]$$

For all particles (Fig. 15) the equations of motion are written:

$$m\frac{d^2\overline{r}_i}{dt^2} = \overline{F}_{T-H_2}(\overline{r}_i) + \sum_{j\neq i}\overline{F}_{H_2-H_2}\left(\overline{r}_i - \overline{r}_j\right),$$

where,
$\overline{F}_{T-H_2}(\overline{r})$ – force, acting by the CNT, $\overline{F}_{H_2-H_2}(\overline{r}_i - \overline{r}_j)$ – force acting on the i-th molecule from the j-th molecule

The coordinates of the molecules are distributed regularly in the space, the velocity of the molecules are distributed according to the Maxwell equilibrium distribution function according to the temperature of the system:

$$f(u,v,w) = \frac{\beta^3}{\pi^{3/2}}\exp\left(-\beta^2\left(u^2+v^2+w^2\right)\right) \quad \beta = \frac{1}{\sqrt{2RT}}$$

The macroscopic flow parameters are calculated from the distribution of positions and velocities of the molecules:

$$\overline{V} = \left\langle \overline{v}_i \right\rangle = \frac{1}{n}\sum_i \overline{v}_i, \quad \rho = \frac{nm}{V_0}, \quad \frac{3}{2}RT = \frac{1}{2}\left\langle \left|\overline{v}_i^{\prime}\right|^2 \right\rangle . \overline{v}_i^{\prime} = \overline{v}_i - \overline{V},$$

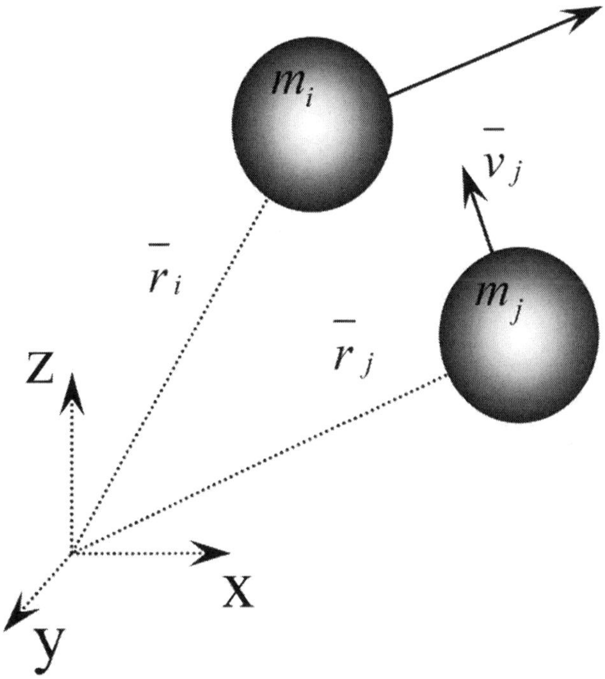

FIGURE 14 Position of Particles.

The resulting system of equations is solved numerically. However, the molecular dynamics method has limitations of applicability:

1. the de Broglie wavelength h/mv (where h – Planck's constant, m – the mass of the particle, v – velocity) of the atom must be much smaller than the interatomic distance;
2. Classical molecular dynamics cannot be applied for modeling systems consisting of light atoms such as helium or hydrogen;
3. at low temperatures, quantum effects become decisive for the consideration of such systems must use quantum chemical methods;
4. necessary that the times at which we consider the behavior of the system were more than the relaxation time of the physical quantities.

In 1873, Van der Waals proposed an equation of state is qualitatively good description of liquid and gaseous systems. It is for one mole (one mole) is:

$$\left(p+\frac{a}{v^2}\right)(v-b)=RT \tag{2}$$

Note that at $p >> \frac{a}{v^2}$ and $v >> b$ this equation becomes the equation of state of ideal gas

$$pv = RT \tag{3}$$

Van der Waals equation can be obtained from the Clapeyron equation of Mendeleev by an amendment to the magnitude of the pressure a/v^2 and the amendment b to the volume, both constant a and b independent of T and v but dependent on the nature of the gas.

The amendment b takes into account:

1. the volume occupied by the molecules of real gas (in an ideal gas molecules are taken as material points, not occupying any volume);
2. so-called "dead space", in which can not penetrate the molecules of real gas during motion, i.e. volume of gaps between the molecules in their dense packing.

Thus, $b = v_{мол} + v_{зaз}$ (Fig. 16). The amendment a/v^2 takes into account the interaction force between the molecules of real gases. It is the internal pressure, which is determined from the following simple considerations. Two adjacent elements of the gas will react with a force proportional to the product of the quantities of substances enclosed in these elementary volumes.

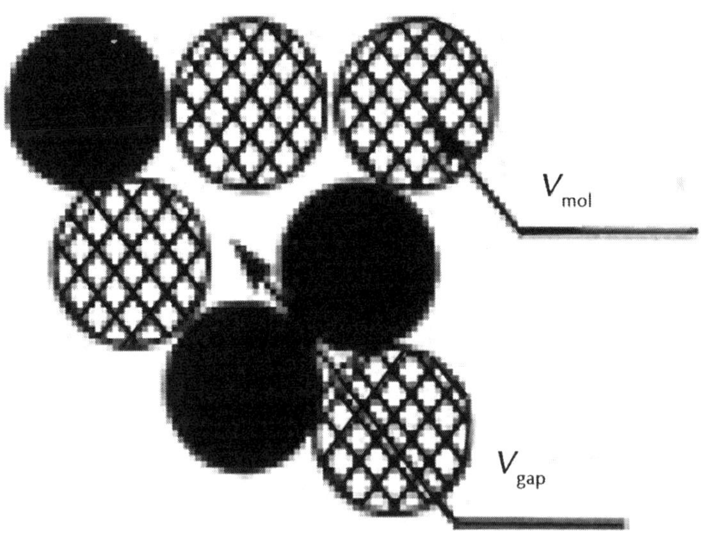

FIGURE 15 Location scheme of molecules in a real gas.

Therefore, the internal pressure p_{BH} is proportional to the square of the concentration n:

$$p_{BH} \sim n^2 \sim \rho^2 \sim \frac{1}{v^2},$$

Where;
ρ – the gas density.

Thus, the total pressure consists of internal and external pressures:

$$p + p_{BH} = p + \frac{a}{v^2}$$

Equation (3) is the most common for an ideal gas. Under normal physical conditions and from Eq. (3) we obtain:

$$R = \frac{R\mu}{\mu} = \frac{8314}{\mu}$$

Knowing $R\mu$ we can find the gas constant for any gas with the help of the value of its molecular mass μ (Table 1):

TABLE 1 The molecular weight of some gases.

Gas	N	Ar	H_2	O_2	CO	CO_2	Ammonia	Air
μ	28	40	2	32	28	44	17	29

For gas mixture with mass M state equation has the form:

$$pv = MR_{cm}T = \frac{8314MT}{\mu_{cm}} \tag{4}$$

where;
R_{cm} – gas constant of the mixture.
The gas mixture can be given by the mass proportions g_i, voluminous r_i or mole fractions n_i respectively, which are defined as the ratio of mass m_i, volume v_i or number of moles N_i of i gas to total mass M, volume v or number of moles N of gas mixture. Mass fraction of component is $g_i = \frac{m_i}{M}$, where $i = 1,n$. It is obvious that $M = \sum_{i=1}^{n} m_i$ and $\sum_{i=1}^{n} g_i = 1$. The volume fraction is $r_i = \frac{v_i}{v_{cm}}$, where v_i – partial volume of component mixtures.

Similarly, we have $\sum_{i=1}^{n} v_i = v_{cm}, \sum_{i=1}^{n} r_i = 1$.

Depending on specificity of tasks the gas constant of the mixture may be determined as follows:

$$R_{cm} = \frac{1}{\sum_{i=1}^{n} r_i R_i^{-1}}; \quad R_{cm} = \frac{1}{\sum_{i=1}^{n} r_i R_i^{-1}}$$

If we know the gas constant R_{cm}, the seeming molecular weight of the mixture is equal to

$$\mu_{cm} = \frac{8314}{R_{cm}} = \frac{8314}{\sum_{i=1}^{n} g_i R_i} = 8314 \sum_{i=1}^{n} r_i R_i^{-1}$$

The pressure of the gas mixture p is equal to the sum of the partial pressures of individual components in the mixture p_i:

$$p = \sum_{i=1}^{n} p_i \tag{5}$$

Partial pressure p_i – pressure that has gas, if it is one at the same temperature fills the whole volume of the mixture ($p_i v_{cm} = RT$).

With various methods of setting the gas mixture partial pressures

$$p_i = p r_i; \; p_i = \frac{p g_i \mu_{cm}}{\mu_i} \tag{6}$$

From the Eq. (6) we see that for the calculation of the partial pressures p_i necessary to know the pressure of the gas mixture, the volume or mass fraction i of the gas component, as well as the molecular weight of the gas mixture μ and the molecular weight of i of gas μ_i.

The relationship between mass and volume fractions are written as follows:

$$g_i = \frac{m_i}{m_{cm}} = \frac{\rho_i v_i}{\rho_{cm} v_{cm}} = \frac{R_{cm}}{R_i} r_i = \frac{\mu_i}{\mu_{cm}} r_i$$

We rewrite Eq. (2) as

$$v^3 - \left(b + \frac{RT}{p} \right) v^2 + \frac{a}{p} v - \frac{ab}{p} = 0 \tag{7}$$

When $p = p_k$ and $T = T_k$, where p_k and T_k – critical pressure and temperature, all three roots of Eq. (7) are equal to the critical volume v_k:

$$v^3 - \left(b + \frac{RT_k}{P_k}\right)v^2 + \frac{a}{p_k}v - \frac{ab}{p_k} = 0 \tag{8}$$

Because the $v_1 = v_2 = v_3 = v_k$, then Eq. (8) must be identical to the equation:

$$(v - v_1)(v - v_2)(v - v_3) = (v - v_k)^3 = v^3 - 3v^2v_k + 3vv_k^2 - v_k^3 = 0 \tag{9}$$

Comparing the coefficients at the equal powers of v in both equations leads to the equalities:

$$b + \frac{RT_k}{p_k} = 3v_k \quad \frac{a}{p_k} = 3v_k^2 \quad \frac{ab}{p_k} = v_k^3 \tag{10}$$

Hence,

$$a = 3v_k^2 p_k; b = \frac{v_k}{3} \tag{11}$$

Considering Eq. (10) as equations for the unknowns p_k, v_k, T_k, we obtain:

$$p_k = \frac{a}{27b^2}; v_k = 3b; T_k = \frac{8a}{27bR} \tag{12}$$

From Eqs. (10) and (11) or (12) we can find the relation:

$$\frac{RT_k}{p_k v_k} = \frac{8}{3} \tag{13}$$

Instead of the variables p, v, T let's introduce the relationship of these variables to their critical values (leaden dimensionless parameters).

$$\pi = \frac{p}{p_k}; \omega = \frac{v}{V_k}; \tau = \frac{T}{T_k} \tag{14}$$

Substituting Eqs. (12) and (14) in Eq. (7) and using Eq. (13), we obtain:

$$\left(\pi p_k + \frac{3v_k^2 p_k}{\omega^2 v_k^2}\right)\left(\omega v_k - \frac{v_k}{3}\right) = RT_k\tau,$$

$$\left(\pi + \frac{3}{\omega^2}\right)(3\omega - 1) = 3\frac{RT_k}{p_k v_k}\tau,$$

$$\left(\pi + \frac{3}{\omega^2}\right)(3\omega - 1) = 8\tau \tag{15}$$

4.3 SLIPPAGE OF THE FLUID PARTICLES NEAR THE WALL

Features of the simulation results of Poiseuille flow in the microtubules, when the molecules at the solid wall and the wall atoms at finite temperature of the wall make chaotic motion lies in the fact that in the intermediate range of Knudsen numbers there is slippage of the fluid particles near the wall.

Researchers describe three possible cases: (1) the liquid can be stable (no slippage), (2) slides relative to the wall (with slippage flow), (3) the flow profile is realized; this is when the friction of the wall is completely absent (complete slippage).

In the framework of classical continuum fluid dynamics, according to the Navier boundary condition the velocity slip is proportional to fluid velocity gradient at the wall:

$$v\big|_{y=0} = L_S\, dv/dy\big|_{y=0} \tag{16}$$

Here and in Fig. 17 L_S represents the "slip length" and has a dimension of length.

Because of the slippage, the average velocity in the channel $\langle v_{pdf} \rangle$ increases.

In a rectangular channel (of width > height h and viscosity of the fluid η) due to an applied pressure gradient – dp/dx the authors of that article obtained:

$$\langle v_{pdf} \rangle = \frac{h^2}{12\eta}\left(-\frac{dp}{dx}\right)\left(1 + \frac{6L_S}{h}\right) \tag{17}$$

The results of molecular dynamics simulation for nanosystems with liquid, with characteristic dimensions of the order of the size of the fluid particles, show that a large slippage lengths (of the order of microns) should occur in the carbon nanotubes of nanometer diameter and, consequently, can increase the flow rate by three orders of ($6L_S / h > 1000$). Thus, the flow with slippage is becoming more and more important for hydrodynamic systems of small size.

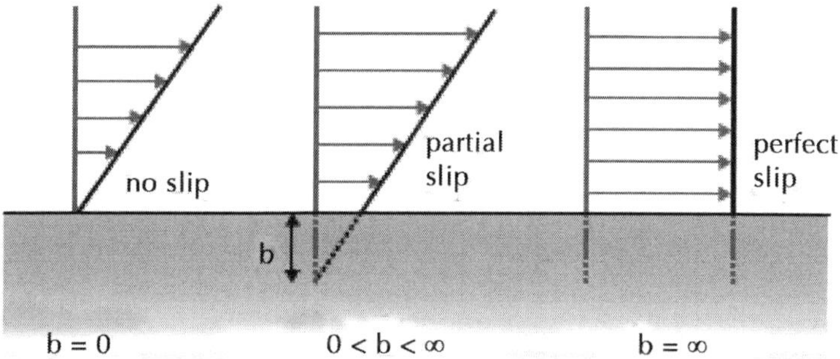

FIGURE 17 Three cases of slip flow past a stationary surface. The slip length b is indicated.

The results of molecular dynamics simulation of unsteady flow of mixtures of water – water vapor, water and nitrogen in a carbon nanotube are reported by many reserchers.

Carbon nanotubes have been considered "zigzag" with chiral vectors (20, 20), (30, 30) and (40, 40), corresponding to pipe diameters of 2.712, 4.068 and 5.424 nm, respectively.

Knowing the value of the flow rate and the system pressure, which varies in the range of 600–800 bars, are high enough to ensure complete filling of the tubes. This pressure can be achieved by the total number of water molecules 736, 904, and 1694, respectively.

The effects of slippage of various liquids on the surface of the nanotube were studied in detail.

The length of slip, can be calculated using the current flow velocity profiles of liquid, shown in Fig. 18, were 11, 13, and 15 nm for the pipes of 2.712, 4.068 and 5.424 nm, respectively. The dotted line marked by theoretical modeling data. The vertical lines indicate the position of the surface of carbon nanotubes.

It was found out that as the diameter decreases, the speed of slippage of particles on the wall of nanotube also decreases. The authors attribute this to the increase of the surface friction.

Experiments with various pressure drops in nanotubes demonstrated slippage of fluid in micro and nanosystems. The most remarkable were the two recent experiments, which were conducted to improve the flow characteristics of carbon nanotubes with the diameters of 2 and 7 nm respectively. In the membranes in which the carbon nanotubes were arranged in parallel, there was a slip of the liquid in the micrometer range. This led to a significant increase in flow rate – up to three – four orders of magnitude.

In the experiments for the water moving in microchannels on smooth hydrophobic surfaces, there are slidings at about 20 nm. If the wall of the channel is not smooth but twisty or rough, and at the same time, hydrophobic, such a structure would lead to an accumulation of air in the cavities and become superhydrophobic (with contact angle greater than 160^0). It is believed that this leads to creation of contiguous areas with high and low slippage, which can be described as "effective slip length". This effective length of the slip occurring on the rough surface can be several tens of microns, which was indeed experimentally confirmed by many researchers.

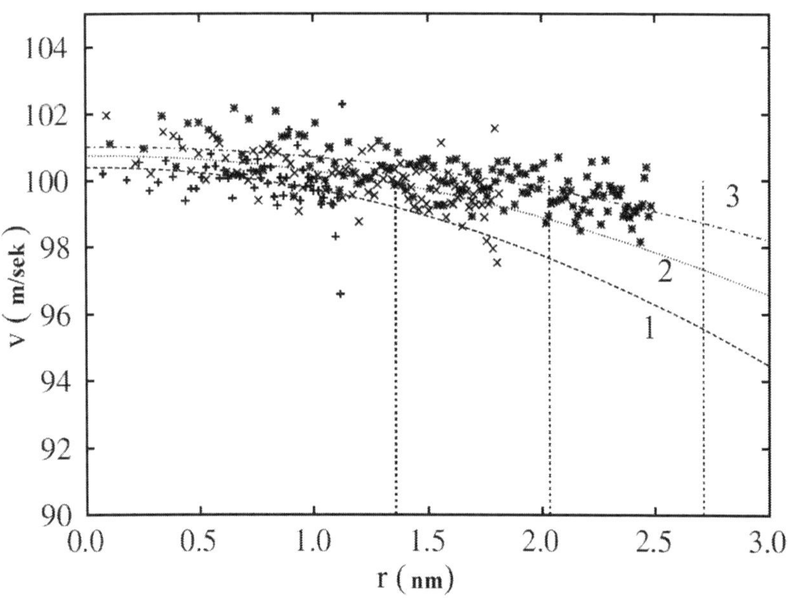

FIGURE 18 Time average streaming velocity profiles of water in a carbon nanotubes of different diameter: 2.712 nm, curve 1; 4.068 nm, curve 2; 5.424 nm, curve 3.

Another possible problem is filling of the hydrophobic systems with liquid. Filling of micron size hydrophobic capillaries is not a big problem, because pressure of less than 1 atm is sufficient. Capillary pressure, however, is inversely proportional to the diameter of the channel, and filling for nanochannels can be very difficult.

4.3.1 THE DENSITY OF THE LIQUID LAYER NEAR A WALL OF CARBON NANOTUBE

Researchers showed radial density profiles of oxygen averaged in time and hydrogen atoms in the "zigzag" carbon nanotube with chiral vector (20, 20) and a radius

$R = 1{,}356$ nm (Fig. 19). The distribution of molecules in the area near the wall of the carbon nanotube indicated a high-density layer near the wall of the carbon nanotube. Such a pattern indicates the presence of structural heterogeneity of the liquid in the flow of the nanotube. In **Figure (3.3)** $\rho^*(r) = \rho(r)/\rho(0)$, 2,712 nm diameter pipe is completely filled with water molecules at $300^0 K$. The overall density $\rho(0) = 1000_{KS/M^3}$. The arrows denote the location of distinguishable layers of the water molecules and the vertical line the position of the CNT wall.

The distribution of molecules in the region of $0{,}95 \le r \le R$ nm indicates a high-density layer near the wall carbon nanotube. Such a pattern indicates the presence of structural heterogeneity of the liquid in the flow of the nanotube.

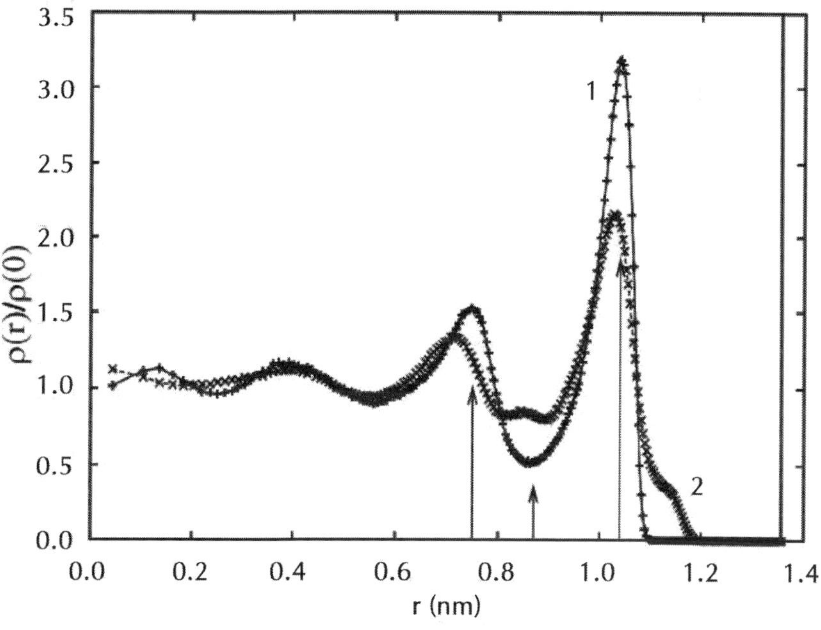

FIGURE 19 Radial density profiles of oxygen (curve 1) and hydrogen (curve 2) atoms for averaged in the time interval.

Many researchers obtained similar results for the flow of the water in the carbon nanotubes using molecular dynamic simulations.

Figure 20 shows the scheme of the initial structure and movement of water molecules in the model carbon nanotube (a), the radial density profile of water molecules inside nanotubes with different radii (b) and the velocity profile of water molecules in a nanotube chirality (60,60) (c).

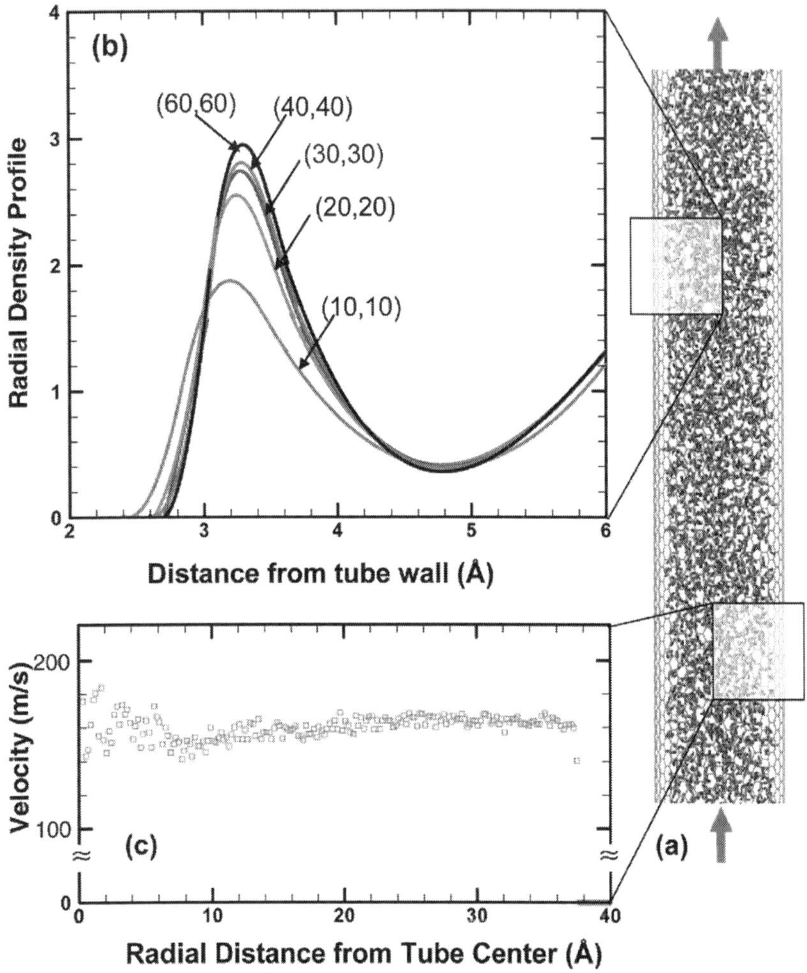

FIGURE 20 Dependences of the density and velocity profiles of water molecules inside the nanotubes with different radii

(a) Schematic of the initial structure and transport of water molecules in a model CNT.

(b) The radial density profile (RDP) of water molecules inside CNTs with different radii.

(c) The representative radial velocity profile (RVP) of water molecules inside a (60,60) nanotube

It is mentioned that the area available for molecules of the liquid is less than the area bounded by a solid wall, primarily due to the van der Waals interactions.

4.3.2 THE EFFECTIVE VISCOSITY OF THE LIQUID IN A NANOTUBE

Scientists showed a significant increase in the effective viscosity of the fluid in the nano volumes compared to its macroscopic value. It was shown that the effective viscosity of the liquid in a nanotube depends on the diameter of the nanotube.

The effective viscosity of the liquid in a nanotube is defined as follows.

Let's establish a conformity nanotubes filled with liquid, possibly containing crystallites with the same size tube filled with liquid, considered as a homogeneous medium (i.e., without considering the crystallite structure), in which Poiseuille flow is realized at the same pressure drop and consumption rate. The viscosity of a homogeneous fluid, which ensures the coincidence of these parameters, will be called the effective viscosity of the flow in the nanotube.

Researchers showed that while flowing in the narrow channels of width less than 2 nm, water behaves like a viscous liquid. In the vertical direction water behaves as a rigid body, and in a horizontal direction it maintains its fluidity.

It is known that at large distances the van der Waals interaction has a magnet ting tendency and occurs between any molecules like polar as well as nonpolar. At small distances it is compensated by repulsion of electron shells. Van der Waals interaction decreases rapidly with distance. Mutual convergence of the particles under the influence magnetting forces continues until these forces are balanced with the increasing forces of repulsion.

Knowing the deceleration of the flow (Fig. 20) of water a, the authors of the cited article calculates the effective shear stress between the wall of the pipe length l and water molecules by the formula

$$\tau = Nma/(2\pi Rl) \tag{18}$$

Here, the shear stress is a function of tube radius and flow velocity $\bar{v}$, m – mass of water molecules, the average speed is related to volumetric flow $\bar{v} = Q/(\pi R^2)$.

Denoting n_0 the density of water molecules number, we can calculate the shear stress in the form of:

$$\tau|_{r=R} = n_0 mRa/2 \tag{19}$$

Figure 21 shows the results of calculations of the authors of the cited article – influence of the size of the tube R_0 on the effective viscosity (squares) and shear stress τ (triangles), when the flow rate is approximately 165 m/sec.

According to classical mechanics of liquid flow at different pressure drops Δp along the tube length l is given by Poiseuille formula:

$$Q_P = \frac{\pi R^4 \Delta p}{8\eta l} \quad Q_P = \pi R^2 \bar{v} \tag{20}$$

$$\tau = \frac{\Delta p R}{2l} \tag{21}$$

and the effective viscosity of the fluid can be estimated as $\eta = \tau \cdot R/(4\bar{v})$.

The change in the value of shear stress directly causes the dependence of the effective viscosity of the fluid from the pipe size and flow rate. In this case the effective viscosity of the transported fluid can be determined from Eqs. (20) and (21) as:

$$\eta = \frac{\tau \cdot R}{4\bar{v}} \tag{22}$$

With increasing radius τ according to Eq. (21) increases.

Calculations showed that the magnitude of the shear stress τ is relatively small in the range of pipe sizes considered. This indicates that the surface of carbon nanotubes is very smooth and the water molecules can easily slide through it.

In fact, shear stress is primarily due to van der Waals interaction between the solid wall and the water molecules. It is noted that the characteristic distance between the near-wall layer of fluid and pipe wall depends on the equilibrium distance between atoms O and C, the distribution of the atoms of the solid wall and bend of the pipe.

From Fig. 20, we can see, that the effective viscosity η increases by two orders of magnitude when R_0 changes from 0.67 to 5.4 nm. The value of the calculated viscosity of water in the tube (10,10) is $8{,}5 \times 10^{-8}$ *Pa/s*, roughly 4 orders of magnitude lower than the viscosity of a large mass of water.

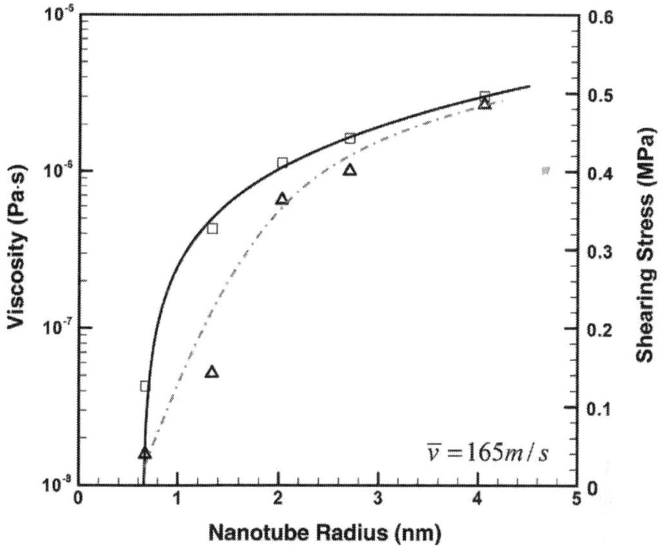

FIGURE 20 Size effect of shearing stress (triangle) and viscosity (square), with $\bar{v}$ =165м/c.

According to Eqs. (20)–(22), the effective viscosity can be calculated as $\eta = \frac{\pi R^4 \Delta p}{8QL}$. The results of calculations are shown in Fig. 21.

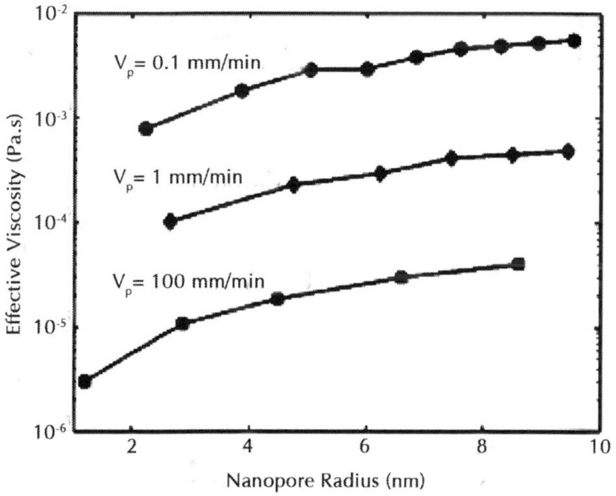

FIGURE 21 Effective viscosity as a function of the nanopore radius and the loading rate.

The dependence of the shear stress on the flow rate is illustrated in Fig. 22. From the example of the tube (20,20) it's clear that τ increases with $\bar{v}$. The growth rate slowed down at higher values $\bar{v}$.

At high speeds $\bar{v}$, while water molecules are moving along the surface of the pipe, the liquid molecules do not have enough time to fully adjust their positions to minimize the free energy of the system. Therefore, the distance between adjacent carbon atoms and water molecules may be less than the equilibrium van der Waals distances. This leads to an increase in van der Waals forces of repulsion and leads to higher shear stress.

Scientists showed that, even though the equation for viscosity is based on the theory of the continuum, it can be extended to a complex flow to determine the effective viscosity of the nanotube.

Figure 22 also shows a dependence η on $\bar{v}$ on the inside of the nanotube (20,20). It is seen that η decreases sharply with increasing flow rate and begins to asymptotically approaches a definite value when $\bar{v}$>150 m/sec. For the current pipe size and flow rate ranges $\eta \sim 1/\sqrt{\bar{v}}$, this trend is the result of addiction $\tau — \bar{v}$, contained the same Fig. 22. According to Fig. 21, high-speed effects are negligible.

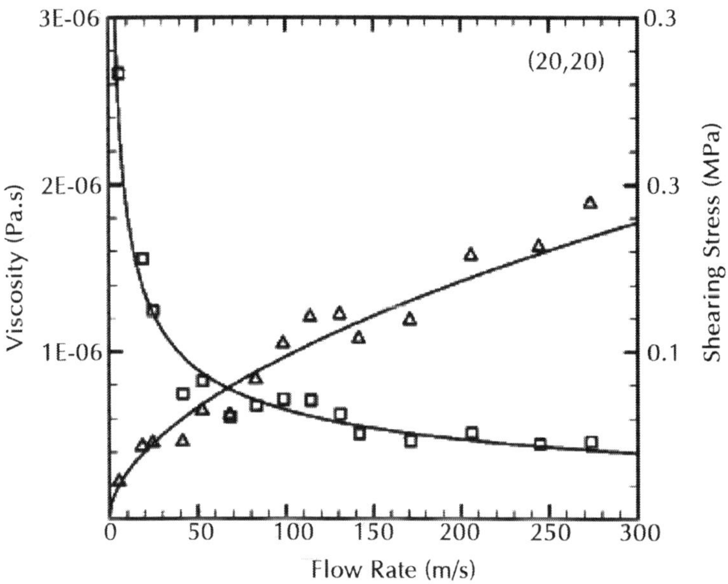

FIGURE 22 Flow rate effect of shearing stress (triangle) and viscosity (square), with R_0 =1.336 нм.

One can easily see that the dependence of viscosity on the size and speed is consistent qualitatively with the results of molecular dynamic simulations. In all studied cases, the viscosity is much smaller than its macroscopic analogy. When the radius of the pores varies from about 1 nm to 10 nm, the value of the effective viscosity increases by an order of magnitude respectively. A more significant change occurs when the speed increases from 0.1 mm/min up to 100 mm/min. This results in a change in the value of viscosity η, respectively, by 3–4 orders. The discrepancy between simulation and test data can be associated with differences in the structure of the nanopores and liquid phase.

Figure 23 shows the viscosity dependence of water, calculated by the method of DM, the diameter of the CNT. The viscosity of water, as shown in the Fig. 23, increases monotonically with increasing diameter of the CNT.

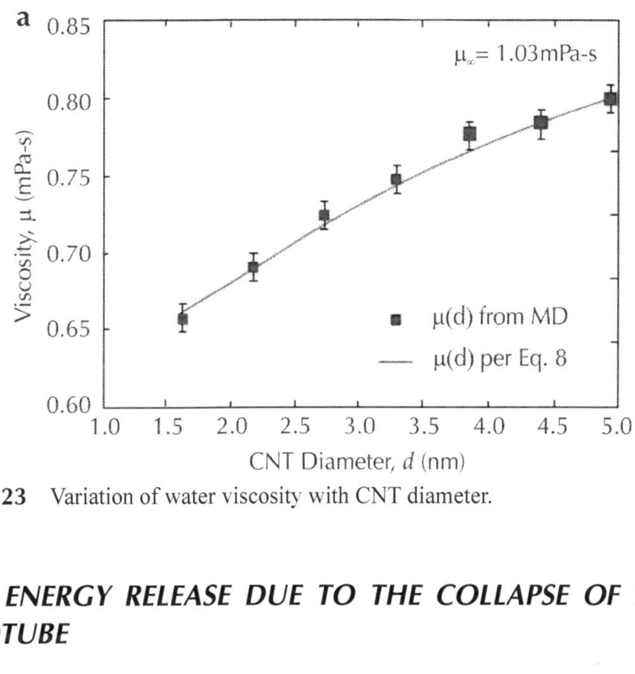

FIGURE 23 Variation of water viscosity with CNT diameter.

4.3.3 ENERGY RELEASE DUE TO THE COLLAPSE OF THE NANOTUBE

Scientists theoretically predicted the existence of a "domino effect" in single-walled carbon nanotube.

Squashing it at one end by two rigid moving to each other by narrow graphene planes (about 0.8 nm in width and 8.5 nm in length), one can observe it rapidly (at a rate exceeding 1 km/s) release its stored energy by collapsing along its length like a row of dominoes. The effect resembles a tube of toothpaste squeezing itself (Fig. 24).

The structure of a single-walled carbon nanotube has two possible stable states: circular or collapsed. Chang realized that for nanotubes wider than 3.5 nm, the circular state stores more potential energy than the collapsed state as a result of van der Waal's forces. He performed molecular dynamics simulations to find out what would happen if one end of a nanotube was rapidly collapsed by clamping it between two graphene bars.

This phenomenon occurs with the release of energy, and thus allows for the first time to talk about carbon nanotubes as energy sources. This effect can also be used as a accelerator of molecules.

The tube does not collapse over its entire length at the same time, but sequentially, one after the other carbon ring, starting from the end, which is tightened (Fig. 24). It happens just like a domino collapses, arranged in a row (this is known as the "domino effect"). Only here the role of bone dominoes is performed by a ring of carbon atoms forming the nanotube, and the nature of this phenomenon is quite different.

Recent studies have shown that nanotubes with diameters ranging from 2 to 6 nm, there are two stable equilibrium states – cylindrical (tube no collapses) and compressed (imploded tube) – with difference values of potential energy, the difference between which and can be used as an energy source.

Researchers found that switching between these two states with the subsequent release of energy occurs in the form of arising domino effect wave. The scientists have shown that such switching is not carried out in carbon nanotubes with diameters of 2 nm and not more, as evident from previous studies, but with a little more, starting from 3.5 nm.

A theoretical study of the "domino effect" was conducted using a special method of classical molecular dynamics, in which the interaction between carbon atoms was described by van der Waals forces (intermolecular forces), mathematically expressed by the Lennard-Jones potential.

The main reason for the observed effect, in author's opinion, is the competition of the potential energy of the van der Waals interactions, which "collapses" the nanotube with the energy of elastic deformation, which seeks to preserve the geometry of carbon atoms, which eventually leads to a bistable (collapsed and no collapsed) configuration of the carbon nanotube.

For small diameter tubes the dominant is the energy of elastic deformation, the cylindrical shape of such a nanotube is stable. For nanotubes with sufficiently large diameter the van der Waals interaction energy is dominant. This means more stability and less compressed nanotube stability, or, as physicists say, metastability (i.e., apparent stability) of its cylindrical shape.

Thus, "domino effect" wave can be produced in a carbon nanotube with a relatively large diameter (more than 3.5 nm, as the author's calculations), because only in such a system the potential energy of collapsing structures may be less than the potential energy of the "normal" nanotube. In other words, the cylindrical and collapsing structure of a nanotubes with large diameters are, respectively, its metastable and stable states.

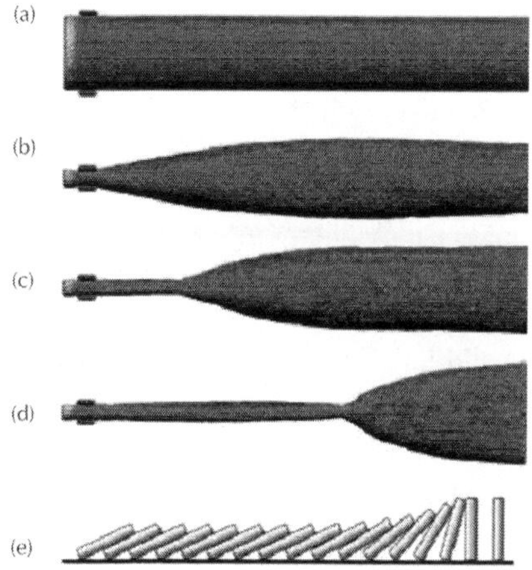

FIGURE 24 "Domino effect" in a carbon nanotube. (a) The initial form of carbon nanotubes – cylindrical. (b) One end of the tube is squeezed. (c), (d) Propagation of domino waves – the configuration of the nanotube 15 and 25 picoseconds after the compression of its end. (e) Schematic illustration of the "domino effect" under the influence of gravity.

Change of the potential energy of a carbon nanotube with a propagating "domin effect" wave in it with time is represented as a graph in Fig. 25a.

This chart shows three sections of features in the change of potential energy. The first (from 0 ps to 10 ps) are composed of elastic strain energy, which appears due to changes in the curvature of the walls of the nanotubes in the process of collapsing, the energy change of van der Waals interactions occurring between the opposite walls of the nanotubes, as well as the interaction between the tube walls and graphene planes, compressing her end.

The second region (from 10 ps to 35 ps) corresponds to the "domino effect"–"domino effect" wave spreads along the surface of the carbon nanotube. Energetically, it looks like this: at every moment, when carbon ring collapses, some of the potential energy of the van der Waals is converted to kinetic energy (the rest to the energy of elastic deformation), which is kind of stimulant to support and "falling domino" – following the collapse of the rings, which form the nanotube, with each coagulated ring reducing the total potential energy of the system.

Finally, the third segment (from 35 ps to 45 ps) corresponds to the ended "domino" process – carbon tube collapsed completely. We emphasize that the nanotube, which collapsed (as seen from the Fig. 25a), has less potential energy than it was before beginning of the "domino effect".

In other words, the spread "domino effect" waves – a process that goes with the release of energy: about 0.01 eV per atom of carbon. This is certainly not comparable in any way with the degree of energy yield in nuclear reactions, but the fact of power generation carbon nanotube is obvious.

Later scientists analyzed the kinematic characteristics of the process – what the rate of propagation of the wave of destruction or collapse of carbon nanotubes and their characteristics is it determined?

Calculations show that the wave of dominoes in a tube diameter of 4–5 nm is about 1 km/s (as seen from the Fig. 25b) and depends on its geometry – the diameter and chirality in a nonlinear manner. The maximum effect should be observed in the tube with a diameter slightly less than 4.5 nm – it will be carbon rings to collapse at a speed of 1.28 km/sec. The theoretical dependence obtained by the author, shows the blue solid line. And now an example of how energy is released in such a system with a "domino effect" can be used in nanodevices. The author offers an original way to use – "nanogun" (Fig. 26a). Imagine that at our disposal there is a carbon nanotube with chirality of (55.0) and corresponding to observation of the dominoes diameter.

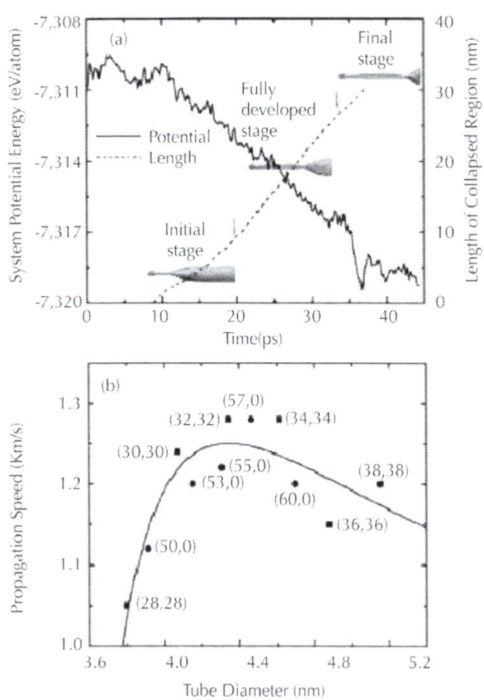

FIGURE 25 (a) Time dependences of potential energy and the length of the collapsed nanotube, (b) The velocity of propagation of the domino wave for carbon nanotubes with different diameters and chirality.

Put a C_{60} fullerene inside a nanotube. A little imagination, and a carbon nanotube can be considered as the gun trunk, and the C_{60} molecule – as its shell.

The molecule located inside will extruded from it into the other, open end under the influence of squeezing nanotube (see fig. 5.6).

The question is, what is the speed of the "core"? Chang estimated that, depending on the initial position of the fullerene molecule when leaving a nanotube, it can reach speeds close to the velocity of "domino effect" waves – about 1 km/s (Fig. 26b). Interestingly, this speed is reached by the "core" for just 2 picoseconds and at a distance of 1 nm. It is easy to calculate that the observed acceleration is of great value $0,5 \cdot 10^{15} m/c^2$. For comparison, the speed of bullets in an AK-47 is 1.5 times lower than the rate of fullerene emitted from a gun.

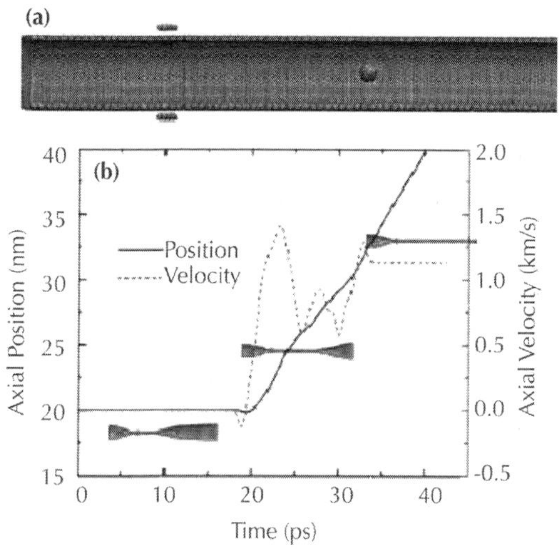

FIGURE 26 Nano cannon scheme acting on the basis of "domino effect" in the incision. (a) Inside a carbon nanotube (55.0) is the fullerene molecule C_{60}. (b) The initial position and velocity of the departure of the "core" (a fullerene molecule), depending on the time. The highest rate of emission of C_{60} (1.13 km /s) comparable to the velocity of the domino wave.

Necessary be noted that the simulation took place in nanogun at assumption of zero temperature in Kelvin. However, this example is not so abstract and may be used in the injecting device.

Thus, for the first time been demonstrated, albeit only in theory, the use of single-walled carbon nanotubes as energy sources.

4.4 FLUID FLOW IN NANOTUBES

The friction of surface against the surface in the absence of the interlayer between the liquid material (so-called dry friction) is created by irregularities in the given surfaces that rub one another, as well as the interaction forces between the particles that make up the surface.

As part of their study, the researchers built a computer model that calculates the friction force between nano-surfaces (Fig. 27). In the model, these surfaces were presented simply as a set of molecules for which forces of intermolecular interactions were calculated.

As a result, scientists were able to establish that the friction force is directly proportional to the number of interacting particles. The researchers propose to consider this quantity by analog of so-called true macroscopic contact area. It is known that the friction force is directly proportional to this area (it should not be confused with common area of the contact surfaces of the bodies).

In addition, the researchers were able to show that the friction surface of the nano-surfaces can be considered within the framework of the classical theories of friction of nonsmooth surfaces.

A literature review shows that nowadays molecular dynamics and mechanics of the continuum in are the main methods of research of fluid flow in nanotubes.

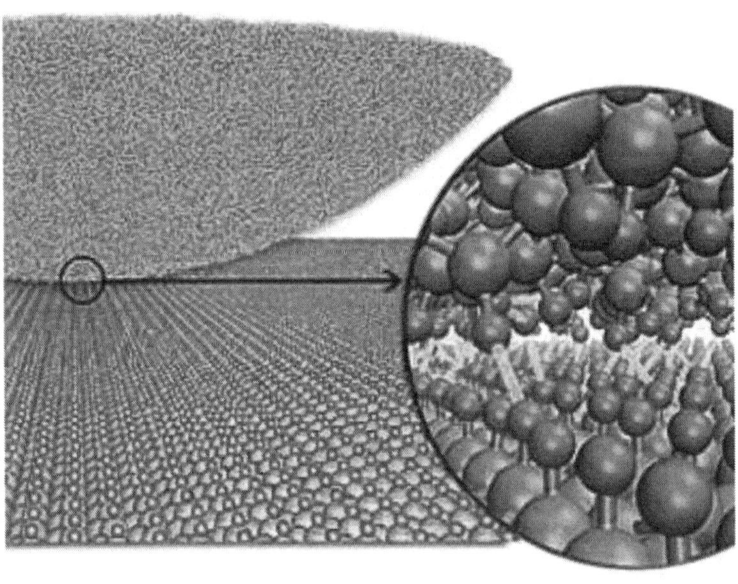

FIGURE 27 Computer model of friction at the nanoscale (Right shows the surfaces of interacting particles).

Although, the method of molecular dynamics simulations is effective, it at the same time requires enormous computing time especially for large systems. Therefore, simulation of large systems is more reasonable to carry out nowadays by the method of continuum mechanics.

The fluid flow in the channel can also be considered in the framework of the continuum hypothesis. The Navier-Stokes equation can be used and the velocity profile can be determined for Poiseuille flow.

However, the water flow by means of pressure differential through the carbon nanotubes with diameters ranging from 1.66 to 4.99 nm is studies using molecular dynamics simulation study. For each nanotube the value enhancement predicted by the theory of liquid flow in the carbon nanotubes is calculated. This formula is defined as a ratio of the observed flow in the experiments to the theoretical values without considering slippage on the model of Hagen-Poiseuille. The calculations showed that the enhancement decreases with increasing diameter of the nanotube.

Important conclusion is that by constructing a functional dependence of the viscosity of the water and length of the slippage on the diameter of carbon nanotubes, the experimental results in the context of continuum fluid mechanics can easily be described. The aforementioned is true even for carbon nanotubes with diameters of less than 1.66 nm.

The theoretical calculations use the following formula for the steady velocity profile of the viscosity η of the fluid particles in the CNT under pressure gradient $\partial p/\partial z$:

$$v(r)=\frac{R^2}{4\eta}\left[1-\frac{r^2}{R^2}+\frac{2L_S}{R}\right]\frac{\partial p}{\partial z} \tag{23}$$

The length of the slip, which expresses the speed heterogeneity at the boundary of the solid wall and fluid is defined as:

$$L_S=\left.\frac{v(r)}{dv/dr}\right|_{r=R} \tag{24}$$

Then the volumetric flow rate, taking into account the slip Q_S is defined as:

$$Q_S=\int_0^R 2\pi r\cdot v(r)dr=\frac{\pi\left[(d/2)^4+4(d/2)^3\cdot L_S\right]}{8\eta}\cdot\frac{\partial p}{\partial z} \tag{25}$$

Equation (4.3) is a modified Hagen-Poiseuille equation, taking into account slippage. In the absence of slip $L_S=0$ (4.3) coincides with the Hagen-Poiseuille flow, Eq. (20) for the volumetric flow rate without slip Q_P. In some works the parameter enhancement flow ε is also introduced. It is defined as the ratio of the calculated volumetric flow rate of slippage to Q_P (calculated using the effective viscosity and the diameter of the CNT). If the measured flux is modeled using Eq. (25), the degree of enhancement takes the form:

$$\varepsilon = \frac{Q_S}{Q_P} = \left[1 + 8\frac{L_S(d)}{d}\right]\frac{\eta_\infty}{\eta(d)} \tag{26}$$

where $d = 2R$ – diameter of CNT, η_∞ – viscosity of water, $L_S(d)$ – CNT slip length depending on the diameter, $\eta(d)$ – the viscosity of water inside CNTs depending on the diameter.

If $\eta(d)$ finds to be equal to η_∞, then the influence of the effect of slip on ε is significant, if $L_S(d) \geq d$. If $L_S(d) < d$ and $\eta(d) = \eta_\infty$, then there will be no significant difference compared to the Hagen-Poiseuille flow with no slip.

Table 2 shows the experimentally measured values of the enhancement water flow. Enhancement flow factor and the length of the slip were calculated using the equations given above.

TABLE 2 Experimentally measured values of the enhancement water flow.

Nanosystems	Diameter (nm)	Enhancement, ε	Slip length, L_S, (nm)
carbon nanotubes	300–500	1	0
	44	22–34	113–177
carbon nanotubes	7	$10^4 - 10^5$	3900–6800
	1.6	560–9600	140–1400

Figure 28 depicts the change in viscosity of the water and the length of the slip in diameter. As can be seen from the Fig. 28, the dependence of slip length to the diameter of the nanotube is well described by the empirical relation:

$$L_S(d) = L_{S,\infty} + \frac{C}{d^3} \tag{27}$$

where $L_{S,\infty}$ =30 nm – slip length on a plane sheet of graphene, a C – const.

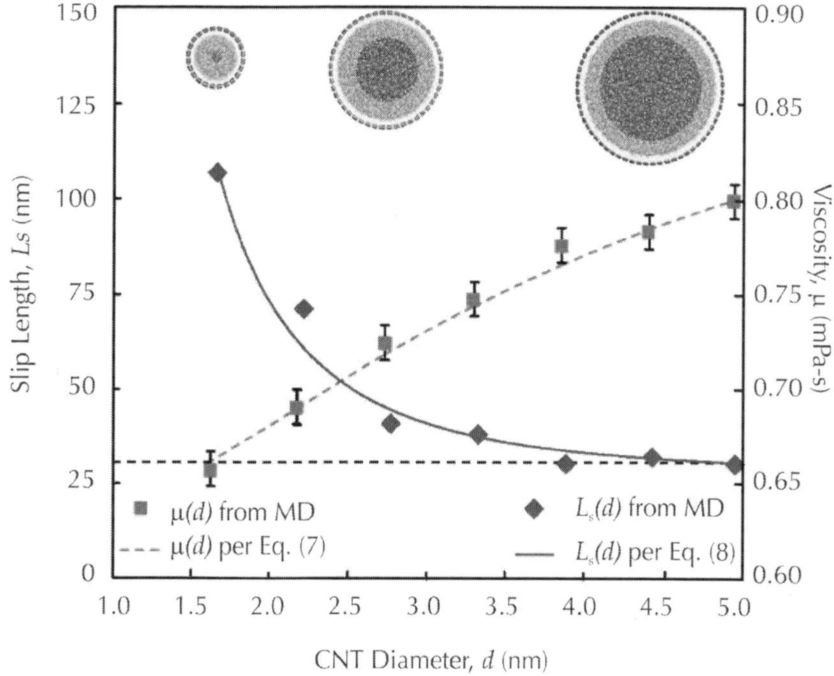

FIGURE 28 Variation of viscosity and slip length with CNT diameter

Figure 29 shows dependence of the enhancement of the flow rate ε on the diameter for all seven CNTs.

There are three important features in the results. First, the enhancement of the flow decreases with increasing diameter of the CNT. Second, with increasing diameter, the value tends to the theoretical value of Eqs. (26) and (27) with a slip $L_{s\infty}$ = 30 nm and the effective viscosity $\mu(d)=\mu_{\infty}$. The dotted line shown the curve of 15% in the second error in the theoretical data of viscosity and slip length. Third, the change ε in diameter of CNTs cannot be explained only by the slip length.

To determine the dependence the volumetric flow of water from the pressure gradient along the axis of single-walled nanotube with the radii of 1.66, 2.22, 2.77, 3.33, 3.88, 4.44 and 4.99 nm in the method of molecular modeling was used. Snapshot of the water-CNT is shown in Fig. 30.

Figure 30 shows the results of calculations to determine the pressure gradient along the axis of the nanotube with the diameter of 2.77 nm and a length of 20 nm. Change of the density of the liquid in the cross sections was less than 1%.

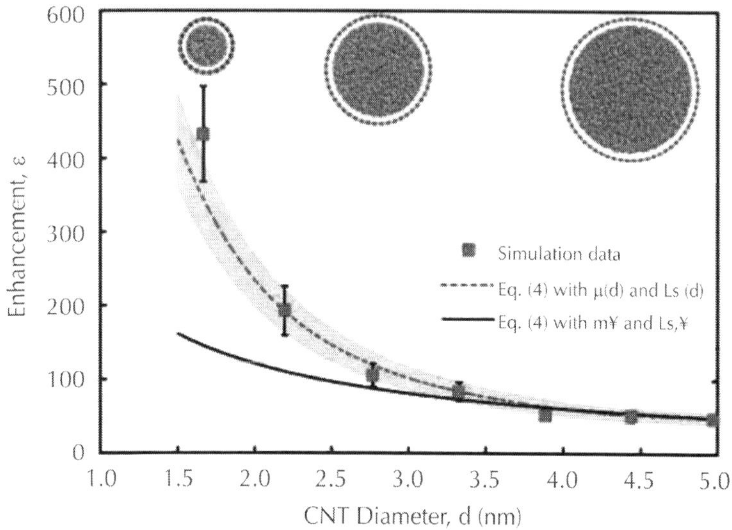

FIGURE 29 Flow enhancement as predicted from MD simulations.

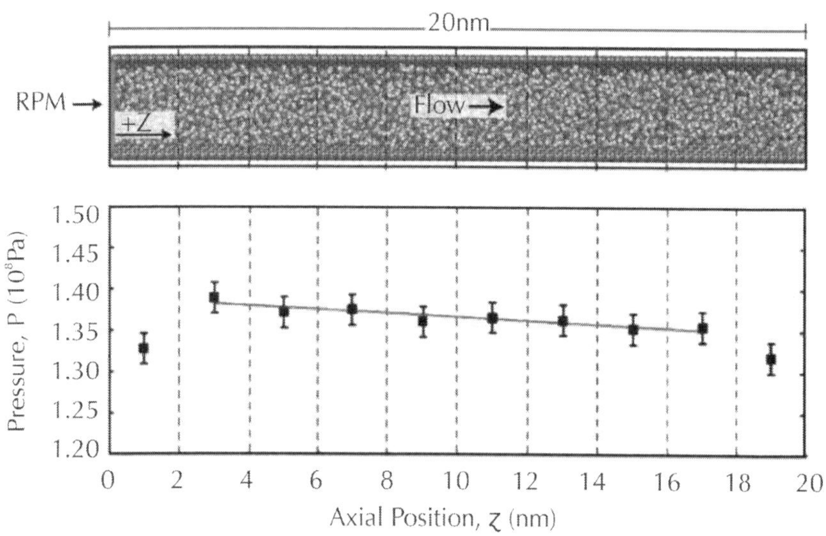

FIGURE 30 Axial pressure gradient inside the 2.77 nm diameter CNT.

Figure 30 shows the dependence of the volumetric flow rate from the pressure gradient for all seven CNTs. The flow rate ranged from 3–14 m/sec. In the range considered here the pressure gradient $(0-3)\times10^9 \, atm/m \;\; Q$

($pl / sek = 10^{-15} m^3 / sek$) is directly proportional $\partial p / \partial z$. Coordinates of chirality for each CNT are indicated in the figure legend. The linearity of the relations between flow and pressure gradient confirms the validity of calculations of the Eq. (29).

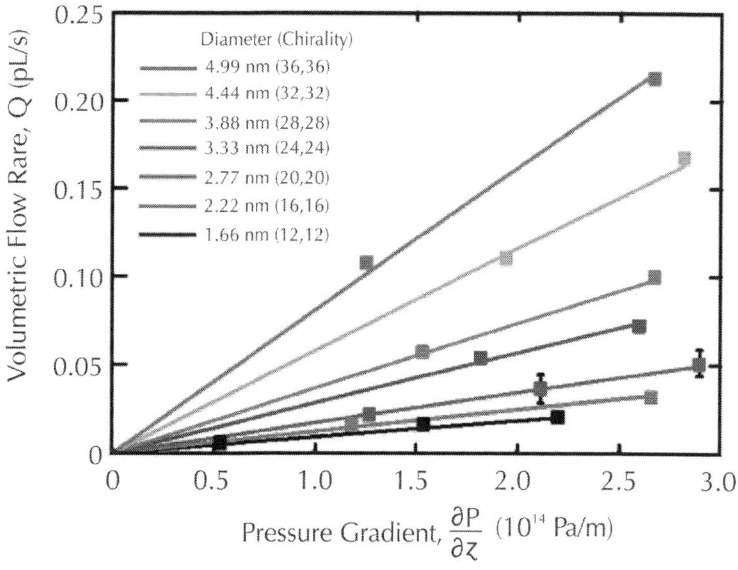

FIGURE 31 Relationship between volumetric liquid flow rate in carbon nanotubes with different diameters and applied pressure gradient

Figure 32 shows the profile of the radial velocity of water particles in the CNT with diameter 2.77 nm. The vertical dotted line at 1.38 nm marked surface of the CNT. It is seen that the velocity profile is close to a parabolic shape.

Researchers consider the flow of water under a pressure gradient in the single-walled nanotubes of "chair" type of smaller radii: 0.83, 0.97, 1.10, 1.25, 1.39 and 1.66 nm.

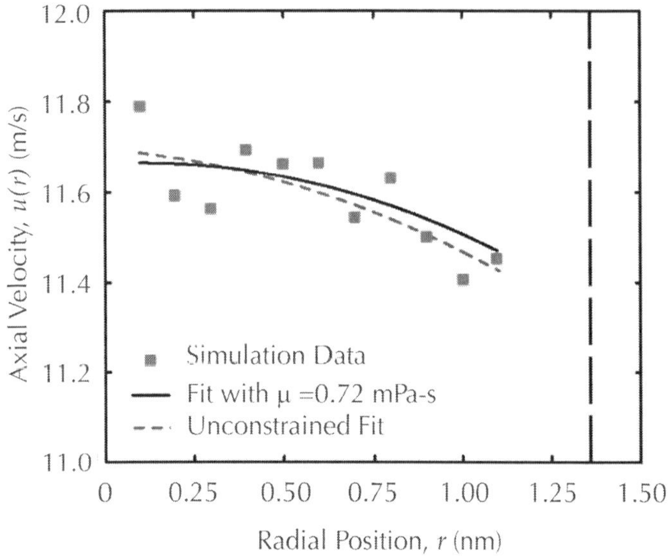

FIGURE 32 Radial velocity profile inside 2.77 nm diameter CNT.

Figure 33 shows the dependence of the mean flow velocity $\bar{v}$ on the applied pressure gradient $\Delta P / L$ in the long nanotubes – 75 nm at 298 K. A similar picture pattern occurs in the tube with the length of 150 nm.

As we can see, there is conformance with the Darcy law, the average flow rate for each CNT increases with increasing pressure gradient. For a fixed value of $\Delta p / L$, however, the average flow rate does not increase monotonically with increasing diameter of the CNTs, as follows from Poiseuille equation. Instead, when at the same pressure gradient, decrease of the average speed in a CNT with the radius of 0.83 nm to a CNT with the radius of 1.10 nm, similar to the CNTs 1.10 and 1.25 nm, then increases from a CNT with the radius of 1.25 nm to a CNT of 1.66 nm.

The nonlinearity of the relationship between $\bar{v}$ and $\Delta P / L$ are the result of inertia losses (i.e., insignificant losses) in the two boundaries of the CNT. Inertial losses depend on the speed and are caused by a sudden expansion, abbreviations, and other obstructions in the flow.

We note the important conclusion of the work of scientists in which the method of molecular modeling shows that the Eq. (23) (Poiseuille parabola) correctly describes the velocity profile of liquid in a nanotube when the diameter of a flow is 5–10 times more than the diameter of the molecule (≈ 0.17 nm for water).

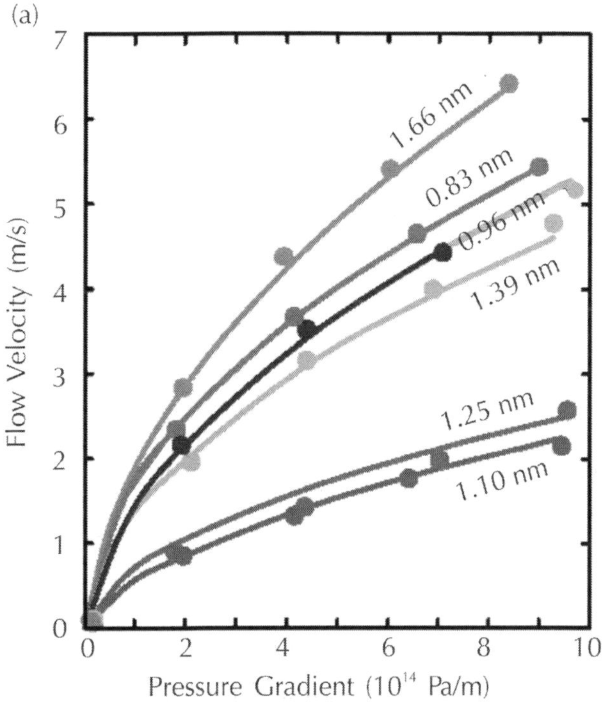

FIGURE 33 Relationship between average flow velocity and applied pressure gradient for the 75 nm long CNTs

In Fig. 34, we can see the effect of slip on the velocity profile at the boundary of radius R of the pipe and fluid. When $L_S = 0$ the fluid velocity at the wall vanishes, and the maximum speed (on the tube axis) exceeds flow speed twice.

The Fig. 34 shows the velocity profiles for Poiseuille flow without slip ($L_S = 0$) and with slippage $L_S = 2R$. The flow rate is normalized to the speed corresponding to the flow without slip. Thick vertical lines indicate the location of the pipe wall. The thick vertical lines indicate the location of the tube wall. As the length of the slip, the flow rate increases, decreases the difference between maximum and minimum values of the velocity and the velocity profile becomes more like a plug.

Velocity of the liquid on a solid surface can also be quantified by the coefficient of slip L_c. The coefficient of slippage – there is a difference between the radial position in which the velocity profile would be zero and the radial position of the solid surface. Slip coefficient is equal to $L_C = \sqrt{R^2 + 2RL_S} = \sqrt{5}R$.

For linear velocity profiles (e.g., Couette flow), the length of the slip and slip rate are equal. These values are different for the Poiseuille flow.

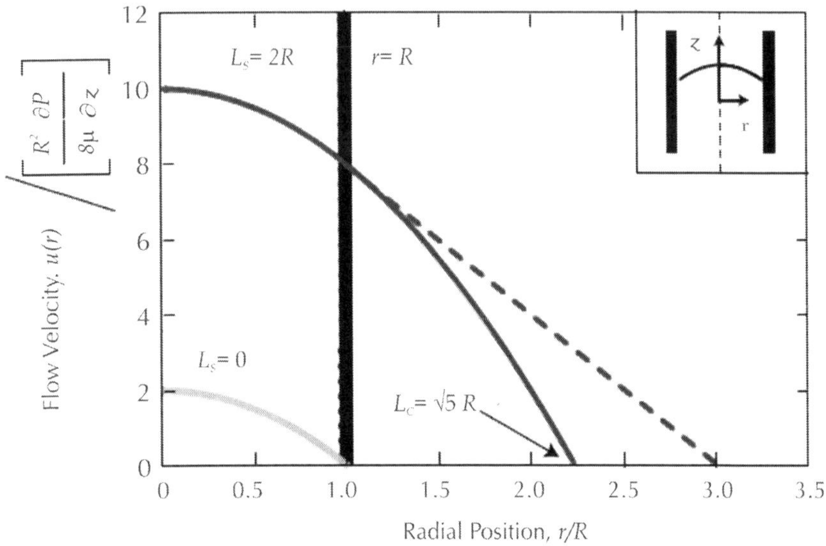

FIGURE 34 No-slip Poiseuille flow and slip Poiseuille flow through a tube.

Figure 35 shows dependence of the volumetric flow rate Q from the pressure gradient $\partial p/\partial z$ in long nanotubes with diameters between 1.66 nm and 6.93 nm. As can be seen in the studied range of the pressure gradient Q is proportional to $\partial p/\partial z$. As in the Poiseuille flow, volumetric flow rate increases monotonically with the diameter of CNT at a fixed pressure gradient. Magnitude of calculations error for all the dependencies are similar to the error for the CNT diameter 4.44 nm, marked in the Fig. 35.

Many researchers considered steady flow of incompressible fluids in a channel width $2h$ under action of the force of gravity ρg or pressure gradient $\partial p/\partial y$, which is described by the Navier-Stokes equations. The velocity profile has a parabolic form:

$$U_y(z) = \frac{\rho g}{2\eta} \cdot \left[(\delta + h)^2 - z^2 \right]$$

where δ – length of the slip, which is equal to the distance from the wall to the point at which the velocity extrapolates to zero.

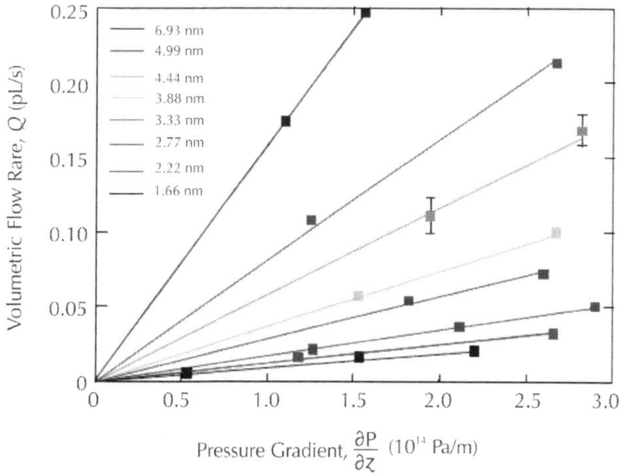

FIGURE 35 Volumetric flow rate in CNTs versus pressure gradient.

4.4.1 SOME OF THE IDEAS AND APPROACHES FOR MODELING IN NANOHYDROMECANICS

Let's consider the fluid flow through the nanotube. Molecules of a substance in a liquid state are very close to each other (Fig. 36).

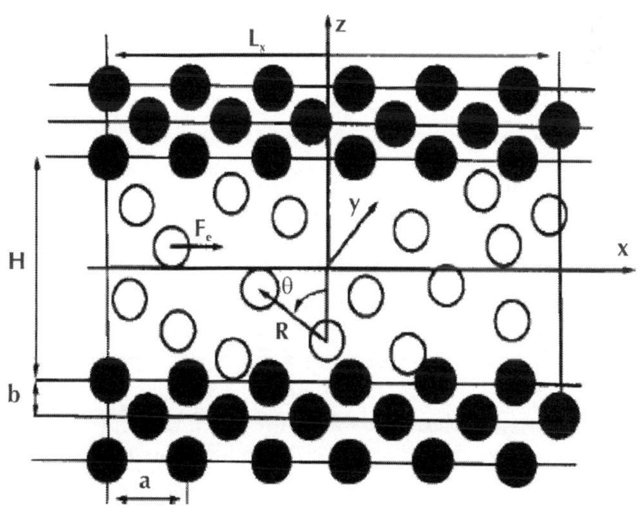

FIGURE 36 The fluid flow through the nanotube.

Most liquid's molecules have a diameter of about 0.1 nm. Each molecule of the fluid is "squeezed" on all sides by neighboring molecules and for a period of time $\left(10^{-10} - 10^{-13} s\right)$ fluctuates around certain equilibrium position, which itself from time to time is shifted in distance commensurating with the size of molecules or the average distance between molecules l_{cp}:

$$l_{cp} \approx \sqrt[3]{\frac{1}{n_0}} = \sqrt[3]{\frac{\mu}{N_A \rho}},$$

where,

n_0 – number of molecules per unit volume of fluid,

N_A – Avogadro's number,

ρ – fluid density,

μ – molar mass.

Estimates show that one cubic of nano water contains about 50 molecules. This gives a basis to describe the mass transfer of liquid in a nanotube-based continuum model. However, the specifics of the complexes, consisting of a finite number of molecules, should be kept in mind. These complexes, called clusters in literatures, are intermediately located between the bulk matter and individual particles – atoms or molecules. The fact of heterogeneity of water is now experimentally established.

There are groups of molecules in liquid—"microcrystals" containing tens or hundreds of molecules. Each microcrystal maintains solid form. These groups of molecules, or "clusters" exist for a short period of time, then break up and are recreated again. Besides, they are constantly moving so that each molecule does not belong at all times to the same group of molecules, or "cluster".

Modeling predicts that gas molecules bounce off the perfectly smooth inner walls of the nanotubes as billiard balls, and water molecules slide over them without stopping. Possible cause of unusually rapid flow of water is maybe due to the small-diameter nanotube molecules move on them orderly, rarely colliding with each other. This "organized" move is much faster than usual chaotic flow. However, while the mechanism of flow of water and gas through the nanotubes is not very clear, and only further experiments and calculations can help understand it.

The model of mass transfer of liquid in a nanotube proposed in this paper is based on the availability of nanoscale crystalline clusters in it.

A similar concept was developed in which the model of structured flow of fluid through the nanotube is considered. It is shown that the flow character in the nanotube depends on the relation between the equilibrium crystallite size and the diameter of the nanotube.

Figure 37 shows the results of calculations by the molecular dynamics of fluid flow in the nanotube in a plane (a) and three-dimensional (b) statement. The figure shows the ordered regions of the liquid.

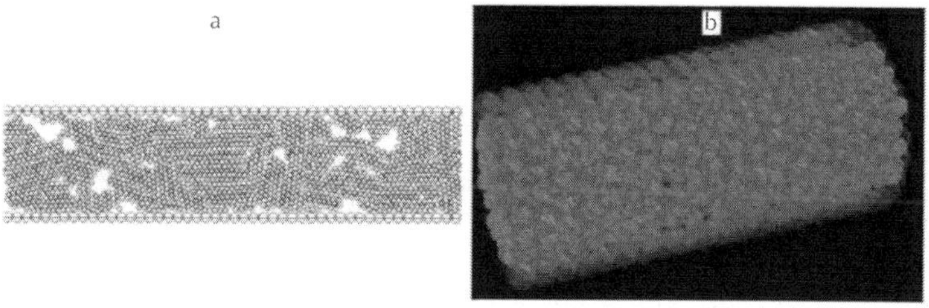

FIGURE 37 The results of calculations of fluid flow in the nanotube.

The typical size of crystallite is 1–2 nm, that is, compared, for example, with a diameter of silica nanotubes of different composition and structure. The flow model proposed in the present chapter is based on the presence of "quasi-solid" phase in the central part of the nanotube and liquid layer, nonautonomous phases.

Consideration of such a structure that is formed when fluid flows through the nanotube is also justified by the aforementioned results of the experimental studies and molecular modeling.

When considering the fluid flow with such structure through the nanotube, we will take into account the aspect ratio of "quasi-solid" phase and the diameter of the nanotube so that a character of the flow is stable and the liquid phase can be regarded as a continuous medium with viscosity η.

Let's establish relationship between the volumetric flow rate of liquid Q flowing from a liquid layer of the nanotube length l, the radius R and the pressure drop $\Delta p/l$, $\Delta p = p - p_0$, where p_0 is the initial pressure in the tube (Fig. 38).

Let R_0 be a radius of the tube from the "quasi-solid" phase, v – velocity of fluid flow through the nanotube.

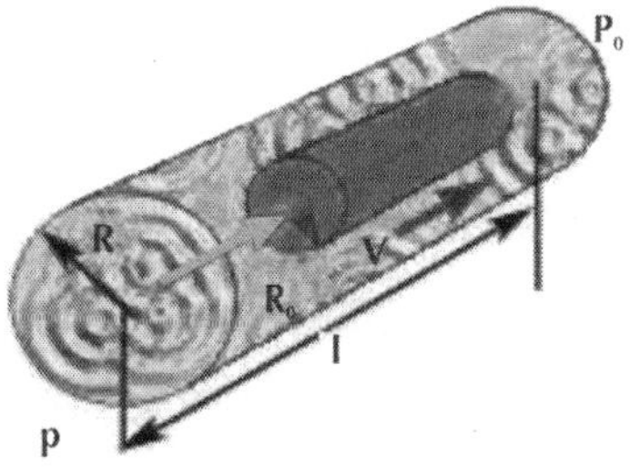

FIGURE 38 Flow through liquid layer of the nanotube.

Structural regime of fluid flow (Fig. 39) implies the existence of the continuous laminar layer of liquid (the liquid layer in the nanotube) along the walls of a pipe. In the central part of a pipe a core of the flow is observed, where the fluid moves, keeping his former structure, that is, as a solid ("quasi-solid" phase in the nanotube). The velocity slip is indicated in Fig. 39a through v_0.

Let's find the velocity profile $v(r)$ in a liquid interlayer $R_0 \le r \le R$ of the nanotube. We select a cylinder with radius r and length l in the interlayer, located symmetrically to the centerline of the pipe (see Fig. 39b).

At the steady flow, the sum of all forces acting on all the volumes of fluid with effective viscosity η, is zero.

The following forces are applied on the chosen cylinder: the pressure force and viscous friction force affects the side of the cylinder with radius r, calculated by the Newton formula.

Thus,

$$(p - p_0)\pi r^2 = -\eta \frac{dv}{dr} 2\pi r l \tag{28}$$

Integrating Eq. (28) between r to R with the boundary conditions $r = R$: $v = v_0$, we obtained a formula to calculate the velocity of the liquid layers located at a distance r from the axis of the tube:

$$v(r) = (p - p_0)\frac{R^2 - r^2}{4\eta l} + v_0 \tag{29}$$

Maximum speed $v_Я$ has the core of the nanotube $0 \le r \le R_0$ and is equal to:

$$v_Я = (p - p_0)\frac{R^2 - R_0^2}{4\eta l} + v_0 \tag{30}$$

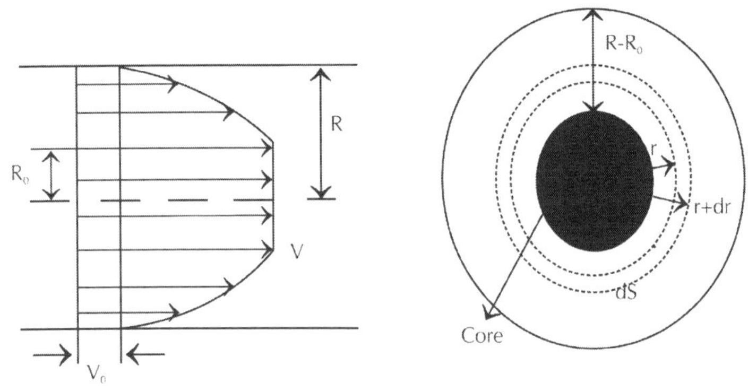

FIGURE 39 Structure of the flow in the nanotube.

Such structure of the liquid flow through nanotubes considering the slip is similar to a behavior of viscoplastic liquids in the tubes. Indeed, as we know, for viscoplastic fluids a characteristic feature is that they are to achieve a certain critical internal shear stresses τ_0 behave like solids, and only when internal stress exceeds a critical value above begin to move as normal fluid. Scientists shown that liquid behave in the nanotube the similar way. A critical pressure drop is also needed to start the flow of liquid in a nanotube.

Structural regime of fluid flow requires existence of continuous laminar layer of liquid along the walls of pipe. In the central part of the pipe is observed flow with core radius R_{00}, where the fluid moves, keeping his former structure, that is, as a solid.

The velocity distribution over the pipe section with radius R of laminar layer of viscoplastic fluid is expressed as follows:

$$v(r)=\frac{\Delta p}{4\eta l}\left(R^2-r^2\right)-\frac{\tau_0}{\eta}(R-r) \quad (31)$$

The speed of flow core in $0 \le r \le R_{00}$ is equal.

Let's calculate the flow or quantity of fluid flowing through the nanotube cross-section S at a time unit. The liquid flow dQ for the inhomogeneous velocity field flowing from the cylindrical layer of thickness dr, which is located at a distance r from the tube axis is determined from the relation

$$v_{Я}=\frac{\Delta p}{4\eta l}\left(R^2-R_{00}^2\right)-\frac{\tau_0}{\eta}\left(R-R_{00}\right)$$

where dS – the area of the cross-section of cylindrical layer (between the dotted lines in Fig. 39).

Let's place Eq. (29) in Eq. (33), integrate over the radius of all sections from R_0 to R and take into account that the fluid flow through the core flow is determined from the relationship $Q_{Я}=\pi R_0^2 v_{Я}$. Then we get the formula for the flow of liquid from the nanotube:

$$Q=\pi R^2 v_0+Q_P\left[1-\left(\frac{R_0}{R}\right)^4\right] \quad (34)$$

If $(R_0/R)^4 < 1$ (no nucleus) and $v_0 \Delta p R^2/8l\eta << 1$ (no slip), then Eq. (34) coincides with Poiseuille Eq. (20). When $R_0 \approx R$ (no of a viscous liquid interlayer in the nanotube), the flow rate Q is equal to volumetric flow $Q \approx \pi R^2 v_0$ of fluid for a uniform field of velocity (full slip).

Accordingly, flow rate of the viscoplastic fluid flowing with a velocity Eq. (29), is equal to:

$$Q = -\frac{\pi R^3 \tau_0}{3\eta}\left[1-\left(\frac{R_{00}}{R}\right)^3\right] + Q_P\left[1-\left(\frac{R_{00}}{R}\right)^4\right] \tag{35}$$

Comparing Eqs. (29)–(32) and (34), (35), we can see that the structure of the flow of the liquid through the nanotubes considering the slippage, is similar to that of the flow of viscoplastic fluid in a pipe of the same radius R.

Given that the size of the central core flow of viscoplastic fluid (radius R_{00}) is defined by:

$$R_{00} = \frac{2\tau_0 l}{\Delta p} \tag{36}$$

for viscoplastic fluid flow we obtain Buckingham formula:

$$Q = Q_P\left[1+\frac{1}{3}\left(\frac{2l\tau_0}{R\Delta p}\right)^4 - \frac{4}{3}\left(\frac{2l\tau_0}{R\Delta p}\right)\right] \tag{37}$$

We'll establish conformity of the pipe, which implements the flow of a viscoplastic fluid with a fluid-filled nanotube, the same size and with the same pressure drop. We say that an effective internal critical shear stress τ_{0ef} of viscoplastic fluid flow, which ensures the coincidence rate with the flow of fluid in the nanotube. Then from Eq. (37) we obtain equation of fourth order to determine τ_{0ef}:

$$\left(\frac{2l\tau_{0ef}}{R\Delta p}\right)^4 - 4\left(\frac{2l\tau_{0ef}}{R\Delta p}\right) = A;\ A = 3(\varepsilon - 1);\ \varepsilon = Q/Q_P \tag{38}$$

The solution of Eq. (38) can be found, for example, the iteration method of Newton:

$$\bar{\tau}_{0ef\,n} = \bar{\tau}_{0ef\,n-1} - \frac{\bar{\tau}^{4}_{0ef\,n-1} - 4\bar{\tau}_{0ef\,n-1} - A}{4\bar{\tau}^{3}_{0ef\,n-1} - 4}\quad \bar{\tau}_{0ef} = \frac{2l\tau_{0ef}}{R\Delta p} \tag{39}$$

The first component in Eq. (34) represents the contribution to the fluid flow due to the slippage, and it becomes clear that the slippage significantly enhances the flow rate in the nanotube, when $l\eta v_0 >\approx \Delta p R^2$.

This result is consistent with experimental and theoretical results (which show that water flow in nanochannels can be much higher than under the same conditions, but for the liquid continuum.

In the absence of slippage $\varepsilon = 1$ the Eq. (38) has a trivial solution $\bar{\tau}_{0ef} = 0$.

4.4.2 RESULTS

Let's determine the dependence of the effective critical inner shear stress τ_{0ef} on the radius of the nanotubes, by taking necessary values for calculations $\varepsilon = Q/Q_P$. The results of calculations at $\Delta p / l = 2{,}1 \cdot 10^{14}$ *Pa/m* are in the table below:

R, м	τ_{0ef} (Па)	$\varepsilon = Q/Q_P$
$0{,}8 \times 10^{-9}$	498498	350
$1{,}11 \times 10^{-9}$	577500	200
$1{,}385 \times 10^{-9}$	632599	114
$1{,}665 \times 10^{-9}$	699300	84
$1{,}94 \times 10^{-9}$	782208	68
$2{,}22 \times 10^{-9}$	855477	57
$2{,}495 \times 10^{-9}$	932631	50

Calculations show that the value of effective internal shear stress depends on the size of the nanotube.

Figure 40. shows the dependence τ_{0ef} on the nanotube radius.

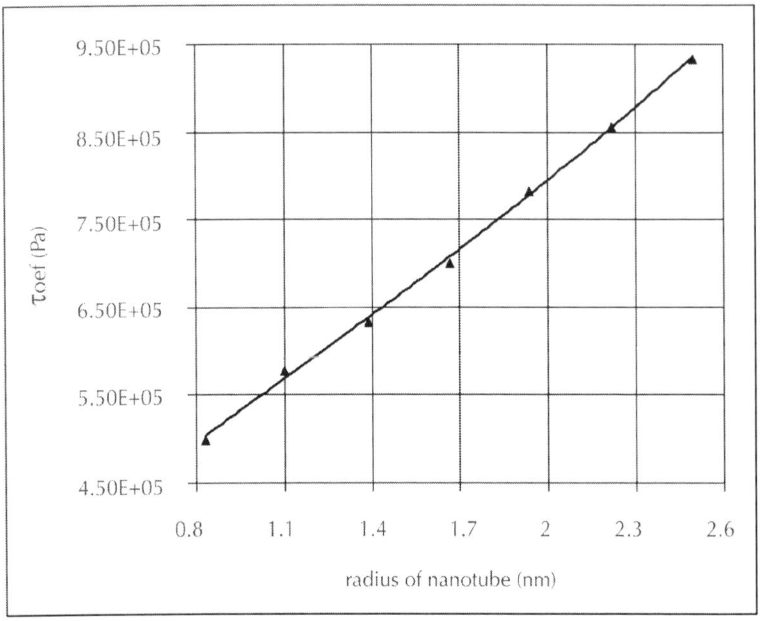

FIGURE 40 Dependence of the effective inner shear stress from the radius of the nanotube.

As you can see, this dependence $\tau_{0ef}(R)$ is almost linear. Within the range of the considered nanotube sizes τ_{0ef} has a relatively low value, which indicates the smoothness of the surface of carbon nanotubes.

4.4.3 THE FLOW OF FLUID WITH AN EMPTY INTERLAYER

The works of many researchers were analyzed in the aforementioned analysis of the structure of liquid flow in carbon nanotubes. The results of the calculations of the cited works showed that during the flow of the liquid particles, an empty layer between the fluid and the nanotube is formed. The area near the walls of the carbon nanotube $R_* \le r \le R$ becomes inaccessible for the molecules of the liquid due to van der Waals repulsion forces of the heterogeneous particles of the carbon and water. Moreover, according to the results of man scientists thicknesses of the layers $R_* \le r \le R$ regardless of radiuses of the nanotubes are practically identical: $R_* / R \approx 0{,}8$.

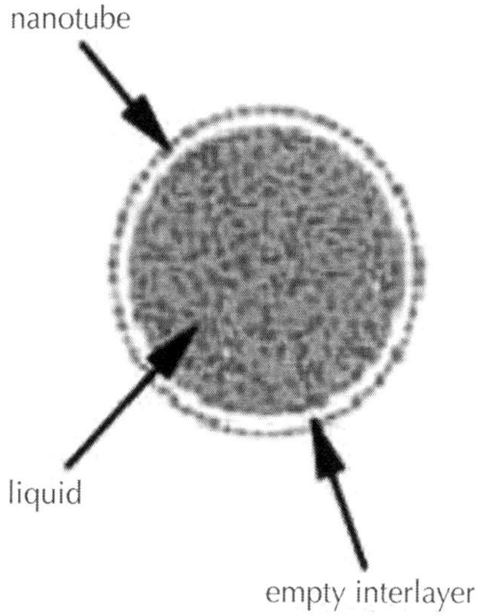

FIGURE 41 The structure of the flow.

A similar result was obtained in by researchers which is an image (Fig. 38) of the configuration of water molecules inside (8, 8) single-walled carbon nanotubes at different temperatures: 298, 325, and 350 ^0K.

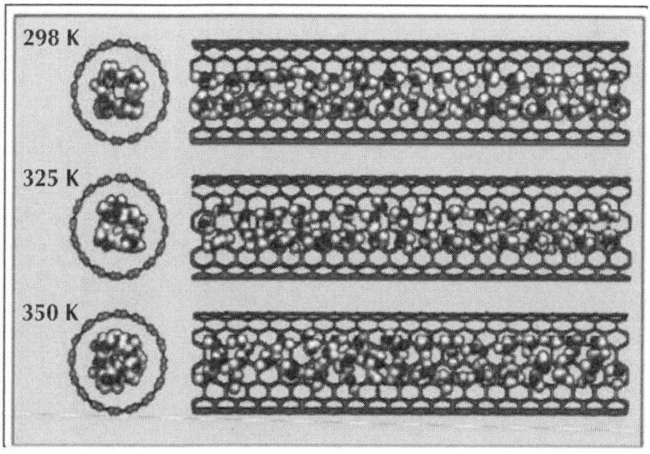

FIGURE 38 The configuration of water molecules inside single-walled carbon nanotubes.

Integrating Eq. (28) between r to R_* at the boundary conditions $r = R_* : v = v_*$, we obtain a formula to calculate the velocity of the liquid layers located at a distance r from the axis of the tube:

$$v(r) = v_* + \frac{2Q_P}{\pi R^2}\left[\left(\frac{R_*}{R}\right)^2 - \left(\frac{r}{R}\right)^2\right] \tag{40}$$

Let's insert Eq. (40) in Eq. (33), integrate over the radius of all sections from 0 to R_*. Then we get a formula for the flow of liquid from the nanotube:

$$Q = Q_P\left[v_*\frac{8l\eta R_*^2}{\Delta p R^4} + \left(\frac{R_*}{R}\right)^4\right] \tag{41}$$

or

$$\varepsilon = \frac{8l\eta v_*}{\Delta p R^2}\left(\frac{R_*}{R}\right)^2 + \left(\frac{R_*}{R}\right)^4$$

from which we can determine the unknown v_*:

$$v_* = \frac{Q_P}{\pi R^2}\left[\varepsilon - \left(\frac{R_*}{R}\right)^4\right]\left(\frac{R}{R_*}\right)^2 \tag{42}$$

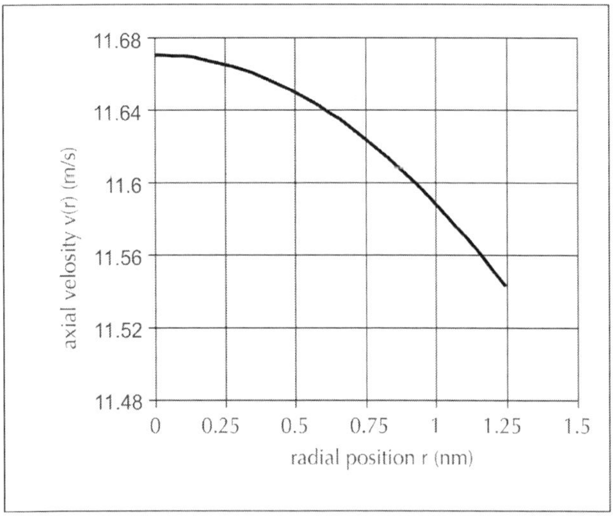

FIGURE 43 Profile of the radial velocity of water in carbon nanotubes.

Figure 43 shows the profile of the radial velocity of water particles in a carbon nanotube with a diameter of 2.77 nm, calculated using the Eq. (44) at $Q_P = 4,75 \cdot 10^{-19} n/m^2$, $\varepsilon = 114$. The velocity at the border v_* is equal to 11,55 m/s. It is seen that the velocity profile is similar to a parabolic shape, and at the same time agrees with the numerical values.

The calculations suggest the following conclusions. Flow of liquid in a nanotube was investigated using synthesis of the methods of the continuum theory and molecular dynamics. Two models are considered. The first is based on the fact that fluid in the nanotube behaves like a viscoplastic. A method of calculating the value of limiting shear stress is proposed, which was dependent on the nanotube radius. A simplified model agrees quite well with the results of the molecular simulations of fluid flow in carbon nanotubes. The second model assumes the existence of an empty interlayer between the liquid molecules and wall of the nanotube. This formulation of the task is based on the results of experimental works known from the literature. The velocity profile of fluid flowing in the nanotube is practically identical to the profile determined by molecular modeling.

As seen from the results of the calculations, the velocity value varies slightly along the radius of the nanotube. Such a velocity distribution of the fluid particles can be explained by the lack of friction between the molecules of the liquid and the wall due to the presence of an empty layer. This leads to an easy slippage of the liquid and, consequently, anomalous increase in flow compared to the Poiseuille flow.

APPENDIX A—MODELING AND SIMULATION TECHNIQUES

A1 MOLECULAR SCALE METHODS

The modeling and simulation methods at molecular level usually employ atoms, molecules or their clusters as the basic units considered. The most popular methods include molecular mechanics (MM), Molecular dynamics (MD) and Monte Carlo (MC) simulation. Modeling of polymer nanocomposites at this scale is predominantly directed toward the thermodynamics and kinetics of the formation, molecular structure and interactions. The diagram in Fig. A1 describes the equation of motion for each method and the typical properties predicted from each of them [1–22]. We introduce here the two widely used molecular scale methods: MD and MC.

A1.1 MOLECULAR DYNAMICS

MD is a computer simulation technique that allows one to predict the time evolution of a system of interacting particles (e.g., atoms, molecules, granules, etc.) and estimate the relevant physical properties [23, 24]. Specifically, it generates such information as atomic positions, velocities and forces from which the macroscopic properties (e.g., pressure, energy, heat capacities) can be derived by means of statistical mechanics. MD simulation usually consists of three constituents: (i) a set of initial conditions (e.g., initial positions and velocities of all particles in the system); (ii) the interaction potentials to represent the forces among all the particles; (iii) the evolution of the system in time by solving a set of classical Newtonian equations of motion for all particles in the system. The equation of motion is generally given by:

$$\vec{F_i}(t) = m_i \frac{d^2 \vec{r_i}}{dt^2} \tag{A1}$$

where $\vec{F_i}(t)$ is the force acting on the ith atom or particle at time t, which is obtained as the negative gradient of the interaction potential U, mi is the atomic mass and $\vec{r_i}$ the atomic position. A physical simulation involves the proper selection of interaction potentials, numerical integration, periodic boundary conditions, and the controls of pressure and temperature to mimicbphysically meaningful thermodynamic ensembles. The interaction potentials together with their parameters, i.e., the so-called force field, describe in detail how the particles in a system interact with each other, i.e., how the potential energy of a system depends on the particle coordinates. Such a force field may be obtained by quantum method (e.g., ab initio), empirical method (e.g., Lennard–Jones, Mores, Born-Mayer) or quantum-empirical method (e.g., embedded atom model, glue model, bond order potential). The criteria for selecting a force field include the accuracy, transferability and computational speed. A typical interaction potential U may consist of a number of bonded and nonbonded interaction terms:

$$U(\vec{r_1},\vec{r_2},\vec{r_3},\ldots,\vec{r_n}) = \sum_{i_{bond}}^{N_{bond}} U_{bond}(i_{bond},\vec{r_a},\vec{r_b}) + \sum_{i_{angle}}^{N_{angle}} U_{angle}(i_{angle},\vec{r_a},\vec{r_b},\vec{r_c})$$

$$+ \sum_{i_{torsion}}^{N_{torsion}} U_{torsion}(i_{torsion},\vec{r_a},\vec{r_b},\vec{r_c},\vec{r_d})$$

$$+ \sum_{i_{inversion}}^{N_{inversion}} U_{inversion}(i_{inversion},\vec{r_a},\vec{r_b},\vec{r_c},\vec{r_d})$$

$$+ \sum_{i=1}^{N-1}\sum_{j>i}^{N} U_{vdw}(i,j,\vec{r_a},\vec{r_b})$$

$$+ \sum_{i=1}^{N-1}\sum_{j>i}^{N} U_{electrostatic}(i,j,\vec{r_a},\vec{r_b}) \quad \text{(A2)}$$

The first four terms represent bonded interactions, that is, bond stretching Ubond, bond-angle bend Uangle and dihedral angle torsion Utorsion and inversion interaction Uinversion, while the last two terms are nonbonded interactions, that is, van der Waals energy Uvdw and electrostatic energy Uelectrostatic. In the equation, $\vec{r}_a,\vec{r}_b,\vec{r}_c,\vec{r}_d$ are the positions of the atoms or particles specifically involved in a given interaction; N_{bond}, N_{angle}, $N_{torsion}$ and $N_{inversion}$ stand for the total numbers of these respective interactions in the simulated system; i_{bond}, i_{angle}, $i_{torsion}$ and $i_{inversion}$ uniquely specify an individual interaction of each type; i and j in the van der Waals and electrostatic terms indicate the atoms involved in the interaction. There are many algorithms for integrating the equation of motion using finite difference methods. The algorithms of varlet, velocity varlet, leap-frog and Beeman, are commonly used in MD simulations [23]. All algorithms assume that the atomic position $\vec{r}$, velocities $\vec{v}$ and accelerations $\vec{a}$ can be approximated by a Taylor series expansion:

$$\vec{r}(t+\delta t) = \vec{r}(t) + \vec{v}(t)\delta t + \frac{1}{2}\vec{a}(t)\delta^2 t + \cdots$$

$$\vec{r}(t+\delta t) = \vec{r}(t) + \vec{v}(t)\delta t + \frac{1}{2}\vec{a}(t)\delta^2 t + \cdots \quad \text{(A3)}$$

$$\vec{v}(t+\delta t) = \vec{v}(t) + \vec{a}(t)\delta t + \frac{1}{2}\vec{b}(t)\delta^2 t + \cdots$$

$$\vec{v}(t+\delta t)=\vec{v}(t)+\vec{a}(t)\delta t+\frac{1}{2}\vec{b}(t)\delta^{2}t+\cdots \tag{A4}$$

$$\vec{a}(t+\delta t)=\vec{a}(t)+\vec{b}(t)\delta t+\cdots \tag{A5}$$

Generally speaking, a good integration algorithm should conserve the total energy and momentum and be time-reversible. It should also be easy to implement and computationally efficient, and permit a relatively long time step. The verlet algorithm is probably the most widely used method. It uses the positions $\overrightarrow{r(t)}$ and accelerations $\vec{a}(t)$ at time t, and the positions $\vec{r}(t-\delta t)$ from the previous step (t–δ) to calculate the new positions $\vec{r}(t+\delta t)$ at (t+δt), we have:

$$\vec{r}(t+\delta t)=\vec{r}(t)+\vec{v}(t)\delta t+\frac{1}{2}\vec{a}(t)\delta t^{2}+\cdots$$

$$\vec{r}(t+\delta t)=\vec{r}(t)+\vec{v}(t)\delta t+\frac{1}{2}\vec{a}(t)\delta t^{2}+\cdots \tag{A6}$$

$$\vec{r}(t-\delta t)=\vec{r}(t)-\vec{v}(t)\delta t+\frac{1}{2}\vec{a}(t)\delta t^{2}+\cdots$$

$$\vec{r}(t-\delta t)=\vec{r}(t)-\vec{v}(t)\delta t+\frac{1}{2}\vec{a}(t)\delta t^{2}+\cdots \tag{A7}$$

$$\vec{r}(t+\delta t)=2\vec{r}(t)-\vec{r}(t-\delta t)+\vec{a}(t)\delta t^{2}+\cdots$$

$$\vec{r}(t+\delta t)=2\vec{r}(t)-\vec{r}(t-\delta t)+\vec{a}(t)\delta t^{2}+\cdots \tag{A8}$$

The velocities at time t and $t+\frac{1}{2\delta t}$ can be respectively estimated.

$$\vec{v}(t)=\frac{\left[\vec{r}(t+\delta t)-\vec{r}(t-\delta t)\right]}{2\delta t} \tag{A9}$$

$$\vec{v}\left(t+\frac{1}{2\delta t}\right)=\frac{\left[\vec{r}(t+\delta t)-\vec{r}(t-\delta t)\right]}{\delta t}$$

$$\vec{v}\left(t+\frac{1}{2\delta t}\right)=\frac{\left[\vec{r}(t+\delta t)-\vec{r}(t-\delta t)\right]}{\delta t} \tag{A10}$$

MD simulations can be performed in many different ensembles, such as grand canonical (μVT), microcanonical (NVE), canonical (NVT) and isothermal–isobaric (NPT). The constant temperature and pressure can be controlled by adding an appropriate thermostat (e.g., Berendsen, Nose, Nose–Hoover and Nose–Poincare) and barostat (e.g., Andersen, Hoover and Berendsen), respectively. Applying MD into polymer composites allows us to investigate into the effects of fillers on polymer structure and dynamics in the vicinity of polymer–filler interface and also to probe the effects of polymer–filler interactions on the materials properties.

A1.2 MONTE CARLO

MC technique, also called Metropolis method [24], is a stochastic method that uses random numbers to generate a sample population of the system from which one can calculate the properties of interest. A MC simulation usually consists of three typical steps. In the first step, the physical problem under investigation is translated into an analogous probabilistic or statistical model. In the second step, the probabilistic model is solved by a numerical stochastic sampling experiment. In the third step, the obtained data are analyzed by using statistical methods. MC provides only the information on equilibrium properties (e.g., free energy, phase equilibrium), different from MD, which gives nonequilibrium, as well as equilibrium properties. In a NVT ensemble with N atoms, one hypothesizes a new configuration by arbitrarily or systematically moving one atom from position i→j. Due to such atomic movement, one can compute the change in the system Hamiltonian ΔH:

$$\Delta H = H(j)\text{-}H(i) \tag{A11}$$

where H(i) and H(j) are the Hamiltonian associated with the original and new configuration, respectively.

This new configuration is then evaluated according to the following rules. If ΔH≥0, then the atomic movement would bring the system to a state of lower energy. Hence, the movement is immediately accepted and the displaced atom remains in its new position. If ΔH≥0, the move is accepted only with a certain probability $P^{i\to j}$ which is given by:

$$P^{i\to j} \propto \exp\left(-\frac{\Delta H}{K_B T}\right) \tag{A12}$$

where K_B is the Boltzmann constant. According to Metropolis et al. [25] one can generate a random number ζ between 0 and 1 and determine the new configuration according to the following rule:

$$\zeta \leq \exp\left(-\frac{\Delta H}{K_B T}\right); \tag{A13}$$

the move is accepted;

$$\zeta > \exp\left(-\frac{\Delta H}{K_B T}\right); \tag{A14}$$

the move is not acccpted.

If the new configuration is rejected, one counts the original position as a new one and repeats the process by using other arbitrarily chosen atoms. In a μVT ensemble, one hypothesizes a new configuration j by arbitrarily choosing one atom and proposing that it can be exchanged by an atom of a different kind. This procedure affects the chemical composition of the system. Also, the move is accepted with a certain probability. However, one computes the energy change ΔU associated with the change in composition. The new configuration is examined according to the following rules. If ΔU≥0, the move of compositional change is accepted. However, if ΔU ≥0, the move is accepted with a certain probability which is given by:

$$P^{i\to j} \propto \exp\left(-\frac{\Delta U}{K_B T}\right) \tag{A15}$$

where ΔU is the change in the sum of the mixing energy and the chemical potential of the mixture. If the new configuration is rejected one counts the original configuration as a new one and repeats the process by using some other arbitrarily or systematically chosen atoms. In polymer nanocomposites, MC methods have been used to investigate the molecular structure at nanoparticle surface and evaluate the effects of various factors.

A2 MICROSCALE METHODS

The modeling and simulation at microscale aim to bridge molecular methods and continuum methods and avoid their shortcomings. Specifically, in nanoparticle–polymer systems, the study of structural evolution (i.e., dynamics of phase separation) involves the description of bulk flow (i.e., hydrodynamic behavior) and the interactions between nanoparticle and polymer components. Note that hydrodynamic behavior is relatively straightforward to handle by continuum methods but is very difficult and expensive to treat by atomistic methods. In contrast, the interactions between components can be examined at an atomistic level but are usually not straightforward to incorporate at the continuum level. Therefore, various simulation methods have been evaluated and extended to study the microscopic structure and phase separation of these polymer nanocomposites, including BD, DPD, LB, time-dependent Ginsburg–Landau (TDGL) theory, and dynamic DFT. In these methods, a polymer system is usually treated with a field description or microscopic particles that incorporate molecular details implicitly. Therefore, they are able to simulate the phenomena on length and time scales currently inaccessible by the classical MD methods.

A2.1 BROWNIAN DYNAMICS

BD simulation is similar to MD simulations [26]. However, it introduces a few new approximations that allow one to perform simulations on the microsecond timescale whereas MD simulation is known up to a few nanoseconds. In BD the explicit description of solvent molecules used in MD is replaced with an implicit continuum solvent description. Besides, the internal motions of molecules are typically ignored, allowing a much larger time step than that of MD. Therefore, BD is particularly useful for systems where there is a large gap of time scale governing the motion of different components. For example, in polymer–solvent mixture, a short time-step is required to resolve the fast motion of the solvent molecules, whereas the evolution of the slower modes of the system requires a larger time step. However, if the detailed motion of the solvent molecules is concerned, they may be removed from the simulation and their effects on the polymer are represented by dissipative ($-\gamma P$) and random ($\sigma\,\zeta(t)$) force terms. Thus, the forces in the governing Eq. (16) is replaced by a Langevin equation,

$$F_i(t) = \sum_{i \neq j} F_{ij}^c - \gamma P_i + \sigma \zeta_i(t) \tag{A16}$$

Where F_{ij}^c is the conservative force of particle j acting on particle i, γ and σ are constants depending on the system, Pi the momentum of particle i, and $\zeta(t)$ a Gaussian random noise term. One consequence of this approximation of the fast degrees of freedom by fluctuating forces is that the energy and momentum are no longer conserved, which implies that the macroscopic behavior of the system will not be hydrodynamic. In addition, the effect of one solute molecule on another through the flow of solvent molecules is neglected. Thus, BD can only reproduce the diffusion properties but not the hydrodynamic flow properties since the simulation does not obey the Navier–Stokes equations.

A2.2 DISSIPATIVE PARTICLE DYNAMICS

DPD was originally developed by Hoogerbrugge and Koelman [27]. It can simulate both Newtonian and nonNewtonian fluids, including polymer melts and blends, on microscopic length and time scales. Like MD and BD, DPD is a particle-based method. However, its basic unit is not a single atom or molecule but a molecular assembly (i.e., a particle). DPD particles are defined by their mass M_i, position r_i and momentum P_i. The interaction force between two DPD particles i and j can be described by a sum of conservative F_{ij}^{c}, dissipative F_{ij}^{D} and random forces F_{ij}^{R} [28–30]:

$$F_{ij} = F_{ij}^{c} + F_{ij}^{D} + F_{ij}^{R} \tag{A17}$$

While the interaction potentials in MD are high-order polynomials of the distance r_{ij} between two particles, in DPD the potentials are softened so as to approximate the effective potential at microscopic length scales. The form of the conservative force in particular is chosen to decrease linearly with increasing r_{ij}. Beyond a certain cut-off separation r_c, the weight functions and thus the forces are all zero. Because the forces are pair wise and momentum is conserved, the macroscopic behavior directly incorporates Navier–Stokes hydrodynamics. However, energy is not conserved because of the presence of the dissipative and random force terms, which are similar to those of BD, but incorporate the effects of Brownian motion on larger length scales. DPD has several advantages over MD, for example, the hydrodynamic behavior is observed with far fewer particles than required in a MD simulation because of its larger particle size. Besides, its force forms allow larger time steps to be taken than those in MD.

A2.3 LATTICE BOLTZMANN

LB [31] is another microscale method that is suited for the efficient treatment of polymer solution dynamics. It has recently been used to investigate the phase separation of binary fluids in the presence of solid particles. The LB method is originated from lattice gas automaton, which is constructed as a simplified, fictitious molecular dynamic in which space, time and particle velocities are all discrete. A typical lattice gas automaton consists of a regular lattice with particles residing on the nodes. The main feature of the LB method is to replace the particle occupation variables (Boolean variables), by single-particle distribution functions (real variables) and neglect individual particle motion and particle–particle correlations in the kinetic equation. There are several ways to obtain the LB equation from either the discrete velocity model or the Boltzmann kinetic equation, and to derive the macroscopic Navier–Stokes equations from the LB equation. An important advantage of the LB method is that microscopic physical interactions of the fluid particles can be conveniently incorporated into the numerical model. Compared with the Navier– Stokes equations, the LB method can handle the interactions among fluid particles and reproduce the microscale mechanism of hydrodynamic behavior. Therefore it belongs to the MD in nature and bridges the gap between the molecular level and macroscopic level. However, its main disadvantage is that it is typically not guaranteed to be numerically stable and may lead to physically unreasonable results, for instance, in the case of high forcing rate or high interparticle interaction strength.

A2.4 TIME-DEPENDENT GINZBURG–LANDAU METHOD

TDGL is a microscale method for simulating the structural evolution of phase-separation in polymer blends and block copolymers. It is based on the Cahn–Hilliard–Cook (CHC) nonlinear diffusion equation for a binary blend and falls under the more general phase-field and reaction-diffusion models [32–34]. In the TDGL method, a free-energy function is minimized to simulate a temperature quench from the miscible region of the phase diagram to the immiscible region. Thus, the resulting time-dependent structural evolution of the polymer blend can be investigated by solving the TDGL/CHC equation for the time dependence of the local blend concentration. Glotzer and co-workers have discussed and applied this method to polymer blends and particle-filled polymer systems [35]. This model reproduces the growth kinetics of the TDGL model, demonstrating that such quantities are insensitive to the precise form of the double-well potential of the bulk free-energy term. The TDGL and CDM methods have recently been used to investigate the phase-separation of polymer nanocomposites and polymer blends in the presence of nanoparticles [36–40].

A2.5 DYNAMIC DFT METHOD

Dynamic DFT method is usually used to model the dynamic behavior of polymer systems and has been implemented in the software package Mesodyn™ from Accelrys [41]. The DFT models the behavior of polymer fluids by combining

Gaussian mean-field statistics with a TDGL model for the time evolution of conserved order parameters. However, in contrast to traditional phenomenological free-energy expansion methods employed in the TDGL approach, the free energy is not truncated at a certain level, and instead retains the full polymer path integral numerically. At the expense of a more challenging computation, this allows detailed information about a specific polymer system beyond simply the Flory–Huggins parameter and mobilities to be included in the simulation. In addition, viscoelasticity, which is not included in TDGL approaches, is included at the level of the Gaussian chains. A similar DFT approach has been developed by Doi and co-workers [42, 43] and forms the basis for their new software tool Simulation Utilities for Soft and Hard Interfaces (SUSHI), one of a suite of molecular and mesoscale modeling tools (called OCTA) developed for the simulation of polymer materials [44]. The essence of dynamic DFT method is that the instantaneous unique conformation distribution can be obtained from the off-equilibrium density profile by coupling a fictitious external potential to the Hamiltonian. Once such distribution is known, the free energy is then calculated by standard statistical thermodynamics. The driving force for diffusion is obtained from the spatial gradient of the first functional derivative of the free energy with respect to the density. Here, we describe briefly the equations for both polymer and particle in the diblock polymer–particle composites [38].

A3 MESOSCALE AND MACROSCALE METHODS

Despite the importance of understanding the molecular structure and nature of materials, their behavior can be homogenized with respect to different aspects, which can be at different scales. Typically, the observed macroscopic behavior is usually explained by ignoring the discrete atomic and molecular structure and assuming that the material is continuously distributed throughout its volume. The continuum material is thus assumed to have an average density and can be subjected to body forces such as gravity and surface forces. Generally speaking, the macroscale methods (or called continuum methods hereafter) obey the fundamental laws of: (i) continuity, derived from the conservation of mass; (ii) equilibrium, derived from momentum considerations and Newton's second law; (iii) the moment of momentum principle, based on the model that the time rate of change of angular momentum with respect to an arbitrary point is equal to the resultant moment; (iv) conservation of energy, based on the first law of thermodynamics; and (v) conservation of entropy, based on the second law of thermodynamics. These laws provide the basis for the continuum model and must be coupled with the appropriate constitutive equations and the equations of state to provide all the equations necessary for solving a continuum problem. The continuum method relates the deformation of a continuous medium to the external forces acting on the medium and the resulting internal stress and strain. Computational approaches range from simple closed-form analytical expressions to micromechanics and complex structural mechanics calculations based on beam and shell theory. In this section, we introduce some continuum methods that have been used in polymer nanocomposites, including micromechanics models (e.g., Halpin–Tsai model, Mori–Tanaka model), equivalent-continuum model, self-consistent model and finite element analysis.

A4 MICROMECHANICS

Since the assumption of uniformity in continuum mechanics may not hold at the microscale level, micromechanics methods are used to express the continuum quantities associated with an infinitesimal material element in terms of structure and properties of the micro constituents. Thus, a central theme of micromechanics models is the development of a representative volume element (RVE) to statistically represent the local continuum properties. The RVE is constructed to ensure that the length scale is consistent with the smallest constituent that has a first-order effect on the macroscopic behavior. The RVE is then used in a repeating or periodic nature in the full-scale model. The micromechanics method can account for interfaces between constituents, discontinuities, and coupled mechanical and nonmechanical properties. Ourpurpose is to review the micromechanics methods used for polymer nanocomposites. Thus, we only discuss here some important concepts of micromechanics as well as the Halpin–Tsai model and Mori–Tanaka model.

A4.1 BASIC CONCEPTS

When applied to particle reinforced polymer composites, micromechanics models usually follow such basic assumptions as (i) linear elasticity of fillers and polymer matrix; (ii) the fillers are axisymmetric, identical in shape and size, and can be characterized by parameters such as aspect ratio; (iii) well-bonded filler–polymer interface and the ignorance of interfacial slip, filler–polymer debonding or matrix cracking. The first concept is the linear elasticity, i.e., the linear relationship between the total stress and infinitesimal strain tensors for the filler and matrix as expressed by the following constitutive equations:

For filler $\sigma^f = C^f \varepsilon^f$

For matrix $\sigma^m = C^m \varepsilon^m$

where C is the stiffness tensor. The second concept is the average stress and strain. Since the point wise stress field $\sigma(x)$ and the corresponding strain field $\varepsilon(x)$ are usually nonuniform in polymer composites, the volume–average stress $\bar{\sigma}$ and strain $\vec{\varepsilon}$ are then defined over the representative averaging volume V, respectively.

$$\bar{\sigma} = \frac{1}{V} \int \sigma(x) dv$$

$$\bar{\varepsilon} = \frac{1}{V} \int \varepsilon(x) dv$$

Therefore, the average filler and matrix stresses are the averages over the corresponding volumes v_f and v_m, respectively.

$$\overline{\sigma_f} = \frac{1}{V_f} \int \sigma(x) dv$$

$$\overline{\sigma_m} = \frac{1}{V_m} \int \sigma(x) dv$$

The average strains for the fillers and matrix are defined, respectively, as

$$\overline{\varepsilon_f} = \frac{1}{V_f} \int \varepsilon(x) dv$$

$$\overline{\varepsilon_m} = \frac{1}{V_m} \int \varepsilon(x) dv$$

Based on the above definitions, the relationships between the filler and matrix averages and the overall averages can be derived as follows:

$$\bar{\sigma} = \overline{\sigma_f} v_f + \overline{\sigma_m}\, v_m \quad \text{(A18)}$$

$$\bar{\varepsilon} = \overline{\varepsilon_f} v_f + \overline{\varepsilon_m}\, v_m \quad \text{(A19)}$$

where v_f, v_m are the volume fractions of the fillers and matrix, respectively.

The third concept is the average properties of composites, which are actually the main goal of a micromechanics model. The average stiffness of the composite is the tensor C that maps the uniform strain to the average stress.

$$\bar{\sigma} = \bar{\varepsilon} C$$

The average compliance S is defined in the same way:

$$\bar{\varepsilon} = \bar{\sigma} S$$

Another important concept is the strain–concentration and stress–concentration tensors A and B, which are basically the ratios between the average filler strain (or stress) and the corresponding average of the composites.

$$\overline{\varepsilon_f} = \bar{\varepsilon} A$$

$$\overline{\sigma_f} = \bar{\sigma} B$$

Using the above concepts and equations, the average composite stiffness can be obtained from the strain concentration tensor A and the filler and matrix properties:

$$C = C_m + v_f (C_f - C_m) A \quad \text{(A19)}$$

A4.2 HALPIN–TSAI MODEL

The Halpin–Tsai model is a well-known composite theory to predict the stiffness of unidirectional composites as a functional of aspect ratio. In this model, the longitudinal E11 and transverse E22 engineering moduli are expressed in the following general form:

$$\frac{E}{E_m} = \frac{1 + \zeta \eta v_f}{1 - \eta v_f} \quad \text{(A20)}$$

where E and E_m represent the Young's modulus of the composite and matrix, respectively, v_f is the volume fraction of filler, and η is given by:

$$\eta = \frac{\frac{E}{E_m} - 1}{\frac{E_f}{E_m} + \zeta_f}$$

where E_f represents the Young's modulus of the filler and ζ_f the shape parameter depending on the filler geometry and loading direction. When calculating longitudinal modulus E_{11}, ζ_f is equal to l/t, and when calculating transverse modulus E_{22}, ζ_f is equal to w/t. Here, the parameters of l, w and t are the length, width and thickness of the dispersed fillers, respectively. If $\zeta_f \rightarrow 0$, the Halpin–Tsai theory converges to the inverse rule of mixture (lower bound):

$$\frac{1}{E} = \frac{v_f}{E_f} + \frac{1 - v_f}{E_m} \quad \text{(A21)}$$

Conversely, if $\zeta_f \rightarrow \infty$, the theory reduces to the rule of mixtures (upper bound),

$$E = E_f v_f + E_m \left(1 - v_f\right) \tag{A22}$$

A4.3 MORI–TANAKA MODEL

The Mori–Tanaka model is derived based on the principles of Eshelby's inclusion model for predicting an elastic stress field in and around ellipsoidal filler in an infinite matrix. The complete analytical solutions for longitudinal E_{11} and transverse E_{22} elastic moduli of an isotropic matrix filled with aligned spherical inclusion are [45, 46]:

$$\frac{E_{11}}{E_m} = \frac{A_0}{A_0 + v_f(A_1 + 2v_0 A_2)} \tag{A23}$$

$$\frac{E_{22}}{E_m} = \frac{2A_0}{2A_0 + v_f(-2A_3 + (1 - v_0 A_4) + (1 + v_0)A_5 A_0)} \tag{A24}$$

where E_m represents the Young's modulus of the matrix, v_f the volume fraction of filler, v_0 the Poisson's ratio of the matrix, parameters, A0, A1, …, A5 are functions of the Eshelby's tensor and the properties of the filler and the matrix, including Young's modulus, Poisson's ratio, filler concentration and filler aspect ratio [45].

A4.4 EQUIVALENT-CONTINUUM AND SELF-SIMILAR APPROACHES

Numerous micromechanical models have been successfully used to predict the macroscopic behavior of fiber-reinforced composites. However, the direct use of these models for nanotube-reinforced composites is doubtful due to the significant scale difference between nanotube and typical carbon fiber. Recently, two methods have been proposed for modeling the mechanical behavior of single walled carbon nanotube (SWCN) composites: equivalent-continuum approach and self-similar approach [47]. The equivalent-continuum approach was proposed by Odegard, et al. [48]. In this approach, MD was used to model the molecular interactions between SWCN–polymer and a homogeneous equivalent-continuum reinforcing element (e.g., (a) SWCN surrounded (b) polymer) were constructed as shown in Fig. A2. Then, micromechanics are used to determine the effective bulk properties of the equivalent-continuum reinforcing element embedded in a continuous polymer. The equivalent-continuum approach consists of four major steps, as briefly described below.

Step 1: MD simulation is used to generate the equilibrium structure of a SWCN–polymer composite and then to establish the RVE of the molecular model and the equivalent-continuum model.

Step 2: The potential energies of deformation for the molecular model and effective fiber are derived and equated for identical loading conditions. The bonded and nonbonded interactions within a polymer molecule are quantitatively described by MM. For the SWCN/polymer system, the total potential energy U^m of the molecular model is:

$$U^m = \sum U^r(k_r) + \sum U^\theta(k_\theta) + \sum U^{vdw}(k_{vdw}) \tag{A25}$$

where U^r, U^θ and U^{vdw} are the energies associated with covalent bond stretching, bond-angle bending, and van der Waals interactions, respectively. An equivalent-truss model of the RVE is used as an intermediate step to link the molecular and equivalent-continuum models. Each atom in the molecular model is represented by a pin-joint, and each truss element represents an atomic bonded or nonbonded interaction. The potential energy of the truss model is

$$U^t = \sum U^a(E^a) + \sum U^b(E^b) + \sum U^c(E^c) \tag{A26}$$

where U^a, U^b and U^c are the energies associated with truss elements that represent covalent bond stretching, bond-angle bending, and van der Waals interactions, respectively. The energies of each truss element are a function of the Young's modulus, E.

Step 3: A constitutive equation for the effective fiber is established. Since the values of the elastic stiffness tensor components are not known a priori, a set of loading conditions are chosen such that each component is uniquely determined from

$$U^f = U^t = U^m$$

Step 4: Overall constitutive properties of the dilute and unidirectional SWCN/polymer composite are determined with Mori–Tanaka model with the mechanical properties of the effective fiber and the bulk polymer. The layer of polymer molecules that are near the polymer/nanotube interface (Fig. 2) is included in the effective fiber, and it is assumed that the matrix polymer surrounding the effective fiber has mechanical properties equal to those of the bulk polymer. The self-similar approach was proposed by Pipes and Hubert [49], which consists of three major steps:

First, a helical array of SWCNs is assembled. This array is termed as the SWCN nanoarray where 91 SWCNs make up the cross-section of the helical nanoarray. Then, the SWCN nanoarrays is surrounded by a polymer matrix and assembled into a second twisted array, termed as the SWCN nanowire Finally, the SWCN nanowires are further impregnated with a polymer matrix and assembled into the final helical array—the SWCN microfiber. The self-similar geometries described in the nanoarray, nanowire and microfiber (Fig. A3) allow the use of the same mathematical and geometric model for all three geometries [49].

A4.5 FINITE ELEMENT METHOD

FEM is a general numerical method for obtaining approximate solutions in space to initial-value and boundary-value problems including time-dependent processes. It employs preprocessed mesh generation, which enables the model to fully capture the spatial discontinuities of highly inhomogeneous materials. It also allows complex, nonlinear tensile relationships to be incorporated into the analysis. Thus, it has been widely used in mechanical, biological and geological systems. In FEM, the entire domain of interest is spatially discretized into an assembly of simply shaped subdomains (e.g., hexahedra or tetrahedral in three dimensions, and rectangles or triangles in two dimensions) without gaps and without overlaps. The subdomains are interconnected at joints (i.e., nodes). The implementation of FEM includes the important steps shown in Fig. A.1. The energy in FEM is taken from the theory of linear elasticity and thus the input parameters are simply the elastic moduli and the density of the material. Since these parameters are in agreement with the values computed by MD, the simulation is consistent across the scales. More specifically, the total elastic energy in the absence of tractions and body forces within the continuum model is given by [50]:

$$U = U_v + U_k$$

$$U_k = \frac{1}{2} \int dr \, p(r) \| \dot{U}_r \|^2$$

$$U_v = \frac{1}{2} \int dr \sum_{\mu,\upsilon,\lambda,\sigma=1}^{3} \varepsilon_{\mu\upsilon}(r) C_{\mu\upsilon\lambda\sigma^3\lambda\sigma}(r)$$

where U_v is the Hookian potential energy term, which is quadratic in the symmetric strain tensor e, contracted with the elastic constant tensor C. The Greek indices (i.e., m, n, l, s) denote Cartesian directions. The kinetic energy U_k involves the time rate of change of the displacement field $\dot{U}$, and the mass density ρ.

These are fields defined throughout space in the continuum theory. Thus, the total energy of the system is an integral of these quantities over the volume of the sample dυ. The FEM has been incorporated in some commercial software packages and open source codes (e.g., ABAQUS, ANSYS, Palmyra and OOF) and widely used to evaluate the mechanical properties of polymer composites. Some attempts have recently been made to apply the FEM to nanoparticle-reinforced polymer nanocomposites. In order to capture the multiscale material behaviors, efforts are also underway to combine the multiscale models spanning from molecular to macroscopic levels [51, 52].

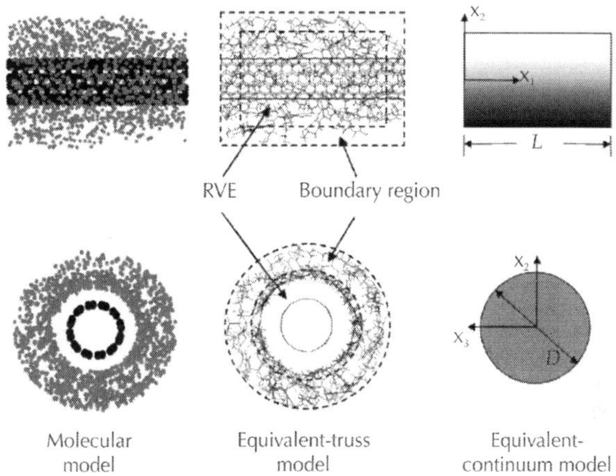

FIGURE A1 Equivalent-continuum modeling of effective fiber.

A5 MULTI SCALE MODELING OF MECHANICAL PROPERTIES

In Odegard's study [48], a method has been presented for linking atomistic simulations of nano-structured materials to continuum models of the corresponding bulk material. For a polymer composite system reinforced with SWNT, the method provides the steps whereby the nanotube, the local polymer near the nanotube, and the nanotube/polymer interface can be modeled as an effective continuum fiber by using an equivalent-continuum model. The effective fiber retains the local molecular structure and bonding information, as defined by molecular dynamics, and serves as a means for linking the equivalent-continuum and micromechanics models. The micromechanics method is then available for the prediction of bulk mechanical properties of SWNT/polymer composites as a function of nanotube size, orientation, and volume fraction. The utility of this method was examined by modeling tow composites that both having a interface. The elastic stiffness constants of the composites were determined for both aligned and three-dimensional randomly oriented nanotubes, as a function of nanotube length and volume fraction. They used Mori– Tanaka model [53] for random and oriented fibers position and compare their model with mechanical properties, the interface between fiber and matrix was assumed perfect. Motivated by micrographs showing that embedded nanotubes often exhibit significant curvature within the polymer, Fisher et al.[54] have developed a model combining finite element results and micromechanical methods (Mori-Tanaka)to determine the effective reinforcing modulus of a wavy embedded nanotube with perfect bonding and random fiber orientation assumption. This effective reinforcing modulus (ERM) is then used within a multiphase micromechanics model to predict the effective modulus of a polymer reinforced with a distribution of wavy nanotubes. We found that even slight nanotube curvature significantly reduces the effective reinforcement when compared to straight nanotubes. These results suggest that nanotube waviness may be an additional mechanism limiting the modulus enhancement of nanotube-reinforced polymers. Bradshaw et al. [55] investigated the degree to which the characteristic waviness of nanotubes embedded in polymers can impact the effective stiffness of these materials. A 3D finite element model of a single infinitely long sinusoidal fiber within an infinite matrix is used to numerically compute the dilute strain concentration tensor. A Mori–Tanaka model uses this tensor to predict the effective modulus of the material with aligned or randomly oriented inclusions. This hybrid finite element micromechanical modeling technique is a powerful extension of general micromechanics modeling and can be applied to any composite microstructure containing nonellipsoidal inclusions. The results demonstrate that nanotube waviness results in a reduction of the effective modulus of the composite relative to straight nanotube reinforcement. The degree of reduction is dependent on the ratio of the sinusoidal wavelength to the nanotube diameter. As this wavelength ratio increases, the effective stiffness of a composite with randomly oriented wavy nanotubes converges to the result obtained with straight nanotube inclusions.

The effective mechanical properties of carbon nanotube-based composites are evaluated by Liu & Chen [56] using a 3-D nanoscale RVE based on 3-D elasticity theory and solved by the finite element method. Formulas to extract the material constants from solutions for the RVE under three loading cases are established using the elasticity. An extended rule of mixtures, which can be used to estimate the Young's modulus in the axial direction of the RVE and to validate the numerical solutions for short CNTs, is also derived using the strength of materials theory. Numerical examples using the FEM to evaluate the effective material constants of a CNT-based composites are presented, which demonstrate that the reinforcing capabilities of the CNTs in a matrix are significant. With only about 2% and 5% volume fractions of the CNTs in a matrix, the stiffness of the composite in the CNT axial direction can increase as many as 0.7 and 9.7 times for the cases of short and long CNT fibers, respectively. These simulation results, which are believed to be the first of its kind for CNT-based composites, are consistent with the experimental results reported in the literature [57–59]. The developed extended rule of mixtures is also found to be quite effective in evaluating the stiffness of the CNT-based composites in the CNT axial direction. Many research issues need to be addressed in the modeling and simulations of CNTs in a matrix material for the development of nanocomposites. Analytical methods and simulation models to extract the mechanical properties of the CNT-based nanocomposites need to be further developed and verified with experimental results. The analytical method and simulation approach developed in this paper are only a preliminary study. Different type of RVEs, load cases and different solution methods should be investigated. Different interface conditions, other than perfect bonding, need to be investigated using different models to more accurately account for the interactions of the CNTs in a matrix material at the nanoscale. Nanoscale interface cracks can be analyzed using simulations to investigate the failure mechanism in nanomaterials. Interactions among a large number of CNTs in a matrix can be simulated if the computing power is available. Single-walled and multiwalled CNTs as reinforcing fibers in a matrix can be studied by simulations to find out their advantages and disadvantages. Finally, large multiscale simulation models for CNT-based composites, which can link the models at the nano, micro and macro scales, need to be developed, with the help of analytical and experimental work [56]. The three RVEs proposed in Ref. [60] and shown in Fig. A2 are relatively simple regarding the models and scales and pictures in Fig. A3 are three loading cases for the cylindrical RVE. However, this is only the first step toward more sophisticated and large-scale simulations of CNT-based composites. As the computing power and confidence in simulations of CNT-based composites increase, large scale 3-D models containing hundreds or even more CNTs, behaving linearly or nonlinearly, with coatings or of different sizes, distributed evenly or randomly, can be employed to investigate the interactions among the CNTs in a matrix and to evaluate the effective material properties. Other numerical methods can also be attempted for the modeling and simulations of CNT-based composites, which may offer some advantages over the FEM approach. For example, the boundary element method, [60, 61], accelerated with the fast multipole techniques, [62, 63], and the mesh free methods [64] may enable one to model an RVE with thousands of CNTs in a matrix on a desktop computer. Analysis of the CNT-based composites using the boundary element method is already underway and will be reported subsequently.

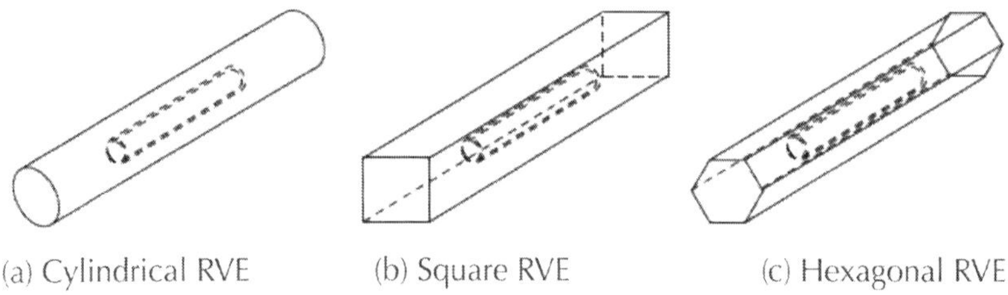

FIGURE A2 Three nanoscale representative volume elements for the analysis of CNT-based nanocomposites.

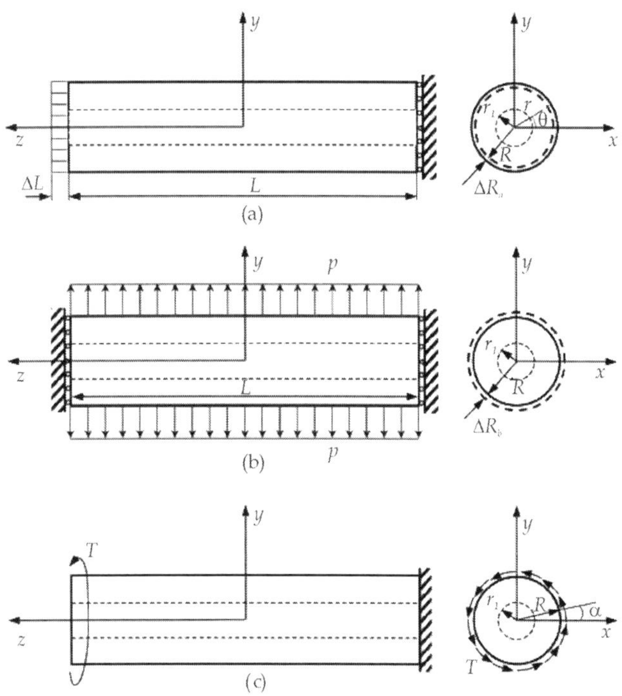

FIGURE A3 Three loading cases for the cylindrical RVE used to evaluate the effective material properties of the CNT-based composites. (a)Under axial stretch DL; (b) under lateral uniform load P; (c) under torsional load T.

The effective mechanical properties of CNT based composites are evaluated using square RVEs based on 3-D elasticity theory and solved by the FEM. Formulas to extract the effective material constants from solutions for the square RVEs under two loading cases are established based on elasticity. Square RVEs with multiple CNTs are also investigated in evaluating the Young's modulus and Poisson's ratios in the transverse plane. Numerical examples using the FEM are presented, which demonstrate that the load-carrying capabilities of the CNTs in a matrix are significant. With the addition of only about 3.6% volume fraction of the CNTs in a matrix, the stiffness of the composite in the CNT axial direction can increase as much as 33% for the case of long CNT fibers [65]. These simulation results are consistent with both the experimental ones reported in the literature [56–59, 66]. It is also found that cylindrical RVEs tend to overestimate the effective Young's moduli due to the fact that they overestimate the volume fractions of the CNTs in a matrix. The square RVEs, although more demanding in modeling and computing, may be the preferred model in future simulations for estimating the effective material constants, especially when multiple CNTs need to be considered. Finally, the rules of mixtures, for both long and short CNT cases, are found to be quite accurate in estimating the effective Young's moduli in the CNT axial direction. This may suggest that 3-D FEM modeling may not be necessary in obtaining the effective material constants in the CNT direction, as in the studies of the conventional fiber reinforced composites. Efforts in comparing the results presented in this paper using the continuum approach directly with the MD simulations are underway. This is feasible now only for a smaller RVE of one CNT embedded in a matrix. In future research, the MD and continuum approach should be integrated in a multiscale modeling and simulation environment for analyzing the CNT-based composites. More efficient models of the CNTs in a matrix also need to be developed, so that a large number of CNTs, in different shapes and forms (curved or twisted), or randomly distributed in a matrix, can be modeled. The ultimate validation of the simulation results should be done with the nanoscale or microscale experiments on the CNT reinforced composites [64].

Griebel and Hamaekers [67] reviewed the basic tools used in computational nanomechanics and materials, including the relevant underlying principles and concepts. These tools range from subatomic ab initio methods to classical molecular dynamics and multiple-scale approaches. The energetic link between the quantum mechanical and classical systems has been discussed, and limitations of the standing alone molecular dynamics simulations have been shown on a series of illustrative examples. The need for multiscale simulation methods to tackle nanoscale aspects of material behavior was therefore emphasized; that was followed by a review and classification of the mainstream and emerging multiscale methods. These simulation methods include the broad areas of quantum mechanics, molecular dynamics and multiple-scale approaches, based on coupling the atomistic and continuum models. They summarize the strengths and limitations of currently available multiple-scale techniques, where the emphasis is made on the latest perspective approaches, such as the bridging scale method, multiscale boundary conditions, and multiscale fluidics. Example problems, in which multiple-scale simulation methods yield equivalent results to full atomistic simulations at fractions of the computational cost, were shown. They compare their results with Odegard, et al. [48], the micromechanic method was BEM Halpin-Tsai Eq. [68] with aligned fiber by perfect bonding.

The solutions of the strain-energy-changes due to a SWNT embedded in an infinite matrix with imperfect fiber bonding are obtained through numerical method by Wan, et al. [69]. A "critical" SWNT fiber length is defined for full load transfer between the SWNT and the matrix, through the evaluation of the strain-energy-changes for different fiber lengths The strain-energy-change is also used to derive the effective longitudinal Young's modulus and effective bulk modulus of the composite, using a dilute solution. The main goal of their research was investigation of strain-energy-change due to inclusion of SWNT using FEM. To achieve full load transfer between the SWNT and the matrix, the length of SWNT fibers should be longer than a "critical" length if no weak interphase exists between the SWNT and the matrix [69].

A hybrid atomistic/continuum mechanics method is established in the Feng, et al. study [70] the deformation and fracture behaviors of carbon nanotubes (CNTs) in composites. The unit cell containing a CNT embedded in a matrix is divided in three regions, which are simulated by the atomic-potential method, the continuum method based on the modified Cauchy–Born rule, and the classical continuum mechanics, respectively. The effect of CNT interaction is taken into account via the Mori–Tanaka effective field method of micromechanics. This method not only can predict the formation of Stone–Wales (5-7–7-5) defects, but also simulate the subsequent deformation and fracture process of CNTs. It is found that the critical strain of defect nucleation in a CNT is sensitive to its chiral angle but not to its diameter. The critical strain of Stone–Wales defect formation of zigzag CNTs is nearly twice that of armchair CNTs. Due to the constraint effect of matrix, the CNTs embedded in a composite are easier to fracture in comparison with those not embedded. With the increase in the Young's modulus of the matrix, the critical breaking strain of CNTs decreases.

Estimation of effective elastic moduli of nanocomposites was performed by the version of effective field method developed in the framework of quasi-crystalline approximation when the spatial correlations of inclusion location take particular ellipsoidal forms [71]. The independent justified choice of shapes of inclusions and correlation holes provide the formulae of effective moduli, which are symmetric, completely explicit, and easily to use. The parametric numerical analyzes revealed the most sensitive parameters influencing the effective moduli which are defined by the axial elastic moduli of nanofibers rather than their transversal moduli as well as by the justified choice of correlation holes, concentration and prescribed random orientation of nanofibers [72].

Li and Chou [73, 74] have reported a multiscale modeling of the compressive behavior of carbon nanotube/polymer composites. The nanotube is modeled at the atomistic scale, and the matrix deformation is analyzed by the continuum finite element method. The nanotube and polymer matrix are assumed to be bonded by van der Waals interactions at the interface. The stress distributions at the nanotube/polymer interface under isostrain and isostress loading conditions have been examined. They have used beam elements for SWCNT using molecular structural mechanics, truss rod for vdW links and cubic elements for matrix, the rule of mixture was used as for comparison in this research. The buckling forces of nanotube/ polymer composites for different nanotube lengths and diameters are computed. The results indicate that continuous nanotubes can most effectively enhance the composite buckling resistance.

Anumandla and Gibson [75] describes an approximate, yet comprehensive, closed form micromechanics model for estimating the effective elastic modulus of carbon nanotube-reinforced composites. The model incorporates the typically observed nanotube curvature, the nanotube length, and both 1D and 3D random arrangement of the nanotubes. The analytical results obtained from the closed form micromechanics model for nanoscale representative volume elements and results from an equivalent finite element model for effective reinforcing modulus of the nanotube reveal that the reinforcing modulus is strongly dependent on the waviness, wherein, even a slight change in the nanotube curvature can induce a prominent change in the effective reinforcement provided. The micromechanics model is also seen to produce reasonable agreement with experimental data for the effective tensile modulus of composites reinforced with multiwalled nanotubes (MWNTs) and having different MWNT volume fractions.

Effective elastic properties for carbon nanotube reinforced composites are obtained through a variety of micromechanics techniques [76]. Using the in-plane elastic properties of graphene, the effective properties of carbon nanotubes are calculated using a composite cylinders micromechanics technique as a first step in a two-step process. These effective properties are then used in the self-consistent and Mori–Tanaka methods to obtain effective elastic properties of composites consisting of aligned single or multiwalled carbon nanotubes embedded in a polymer matrix. Effective composite properties from these averaging methods are compared to a direct composite cylinders approach extended from the work of Z. Hashin and B. Rosen [77] and R. Christensen and K. Lo [78]. Comparisons with finite element simulations are also performed. The effects of an interphase layer between the nanotubes and the polymer matrix as result of functionalization is also investigated using a multilayer composite cylinders approach. Finally, the modeling of the clustering of nanotubes into bundles due to interatomic forces is accomplished herein using a tessellation method in conjunction with a multiphase Mori–Tanaka technique. In addition to aligned nanotube composites, modeling of the effective elastic properties of randomly dispersed nanotubes into a matrix is performed using the Mori–Tanaka method, and comparisons with experimental data are made.

Selmi, et al. [79] deal with the prediction of the elastic properties of polymer composites reinforced with single walled carbon nanotubes. Our contribution is the investigation of several micromechanical models, while most of the papers on the subject deal with only one approach. They implemented four homogenization schemes, a sequential one and three others based on various extensions of the Mori–Tanaka (M–T) mean-field homogenization model: two-level (M–T/M–T), two-step (M–T/M–T) and two-step (M–T/Voigt). Several composite systems are studied, with various properties of the matrix and the graphene, short or long nanotubes, fully aligned or randomly oriented in 3D or 2D. Validation targets are experimental data or finite element results, either based on a 2D periodic unit cell or a 3D representative volume element. The comparative study showed that there are cases where all micromechanical models give adequate predictions, while for some composite materials and some properties, certain models fail in a rather spectacular fashion. It was found that the two-level (M–T/M–T) homogenization model gives the best predictions in most cases. After the characterization of the discrete nanotube structure using a homogenization method based on energy equivalence, the sequential, the two-step (M–T/M–T), the two-step (M–T/Voigt), the two-level (M–T/M–T) and finite element models were used to predict the elastic properties of SWNT/polymer composites. The data delivered by the micromechanical models are compared against those obtained by finite element analyzes or experiments. For fully aligned, long nanotube polymer composite, it is the sequential and the two-level (M–T/M–T) models, which delivered good predictions. For all composite morphologies (fully aligned, two-dimensional in-plane random orientation, and three-dimensional random orientation), it is the two-level (M–T/M–T) model, which gave good predictions compared to finite element and experimental results in most situations. There are cases where other micromechanical models failed in a spectacular way.

Luo, et al. [80] have used multiscale homogenization (MH) and FEM for wavy and straight SWCNTs, they have compare their results with Mori-Tanaka, Cox, Halpin-Tsai, Fu, et al. [81], Lauke [82]. Trespass, et al. [83] used 3D elastic beam for C-C bond and, 3D space frame for CNT and progressive fracture model for prediction of elastic modulus, they used rule of mixture for compression of their results. Their assumption was embedded a single SWCNT in polymer with Perfect bonding. The multiscale modeling, Monte Carlo, FEM and using equivalent continuum method was used by Spanos and Kontsos [84] and compared with Zhu, et al. [85] and Paiva, et al. [86]'s results.

The effective modulus [87] of CNT/PP composites is evaluated using FEA of a 3D RVE which includes the PP matrix, multiple CNTs and CNT/PP interphase and accounts for poor dispersion and non homogeneous distribution of CNTs within the polymer matrix, weak CNT/polymer interactions, CNT agglomerates of various sizes and CNTs orientation and waviness. Currently, there is no other model, theoretical or numerical, that accounts for all these experimentally observed phenomena and captures their individual and combined effect on the effective modulus of nanocomposites. The model is developed using input obtained from experiments and validated against experimental data. CNT reinforced PP composites manufactured by extrusion and injection molding are characterized in terms of tensile modulus, thickness and stiffness of CNT/PP interphase, size of CNT agglomerates and CNT distribution using tensile testing, AFM and SEM, respectively. It is concluded that CNT agglomeration and waviness are the two dominant factors that hinder the great potential of CNTs as polymer reinforcement. The proposed model provides the upper and lower limit of the modulus of the CNT/PP composites and can be used to guide the manufacturing of composites with engineered properties for targeted applications. CNT agglomeration can be avoided by employing processing techniques such as sonication of CNTs, stirring, calendaring etc., whereas CNT waviness can be eliminated by increasing the injection pressure during molding and mainly by using CNTs with smaller aspect ratio. Increased pressure during molding can also promote the alignment of CNTs along the applied load direction. The 3D modeling capability presented in this study gives an insight on the upper and lower bound of the CNT/PP composites modulus quantitatively by accurately capturing the effect of various processing parameters. It is observed that when all the experimentally observed factors are considered together in the FEA the modulus prediction is in good agreement with the modulus obtained from the experiment. Therefore, it can be concluded that the FEM models proposed in this study by systematically incorporating experimentally observed characteristics can be effectively used for the determination of mechanical properties of nanocomposite materials. Their result is in agreement with the results reported in Ref. [88], The theoretical micromechanical models, shown in Fig. A4, are used to confirm that our FEM model predictions follow the same trend with the one predicted by the models as expected.

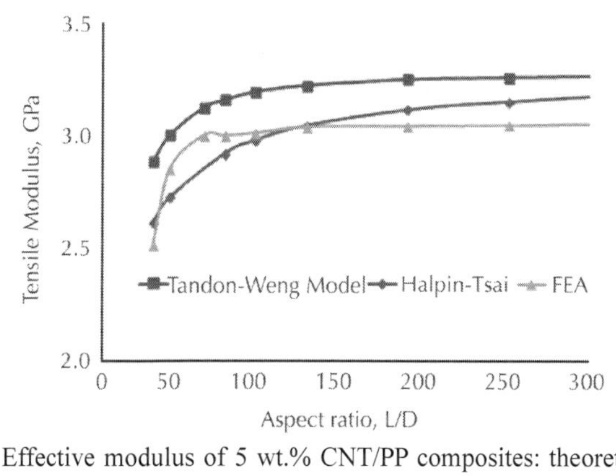

FIGURE A4 Effective modulus of 5 wt.% CNT/PP composites: theoretical models vs. FEA.

For reasons of simplicity and in order to minimize the mesh dependency on the results the hollow CNTs are considered as solid cylinders of circular cross-sectional area with an equivalent average diameter, shown in Fig. A.5calculated by equating the volume of the hollow CNT to the solid one [87].

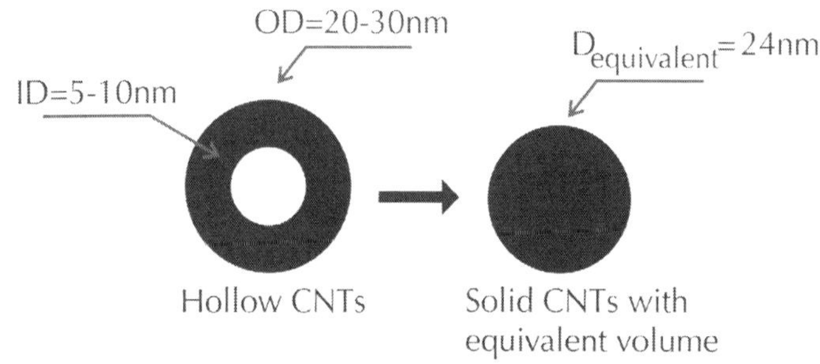

FIGURE A5 Schematic of the CNTs considered for the FEA.

The micromechanical models used for the comparison were Halpin–Tsai (H–T) [89] and Tandon–Weng (T–W) [90] model and the comparison was performed for 5 wt.% CNT/PP. It was noted that the H–T model results to lower modulus compared to FEA because H–T equation does not account for maximum packing fraction and the arrangement of the reinforcement in the composite. A modified H–T model that account for this has been proposed in the literature [91]. The effect of maximum packing fraction and the arrangement of the reinforcement within the composite become less significant at higher aspect ratios [92].

A finite element model of carbon nanotube, interphase and its surrounding polymer is constructed to study the tensile behavior of embedded short carbon nanotubes in polymer matrix in presence of vdW interactions in interphase region by Shokrieh and Rafiee [93]. The interphase is modeled using nonlinear spring elements capturing the force-distance curve of vdW interactions. The constructed model is subjected to tensile loading to extract longitudinal Young's modulus. The obtained results of this work have been compared with the results of previous research of the same authors [94] on long embedded carbon nanotube in polymer matrix. It shows that the capped short carbon nanotubes reinforce polymer matrix less efficient than long CNTs.

Despite the fact that researches have succeeded to grow the length of CNTs up to 4 cm as a world record in US Department of Energy Los Alamos National Laboratory [95] and also there are some evidences on producing CNTs with lengths up to millimeters [96,97], CNTs are commercially available in different lengths ranging from 100 nm to approximately 30 lm in the market based on employed process of growth [98–101]. Chemists at Rice University have identified a chemical process to cut CNTs into short segments [102]. As a consequent, it can be concluded that the SWCNTs with lengths smaller than 1000 nm do not contribute significantly in reinforcing polymer matrix. On the other hand, the efficient length of reinforcement for a CNT with (10, 10) index is about 1.2 lm and short CNT with length of 10.8 lm can play the same role as long CNT reflecting the uppermost value reported in our previous research [94]. Finally, it is shown that the direct use of Halpin–Tsai equation to predict the modulus of SWCNT/composites overestimates the results. It is also observed that application of previously developed long equivalent fiber stiffness [94] is a good candidate to be used in Halpin–Tsai equations instead of Young's modulus of CNT. Halpin–Tsai equation is not an appropriate model for smaller lengths, since there is not any reinforcement at all for very small lengths.

Earlier, a nano-mechanical model has been developed by Chowdhury et al. [103] to calculate the tensile modulus and the tensile strength of randomly oriented short carbon nanotubes (CNTs) reinforced nanocomposites, considering the statistical variations of diameter and length of the CNTs. According to this model, the entire composite is divided into several composite segments which contain CNTs of almost the same diameter and length. The tensile modulus and tensile strength of the composite are then calculated by the weighted sum of the corresponding modulus and strength of each composite segment. The existing micromechanical approach for modeling the short fiber composites is modified to account for the structure of the CNTs, to calculate the modulus and the strength of each segmented CNT reinforced composites. Multi-walled CNTs with and without intertube bridging (see Fig. A6) have been considered. Statistical variations of the diameter and length of the CNTs are modeled by a normal distribution. Simulation results show that CNTs intertube bridging, length and diameter affect the nanocomposites modulus and strength. Simulation results have been compared with the available experimental results and the comparison concludes that the developed model can be effectively used to predict tensile modulus and tensile strength of CNTs reinforced composites.

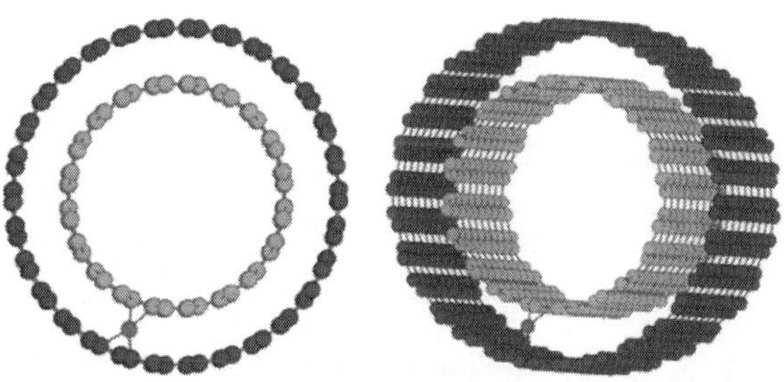

(a) (b)

FIGURE A6. Schematic of MWNT with intertube bridging. (a) Top view and (b) oblique view.

The effective elastic properties of carbon nanotube-reinforced polymers have been evaluated by K.I. Tserpes and A. Chanteli [104] as functions of material and geometrical parameters using a homogenized RVE. The RVE consists of the polymer matrix, a multiwalled carbon nanotube (MWCNT) embedded into the matrix and the interface between them. The parameters considered are the nanotube aspect ratio, the nanotube volume fraction as well as the interface stiffness and thickness. For the MWCNT, both isotropic and orthotropic material properties have been considered. Analyzes have been performed by means of a 3D FE model of the RVE. The results indicate a significant effect of nanotube volume fraction. The effect of nanotube aspect ratio appears mainly at low values and diminishes after the value of 20. The interface mostly affects the effective elastic properties at the transverse direction. Having evaluated the effective elastic properties of the MWCNT–polymer at the microscale, the RVE has been used to predict the tensile modulus of a polystirene specimen reinforced by randomly aligned MWCNTs for which experimental data exist in the literature. A very good agreement is obtained between the predicted and experimental tensile moduli of the specimen. The effect of nanotube alignment on the specimen's tensile modulus has been also examined and found to be significant since as misalignment increases the effective tensile modulus decreases radically. The proposed model can be used for the virtual design and optimization of CNT– polymer composites since it has proven capable of assessing the effects of different material and geometrical parameters on the elastic properties of the composite and predicting the tensile modulus of CNT-reinforced polymer specimens [105–144].

APPENDIX B

The notion "cluster" assumes energy-wise compensated nucleus with a shell, the surface energy of which is rather small, as a result under given conditions the cluster represents a stable formation. Nanocomposites of various shapes can be obtained either by the dispersion of a substance in a definite medium, or condensation or synthesis from low-molecular chemical particles (Fig. B1). Thus, the techniques for obtaining nanostructures can be classified by the mechanism of their formation. Other features, by which the methods for nanostructure production can be classified, comprise the variants of nanoproduct formation process by the change of energy consumption. A temperature or an energy factor is usually evident here. Besides, a so-called apparatus factor plays an important role together with the aforesaid features. At present, nanostructural "formations" of various shapes and compositions are obtained in a rather wide region of actions upon the chemical particles and substances.

In chemical literature a cluster is equated with a complex compound containing a nucleus and a shell. Usually a nucleus consists of metal atoms combined with metallic bond, and a shell of ligands. Manganese carbonyls $[(Co)_5 MnMn(Co)_5]$ and cobalt carbonyls $[Co_6(Co)_{18}]$, nickel pentadienyls $[Ni_6(C_5H_6)_6]$ belong to elementary clusters.

In recent years, the notion "cluster" has got an extended meaning. At the same time the nucleus can contain not only metals or not even contain metals. In some clusters, for instance, carbon ones there is no nucleus at all. In this case their shape can be characterized as a sphere (icosahedron, to be precise)—fullerenes, or as a cylinder—fullerene tubules. Surely a certain force field is formed by atoms on internal walls inside such particles. It can be assumed that electrostatic, electromagnetic, and gravitation fields conditioned by corresponding properties of atoms contained in particle shells can be formed inside tubules and fullerenes. If analyze papers recently published, it should be noted that a considerable exceeding of surface size over the volume and, consequently, a relative growth of the surface energy in comparison with the growth of volume and potential energy is the main feature of clusters. If particle dimensions (diameters of "tubes" and "spheres") change from 1 up to several hundred nanometers, they would be called nanoparticles. In some papers, the area of nanoclusters existence is within 1–10 nm.

Based on classical definitions, in given paper, metal nanoparticles and nanocrystals are referred to as nanoclusters. Apparently, the difference of nanoparticles from other particles (smaller or larger) is determined by their specific characteristics. The search of nanoworld distinctions from atomic-molecular, micro and macroworld can lead to finding analogies and coincidences in colloid chemistry, chemistry of polymers, and coordination compounds. Firstly, it should be noted that nanoparticles usually represent a small collective aggregation of atoms being within the action of adjacent atoms, thus conditioning the shape of nanoparticles. A nanoparticle shape can vary depending upon the nature of adjacent atoms and character of formation medium. Obviously, the properties of separate atoms and molecules (of small size) are determined by their energy and geometry characteristics, the determinative role being played by electron properties. In particular, electron interactions determine the geometry of molecules and atomic structures of small size and mobility of these chemical particles in media, as well as their activity or reactivity.

When the number of atoms in chemical particle exceeds 30, a certain stabilization of its shape being also conditioned by collective influence of atoms constituting the particle is observed. Simultaneously, the activity of such a particle remains high but the processes with its participation have a directional character. The character of interactions with the surroundings of such structures is determined by their formation mechanism.

During polymerization or copolymerization the influence of macromolecule growth parameters changes with the increase of the number of elementary acts of its growth. According to scientists, after 7–10 acts the shape or geometry of nanoparticles formed becomes the main determinative factor providing the further growth of macromolecule (chain development). A nanoparticle shape is usually determined not only by its structural elements but also by its interactions with surrounding chemical particles.

Due to the overlapping of different classification features, it is appropriate to present a set of diagrams by main features. For instance, the methods for nanoparticle formation by substance dispersion and chemical particle condensation can be identified with physical and chemical methods, though such decision is incorrect, since substance destruction methods can contain both chemical and physical impacts. In turn, when complex nanostructures are formed from simple ones, both purely physical and chemical factors are possible. However, in the process of substance dispersion high-energy sources, such as electric arc, laser pyrolysis, plasma sources, mechanical crushing or grinding should be applied.

From the aforesaid, it can be concluded that the possibility of self-organization of nanoparticles with the formation of corresponding nanosystems and nanomaterials is the main distinction of nanoworld from pico, micro, and macroworld. Recently, much attention has been paid to synergetics or the branch of science dealing with self-organization processes since these processes, in many cases, proceeds with small energy consumption and, consequently, is more ecologically clear in comparison with existing technological processes.

In turn, nanoparticle dimensions are determined by its formation conditions. When the energy consumed for macroparticle destruction or dispersion over the surface increases, the dimensions of nanomaterials are more likely to decrease. The notion "nanomaterial" is not strictly defined. Several researchers consider nanomaterials to be aggregations of nanocrystals, nanotubes, or fullerenes. Simultaneously, there is a lot of information available that nanomaterials can represent materials containing various nanostructures. The most attention researchers pay to metallic nanocrystals. Special attention is paid to metallic nanowires and nanofibers with different compositions.

Here are some names of nanostructures:

(1) fullerenes, (2) gigantic fullerenes, (3) fullerenes filled with metal ions, (4) fullerenes containing metallic nucleus and carbon (or mineral) shell, (5) one-layer nanotubes, (6) multilayer nanotubes, (7) fullerene tubules, (8) "scrolls", (9) conic nanotubes, (10) metal-containing tubules, (11) "onions", (12) "Russian dolls", (13) bamboo-like tubules, (14) "beads", (15) welded nanotubes, (16) bunches of nanotubes, (17) nanowires, (18) nanofibers, (19) nanoropes, 20) nanosemispheres (nanocups), (21) nanobands and similar nanostructures, as well as various derivatives from enlisted structures. It is quite possible that a set of such structures and notions will be enriched.

In most cases nanoparticles obtained are bodies of rotation or contain parts of bodies of rotation. In natural environment there are minerals containing fullerenes or representing thread-like formations comprizing nanometer pores or structures. In the first case, it is talked about schungite that is available in quartz rock in unique deposit in Prionezhje. Similar mineral can also be found in the river Lena basin, but it consists of micro and macrodimensional cones, spheroids, and complex fibers. In the second case, it is talked about kerite from pegmatite on Volyn (Ukraine) that consists of polycrystalline fibers, spheres, and spirals mostly of micron dimensions, or fibrous vetcillite from the state of Utah (USA); globular anthraxolite and asphaltite.

Diameters of some internal channels are up to 20–50 nm. Such channels can be of interest as nanoreactors for the synthesis of organic, carbon, and polymeric substances with relatively low energy consumption. In case of directed location of internal channels in such matrixes and their inner-combinations the spatial structures of certain purpose can be created. Terminology in the field of nanosystems existence is still being developed, but it is already clear that nanoscience obtains qualitatively new knowledge that can find wide application in various areas of human practice thus, significantly decreasing the danger of people's activities for themselves and environment.

The system classification by dimensional factor is known, based on which we consider the following:

- microobjects and microparticles 10^{-6}–10^{-3} m in size;
- nanoobjects and nanoparticles 10^{-9}–10^{-6} m in size;
- picoobjects and picoparticles 10^{-12}–10^{-9} m in size.

Assuming that nanoparticle vibration energies correlate with their dimensions and comparing this energy with the corresponding region of electromagnetic waves, we can assert that energy action of nanostructures is within the energy region of chemical reactions. System self-organization refers to synergetics. Quite often, especially recently, the papers are published, for example, it is considered that nanotechnology is based on self-organization of metastable systems. As assumed, self-organization can proceed by dissipative (synergetic) and continual (conservative) mechanisms. Simultaneously, the system can be arranged due to the formation of new stable ("strengthening") phases or due to the growth provision of the existing basic phase. This phenomenon underlies the arising nanochemistry. Below is one of the possible definitions of nanochemistry.

Nanochemistry is a science investigating nanostructures and nanosystems in metastable ("transition") states and processes flowing with them in near-"transition" state or in "transition" state with low activation energies.

To carry out the processes based on the notions of nanochemistry, the directed energy action on the system is required, with the help of chemical particle field as well, for the transition from the prepared near-"transition" state into the process product state (in our case–into nanostructures or nanocomposites). The perspective area of nanochemistry is the chemistry in nanoreactors. Nanoreactors can be compared with specific nanostructures representing limited space regions in which chemical particles orientate creating "transition state" prior to the formation of the desired nanoproduct. Nanoreactors have a definite activity, which predetermines the creation of the corresponding product. When nanosized particles are formed in nanoreactors, their shape and dimensions can be the reflection of shape and dimensions of the nanoreactor.

In the last years a lot of scientific information in the field of nanotechnology and science of nanomaterials appeared. Scientists defined the interval from 1 to 1000 nm as the area of nanostructure existence, the main feature of which is to regulate the system self-organization processes. However, later some scientists limited the upper threshold at 100 nm. At the same time, it was not well substantiated. Now many nanostructures varying in shapes and sizes are known. These nanostructures have sizes that fit into the interval, determined by Smally, and are active in the processes of self-organization, and also demonstrate specific properties.

Problems of nanostructure activity and the influence of nanostructure super small quantities on the active media structural changes are explained.

The molecular nanotechnology ideology is analyzed. In accordance with the development tendencies in self-organizing systems under the influence of nanosized excitations the reasons for the generation of self-organization in the range 10^{-6}–10^{-9} m should be determined.

Based on the law of energy conservation the energy of nanoparticle field and electromagnetic waves in the range 1–1000 nm can transfer, thus corresponding to the range of energy change from soft X-ray to near IR radiation. This is the range of energies of chemical reactions and self-organization (structuring) of systems connected with them.

Apparently the wavelengths of nanoparticle oscillations near the equilibrium state are close or correspond to their sizes. Then based on the concepts of ideologists of nanotechnology in material science the definition of nanotechnology can be as follows:

Nanotechnology is a combination of knowledge in the ways and means of conducting processes based on the phenomenon of nano-sized system self-organization and utilization of internal capabilities of the systems that results in decreasing the energy consumption required for obtaining the targeted product while preserving the ecological cleanness of the process.

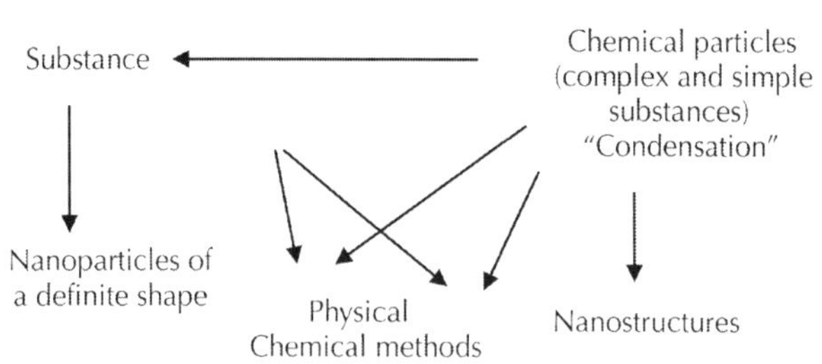

FIGURE B1 Classification diagram of nanostructure formation techniques by the features "dispersion" and "condensation".

At the same time, the conceptions "dispersion" and "condensation" are conditional and can be explained in various ways (Figs. B2 and B3). Among the physical methods of "dispersion" high-temperature methods and methods with relatively low temperatures or high-energy and low-energy ones are distinguished.

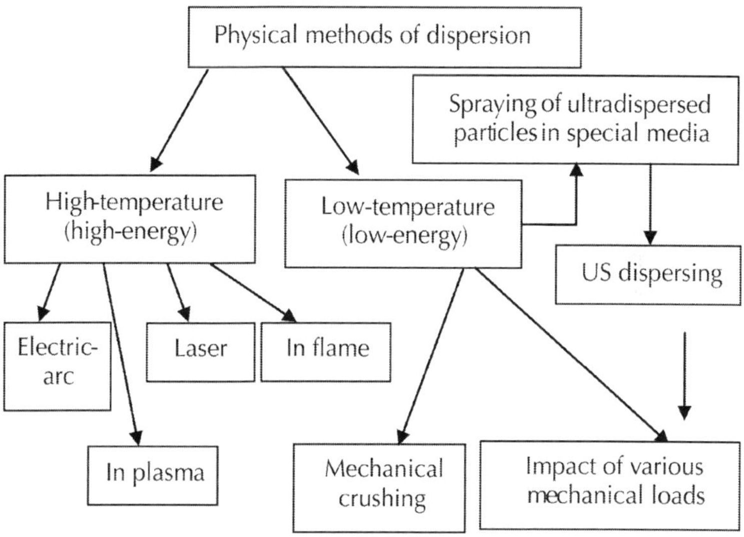

FIGURE B2 Classification diagram of physical methods of dispersion.

Considerably fewer chemical methods of dispersion are known, though, in the physical methods listed chemical processes are surely present, since it is difficult to imagine spraying and grinding of a substance without chemical reactions of destruction. Therefore, it is more appropriate to speak about physical methods of impact upon the substance that lead to their dispersion (decomposition, destruction at high temperatures, radiations or mechanical loads). Then, chemical methods are identified with methods of impact upon the substances of chemical particles and media.

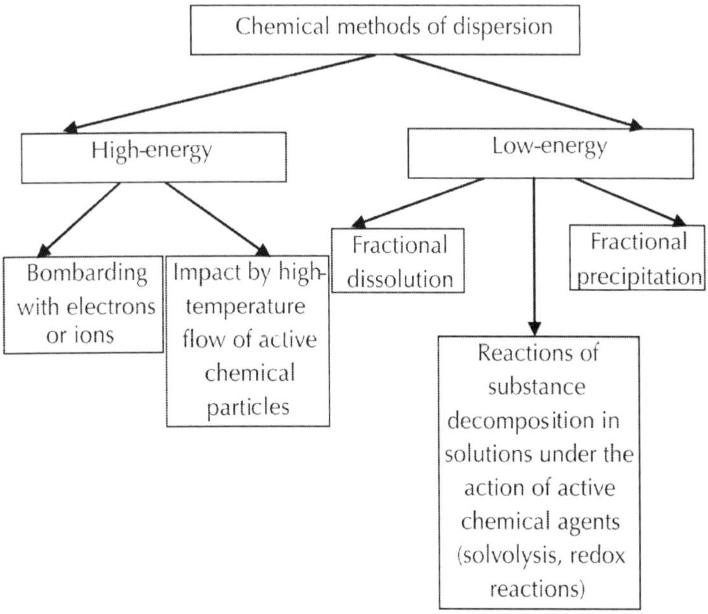

FIGURE B3 Classification diagram of dispersion chemical methods.

Basically nanostructures are obtained from picosized chemical particles by means of chemical or physical-chemical techniques, in which the activity of a chemical particle but not its activation during "condensation" is determinant (Fig. B4). Therefore, the classification diagram of chemical techniques for nanostructure formation under the action of chemical media can be given as follows:

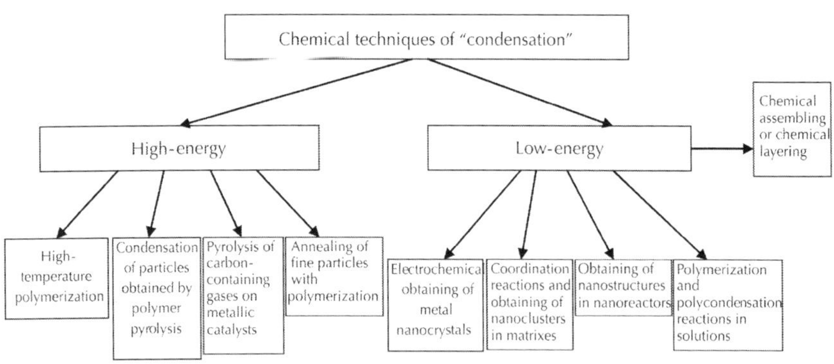

FIGURE B4 Classification diagram of chemical techniques of "condensation".

Actually, such diagrams are approximate in the same way as conditional are separate points of separation and difference between them. Several techniques can be referred to combined or physical methods of impact upon active chemical particles. For instance, CVD method comprises high-energy technique that leads to the formation of gaseous phase that can be attributed to chemical methods of dispersion, and then active chemical particles formed during the pyrolysis "transform" into nanostructures. The polymerization processes of gaseous phase particles are carried out by means of probe technological stations, and this can be referred to the physical techniques of "condensation." So, under the physical methods of formation of various nanostructures, including nanocrystals, nanoclusters, fullerenes, nanotubes, and so on, it means the techniques in which physical impact results in the formation of nanostructures from active picosized chemical particles (Fig. B5).

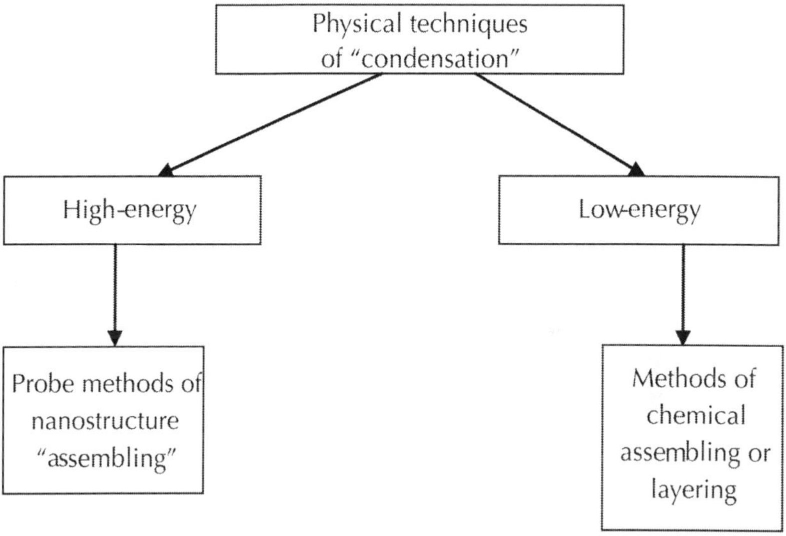

FIGURE B5 Physical impact results in the formation of nanostructures from active picosized chemical particles.

The proposed classification does not reflect multivicissitude of methods for nanostructure formation. Usually nanostructure production comprises the following:

1. preparation of "embryos" or precursors of nanostructures;
2. production of nanostructures;
3. isolation and refining of nanostructures of a definite shape.

The production of nanostructures by various methods, including mechanical ones, for instance, extrusion, grinding and similar operations can proceed in several stages.

Mechanical crushing, combined methods for grinding in media and influence of action power and medium, where the substance is crushed and sprayed, upon the size and shape of nanostructures formed will be discussed in this chapter. At the same time, multistepped combined methods for obtaining nanostructures with different shapes and sizes have been applied in recent years. Ways of substance dispersion by mechanical methods till nanoproducts are obtained (fine powders or ultradispersed particles) can be given in the following diagram (Fig. B6):

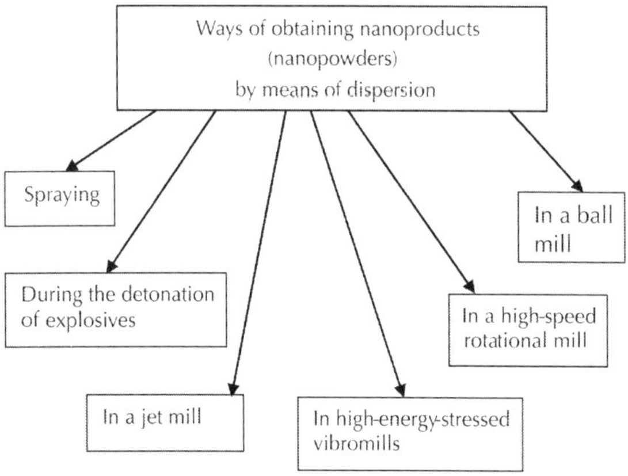

FIGURE B6 Diagram of classification of ways for obtaining powder-like nanoproducts by means of dispersion.

One of the most widely applied methods used for grinding different substances is the crushing and grinding in mills: ball, rod, colloid combined with the action of shock force, abrasion force, centrifugal force and vibration upon the materials. Since, initially, carbon nanostructures, fullerenes and nanotubes were extracted from the carbon dust obtained in electrical charge, the possibility of the formation of corresponding nanostructures when using conventional mechanical methods of action upon substances seems doubtful. However, the investigation of the products obtained after graphite crushing in ball mills allows making the conclusion that the method proposed can be quite competitive high-temperature way for forming nanostructures when the mechanical power provides practically the same conditions as the heat. In this case, it is difficult to imagine a directed action of the combined forces upon the substance. However, at the set speed of the mill drum the directed action of the "stream" of steel balls upon the material can be predicted. As an example, let us give the description of one of such processes during graphite grinding. Carbon nanostructures resembling "onions" by their shape are obtained when grinding graphite in ball mills. The product is ground in planetary ball mill in the atmosphere of pure argon. Mass ratio of steel balls to the powder of pure graphite equals 40. The rotation speed of the drum is 270 rpm. The grinding time changed from 150 to 250 h. It was observed that after 150 h of grinding the nanoproduct obtained resembles by characteristics and appearance the nanoparticles obtained in electric-arc method.

Iron-containing nanostructures were also obtained in planetary ball mills after milling the iron powders in heptane adding oleic acid.

The distinctive feature of the considered technique for producing metal/carbon nanocomposites is a wide application of independent, modern, experimental and theoretical analysis methods to substantiate the proposed technique and investigation of the composites obtained (quantum-chemical calculations, methods of transmission electron microscopy and electron diffraction, method of X-ray photoelectron spectroscopy, X-ray phase analysis, etc.). The technique developed allows synthesizing a wide range of metal/carbon nanocomposites by composition, size and morphology depending on the process conditions. In its application it is possible to use secondary metallurgical and polymer raw materials. Thus, the nanocomposite structure can be adjusted to extend the function of its application without prefunctionalization. Controlling the sizes and shapes of nanostructures by changing the metal-containing phase, to some extent, apply completely new, practicable properties to the materials which sufficiently differ from conventional materials.

The essence of the method consists in coordination interaction of functional groups of polymer and compounds of 3d-metals as a result of grinding of metal-containing and polymer phases Further, the composition obtained undergoes thermolysis following the temperature mode set with the help of thermogravimetric and differential thermal analyzes. At the same time, one observes the polymer carbonization, partial or complete reduction of metal compounds and structuring of carbon material in the form of nanostructures with different shapes and sizes.

Metal/carbon nanocomposite (Me/C) represents metal nanoparticles stabilized in carbon nanofilm structures. In turn, nanofilm structures are formed with carbon amorphous nanofibers associated with metal containing phase. As a result of stabilization and association of metal nanoparticles with carbon phase, the metal chemically active particles are stable in the air and during heating as the strong complex of metal nanoparticles with carbon material matrix is formed.

APPENDIX C

The propagation of ultrashort optical pulses in the array of carbon nanotubes was investigated. The electromagnetic field is considered based on the Maxwell's equations, and the electronic system of carbon nanotubes based on the Boltzmann kinetic equation in the relaxation time approximation. We construct the distribution of the pulse intensity and observed the periodic effect of a partition of pulse maximum into two maxima, and their subsequent merger into a single maximum. The objective of this appendix is to consider the problem of three-dimensional propagation of ultrashort optical pulse in the array of carbon nanotubes. The rapid development of laser and optical technologies, the application of powerful lasers and the unique precision of optical measurements allowed substantial progress in the study of nonlinear phenomena that occur in a wide range of substances_with a very different physical properties. This is due both to the rapid progress of computer technology, and with interest in modern physics the study of nonlinear dynamic processes. The object of the study should be as a substance with pronounced nonlinear properties and important from the viewpoint of practical applications. In the last decade, as a matter of researchers are increasingly attracted to carbon nanotubes (CNTs) are unique macromolecular systems. Nanometer diameter and micron length of CNTs make them attractive for use in nano- and microelectronics, as they allow them to be considered the closest in structure to the ideal one-dimensional systems. The study of optical soliton propagation in CNTs is one of the most promising areas of research. Although these papers were predicted by the existence possibility of electromagnetic solutions and dependence of their characteristics on the parameters of CNTs remained a series of questions requiring clarification. This question related to output beyond the one-dimensional approximation, and the question of the propagation of optical pulses with light diffraction, and the scattering problem of the optical pulse in the multidimensional case the inhomogeneities of various types. These structures are localized in three spatial dimensions, will be called "light bullets", because in this problem localization occurs in three dimensions. All these factors make the problem of the study of nonlinear dynamic processes in CNTs relevant for theory and in practice.

C1 MODEL AND BASIC RELATIONS

We assume in the model that the electric field vector $\mathbf{E}(x, y, z, t)$ directed along the tube axis x, and the electromagnetic wave is moving in the transverse direction (Fig. 1). Carbon nanotubes are considered to be ideal to have a structural modification of the "zigzag" and located on the same interatomic distances equal to 0.34 nm [3] to simplify the calculations. Method of packing CNTs in the array is not important, because it does not take into account the interaction between the CNTs, and therefore used a tetragonal packing of single-walled carbon nanotubes in two-dimensional array with space group P42/mc(D_{4h}^{9}) in Fig. C1.

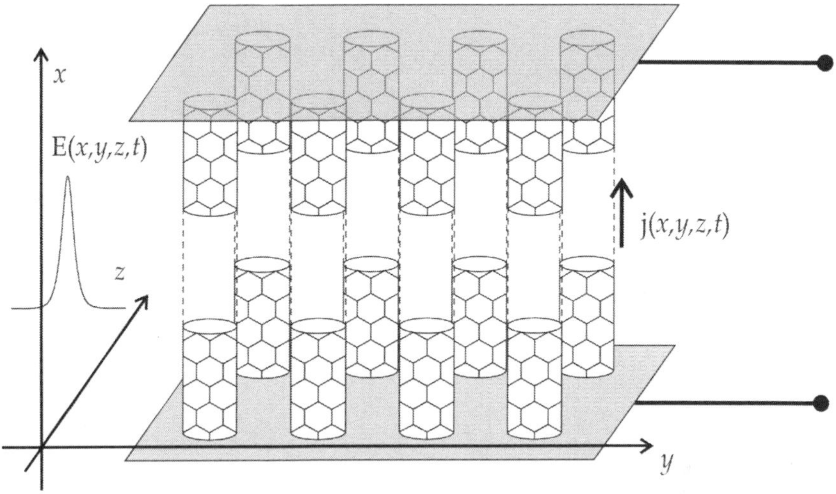

FIGURE C1 The geometry of the problem. Ultrashort optical pulse with the electric field **E**(x, y, z, t) is directed along the tube axis x, moving in the transverse direction along the axis z.

In this appendix, we ignore the influence of the substrate on which the nanotubes are grown. This can lead particularly to the appearance of surface waves of these system, we will not considered these features in such article.

Electronic structure of "zigzag" carbon nanotubes is characterized by chiral indices (n, 0) and described by the well-known dispersion relation which obtained in the framework of Huckel π-electron approximation:

$$E(\mathbf{p}) = \pm\gamma\sqrt{1+4\cos\left(ap_x\right)\cos\left(\pi s/n\right)+4\cos^2\left(\pi s/n\right)} \tag{27}$$

where $a = 3d/2\hbar$, $d = 0.142$ nm is the distance between adjacent carbon atoms in graphene, $\mathbf{p} = (p_x, s)$ is the quasimomentum of the electrons in graphene, p_x is the parallel component of the graphene sheet of the quasimomentum and $s = 1, 2, \ldots, n$ are the quantization numbers of the momentum components depending on the width of the graphene ribbon. Different signs are related to the conductivity band and to the valence band accordingly.

The electromagnetic field pulse is described classically on the basis of Maxwell's equations. We are considering the dielectric and magnetic properties of CNTs and two-dimensionality of the problem the Maxwell equations for the vector potential **A** in the gauge $\mathbf{E} = -\dfrac{1}{c}\dfrac{\partial \mathbf{A}}{\partial t}$ will be:

$$\frac{\partial^2 \mathbf{A}}{\partial x^2}+\frac{\partial^2 \mathbf{A}}{\partial y^2}+\frac{\partial^2 \mathbf{A}}{\partial z^2}-\frac{1}{c}\frac{\partial^2 \mathbf{A}}{\partial t^2}=-\frac{4\pi}{c}\mathbf{j} \tag{28}$$

We ignore the diffraction spreading of the laser beam in the x-direction in Eq. (28). The vector potential **A** is chosen as $\mathbf{A} = (Ax\ (x, y, z, t),\ 0,\ 0)$. Using the apparatus shown in Refs. [4–9], Eq. (27), we obtain an effective equation for the vector potential:

$$\frac{\partial^2 A_z}{\partial x^2}+\frac{\partial^2 A_z}{\partial y^2}+\frac{\partial^2 A_z}{\partial z^2}-\frac{1}{c}\frac{\partial^2 A_z}{\partial t^2}+\frac{q}{\pi\hbar}\sum_m c_m \sin\left(\frac{maq}{c}A_z(t)\right)=0 \tag{29}$$

where F0 is the equilibrium Fermi distribution function, q is the elementary electron charge, q0 is the pulse on the boundary of the Brillouin zone $q_0 = \frac{2\pi\hbar}{3b}$. The coefficients a_{ms} are determined by Fourier series expansion of velocity component $v_x\ (s, p)$ obtained from (1) as $v_x = \partial E(p)/\partial p_x$.

These simplifications have been made: (a) interband transitions are not counted. This assumption imposes a restriction on the maximum frequency of laser pulses, which for the CNT is in the near infrared. (b) the solution for the field component $A_x(t)$ is in the class of rapidly decreasing functions. (c) the relaxation time τ is large enough for the common durations of ultrashort laser pulses.

Equation (3) were solved numerically using the direct difference scheme “cross”. Steps to time and coordinate were determined from the standard conditions of stability and decreased successively in 2 times as long as the solution is not changed in the eighth digit after dot. These calculations for the one-dimensional case are presented in Refs. [4–9]. The evolution of the initial conditions considered in time (the variable t). The initial pulse profile has a Gaussian form:

$$E(x,y,z,t)=E_{\max}\exp\left(-\frac{(x-x_0)^2}{\mu_x}\right)\exp\left(-\frac{(y-y_0)^2}{\mu_y}\right)\exp\left(-\frac{(z-z_0)^2}{\mu_z}\right) \tag{30}$$

where E_{max} is the maximum field amplitude, x_0 is the coordinate of the maximum intensity the axis of x, y_0 is the coordinate of the maximum intensity the axis of y, z_0 is the coordinate of the maximum intensity the axis of z, μ_x, μ_y and μ_z are the parameters defining pulse width to the coordinates x, y and z respectively.

This appendix focused on the results of the impulse, which localized in three spatial dimensions. In the evolution of ultrashort pulse periodically split into two pulses and then combined into one, while the amplitude of the pulses is changed. Figures C2 and C3 shows the results of numerical simulation of the pulse evolution in the three-dimensional CNTs array such as (8.0) and (20.0), respectively.

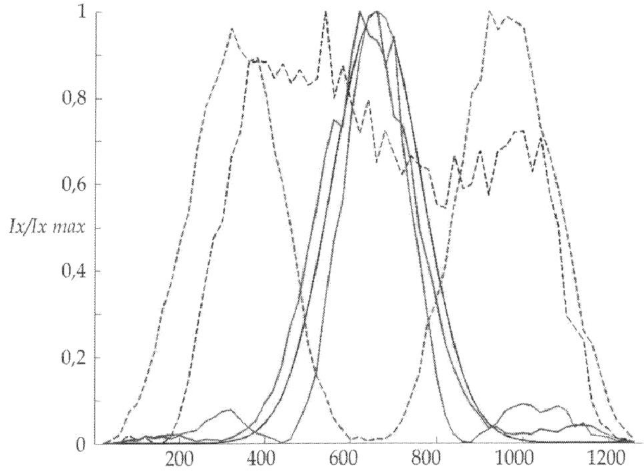

FIGURE C2 Profiles of the intensities I_x/I_{xmax} bullets in the cross section parallel to the plane *ZOY* and passing through a maximum pulse propagating in the array of CNTs such as (8.0). The ordinate shows the ratio I_x/I_{xmax}. Type of pulse: 1 – initial pulse, 2 – pulse after $t = 0.8 \cdot 10^{-12} s$, 3 – pulse after $t = 1.6 \cdot 10^{-12} s$, 4 – pulse after $t = 2.4 \cdot 10^{-12} s$, 5 – pulse after $t = 2.9 \cdot 10^{-12} s$.

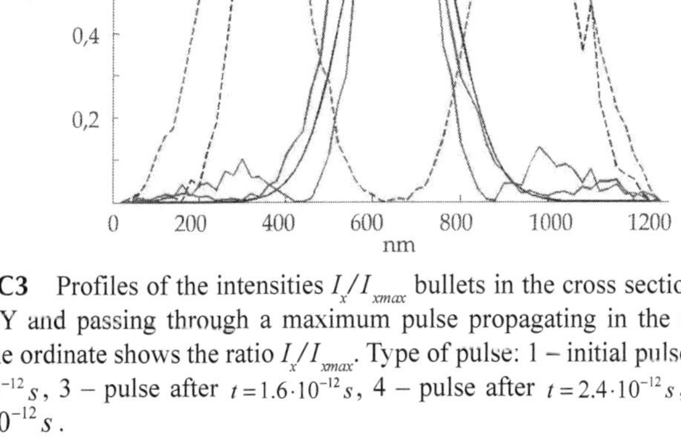

FIGURE C3 Profiles of the intensities I_x/I_{xmax} bullets in the cross section parallel to the plane ZOY and passing through a maximum pulse propagating in the array CNT type (20.0). The ordinate shows the ratio I_x/I_{xmax}. Type of pulse: 1 – initial pulse, 2 – pulse after $t = 0.8 \cdot 10^{-12} s$, 3 – pulse after $t = 1.6 \cdot 10^{-12} s$, 4 – pulse after $t = 2.4 \cdot 10^{-12} s$, 5 – pulse after $t = 2.9 \cdot 10^{-12} s$.

Figures C2 and C3 shows the distribution of intensities $I_x(x, y, z, t) = E_x^2(x, y, z, t)$ at different times. Analysis of the data graphs show that the stable propagation of pulses in the array of CNTs is possible, which are localized in three dimensions. These pulses are often called in literature as "light bullets". The interesting effect of a periodic partition of pulse maximum into two maxima, and their subsequent merger into a single maximum was discovered. The pulse shape varies significantly during the evolution of the pulse. We consider that this effect is akin to the dynamics of internal modes and consists in the excitation of internal vibration modes of the "light bullet" when passing a pulse through the array of CNTs, which leads to periodic separation of the pulse maximum and their interference with the original "bullet." This fact suggests that there may be bumeron analogs in the other highly nonlinear three-dimensional systems.

1. The model and the effective equation describing the dynamics of an ultrashort pulse laser beams in the CNT are obtained. The approximations, which used in this model were formulated.
2. Numerical calculations showed that in three-dimensional case are possible stable nonlinear waves that are localized in three directions of light pulses, which are analogs of "light bullets".
3. The periodic separation of peak pulse into two pulses and subsequent integration occur in the case of scattering when light pulse passing through the array of CNTs. The pulse shape at the same time varies considerably.

KEYWORDS

- **domino effect**
- **Hagen-Poiseuille**
- **light bullets**
- **nanogun**
- **nanostructures architecture**
- **quasi-solid phase**

REFERENCES

1. Iijima, S.; Helical microtubules of graphitic carbon. Nature **1991**, *354*, 6348, 56.
2. Dresselhaus, M. S.; Dresselhaus, G.; Eklund, P. C.;Science of fullerenes and carbon nanotubes. New York: Academic Press; **1996**.

3. Saito, R.; Dresselhaus, G.; Dresselhaus, M. S.; Physical properties of carbon nanotubes. Imperial College Press; **1998**.
4. Harris, P.; J. F.; Carbon nanotubes and related structures: new materials for the 20-first century. Cambridge, United Kingdom: Cambridge University Press; **1999,** 279.
5. Wagner, H. D.; Vaia, R. A.; Nanocomposites: issues at the interface. Mater Today **2004**, *7(11),* 38.
6. Zeng, Q. H.; Yu, A. B.; Lu, M. G. Q. ultiscale modeling and simulation of polymer nanocomposites, Prog. Polym. Sci. **2008,** *33,* 191–269.
7. Iijima, S.; Helical microtubules of graphitic carbon. Nature **1991**, *354,* 56–58.
8. Bethune, D. S.; Klang, C. H.; De Vries, M. S.; et al. Cobalt-catalyzed growth of carbon nanotubes with single-atomic-layer walls. Nature **1993**, *363,* 605–607.
9. Dresselhaus, M. S.; Dresselhaus, G.; Saito, R.; Physics of carbon nanotubes. Carbon **1995,** *33,* 883–891.
10. Thostenson, E. T.; Ren, Z. F.; Chou, T. W.; Advances in the science and technology of CNTs and their composites: a review. Compos Sci Technol **2001,** *61,* 1899–912.
11. Yakobson, B. I.; Avouris, P.; Mechanical properties of carbon nanotubes. Top Appl Phys **2001,** *80,* 287–327.
12. Ajayan, P. M.; Schadler, L. S.; Braun, P. V.; Nanocomposite science and technology. Weinheim: Wiley-VCH; **2003,** p. 77–80.
13. Li J, Ma, P.; C, Chow, W. S.; To C. K, Tang, B. Z.; Kim J. K.; Correlations between percolation threshold, dispersion state and aspect ratio of carbon nanotube, Adv Funct Mater **2007,** *17,* 3207–15.
14. Ajayan, P. M.; Stephan, O.; Colliex, C.; Trauth, D.; Aligned carbon nanotube arrays formed by cutting a polymer resin-nanotube composite. Science **1994,** *265,* 1212–4.
15. Summary of Searching Results. <http: //www.scopus.com> (accessed January 2010).
16. Du, J. H. ; Bai, J.; Cheng, H. M.;The present status and key problems of carbon nanotube based polymer composites. Express Polym Lett **2007,** *1,* 253–73.
17. Lee, J. Y.; Baljon, A. R. C.; Loring, R. F.; Panagiotopoulos, A. Z.; Simulation of polymer melt intercalation in layered nanocomposites. J Chem Phys **1998,** *109,* 10321–30.
18. Smith, G. D.; Bedrov, D.; Li, L.; W, Byutner, O.; A molecular dynamics simulation study of the viscoelastic properties of polymer nanocomposites. J Chem Phys **2002,** *117,* 9478–89.
19. Smith, J. S.; Bedrov, D.; Smith, G. D. A molecular dynamics simulation study of nanoparticle interactions in a model polymer–nanoparticle composite. Compos Sci Technol **2003,** *63,* 1599–605.
20. Zeng, Q. H.; Yu, A. B. ; Lu, G. Q.; Standish, R. K.; Molecular dynamics simulation of organic-inorganic nanocomposites: layering behavior and interlayer structure of organoclays. Chem Mater **2003,** *15,* 4732–8.
21. Vacatello, M.; Predicting the molecular arrangements in polymer-based nanocomposites. Macromol Theory Simul **2003,** *12,* 86–91.
22. Zeng, Q. H.; Yu, A. B.; Lu, G. Q.; Interfacial interactions and structure of polyurethane intercalated nanocomposite. Nanotechnology **2005,** *16,* 2757–63.
23. Allen, M. P.; Tildesley, D. J.; Computer simulation of liquids.Oxford: Clarendon Press; 1989.

24. Frenkel, D.; Smit, B.; Understanding molecular simulation: from algorithms to applications. 2nd Ed. San Diego: Academic Press; 2002.
25. Metropolis, N.; Rosenbluth, A. W.; Marshall, N.; Rosenbluth, M. N.; Teller, A. T.;Equation of state calculations by fast computing machines. J Chem Phys **1953,** *21,* 1087–92.
26. Carmesin, I.; Kremer, K.; The bond fluctuation method: a new effective algorithm for the dynamics of polymers in all spatial dimensions. Macromolecules **1988,** *21,* 2819–23.
27. Hoogerbrugge, P. J.; Koelman, J. M. V. A.; Simulating microscopic hydrodynamic phenomena with dissipative particle dynamics. Europhys Lett **1992,** *19,* 155–60.
28. Gibson, J. B.; Chen, K.; Chynoweth, S.; Simulation of particle adsorption onto a polymer-coated surface using the dissipative particle dynamics method. J Colloid Interface Sci **1998,** *206,* 464–74.
29. Dzwinel V.; Yuen, D. A.; A two-level, discrete particleapproach for large-scale simulation of colloidal aggregates. Int J Mod Phys C **2000,** *11,* 1037–61.
30. Dzwinel, W.; Yuen, D. A.; A two-level, discrete-particle approach for simulating ordered colloidal structures. J Colloid Interface Sci **2000,** *225,* 179–90.
31. Chen, S.; Doolen, G. D.; Lattice Boltzmann method for fluid flows. Annu Rev Fluid Mech **1998,** *30,* 329–64.
32. Cahn, J. W.; On spinodal decomposition. Acta Metall **1961,** *9,* 795–801.
33. Cahn, J. W.; Hilliard, J. E.; Spinodal decomposition: a reprise. Acta Metall **1971,** *19,* 151–61.
34. Cahn, J. W.; Free energy of a nonuniform system. II.; Thermodynamic basis. J Chem Phys **1959,** *30,* 1121–4.
35. Lee, B. P.; Douglas, J. F.; Glotzer, S. C.; Filler-induced composition waves in phase-separating polymer blends. Phys Rev E **1999,** *60,* 5812–22.
36. Ginzburg, V. V.; Qiu, F.; Paniconi, M.; Peng, G. W.; Jasnow, D.; Balazs, A. C.; Simulation of hard particles in a phaseseparating binary mixture. Phys Rev Lett **1999,** *82,* 4026–9.
37. Qiu, F.; Ginzburg, V. V.; Paniconi, M.; Peng, G. W.; Jasnow, D.; Balazs, A. C.; Phase separation under shear of binary mixtures containing hard particles. Langmuir **1999,** *15,* 4952–6.
38. Ginzburg, V. V.; Gibbons, C.; Qiu, F.; Peng, G. W.; Balazs, A. C.; Modeling the dynamic behavior of diblock copolymer/ particle composites. Macromolecules **2000,** *33,* 6140–7.
39. Ginzburg, V. V.; Qiu, F.; Balazs, A. C.; Three-dimensional simulations of diblock copolymer/particle composites.Polymer **2002,** *43,* 461–6.
40. He G, Balazs, A. C.; Modeling the dynamic behavior of mixtures of diblock copolymers and dipolar nanoparticles.J Comput Theor Nanosci **2005,** *2,* 99–107.
41. Altevogt, P.; Ever, O. A.; Fraaije J. G. E. M.; Maurits, N. M.; van Vlimmeren, B. A. C.; The MesoDyn project: software for mesoscale chemical engineering. J Mol Struct **1999,** *463,* 139–43.
42. Kawakatsu, T.; Doi, M.; Hasegawa, A.; Dynamic density functional approach to phase separation dynamics of polymer systems. Int J Mod Phys C **1999,** *10,* 1531–40.

43. Morita, H.; Kawakatsu, T.; Doi, M.; Dynamic density functional study on the structure of thin polymer blend films with a free surface. Macromolecules **2001,** *34,* 8777–83.
44. Doi, M.; OCTA-a free and open platform and softwares of multiscale simulation for soft materials /http://octa.jp/S.2002.
45. Tandon, G. P.; Weng, G. J.; The effect of aspect ratio of inclusions on the elastic properties of unidirectionally aligned composites. Polym Compos **1984,** *5,* 327–33.
46. Fornes, T. D.; Paul, D. R.; Modeling properties of nylon 6/clay nanocomposites using composite theories. Polymer **2003,** *44,* 4993–5013.
47. Odegard, G. M.; Pipes, R. B.; Hubert, P.; Comparison of two models of SWCN polymer composites. Compos Sci Technol **2004,** *64,* 1011–20.
48. Odegard, G. M.; Gates, T. S.; Wise, K. E.; Park, C.; Siochi, E. J.; Constitutive modeling of nanotube-reinforced polymer composites. Compos Sci Technol **2003,** *63,* 1671–87.
49. Pipes, R. B.; Hubert, P.; Helical carbon nanotube arrays: mechanical properties. Compos Sci Technol **2002,** *62,* 419–28.
50. Rudd, R. E.; Broughton, J. Q.; Concurrent coupling of length scales in solid state systems. Phys Stat Sol B **2000,** *217,* 251–91.
51. Starr, F. W.; Glotzer, S. C.; Simulations of filled polymers on multiple length scales, In: Nakatani, A. I.; Hjelm, R. P.; Gerspacher, M.; Krishnamoorti, R.; editors. Filled and nanocomposite polymer materials, Materials research symposium proceedings. Warrendale: Materials Research Society, **2001,** p. KK4.1.1–KK4.1.13.
52. Glotzer, S. C.; Starr, F. W.; Towards multiscale simulations of filled and nanofilled polymers, In: Cummings, P. T.; Westmoreland, P. R.; Carnahan, B.; editors. Foundations of molecular modeling and simulation: Proceedings of the 1st international conference on molecular modeling and simulation. Keystone: American Institute of Chemical Engineers, **2001,** p. 44–53.
53. Mori, T.; Tanaka, K.; Average stress in matrix and average elastic energy of materials with misfitting inclusions. Acta Metallurgica **1973,** *21,* 571–575.
54. Fisher, F. T.; Bradshaw, R. D.; Brinson, L. C.; Fiber waviness in nanotube-reinforced polymer composites—I: modulus predictions using effective nanotube properties, Comp Sci and Tech **2003,** *63,* 1689–1703.
55. Fisher, F. T.; Bradshaw, R. D.; Brinson LC Fiber waviness in nanotube-reinforced polymer composites—II: modeling via numerical approximation of the dilute strain concentration tensor. Comp Sci and Tech **2003,** *63,* 1705–1722.
56. Liu, Y. J.; Chen, X. L.; Evaluations of the effective material properties of carbon nanotube-based composites using a nanoscale representative volume element, Mech of Mat **2003,** *35,* 69–81.
57. Schadler, L. S.; Giannaris, S. C.; Ajayan, P. M.; Load transfer in carbon nanotube epoxy composites. Applied Physics Letters **1998,** *73 (26),* 3842–3844.
58. Wagner, H. D.; Lourie, O.; Feldman, Y.; Tenne, R.; Stress-induced fragmentation of multiwall carbon nanotubes in a polymer matrix. Applied Physics Letters **1998,** *72 (2),* 188–190.
59. Qian, D.; Dickey, E. C.; Andrews, R.; Rantell, T.; Load transfer and deformation mechanisms in carbon nanotube polystirene composites. Applied Physics Letters **2000,** *76 (20),* 2868–2870.

60. Liu, Y. J.; Chen, X. L.; Modeling and analysis of carbon nanotube-based composites using the FEM BEM.; Submitted to CMES: Computer Modeling in Engineering and Science. **2002.**
61. Liu, Y. J.; Xu, N.; Luo, J. F.; Modeling of inter phases in fiber-reinforced composites under transverse loading using the boundary element method. Journal of Applied Mechanics **2000,** *67 (1),* 41–49.
62. Fu, Y.; Klimkowski, K. J.; Rodin, G. J.; Berger, E.; et al.; A fast solution method for three-dimensional many-particle problems of linear elasticity. International Journal for Numerical Methods in Engineering **1998,** *42,* 1215–1229.
63. Nishimura, N.; Yoshida, K.-I.; Kobayashi, S.; A fast multipole boundary integral equation method for crack problems in 3D Engineering Analysis with Boundary Elements **1999,** *23,* 97–105.
64. Qian, D.; Liu, W. K.; Ruoff, R. S.; Mechanics of C60 in nanotubes. The Journal of Physical Chemistry B **2001,** *105 (44),* 10753–10758.
66. Chen, X. L.; Liu, Y. J.; Square representative volume elements for evaluating the effective material properties of carbon nanotube-based composites, Comput Mater Sci **2004,** *29,* 1–11.
67. C.; Bower, R.; Rosen, L.; Jin, J.; Han, O.; Zhou, Deformation of carbon nanotubes in nanotube–polymer composites, Applied Physics Letters **1999,** *74,* 3317–3319.
68. Gibson, R. F.; Principles of composite material mechanics, CRC Press, 2nd edition. **2007**, 97–134.
69. Wan, H.; Delale, F.; Shen, L.; Effect of CNT length and CNT-matrix interphase in carbon nanotube (CNT) reinforced composites, Mech Res Commun **2005,** *32,* 481–489.
70. Shi, D.; Feng, X.; Jiang, H.; Huang, Y. Y.; Hwang, K.; Multiscale analysis of fracture of carbon nanotubes embedded in composites, Int J of Fract **2005,** *134,* 369–386.
71. Buryachenko, V. A.; Roy, A.; Effective elastic moduli of nanocomposites with prescribed random orientation of nanofibers, Comp: Part B **2005,** *36(5),* 405–416.
72. Buryachenko, V. A.; Roy, A.; Lafdi, K.; Andeson, K. L.; Chellapilla, S.; Multi-scale mechanics of nanocomposites including interface: experimental and numerical investigation, Comp Sci and Tech **2005,** *65,* 2435–246.
73. Li C, Chou, T. W.; Multiscale modeling of carbon nanotube reinforced polymer composites, J of Nanosci Nanotechnol **2003,** *3,* 423–430.
74. Li C, Chou, T. W.; Multiscale modeling of compressive behavior of carbon nanotube/polymer composites, Comp Sci and Tech **2006,** *66,* 2409–2414.
75. Anumandla, V.; Gibson, R. F.; A comprehensive closed form micromechanics model for estimating the elastic modulus of nanotube-reinforced composites, Com: Part A **2006,** *37,* 2178–2185.
76. Seidel, G. D.; Lagoudas, D. C.; Micromechanical analysis of the effective elastic properties of carbon nanotube reinforced composites, Mech of Mater **2006,** *38,* 884–907.
77. Hashin Z.; Rosen, B.; The elastic moduli of fiber-reinforced materials. Journal of Applied Mechanics **1964,** *31,* 223–232
78. R.; Christensen and, K.; Lo. Solutions for effective shear properties in three phase sphere and cylinder models. Journal of the Mechanics and Physics of Solids, **1979,** *27*,315–330.

79. Selmi, A.; Friebel, C.; Doghri, I.; Hassis, H.; Prediction of the elastic properties of single walled carbon nanotube reinforced polymers: A comparative study of several micromechanical models, Comp Sci and Tech **2007,** *67,* 2071–2084
80. Luo, D.; Wang, W. X.; Takao, Y.; Effects of the distribution and geometry of carbon nanotubes on the macroscopic stiffness and microscopic stresses of nanocomposites, Comp Sci and Tech **2007,** *67,* 2947–2958.
81. Fu SY, Yue, C. Y.; Hu X, Mai, Y. W.; On the elastic transfer and longitudinal modulus of unidirectional multishort-fiber composites, Compos Sci Technol **2000,** *60,* 3001–3013.
82. Lauke, B.; Theoretical considerations on deformation and toughness of short-fiber reinforced polymers, J Polym Eng **1992,** *11,* 103–154.
83. Tserpes, K. I.; Panikos, P.; Labeas, G.; Panterlakis SpG.; Multi-scale modeling of tensile behavior of carbon nanotube-reinforced composites, Theoret and Appl Fract Mech, **2008,** *49,* 51–60.
84. Spanos, P. D.; Kontsos, A.; A multiscale monte carlo finite element method for determining mechanical properties of polymer nanocomposites, Prob Eng Mech **2008,** doi: 10.1016/j.probengmech.2007.09.002
85. Paiva, M. C.; Zhou, B.; Fernando, K. A. S.; Lin, Y.; Kennedy, J. M.; Sun, Y-P.; Mechanical and morphological characterization of polymer-carbon nanocomposites from functionalized carbon nanotubes. Carbon **2004,** *42,* 2849–54.
86. Zhu, J.; Peng, H.; Rodriguez-Macias, F.; Margrave, J.; Khabashesku, V.; Imam, A.; Lozano, K.; Barrera, E.; "Reinforcing epoxy polymer composites through covalent integration of functionalized nanotubes", Advanced Functional Materials **2004,** *14 (7),* 643–648.
87. Bhuiyan Md. A.; Raghuram, V.; Pucha, A.; Johnny Worthy, B.; Mehdi Karevan, A.; Kyriaki Kalaitzidou, Defining the lower and upper limit of the effective modulus of CNT/polypropylene composites through integration of modeling and experiments. Composite Structures **2013,** *95,* 80–87.
88. Papanikos, P.; Nikolopoulos, D. D.; Tserpes, K. I.; Equivalent beams for carbon nanotubes. Comput Mater Sci **2008,** *43(2),* 345–52.
89. Affdl, J. C. H.; Kardos, J. L.; The Halpin–Tsai equations: a review. Polym Eng Sci **1976,** *16(5),* 344–52.
90. Tandon, G. P.; Weng, G. J.; The effect of aspect ratio of inclusions on the elastic properties of unidirectionally aligned composites. Polym Composite **1984,** *5(4),* 327–33.
91. Nielsen, L. E.; Mechanical properties of polymers and composites, 2. New York: Marcel Dekker; 1974.
92. Tucker III CL, Liang, E.; Stiffness predictions for unidirectional short-fiber composites: review and evaluation. Compos Sci Technol **1999,** *59(5),* 655–71.
93. Shokrieh, M. M.; Rafiee, R.; Investigation of nanotube length effect on the reinforcement efficiency in carbon nanotube based composites Composite Structures 92 **2010,** 2415–2420.
94. Shokrieh, M. M.; Rafiee, R.; On the tensile behavior of an embedded carbon nanotube in polymer matrix with nonbonded interphase region. J Compos Struct **2009,** *22,* 23–5.
95. Press Release. US Consulate. World-record-length carbon nanotube grown at US Laboratory. Mumbai-India; September 15, 2004.

96. Evans, J.; Length matters for carbon nanotubes: long carbon nanotubes hold promise for new composite materials. Chemistry World News **2004,** <http://www/rsc.org/chemistryworld/news>.
97. Pan, Z.; Xie, S. S.; Chang, B.; Wang, C.; Very long carbon nanotubes. Nature **1998,** *394,* 631–2.
98. http: //www.carbonsolution.com.
99. http: //www.fibermax.eu/shop/.
100. http: //www.nanoamor.com.
101. www.thomas-swan.co.uk.
102. Rice University's chemical 'Scissors' yield short carbon nanotubes. New process yields nanotubes small enough to migrate through cells. Science Daily [July 2003. <http: //www.sciencedaily.com/releases/2003/07/030723083644.htm>.
103. Chowdhury, S. C.; B. Z. (Gama) Haque, T.; Okabe, J. W.; Gillespie Jr.; Modeling the effect of statistical variations in length and diameter of randomly oriented CNTs on the properties of CNT reinforced nanocomposites, Composites: Part B **2012,** *43,* 1756–1762
104. Tserpes, K. I. Chanteli, A.; Parametric numerical evaluation of the effective elastic properties of carbon nanotube-reinforced polymers, Composite Structures **2013,** *99,* 366–374.
105. Gou, J.; Minaie, B.; Wang, B.; Liang, Z.; Zhang, C.; Computational and experimental study of interfacial bonding of single-walled nanotube reinforced composites. Comput Mater Sci **2004,** *31,* 225–36.
106. Frankland, S. J. V.; Harik, V. M.; Analysis of carbon nanotube pull-out from a polymer matrix. Surf Sci **2003,** *525,* 103–8.
107. Natarajan, U.; Misra, S.; Mattice, W. L.; Atomistic simulation of a polymer-polymer interface: interfacial energy and work of adhesion. Comput Theor Polym Sci **1998,** *8,* 323–9.
108. Lordi V, Yao, N.; Molecular mechanics of binding in carbon-nanotube polymer composites. J Mater Res **2000,** *15,* 2770–9.
109. Wong, M.; Paramsothy, M.; Xu XJ, Ren, Y.; Li S, Liao, K.; Physical interactions at carbon nanotube-polymer interface. Polymer **2003,** *44,* 7757–64.
110. Qian, D.; Liu, W. K.; Ruoff, R. S.; Load transfer mechanism in carbon nanotube ropes. Compos Sci Technol **2003,** *63,* 1561–9.
111. Liu, Y. J.; Chen, X. L.; Continuum models of carbon nanotube-based composites using the boundary element method. J Boundary Elem **2003,** *1,* 316–35.
112. Chen, X. L.; Liu, Y. J.; Square representative volume elements for evaluating the effective material properties of carbon nanotube-based composites. Comput Mater Sci **2004,** *29,* 1–11.
113. Chen, X. L.; Liu, Y. J.; Evaluations of the effective material properties of carbon nanotube-based composites using a nanoscale representative volume element. Mech Mater **2003,** *35,* 69–81.
114. Qian, D.; Dickey, E. C.; Andrews, R.; Rantell, T.; Load transfer and deformation mechanisms in carbon nanotubepolystirene composites. Appl Phys Lett **2000,** *76,* 2868.

115. Thostenson, E. T.; Chou T-W.; Aligned multiwalled carbon nanotube-reinforced composites: processing and mechanical characterization. J Phys D Appl Phys **2002,** *35,* 77–80.
116. Bower, C.; Rosen, R.; Jin, L.; Han, J.; Zhou, O.; Deformation of carbon nanotubes in nanotube-polymer composites. Appl Phys Lett **1999,** *74,* 3317–9.
117. Cooper, C. A.; Cohen, S. R.; Barber, A. H.; Wagner, H. D.; Detachment of nanotubes from a polymer matrix. Appl Phys Lett **2002,** *81,* 3873–5.
118. Qian, D.; Dickey, E. C.; In-situ transmission electron microscopy studies of polymer-carbon nanotube composite deformation. J Microsc **2001,** *204,* 39–45.
119. Schadler, L. S.; Giannaris, S. C.; Ajayan, P. M.; Load transfer in carbon nanotube epoxy composites. Appl Phys Lett **1998,** *73,* 3842.
120. Wagner, H. D.; Nanotube-polymer adhesion: a mechanics approach. Chem Phys Lett **2002,** *361,* 57–61.
121. Ajayan, P. M.; Schadler, L. S.; Giannaris, C.; Rubio, A.; Single-walled carbon nanotube-polymer composites: strength and weakness. Adv Mater **2000,** *12,* 750–3.
122. Cooper, C. A.; Young, R. J.; Halsall, M.; Investigation into the deformation of carbon nanotubes and their composites through the use of Raman spectroscopy. Compos Part A Appl Sci Manuf **2001,** *32,* 401–11.
123. Hadjiev, V. G.; Iliev, M. N.; Arepalli, S.; Nikolaev, P.; Files, B. S.; Raman scattering test of single-wall carbon nanotube composites. Appl Phys Lett **2001,** *78,* 3193.
124. Paipetis, A.; Galiotis, C.; Liu, Y. C.; Nairn, J. A.; Stress transfer from the matrix to the fiber in a fragmentation test: Raman experiments and analytical modeling. J Compos Mater **1999,** *33,* 377–99.
125. Valentini, L.; Biagiotti, J.; Kenny, J. M.; Lopez Manchado, M. A.; Physical and mechanical behavior of singlewalled carbon nanotube/polypropylene/ethylene-propylene-diene rubber nanocomposites. J Appl Polym Sci **2003,** *89,* 2657–63.
126. Qian, D.; Wagner, G. J.; Liu, W. K.; Yu M-F, Ruoff, R. S.; Mechanics of carbon nanotubes. Appl Mech Rev **2002,** *55,* 495–532.
127. Yu M-F, Yakobson, B. I.; Ruo, R. S.; Controlled sliding and pulout of nested shells in individual multiwalled nanotubes. J Phys Chem B **2000,** *104,* 8764–7.
128. Qian, D.; Liu, W. K.; Ruoff, R. S.; Load transfer mechanism in carbon nanotube ropes. Compos Sci Technol **2003,** *63,* 1561–9.
129. Liao, K.; Li, S.; Interfacial characteristics of a carbon nanotubepolystirene composite system. Appl Phys Lett **2001,** *79,* 4225–7.
130. Andrews, R.; Weisenberger, M. C.; Carbon nanotube polymer composites. Curr Opin Solid State Mater Sci **2004,** *8,* 31–7.
131. Lordi, V.; Yao, N.; Molecular mechanics of binding in carbonnanotube- polystirene composite system. J Mater Res **2000,** *5,* 2770–9.
132. Wagner, H. D.; Vaia, R. A.; Nanocomposites: issue at the interface. Mater Today **2004,** *7,* 38–42.
133. Wernik, J. M.; Cornwell-Mott, B. J.; Meguid, S. A.; Determination of the interfacial properties of carbon nanotube reinforced polymer composites using atomistic-based continuum model, Inte. Jour. of Sol. and Struct. **2012,** *49,* 1852–1863.
134. Wernik, J. M.; Meguid, S. A.; Multiscale modeling of the nonlinear response of nano-reinforced polymers. Acta Mech. **2011,** *217,* 1–16.

135. Yang, S.; Yu, S.; Kyoung, W.; Han, D. S.; Cho, M.; Multiscale modeling of size-dependent elastic properties of carbon nanotube/polymer nanocomposites with interfacial imperfections Polymer **2012,** *53,* 623–633.
136. Ayatollahi, M. R. Shadlou, S.; Shokrieh, M. M.; Multiscale modeling for mechanical properties of carbon nanotube reinforced nanocomposites subjected to different types of loadingComposite Structures **2011,** *93,* 2250–2259
137. Jiang, L. Y.; Huang, Y.; Jiang, H.; Ravichandran, G.; Gao, H.; Hwang, K. C.; et al. A cohesive law for carbon nanotube/polymer interfaces based on the van der Waals force. J Mech Phys Solids **2006,** *54,* 2436–52.
138. Tan, H.; Jiang, L. Y.; Huang, Y.; Liu, B.; Hwang, K. C.; The effect of van der Waals-based interface cohesive law on carbon nanotube-reinforced composite materials, Composites Science and Technology **2007,** *67,* 2941–2946.
139. Zalamea, L.; Kim, H.; Pipes, R. B.; Stress transfer in multiwalled carbon nanotubes. Compos Sci Technol **2007,** *67(15–16),* 3425–33.
140. Shen, G. A.; Namilae, S.; Chandra, N.; Load transfer issues in the tensile and compressive behavior of multiwall carbon nanotubes. Mater Sci Eng A **2006,** *429(1–2),* 66–73.
141. Gao, X. L.; Li, K.; A shear-lag model for carbon nanotube-reinforced polymer composites. Int J Solids Struct **2005,** *42(5–6),* 1649–67.
142. Li C.; Chou, T. W.; A structural mechanics approach for the analysis of carbon nanotubes. Int J Solids Struct **2003,** *40(10),* 2487–99.
143. Tsai, J. L.; Lu, T. C.; Investigating the load transfer efficiency in carbon nanotubes reinforced nanocomposites. Compos Struct **2009,** *90(2),* 172–9.
144. Lu, T. C.; Tsai, J. L.; Characterizing load transfer efficiency in double-walled carbon nanotubes using multiscale finite element modeling, Composites: Part B **2013,** *44,* 394–402

CHAPTER 5

A DETAILED REVIEW ON FABRICATION AND CHARACTERIZATION OF THE METAL NANO-SIZED BRANCHED STRUCTURES

GUOQIANG XIE, MINGHUI SONG, KAZUO FURUYA, and AKIHISA INOUE

CONTENTS

5.1 Introduction 256
5.2 Fabrications and Characterization of Nano-Sized Branched Structures 257
5.3 Growth Mechanism of Nano-Sized Branched Structures 263
5.4 Effect of Accelerating Voltage on Crystallization of The Nanostructures 266
5.5 Composite Nanostructures of Nanoparticles/Nanodendrites 269
5.6 Conclusion 271
Keywords 272
References 272

5.1 INTRODUCTION

Using an electron-beam-induced deposition (EBID) process in a transmission electron microscope, we fabricated self-standing metal nano-sized branched structures including nanowire arrays, nanodendrites, and nanofractal-like trees, as well as their composite nanostructures with controlled size and position on insulator (SiO_2, Al_2O_3) substrates. The fabricated nanostructures were characterized with high-resolution transmission electron microscopy and X-ray energy dispersive spectroscopy. The growth mechanism was discussed. Effect of the electron beam accelerating voltage on crystallization of the nanostructures was investigated. The nanostructures of the different morphologies were obtained by controlling the intensity of the electron beam during the EBID process. A mechanism for the growth and morphology of the nanostructures was proposed involving charge-up produced on the surface of the substrate, and the movement of the charges to and charges accumulation at the convex surface of the substrate and the tips of the deposits. High-energy electron irradiation enhanced diffusion of the metallic atoms in the nanostructures and hence promoted crystallization. More crystallized metal nano-branched structures were achieved by the EBID process using high-energy electron beams.

Metal nanostructures are of great importance in nanotechnology due to their potential applications as building blocks in optoelectronic devices, catalysis, chemical sensors, and other areas [1–3]. The formation of nanostructures with controlled size and morphology has been the focus of intensive research in recent years. Such nanostructures are important in the development of nanoscale devices and in the exploitation of the properties of nanomaterials [4–7]. Many fabrication methods including chemical vapor decomposition [8], the arc-discharge method [9], evaporation [10], hydrothermal reactions [11], and so on, have been developed for the production of nanomaterials with controlled sizes and shapes. However, up to date, fabrication of position controllable nanostructures at selected positions on a substrate is still a challenge.

Among the methods, electron-beam-induced deposition (EBID) process is one of the most promizing techniques to fabricate small-sized structures on substrates. An advantage of this process is that the deposited position can be controlled. In this approach, an electron beam in a high vacuum chamber is focused on a substrate surface on which precursor molecules, containing the element to be deposited (e.g., organometallic compound or hydrocarbon), are adsorbed. As a result of complex beam-induced surface reactions, the precursor molecules absorbed in or near to the irradiated area, are dissociated into nonvolatile and volatile parts by energetic electrons. The nonvolatile materials are deposited on surface of the substrate, while the volatile components are pumped away. Due to the controllability of electron beam, fabrication of position controllable nanostructures can be realized. Using the technique, a variety of nanometer-sized structures, such as nanodots, nanowires, nanotubes, nanopatterns, two or three dimensional nano-objects, and so on, have been fabricated [12–22]. Due to easy to receive a stable fabrication condition, conductive substrates are generally used in these fabrications [16]. As a result, compact structures are usually fabricated with this process.

In the case using nonconductive substrates, the nanofabrication with the EBID technique is also very important for the technique to be applied in technology. The accumulation of charges readily occurs on an insulator substrate during the EBID process. The deposition of novel structures may occur under this condition. The growth of carbon fractal-like structures has already been observed on insulator substrates under electron-beam irradiation due to the existence of residual pump oil remaining in the atmosphere of the vacuum chamber [23–25], but the growth of metallic nanostructures on insulator substrates using EBID process was first reported recently by Song et al. [26]. By using insulator substrates such as SiO_2 [27–30] and Al_2O_3 [26, 31–35], characteristic morphologies of nanostructures, such as arrays of nanowhiskers (or nanowires), arrays of nanodendrites, and fractal treelike structures, have been fabricated in transmission electron microscopes (TEMs) by the EBID process [26–35]. The typical size of the diameter of a nanowhisker, the tip of a nanodendrite, and the tip of a nanotree is about 3 nm, and are almost completely independent of the size of the electron beam. In the present review we reported the fabrication and characterization of the metal nano-sized branched structures and the composite nanostructures grown on insulator substrates by the EBID process.

5.2 FABRICATIONS AND CHARACTERIZATION OF NANO-SIZED BRANCHED STRUCTURES

Thin films of SiO_2 and Al_2O_3 were irradiated by an electron beam in TEM with organometallic precursor gasses at room temperature. The EBID process was carried out using JEM-2010 or JEM-ARM1000 TEMs made by JEOL Co., Ltd. The energy of the electron beam used was from 200 to 1000 keV. The pressure in the specimen chambers of the TEMs was on the order of 10^{-5} to 10^{-6} Pa. Insulator SiO_2 and Al_2O_3 substrates suitable for TEM observation were used. An organometallic precursor gas was introduced near the substrate using a special designed system comprizing a nozzle with a diameter smaller than 0.1 mm and a reservoir containing the precursor [26]. Tungsten hexacarbonyl ($W(CO)_6$) or (Methylcyclopentadienyl)trimethylplatinum (Me_3 McCpPt) powders were used as the precursors for fabricating tungsten (W)-containing, or platinum (Pt)-containing nanostructures, respectively. The vapor pressures of these precursors are several Pa at room temperature. These precursors have been previously typically used to fabricate small objects on conductive substrates by EBID process [16]. The intensity of the electron beam for the EBID process was from 3.2 to 111.7×10^{18} e cm^{-2} s^{-1} (current density: 0.51 to 17.9 A cm^{-2}), which was estimated by measuring the total intensity and the size of the electron beam under operating conditions. The fabricated structure was characterized cm or after fabrication in a TEM. The experiments were performed at room temperature.

Figure 1 shows a series of micrographs of growth process of the W-nanodendrite structures using the TEM at an accelerating voltage of 1000 kV during the electron-beam irradiation with an electron beam current density of 1.6 A cm^{-2} [29]. Figure 1(a) presents a micrograph before the electron-beam irradiation. No deposits are observed at the SiO_2 substrate. Figure 1(b) shows a micrograph taken 3 s after the beginning of the irradiation at a fluence of 3.0×10^{19} e cm^{-2}. The electron beam irradiated the specimen from a direction perpendicular to the plane of the micrograph. Whisker-like deposits begin to nucleate and grow on the surface of the substrate. The deposits are about 3 nm in diameter and about 5 nm in length. The structures grow self-standing at positions separated from each other at a distance of several nanometers, and in parallel and nearly perpendicular to the surface of the substrate within the irradiated area. The whisker-like deposits grow longer and also denser under further electron beam irradiation, and new nucleation and growth deposits are not observed. Figure 1(c) shows a micrograph taken 3 min after the beginning of irradiation at a fluence of 1.8×10^{21} e cm^{-2}. Similarity in length and an almost even thickness of all deposits are observed. Branching is observed at the tips of the deposits. The fabricated deposits have a nanodendrite structure. With further electron beam irradiation, new branches grew from the tips of the grown branches. This process continued as long as the irradiation continued. Figure 1(d) shows a micrograph taken 10 min after the beginning of irradiation at a fluence of 6.0×10^{21} e cm^{-2}. The nanodendrite structures grow at both the tip and trunk. The diameter of the nanodendrite structures increased with the decrease in distance from the substrate, but is about 3 nm at the tips. The average length of the nanodendrite structures increased to about 51.4 nm. Figure 2 shows the length of nanodendrite structures grown at various electron-beam irradiation fluences (namely, electron-beam irradiation time). The length of nanodendrite structures increases approximated linearly with electron-beam irradiation fluence [29].

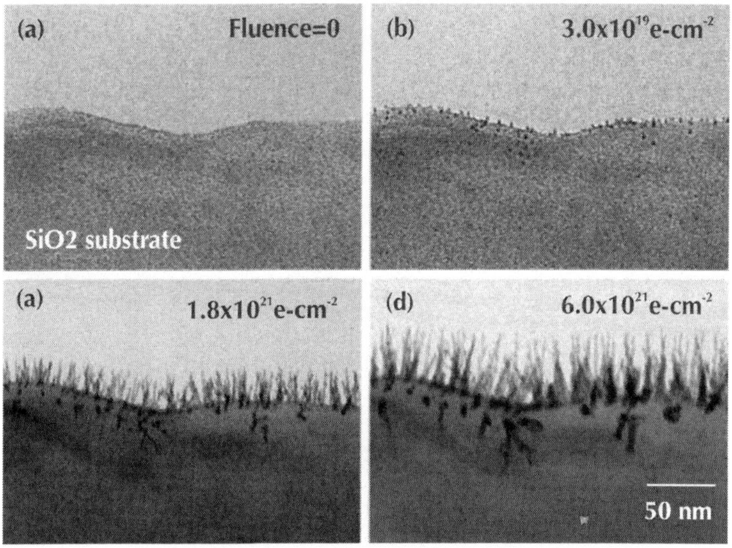

FIGURE 1 Bright-field TEM images of W-nanodendrite structures grown on surface of SiO_2 substrate by the EBID process at an electron beam accelerating voltage of 1000 kV with a current density of 1.6 A cm^{-2} after beginning of electron beam irradiation. (a) 0 s; (b) 3 s; (c) 3 min; and (d) 10 min [29].

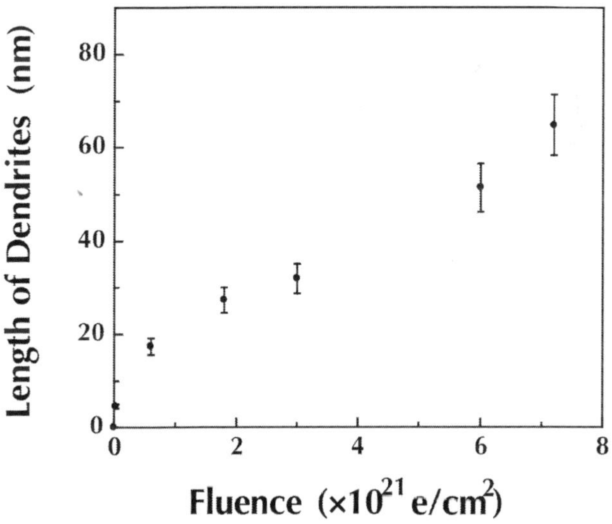

FIGURE 2 Relationship between length of the W-nanodendrite structures grown on surface of SiO_2 substrate by the EBID process at an electron beam accelerating voltage of 1000 kV with a current density of 1.6 A cm^{-2} and electron beam irradiating fluence [29].

Furthermore, the morphology of the nanostructures grown can also be controlled by the intensity of the electron beam [26, 36]. Figure 3(a) shows a micrograph of an array of W nanowhiskers (or nanowires) fabricated on an Al_2O_3 substrate irradiated with a 200 keV electron beam at an intensity of 4.7×10^{18} e cm^{-2} s^{-1} (0.75 A cm^{-2}) to about 120 s [26]. The growth speed was very low. By increasing the current density to a higher value such as 1.6 A cm^{-2}, the growth speed was increased. The nanowhiskers were longer and also denser than those grown at the lower current density with an almost even thickness. Branching is observed to take place at the tips of the nanowhiskers, and the morphology becomes dendritic structures [27, 29]. A further increase in the current density to 17.9 A cm^{-2} resulted in extensive branching and the formation of complicated nanotree structures (Figure 3(c)) [26].

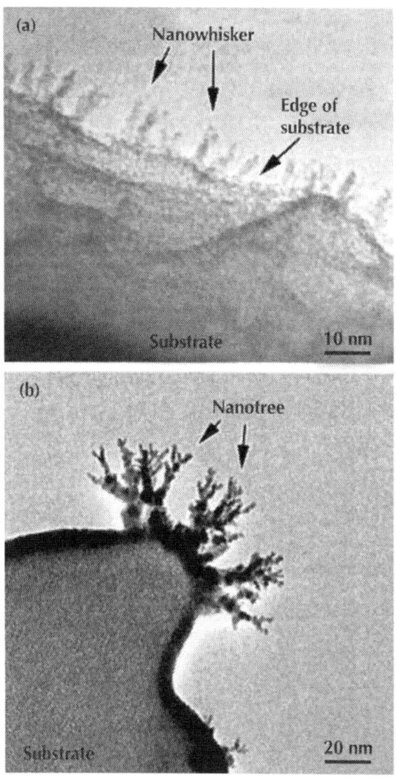

FIGURE 3 (a) W nanowhisker arrays grown on the surface of an Al_2O_3 substrate by the EBID process at an accelerating voltage of 200 kV. The current density of the electron beam was 0.75 A cm^{-2} and the irradiation time was 120 s. (b) Fractal-like W nanotrees grown on the surface of an Al_2O_3 substrate by the EBID process. The current density of the electron beam was 17.9 A cm^{-2} and the irradiation time was 10 s [26].

The composition of the fabricated nanodendrites is important, since they are related to the physical and chemical properties of deposits. The chemical composition of the nanodendrites was examined using an X-ray energy dispersive spectroscopy (EDS). Figure 4 shows the spectra taken using EDS from an as-deposited nanodendrite fabricated at an electron-beam irradiating fluence of 6.0×10^{21} e cm^{-2} after detaching the precursor source [29]. Figure 4(a) presents a spectrum taken from the tips of a nanodendrite structure, and Figure 4(b) shows that taken from its trunk. The size of the electron beam for the analysis using EDS is about 10 nm. The peaks of tungsten dominate these spectra. No obvious differences in the composition between the tip and the trunk of the nanodendrite structure were observed. On the basis of the analyzes using EDS, a relative content of 89.4% W compared with 10.6% C for the spectrum from the tips is obtained. The relative content of W is higher than the reported values for W-deposits fabricated using the same precursor but at a lower voltage of the electron beam (25 kV) by EBID process, which was 75% [17].

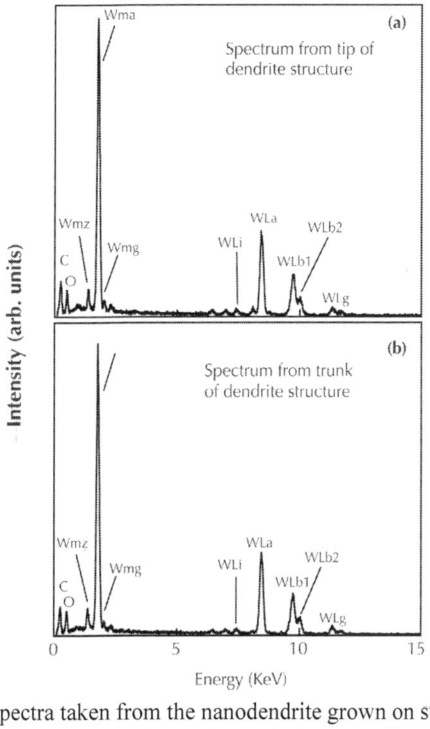

FIGURE 4 EDS spectra taken from the nanodendrite grown on surface of SiO_2 substrate by the EBID process at an accelerating voltage of 1000 kV with a current density of 1.6 A cm^{-2} to an electron-beam irradiating fluence of 6.0×10^{21} e cm^{-2} (a) at tip; and (b) at trunk [29].

Using (Methylcyclopentadienyl)trimethylplatinum (Me_3 MeCpPt) powder as a precursor, one can obtain Pt nanodendritic structures by the EBID process. Figure 5 shows TEM micrographs of Pt nanodendrite-like structures grown on an Al_2O_3 substrate in a 200 kV TEM [34]. The electron beam is defocused to a size of about 600 nm, corresponding to a current density of 0.52 A cm^{-2}. The irradiated time is 1 min. The irradiating fluence of the electron beam is 2.0×10^{20} e cm^{-2}. The edge of the electron beam is indicated by arrows in Fig. 5(a). The contrast inside the irradiated area is obviously dark. The size of dendrite structures is much smaller than the diameter of the electron beam. The dendrites show a tendency to grow at the edge of the substrate. The dendrites have branches at the tip, as observed in Fig. 5(b). The diameter of the nanodendrites become thicker near the substrate, which implies that the deposition takes place at both tip and trunk part. The typical thickness of the tips is less than 3 nm, as observed in Fig. 5(b). TEM observation, diffraction pattern analyzes, and EDS analyzes indicated that a nanodendrite structure with a high Pt content was formed. Figure 6 shows a high resolution TEM (HRTEM) micrograph of the fabricated Pt nanodendrite [34]. It is confirmed that fcc Pt nanoparticles and an amorphous part are contained in the structure. The morphology of the Pt nanodendrite is considerably different from that of the W nanodendrite fabricated on an Al_2O_3 substrate, suggesting that the growth of a nanodendrite depends largely on the properties of the precursor.

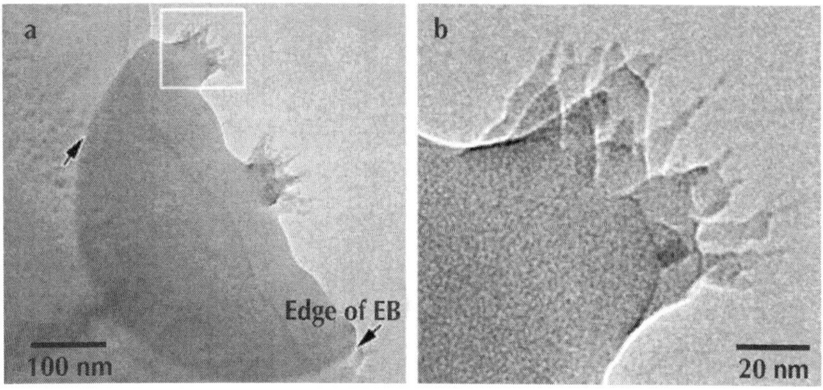

FIGURE 5 (a) Bright-field TEM micrographs of Pt-nanodendrite structures grown on the surface of an Al_2O_3 substrate in a 200 kV TEM with an EBID process 1 min after the start of the electron beam irradiation to a fluence of 2.0×10^{20} e cm^{-2}. (b) Enlargement of the square area in (a), showing the nanodendrites in more detail [34].

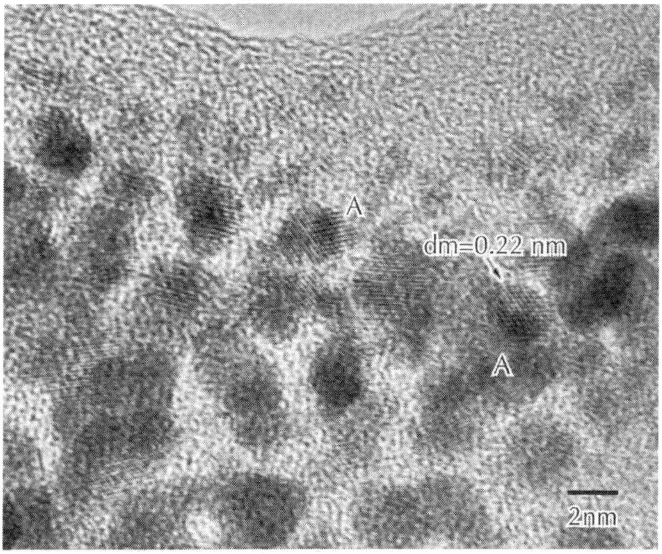

FIGURE 6 An HRTEM micrograph of a nanodendrite structure fabricated on an electron beam irradiating fluence of 2.6×10^{21} e cm^{-2} [34].

5.3 GROWTH MECHANISM OF NANO-SIZED BRANCHED STRUCTURES

It is well known that the EBID process is caused by the dissociation of molecules adsorbed to a surface by energetic electron beam. In this approach, a molecule of a precursor is first adsorbed on surface of a substrate and then decomposed into volatile and nonvolatile parts by further irradiation of the energetic beam. The nonvolatile fraction accumulates to form a deposit, while the volatile component is pumped away by the vacuum system. The dissociation mechanism is complex and not fully understood until the present time because of the huge number of excitation channels available even for small molecules. The details involving the decomposition have been argued to relate to secondary electrons on surface of a specimen produced by incident electron beam and backscattered electrons in an EBID process [17, 19].

If an EBID process is carried out on the surface of an electric grounded conductive substrate, the molecules absorbed may move or not move, but distribute randomly and be decomposed on the surface. Therefore a compact deposit is usually formed. On the other hand, in a case that the deposition is conducted on an insulator substrate, charge-up may take place on the surface due to emission of secondary electrons under energetic beam irradiation. When specimens are exposed to electron bombardments, the molecules absorbed to a surface of a substrate or near the surface in the irradiated area may be polarized by irradiation of an incident electron beam. The irradiated area on the insulator is easily charged, forms local electric potential. It is reasonable to consider that the distribution of charges due to charge-up on the surface may be not even in a nanometer scale. This unevenness may be resulted from the nanoscaled unflatness or atomic steps on surface. Charges may accumulate at some places to some extent. Figure 7(a) shows the growth mechanism of a nanowhisker array [26]. The intensity of the electron beam is weak, but the accumulation of charges occurs. The distribution of the charges on the surface is assumed to be uneven at a nanometer scale. Charges may accumulate at some places (charge centers), where an electric field is generated. The dark dots represent the nonvolatile fraction of the precursor molecules of which the deposits are composed, while the brighter dots represent the volatile fraction. Because the intensity of the electron beam is weak, the adsorbed molecules can move around considerably before being decomposed. The precursor molecules adsorbed on the surface may be attracted to the charge centers since the molecules are easily polarized due to the weak bonding between the atoms of the molecules. The precursor molecules then decompose and form a deposit. After a deposit is formed, the charges on the surface of the substrate or the deposit tend to move and accumulate at the tip of the deposit, since W deposits have been reported to be conductive [17, 19]. Thus, molecules are attracted to the tip, and the deposit increases in length upon further electron beam irradiation. The formation of W nanodot deposits on the surface of a SiO_2 substrate and the growth of W nanowhiskers from these nanodots [30] are consistent with the above mechanism.

Figures 7(b) and (c) schematically show the growth conditions of nanodendrites and fractal-like nanotrees, respectively [26]. When the current density of the electron beam is increased to a definite extent, an adsorbed molecule on the surface may be decomposed in a very short period before it can move. Hence, a compact layer is deposited on the surface. This results in the entire irradiated area becoming conductive. After this stage, the charges on the surface can move a long distance to the substrate edge, particularly when the surface is convex. A strong electric field is thus generated near the convex surface. Once a deposit grows away from the surface at a point, charges then accumulate there. This deposit thus grows preferentially because the charges attract precursor molecules. When the current density is moderate, the electric field generated near a tip may not strongly affect the trajectory of a molecule. Therefore, the deposit grows at both its tip and its main body because the molecules arrive at the tip and the main body at the same time. Therefore, the main body of the deposit increases in thickness with time and develops a dendritic morphology. On the other hand, if the current density of the electron beam is very strong, the generated electric field will become sufficiently strong to affect the trajectory of a molecule near the tip of a deposit. The molecules are thus likely to be attracted to the tip and are not easily adsorbed on the main body of the deposit. Therefore, a fractal-like tree morphology is formed.

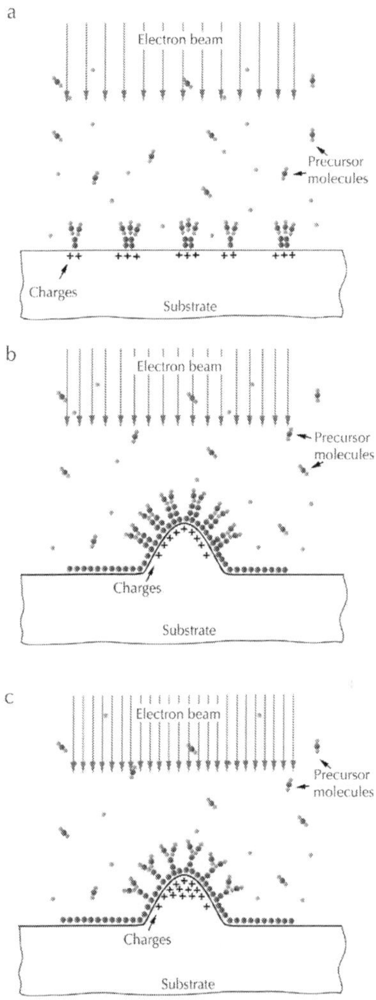

FIGURE 7 Schematic drawings showing the growth process of the nanostructures with different morphologies on insulator substrates by the EBID process. The dark dots represent the nonvolatile fraction of the precursor and the deposit, while the brighter dots represent the volatile fraction of the precursor. (a) The growth of nanowhiskers (or nanowires) inside the area of irradiation with a weak electron beam; (b) the growth of nanodendrites on a convex surface under irradiation with a moderate electron beam; and (c) the growth of fractal-like nanotrees on a convex surface under irradiation with a strong electron beam [26].

5.4 EFFECT OF ACCELERATING VOLTAGE ON CRYSTALLIZATION OF THE NANOSTRUCTURES

The fabricated nanostructure morphologies at various electron beam-accelerating voltages have been investigated with a conventional TEM. Figure 8 shows a set of TEM micrographs of dendrite-like structures grown on SiO_2 substrate at different electron beam accelerating voltages to an electron beam fluence of 6.0×10^{21} e cm^{-2} [27]. The electron beam irradiated the specimen from a direction perpendicular to the plane of the micrograph. The obvious difference in dendrite morphology and their composites for the deposits fabricated by various accelerating voltages is not observed. The length of the nanodendrite structures increases with an increase in electron beam irradiating fluence, and has not obvious effect with the increase of accelerating voltage, as shown in Fig. 9.

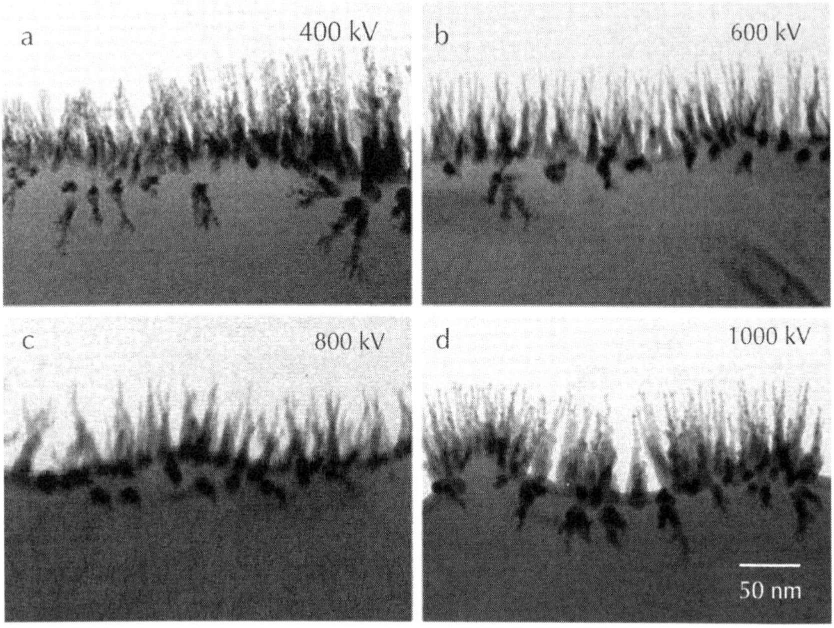

FIGURE 8 A set of bright-field TEM micrographs of the nanodendrite structures grown on SiO_2 substrate irradiated to an electron beam fluence of 6.0×10^{21} e cm^{-2} at different electron beam accelerating voltages. (a) 400 kV; (b) 600 kV; (c) 800 kV; (d) 1000 kV. The current density of electron beam for irradiating the specimens was 1.1 A cm^{-2} for 400 kV, and 1.6 A cm^{-2} for 600, 800 and 1000 kV, respectively [27].

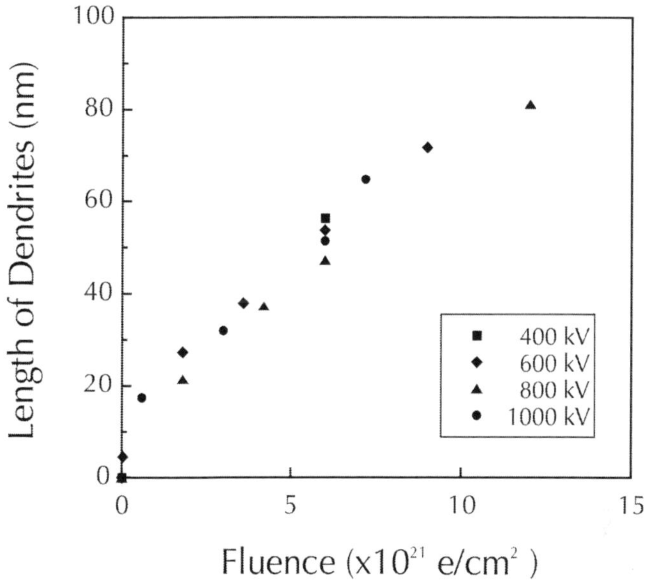

FIGURE 9 Dependence of the length of nanodendrites on electron beam irradiating fluence at different electron beam accelerating voltages [27].

The effect of electron beam accelerating voltage on crystallization of the nanodendrite structures has been investigated with an HRTEM. Figure 10 shows a series of HRTEM micrographs of some branches of nanodendrite structures fabricated with various accelerating voltages to a fluence of 6.0×10^{21} e cm^{-2} [28]. Figure 10(a) is an HRTEM micrograph of some branches of nanodendrites grown with an accelerating voltage of 400 kV. Lattice fringes are observed at the most places. This indicates that they are crystal in several nanometers. The largest and also the most observed lattice spacing measured from micrographs is 0.22 nm. It is close to the lattice spacing, 0.224 nm, of {110} of bcc W crystals with a deviation smaller than 5%. Moreover, lattice fringes with spacing d = 0.22 nm have interfringe angle of 60 degrees (grain A), as well as that of 90 degrees (grain B), as indicated in Fig. 10. They agree to zone axis of [111] and [001] of bcc W structure, respectively. Combined with nanometer-sized area diffraction pattern analyzes and EDS analyzes, it is clarified that these crystals are W grains in bcc structure. Furthermore, lattice fringes cannot be clearly observed in some places, indicating that a large part of them is in amorphous state, as shown by arrows in Fig. 10. Figure 10(b) shows a micrograph of some branches of nanodendrites grown with an accelerating voltage of 600 kV. The fraction of amorphous state in the as-fabricated nanodendrites decreases obviously. For the dendrites fabricated with an accelerating voltage of 800 kV, as shown in Figure 10(c), the amorphous state is observed only in a few places. Figure 10(d) shows an HRTEM micrograph of some branches of nanodendrites grown with an accelerating voltage of 1000 kV. Lattice fringes are observed clearly in almost all of the grains except in thick region, where small crystals in random orientations overlapped each other, so that their lattice fringes cannot be clearly observed. Therefore, it is indicated that the fraction of amorphous state in the as-fabricated nanodendrites decreases with an increase in electron beam accelerating voltage.

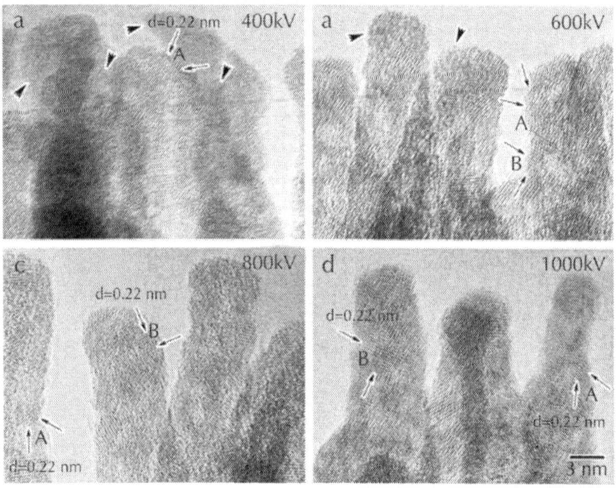

FIGURE 10 A series of HRTEM micrographs of some branches of nanodendrites grown on SiO_2 substrate irradiated to a fluence of 6.0×10^{21} e cm^{-2} with different electron beam accelerating voltages. (a) 400 kV; (b) 600 kV; (c) 800 kV; (d) 1000 kV. Arrows indicate lattice fringes in grains A and B, which have interfringe angles of 60 (A) and 90 (B) degrees, respectively. The current density of electron beam for irradiating the specimens was 1.1 A cm^{-2} for 400 kV, and 1.6 A cm^{-2} for 600, 800 and 1000 kV, respectively [28].

It is known that the growth of a W nanodendrite during the EBID process is controlled by a random accumulation of nonvolatile elements with thermal energy. Therefore the element does not have enough energy to move and form crystalline grains. The electron beam during the EBID process may transfer some energy to the element, but it is not enough for total deposit to transform into crystalline state with a 400 keV electron beam. Therefore, some amorphous structures are remained in deposits (refer to Fig. 10(a)). This also may explain the reason that an amorphous state has often been obtained in the W-deposits fabricated by a lower energy EBID [19, 26]. With an increase in electron beam accelerating voltage, the fraction transformed into crystalline state in the deposits increases, and the fraction of amorphous structure decreases. The results clearly indicate that high-energy electron irradiation enhances crystallization of an amorphous structure.

Song et al. [37] has reported that the effect of 1 MeV electron beam irradiation on crystallization of nanometer-sized W-dendritic structure fabricated on Al_2O_3 substrates with the EBID process at an accelerating voltage of 200 kV. After irradiation at room temperature for 100 min at a current density of 6.4 A cm^{-2}, almost all the grains crystallized. The nanodendrite structure also changes morphology and shrinks its size. In general, there are two important factors in the occurrence of electron irradiation induced crystallization of an amorphous state: (1) the promotion of atomic diffusion by electron irradiation in an amorphous state; and (2) the high stability of crystalline phase against electron irradiation [38]. When the two factors are satisfied simultaneously, electron irradiation induced crystallization of the amorphous structure. The change in morphology and shrinkage in size may be resulted from crystallization and sputtering during the irradiation.

5.5 COMPOSITE NANOSTRUCTURES OF NANOPARTICLES/NANODENDRITES

Complex-shape nanostructures have attracted great interest because advanced functional materials might be emerged if they can be formed with well-defined three-dimensional (3D) architectures [39–41]. Furthermore, the impact is greater for multielement systems, as in the case of advanced nanostructured systems like Au/Cu, Pt/C composite materials [42, 43] and so on, which are used in catalysis, sensors, energy sources [44, 45] and in many other applications. It has been demonstrated that bimetallic bonding can induce significant changes in the properties of the surface, producing in many cases catalysts that have superior activity and/or selectivity [42, 43, 46, 47].

The nanostructures fabricated on insulator substrates by the EBID process are good base materials for the fabrication of complex-shape nanostructures because of their superior features. Pt nanoparticles or Au nanoparticles were deposited on W nanodendrites fabricated on insulator Al_2O_3 or SiO_2 substrates, and a composite nanostructure consisting of Pt nanoparticles and W nanodendrites (Pt-nanoparticle-decorated W nanodendrites, or Pt nanoparticle/W nanodendrite), or Au nanoparticle and W nanodendrites were fabricated by ion-sputtering [33]. Figure 11 shows the Pt nanoparticle/W nanodendrite composite structures. Figure 11(a) is a bright-field TEM image of the as-fabricated composite structure produced with an electron beam irradiation fluence of 5.0×10^{21} cm^{-2} and an ion sputtering time of 40 s. The Pt-nanoparticles are nearly uniformly distributed on the W-nanodendrites. Figure 11(b) shows an HRTEM image of a tip of the composite nanostructure. Lattice fringes are observed in the image. By measuring the lattice spacing and the interfringe angle from this image and other images, one can demonstrate that the as-fabricated composite nanostructures consist of nanocrystals of Pt and W. By using selected-area diffraction (SAD) pattern (Fig. 11(c)), these nanocrystals are identified to be equilibrium phases of fcc Pt and bcc W. Figure 11(d) is an EDS spectrum obtained at the tip corresponding to Fig. 11(b). Pt and W peaks dominate in the spectrum although traces of C and O are also observed. Thus, it is confirmed that Pt has been effectively grown on the nanodendritic W structures. From the HRTEM micrographs, the average nanoparticle size at the present ion sputtering conditions was 2.3 nm. The particle size can be easily controlled by variation of the ion sputtering time.

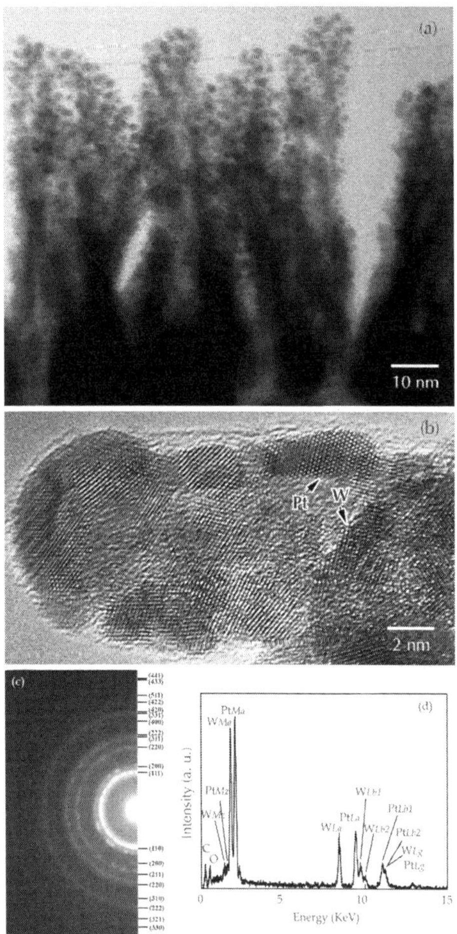

FIGURE 11 (a) Bright-field TEM image of the as-fabricated Pt-nanoparticle/W-nanodendrite composite structures on an Al_2O_3 substrate (b) an HRTEM image, (c) SAD pattern and (d) EDS spectrum taken from a tip of the composite nanostructures [33].

Using this process, various other metal composite nanoparticle/nanodendrite structures, such as Au/W, Mo/W, Au/Pt, and so on, can also be fabricated. Figure 12 shows an example of Au-nanoparticle/W-nanodendrite compound nanostructures grown on an insulator SiO_2 substrate. Therefore, the technique may be easily employed to fabricate metal nanoparticle/nanodendrite composite nanostructures of a wide range of materials. Because the nanodendrite structure possesses a comparatively large specific surface area, it has potential applications in catalysts, sensors, gas storages and so on.

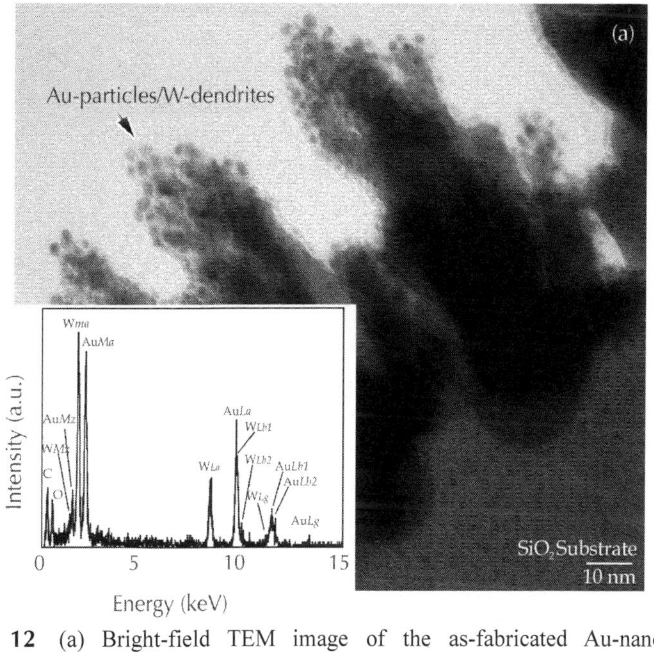

FIGURE 12 (a) Bright-field TEM image of the as-fabricated Au-nanoparticle/W-nanodendrite composite structures on a SiO_2 substrate and (b) an EDS spectrum taken from a tip of the composite nanostructures. The W nanodendrites were fabricated with an electron beam irradiation fluence of 4.4×10^{21} cm^{-2}. Au nanoparticles were deposited by a quick auto coater system (JEOL JFC-1500). The anodic voltage used during sputtering was 1 kV, and anodic current was 7 mA. The ion sputtering time was 7 s. The average Au nanoparticle size was measured to be 2.1 nm [33].

5.6 CONCLUSION

Self-standing metal nano-sized branched structures including nanowire arrays, nanodendrites, and nanofractal-like trees with controlled size and position were grown on insulator (SiO_2, Al_2O_3) substrates by the EBID process. The nanostructures of the different morphologies can be obtained by controlling the intensity of the electron beam during the EBID process. The nucleation and growth of the nano-sized branched structures are proposed to be related to a mechanism involving charge-up produced on the surface of the substrate, and the movement of the charges to and charges accumulation at the convex surface of the substrate and the tips of the deposits. High-energy electron irradiation enhances diffusion of the metallic atoms in the nanostructures and hence promotes crystallization. More crystallized metal nano-branched structures are achieved by the EBID process using high-energy electron beams. The nano-branched structures can be easily decorated by metallic nanoparticles to form composite nano-structures such as the Pt nanoparticles/nanodendrites. Therefore the nano-sized branched structures and their fabrication process may be applied in technology to realize various functional nanomaterials such as catalysts, sensor materials, and emitters.

KEYWORDS

- **electron-beam-induced deposition**
- **nanodendrites**
- **nanofractal-like trees**
- **nanowire arrays**

REFERENCES

1. Shi, J.; Gider, S.; Babcock, K.; Awschalom, D. D.; *Science* **1996,** *271,* 937.
2. Favier, F.; Walter, E.; Zach, M.; Benter, T.; Penner, R. M.; *Science* **2001,** *293,* 2227.
3. Xia, Y. N.; Yang, P. D.; Sun, Y. G.; Wu, Y. Y.; Mayer, B.; Gates, B.; Yin, Y. D.; Kim, F.; Yan, H. Q.; *Adv. Mater.* **2003,** *15,* 353.
4. Lao, Y. L.; Wen, J. G.; Ren, Z. F.; *Nano Lett.* **2002,** *2,* 1287.
5. Manna, L.; Milliron, D. J.; Meisel, A.; Scher, E. C.; Alivisatos, A. P.; *Nature Mater.* **2003,** *2,* 382.
6. Yan, H.; He, R.; Pham, J.; Yang, P.; *Adv. Mater.* **2003,** *15,* 402.
7. Dick, K. A.; Deppert, K.; Larsson, M. W.; Martensson, T.; Seifert, W.; Wallenberg, L. R.; Samuelson, L.; *Nature Mater.* **2004,** *3,* 380.
8. Zhang, H. F.; Wang, C. M.; Buck, E. C.; Wang, L. S.; *Nano Lett.* **2003,** *3,* 577.
9. Shi, Z. J.; Lian, Y. F.; Liao, F. H.; Zhou, X. H.; Gu, Z. N.; Zhang, T.; Iijima, S.; Li, H. D.; Yue, K. T.; Zhang, S. L.; *J. Phys. Chem. Solids* **2000,** *61,* 1031.
10. Dai, Z. R.; Pan, Z. W.; Wang, Z. L.; *Adv. Funct. Mater.* **2003,** *13,* 9.

11. Jin, Y.; Tang, K. B.; Huang, C. H.; An, L. Y.; *J. Cryst. Growth* **2003,** *253,* 429.
12. Mitsuishi, K.; Shimojo, M.; Han, M.; Furuya, K.; *Appl. Phys. Lett.* **2003,** *83,* 2064.
13. Dong, L. X.; Arai, F.; Fukuda, T.; *Appl. Phys. Lett.* **2002,** *81,* 1919.
14. Utke, I.; Hoffmann, P.; Dwir, B.; Leifer, K.; Kapon, E.; Doppelt, P.; *J. Vac. Sci. Technol. B* **2000,** *18,* 3168.
15. Brückl, H.; Kretz, J.; Koops, H. W.; Reiss, G.; *J. Vac. Sci. Technol. B* **1999,** *17,* 1350.
16. H. Koops, W. P.; J.; Kretz, M.; Rudolph, M.; Weber, G.; Dahm, Lee, K. L.; *Jpn. J. Appl. Phys. Part* **1994,** *1–33,* 7099.
17. H. Koops, W. P.; R.; Weiel, Kern, D. P.; Baum, T. H.; *J.; Vac. Sci. Technol. B* **1988,** *6,* 477.
18. Hiroshima, H.; Suzuki, N.; Ogawa, N.; Komuro, M.; *Jpn. J. Appl. Phys.* **1999,** *1–38,* 7135.
19. Hoyle, P. C.; Cleaver, J. R. A.; Ahmed, H.; *J. Vac. Sci. Technol. B* **1996,** *14,* 662.
20. Kohlmann-von Platen, K. T.; Chlebek, J.; Weiss, M.; Reimer, K.; Oertel, H.; Brünger, W. H.; *J. Vac. Sci. Technol. B* **1993,** *11,* 2219.
21. Matsui, S.; Kaito, T.; Fujita, J.; Komura, M.; Kanda, K.; Haruyama, Y.; *J. Vac. Sci. Technol. B* **2000,** *18,* 3181.
22. Liu, Z. Q.; Mitsuishi, K.; Furuya, K.; *Nanotechnology* **2004,** *15,* S414.
23. Banhart, F.; *Phys. Rev. E* **1995,** *52,* 5156.
24. Zhang, J. Z.; X. Y. Ye, Yang, X. J.; Liu, D.; *Phys. Rev. E* **1997,** *55,* 5796.
25. Wang, H. Z.; Liu, X. H.; Yang, X. J.; Wang, X.; *Mat. Sci. Eng. A* **2001,** *311,* 180.
26. Song, M.; Mitsuishi, K.; Tanaka, M.; Takeguchi, M.; Shimojo, K.; Furuya, *Appl. Phys. A* **2004,** *80,* 1431.
27. Xie, G. Q.; Song, M.; Mitsuishi, K.; Furuya, K.; *J.; Nanosci. Nanotechnol.* **2005,** *5,* 615.
28. Xie, G. Q.; Song, M.; Mitsuishi, K.; Furuya, K.; *Physica E* **2005,** *29,* 564.
29. Xie, G. Q.; Song, M.; Mitsuishi, K.; Furuya, K.; *Jpn. J.; Appl. Phys.* **2005,** *44,* 5654.
30. Song, M.; Mitsuishi, K.; Furuya, K.; *Mater. Trans.* **2007,** *48,* 2551.
31. Song, K.; Mitsuishi, M.; Takeguchi, K.; Furuya, *Appl. Surf. Sci.* **2005,** *241,* 107.
32. Song, M.; Mitsuishi, K.; Furuya, K.; *Physica E* **2005,** *29,* 575.
33. Xie, G. Q.; Song, M.; Furuya, K.; Louzguine, D. V.; Inoue, A.; *Appl. Phys. Lett.* **2006,***88,* 263120.
34. Xie, G. Q.; Song, M.; Mitsuishi, K.; Furuya, K.; *J. Mater. Sci.* **2006,** *41,* 2567.
35. Xie, G. Q.; Song, M.; Furuya, K.; *J. Mater. Sci.* **2006,** *41,* 4537.
36. Furuya, K.; Takeguchi, M.; Song, M.; Mitsuishi, K.; Tanaka, M.; *J. Phys. Conf. Ser.* **2008,** *126,* 012024.
37. Song, M.; Mitsuishi, K.; Furuya, K.; *Mater. Sci. Forum* **2005,** *475–479,* 4035.
38. Nagase, T.; Umakoshi, Y.; *Mater. Sci. Eng. A* **2003,** *352,* 251.
39. Gao, P. X.; Ding, Y.; Mai, W.; Hughes, W. L.; Lao, C.; Wang, Z. L.; *Science* **2005,** *309,* 1700.
40. Dick, K. A.; Deppert, K.; Larsson, M. W.; Martensson, T.; Seifert, W.; Wallenberg, L. R.; Samuelson, L. *Nature Mater.* **2004,** *3,* 380.
41. Li, M.; Schnablegger, H.; Mann, S.; *Nature* **1999,** *402,* 393.
42. Pal, U. J.; Ramirez, F. S.; Liu, H. B.; Medina, A.; Ascencio, J. A.; *Appl. Phys. A* **2004,** *79,* 79.

43. Joo, S. H.; Choi, S. J.; Oh, I.; Kwak, J.; Liu, Z.; Terasaki, O.; Ryoo, R.; *Nature* **2001,** *412,* 169.
44. A.; Ruiz, J.; Arbiol, A.; Cirera, A.; Cornet, Morante, J. R.; *Mater. Sci. Eng. C* **2005,** *19,* 105.
45. De Meijer, R. J.; Stapel, C.; Jones, D. G.; Roberts, P. D.; Rozendaal, A.; Macdonald, W. G.; Chen, K. Z.; Zhang, Z. K.; Cui, Z. L.; Zuo, D. H.; Yang, D. Z.; *Nanostruct. Mater.* **1997,** *8,* 205.
46. Wang, A.; Liu, J.; Lin, S.; Lin, T.; Mou, C.; *J.; Catal.* **2005,** *233,* 1486.
47. Liu, P.; Rodriguez, J. A.; Muckerman, J. T.; Hrbek, J.; *Surf. Sci.* **2003,** *530,* L313.

CHAPTER 6

NEW HORIZONS IN NANOTECHNOLOGY

ANAMIKA SINGH, RAJEEV SINGH, KOZLOV GEORGII VLADIMIROVICH, and YANOVSKII YURII GRIGOR'EVICH

CONTENTS

6.1 Nanotherapeutics 276
6.1.1 Introduction 276
6.1.2 Properties of Nanotherapeutics 276
6.1.3 Nanoparticles Types 277
6.1.4 Conclusion 283
6.2 Nanocomposites Structures 283
6.2.1 Introduction 283
6.2.2 Experimental Part 285
6.2.3 Results and Discussion 286
6.2.4 Conclusions 308
Keywords 309
References 309

6.1 NANOTHERAPEUTICS

6.1.1 INTRODUCTION

Nanotechnology expands itself as one of the major area of science [1, 2]. An important and exciting aspect of nanomedicine is the use of nanoparticle for target specific drug delivery systems and it allows a new innovative therapeutic approaches. Due to their small size, these drug delivery systems are promising tools in therapeutic approaches such as selective or targeted drug delivery towards a specific tissue or organ. It also enhances drug transport across biological barriers and intracellular drug delivery. Nanotherapeutic agents work successfully against traditional drugs, which exhibits very less bioavailability and are often associated with gastro-intestinal side effects. Nanotherapeutics improves the delivery of drugs that cannot normally be taken orally and it also improves the safety and efficacy of low molecular weight drugs. It also improves the stability and absorption of proteins that normally cannot be taken orally.

6.1.2 PROPERTIES OF NANOTHERAPEUTICS

A good therapeutic agent or nanoparticle should have ability to: (1) cross one or various biological membranes (e.g., mucosa, epithelium, endothelium) before (2) diffusing through the plasma membrane to (3) finally gain access to the appropriate organelle where the biological target is located.

Different methods of drug delivery are listed below:

1. Oral Drug Delivery
2. Injection Based Drug Delivery
3. Transdermal Drug Delivery
4. Bone Marrow Infusion
5. Control Release Systems
6. Targeted Drug Delivery:
    a. Therapeutical Monoclonal
    b. Antibodies
    c. Liposomes
    d. Microparticles
    e. Modified Blood Cells
    f. Nanoparticles
7. Implant Drug Delivery System

In this chapter, we will discuss about Nanoparticles and its mode of action.

6.1.3 NANOPARTICLES TYPES

6.1.3.1 METAL-BASED NANOPARTICLES

Metallic nanoparticles are either spherical metal or semiconductor particles with nanometer-sized diameters. Similar to nanoparticles "nanocrystal" (NC) or "nanocrystallite" term is used for nanoparticles made up of semiconductors. Generally semiconductor materials from the groups II-VI (CdS, CdSe, CdTe), III-V (InP, InAs) and IV-VI (PbS, PbSe, PbTe) are of particular interest. Metal Based nanoparticle are used for nanoscale transistors, biological sensors, and next generation photovoltaics [3].

6.1.3.2 LIPID-BASED NANOPARTICLES

Lipid based nanoparticles are widely used for drug targeting and drug delivery. Solid lipid nanoparticles (SLN) introduced in 1991 represent an alternative carrier system for drugs [4]. Nanoparticles made from solid lipids are attracting major attention as novel colloidal drug carrier for intravenous applications as they have been proposed as an alternative particulate carrier system. SLN are submicron colloidal carriers ranging from 50 to 1000 nm, which are composed of physiological lipid [5–7].

Lipid based system are many used because:

1. Lipids enhance oral bioavailability and reduce plasma profile variability.
2. Better characterization of lipoid excipients.
3. An improved ability to address the key issues of technology transfers and manufactures scale-up.

6.1.3.3 POLYMER-BASED NANOPARTICLES

Polymeric nanoparticles are nanoparticles, which are prepared from polymers. Polymeric nanoparticles forms (1) the micronization of a material into nanoparticles and (2) the stabilization of the resultant nanoparticles [8]. As for the micronization, one can start with either small monomers or a bulk polymer. The drug is dissolved, entrapped, encapsulated or attached to a nanoparticles and one can obtain different nanoparticles, nanospheres or nanocapsules according to methods of preparation [9]. Gums, Gelatin Sodium alginate Albumin are used for polymer based drug delivery. Polymeric nanoparticles are prepared by Cellulosics, Poly(2-

hydroxy ethyl methacrylate), Poly(N-vinyl pyrrolidone), Poly(vinyl alcohol), Poly(methyl methacrylate), Poly(acrylic acid), Polyacrylamide, Poly(ethylene-covinyl acetate) like polymeric materials. Polymer used in drug delivery must have following qualities like it should be chemically inert, nontoxic and free of leachable impurities [10].

6.1.3.4 BIOLOGICAL NANOPARTICLES

For proper interaction with molecular targets, a biological or molecular coating or layer is required which act as a bioinorganic interface, and it should be attached to the nanoparticle for proper interaction. These biological coating may be antibodies, biopolymers like collagen [11]. A layer of small molecules that make the nanoparticles biocompatible are also used as bioinorganic interface [12]. For detection and analysis these nanoparicles should be optically active and they should either fluoresces or change their color in different intensities of light.

6.1.3.5 NANOCARRIERS IN DRUG DELIVERY

Nanoparticles and Liposomes get absorbed by blood streams for its mode of action. For particular function theses nanoparticles must have specific structure and composition. These particles rapidly get cleared from blood stream by macrophages of reticuloendothelial system (RES) [13, 14].

Nanoparticles can be delivered to targets by Phagocytosis pathway and Non-phagocytosis pathway are described below.

6.1.3.5.1 PHAGOCYTOSIS PATHWAY

Phagocytosis occurs by profession phagocytes and by nonprofessional phagocytes. A professional phagocyte includes macrophages, monocytes, neutrophils and dendritic cell, while nonprofessional phagocytes are fibroblast, epithelial and endothelial cell [15, 16].

Phagocytosis takes place by:

1. Recognition of the opsonized particles in the blood stream.
2. Adhesion of particles to macrophages
3. Ingestion of the particle

Opsonization of the nanoparticles is a major step and it takes place before phagocytosis. In this process nanoparticles get tagged by a protein known as *opsonins*. Due to this tagging nanoparticles make a visible complex for macrophages, this whole process takes place in blood streams. Opsonins are immunoglobins (Ig) G, M and complement components like C3, C4 and C5 [17, 18].

These activated particles then get attached to macrophages and this interaction is similar to receptor-ligand interaction [15, 19]. The most important receptor is Fc receptor (FcR) and its complement receptor (CR). This receptor attachment is mediated by Rho family GTPase and surface extension occurs, which leads to formation of pseudopodia around nanoparticles. Small actin fibers get assembled in it and engulfment of nanoparticles takes place. [20, 21].

As actin is depolymerized from the phagosome, the newly denuded vacuole membrane becomes accessible to early endosomes [15, 22]. Through a series of fusion and fission events, the vacuolar membrane and its contents will mature, fusing with late endosomes and ultimately lysosomes to form a phagolysosome (Fig. 1 of Chapter 3). The rate of these events depends on the surface properties of the ingested particle, typically from half to several hours [15]. The phagolysosomes become acidified due to the vacuolar proton pump ATPase located in the membrane and acquire many enzymes, including esterases and cathepsins [16, 23].

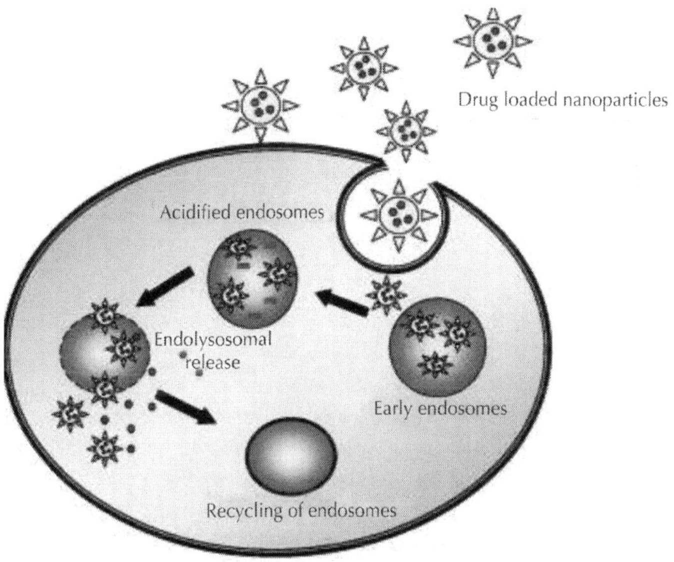

FIGURE 1 Nanoparticles enter the cell through receptor mediated endocytosis and get localized in the endolysosomal compartment.

6.1.3.5.2 NON-PHAGOCYTIC PATHWAY

This is also known as pinocytosis or cell drinking method. It is basically taking of fluids and solutes. Very small nanoparticles get entered in well and absorbed by pinocytosis as phagocytosis is restricted to size of particle. The process of Pinocytosis occurs in specialized cell. It may be clathrin mediated endocytosis, caveolae-mediated endocytosis, macropinocytosis and caveolae-independent endocytosis

6.1.3.5.2.1 CLATHRIN-MEDIATED ENDOCYTOSIS

Endocytosis via clathrin-coated pits, or clathrin-mediated endocytosis (CME), occurs constitutively in all mammalian cells, and fulfills crucial physiological roles, including nutrient uptake and intracellular communication. For most cell types, CME serves as the main mechanism of internalization for macromolecules and plasma membrane constituents. CME via specific receptor-ligand interaction is the best described mechanism, to the extent that it was previously referred to as "receptor-mediated endocytosis" (RME). However, it is now clear that alternative nonspecific endocytosis via clathrin-coated pits also exists (as well as receptor-mediated but clathrin-independent endocytosis). Notably, the CME, either receptor-dependent or independent, causes the endocytosed material to end up in degradative lysosomes. This has an important impact in the drug delivery field since the drug-loaded nanocarriers may be tailored in order to become metabolized into the lysosomes, thus releasing their drug content intracellularly as a consequence of lysosomal biodegradation.

6.1.3.5.2.2 CAVEOLAE-MEDIATED ENDOCYTOSIS

Although CME is the predominant endocytosis mechanism in most cells, alternative pathways have been more recently identified, caveolae-mediated endocytosis (CvME) being the major one. Caveolae are characteristic flask-shaped membrane invaginations, having a size generally reported in the lower end of the 50–100 nm range [24–27], typically 50–80 nm. They are lined by caveolin, a dimeric protein, and enriched with cholesterol and sphingolipids (Fig. 2 of Chapter 3). Caveolae are particularly abundant in endothelial cells, where they can constitute 10–20% of the cell surface [26], but also smooth muscle cells and fibroblasts. CvMEs are involved in endocytosis and trancytosis of various proteins; they also constitute a port of entry for viruses (typically the SV40 virus) [28] and receive increasing attention for drug delivery applications using nanocarriers.

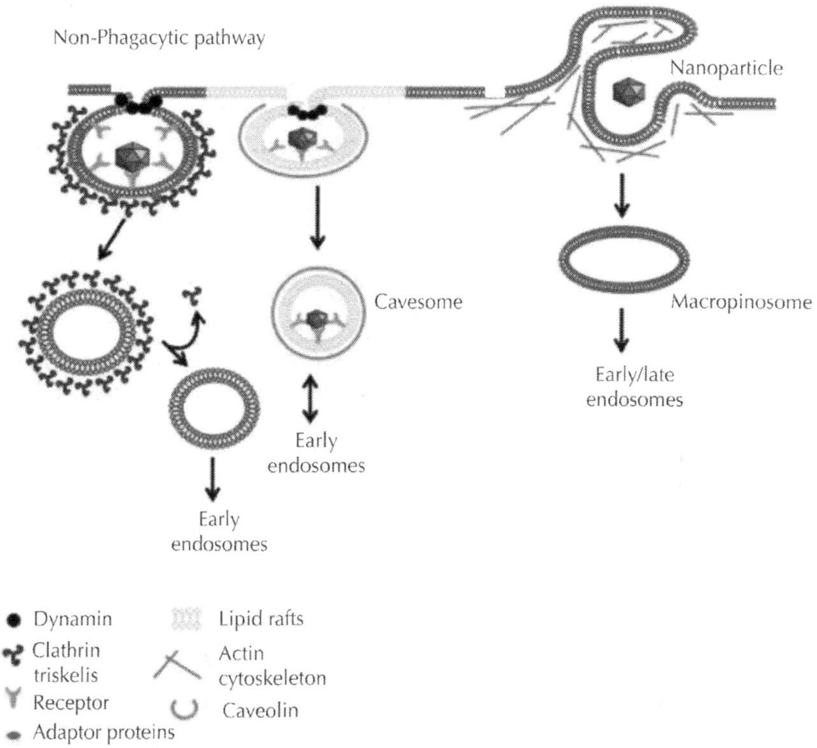

FIGURE 2 Diagrammamtic presentation of (A) Clathrin-mediated endocytosis (B) Caveolae-mediated endocytosis (C) Macropinocytosis

Unlike CME, CvME is a highly regulated process involving complex signaling, which may be driven by the cargo itself [25, 26]. After binding to the cell surface, particles move along the plasma membrane to caveolae invaginations, where they may be maintained through receptor-ligand interactions [25]. Fission of the caveolae from the membrane, mediated by the GTPase dynamin, then generates the cytosolic caveolar vesicle, which does not contain any enzymatic cocktail. Even this pathway is employed by many pathogens to escape degradation by lysosomal enzymes. The use of nanocarriers exploiting CvME may therefore be advantageous to by-pass the lysosomal degradation pathway when the carried drug (e.g., peptides, proteins, nucleic acids, etc.) is highly sensitive to enzymes. On the whole, the uptake kinetics of CvME is known to occur at a much slower rate than that of CME. Ligands known to be internalized by CvME include folic acid, albumin and cholesterol [25].

6.1.3.5.2.3 MACROPINOCYTOSIS

Macropinocytosis is another type of clathrin-independent endocytosis pathway [29], occurring in many cells, including macrophages [24]. It occurs via formation of actin-driven membrane protusions, similarly to phagocytosis. However, in this case, the protusions do not zipper up along the ligand-coated particle; instead, they collapse onto and fuse with the plasma membrane [26] (Fig. 2 of Chapter 3). This generates large endocytic vesicles, called macropinosomes, which sample the extracellular milieu and have a size generally bigger than 1 lm [26] (and sometimes as large as5 lm [24]). The intracellular fate of macropinosomes varies depending on the cell type, but in most cases, they acidify and shrink. They may eventually fuse with lysosomal compartments or recycle their content to the surface [24] (Fig. 2 of Chapter 3). Macropinosomes have not been reported to contain any specific coating, nor do they concentrate receptors. This endocytic pathway does not seem to display any selectivity, but is involved, among others, in the uptake of drug nanocarriers [29].

6.1.3.5.2.4 MAGNETIC NANOPARTICLES (MNPS)

Magnetic nanoparticles (MNPs) have many applications in different areas of biology and medicine. MNPs are used for hyperthermia, magnetic resonance imaging, immunoassay, purification of biologic fluids, cell and molecular separation, tissue engineering. The design of magnetically targeted nanosystems (MNSs) for a smart delivery of drugs to target cells is a promizing direction of nanobiotechnology. They traditionally consist on one or more magnetic cores and biological or synthetic molecules, which serve as a basis for polyfunctional coatings on MNPs surface. The coatings of MNSs should meet several important requirements. They should be biocompatible, protect magnetic cores from influence of biological liquids, prevent MNSs agglomeration in dispersion, provide MNSs localization in biological targets and homogenity of MNSs sizes. The coatings must be fixed on MNPs surface and contain therapeutic products (drugs or genes) and biovectors for recognition by biological systems (Fig. 3 of Chapter 3).

Proteins are promizing materials for creation of coatings on MNPs for biology and medicine. When proteins are used as components of coatings it is of the first importance that they keep their functional activity. Protein binding on MNPs surface is a difficult scientific task. Traditionally bifunctional linkers (glutaraldehyde, carbodiimide) are used for protein cross-linking on the surface of MNPs and modification of coatings by therapeutic products and biovectors. Researchers modified MNPs surface with aminosilanes and performed protein molecules attachment using glutaraldehyde. In the issue bovine serum albumin (BSA) was adsorbed on MNPs surface in the presence of carbodiimide. These works revealed several disadvantages of this way of protein fixing which make it unpromizing. Some of them are clusters formation as a result of linking of protein molecules adsorbed on different MNPs, desorption of proteins from MNSs surface as a result of incomplete linking, uncontrollable linking of proteins in solution (Fig. 4 of Chapter 3). The creation of stable protein coatings with retention of native properties of molecules still is an important biomedical problem.

It is known that proteins can be chemically modified in the presence of free radicals with formation of cross-links. The goals of the work were to create stable protein coating on the surface of individual MNPs using a fundamentally novel approach based on the ability of proteins to form interchain covalent bonds under the action of free radicals and estimate activity of proteins in the coating.

6.1.4 CONCLUSION

Now-a-days *n*-number of drugs are available in market, so nanotherapeutics is particularly very important to minimize health hazards caused by drugs. The most important aspect of nanotherapeutics is that it should be noninvasive and target oriented and it should have very less or no side effects. In-spite of using nonmaterial one can create nanodevices for better functioning. As most of the peptides and proteins have short half-lives so frequent dosing or injections are required, which is not applicable for many situations by using nanotherapeutics one can solve these problems. Thus, nanotherapeutics is expanding day by day in drug delivery system.

6.2 NANOCOMPOSITES STRUCTURES

6.2.1 INTRODUCTION

The modern methods of experimental and theoretical analysis of polymer materials structure and properties allow not only to confirm earlier propounded hypotheses, but also to obtain principally new results. Let us consider some important problems of particulate-filled polymer nanocomposites, the solution of which allows advancing substantially in these materials properties understanding and prediction. Polymer nanocomposites multicomponentness (multiphaseness) requires their structural components quantitative characteristics determination. In this aspect interfacial regions play a particular role, since it has been shown earlier, that they are the same reinforcing element in elastomeric nanocomposites as nanofiller actually [30]. Therefore the knowledge of interfacial layer dimensional characteristics is necessary for quantitative determination of one of the most important parameters of polymer composites in general—their reinforcement degree [31, 32].

The aggregation of the initial nanofiller powder particles in more or less large particles aggregates always occurs in the course of technological process of preparation particulate-filled polymer composites in general [33] and elastomeric nanocomposites in particular [34]. The aggregation process tells on composites (nanocomposites) macroscopic properties [31–33]. For nanocomposites nanofiller aggregation process gains special significance, since its intensity can be the one, that nanofiller particles aggregates size exceeds 100 nm—the value, which is assumed (though conditionally enough [35]) as an upper dimensional limit for nanoparticle. In other words, the aggregation process can result to the situation when primordially supposed nanocomposite ceases to be one. Therefore at present several methods exist, which allow to suppress nanoparticles aggregation process [34, 36]. This also assumes the necessity of the nanoparticles aggregation process quantitative analysis.

It is well known [37, 38] that in particulate-filled elastomeric nanocomposites (rubbers) nanofiller particles form linear spatial structures ("chains"). At the same time in polymer composites, filled with disperse microparticles (microcomposites) particles (aggregates of particles) of filler form a fractal network, which defines polymer matrix structure (analog of fractal lattice in computer simulation) [33]. This results to different mechanisms of polymer matrix structure formation in micro and nanocomposites. If in the first filler particles (aggregates of particles) fractal network availability results to "disturbance" of polymer matrix structure, that is expressed in the increase of its fractal dimension d_f [33], then in case of polymer nanocomposites at nanofiller contents change the value d_f is not changed and equal to matrix polymer structure fractal dimension [39]. As it has been expected, the change of the composites of the indicated classes structure formation mechanism change defines their properties, in particular, reinforcement degree [40, 41]. Therefore, nanofiller structure factuality strict proof and its dimension determination are necessary.

As it is known [42, 43], the scale effects in general are often found at different materials mechanical properties study. The dependence of failure stress on grain size for metals (Holl-Petsch formula) [44] or of effective filling degree on filler particles size in case of polymer composites [45] are examples of such effect. The strong dependence of elasticity modulus on nanofiller particles diameter is observed for particulate-filled elastomeric nanocomposites [5]. Therefore, it is necessary to elucidate the physical grounds of nano- and micromechanical behavior scale effect for polymer nanocomposites.

At present a wide list of disperse material is known, which is able to strengthen elastomeric polymer materials [34]. These materials are very diverse on their surface chemical constitution, but particles small size is a common feature for them. On the basis of this observation the hypothesis was offered that any solid material would strengthen the rubber at the condition that it is in a very dispersed state and it could be dispersed in polymer matrix. Edwards [34] points out, that filler particles small size is necessary and, probably, the main requirement for reinforcement effect realization in rubbers. Using modern terminology, one can say, that for rubbers reinforcement the nanofiller particles, for which their aggregation process is suppressed as far as possible, would be the most effective ones [32, 41]. Therefore, the theoretical analysis of a nanofiller particles size influence on polymer nanocomposites reinforcement is necessary.

This chapter's purpose is to find the solution of the above-considered paramount problems with the help of modern experimental and theoretical techniques on the example of particulate-filled butadiene-stirene rubber.

6.2.2 EXPERIMENTAL PART

The made industrially butadiene-stirene rubber of mark SKS-30, which contains 7.0–12.3% *cis*- and 71.8–72.0% transbonds, with density of 920–930 kg/m^3 was used as matrix polymer. This rubber is fully amorphous one.

Fullerene-containing mineral shungite of Zazhoginsk's deposit consists of ~30% globular amorphous metastable carbon and ~70% high-disperse silicate particles. Besides, industrially made technical carbon of mark № 220 was used as nanofiller. The technical carbon, nano- and microshugite particles average size makes up 20, 40 and 200 nm, respectively. The indicated filler content is equal to 37 mass %. Nano- and microdimensional disperse shungite particles were prepared from industrially output material by the original technology processing. The size and polydispersity analysis of the received in milling process shungite particles was monitored with the aid of analytical disk centrifuge (CPS Instruments, Inc., USA), allowing to determine with high precision size and distribution by the sizes within the range from 2 nm up to 50 mcm.

Nanostructure was studied on atomic-forced microscopes Nano-DST (Pacific Nanotechnology, USA) and Easy Scan DFM (Nanosurf, Switzerland) by semi-contact method in the force modulation regime. Atomic-force microscopy results were processed with the help of specialized software package SPIP (Scanning Probe Image Processor, Denmark). SPIP is a powerful programs package for processing of images, obtained on SPM, AFM, STM, scanning electron microscopes, transmission electron microscopes, interferometers, confocal microscopes, profilometers, optical microscopes and so on. The given package possesses the whole functions number, which are necessary at images precise analysis, in a number of which the following ones are included:

- the possibility of three-dimensional reflecting objects obtaining, distortions automatized leveling, including Z-error mistakes removal for examination of separate elements and so on;
- quantitative analysis of particles or grains, more than 40 parameters can be calculated for each found particle or pore: area, perimeter, mean diameter, the ratio of linear sizes of grain width to its height, distance between grains, coordinates of grain center of mass a.a. can be presented in a diagram form or in a histogram form.

The tests on elastomeric nanocomposites nanomechanical properties were carried out by a nanointentation method [46] on apparatus Nano Test 600 (Micro Materials, Great Britain) in loads wide range from 0.01 mN up to 2.0 mN. Sample indentation was conducted in 10 points with interval of 30 mcm. The load was increased with constant rate up to the greatest given load reaching (for the rate 0.05 mN/s-1 mN). The indentation rate was changed in conformity with the greatest load value counting, that loading cycle should take 20 s. The unloading was conducted with the same rate as loading. In the given experiment the "Berkovich indentor" was used with the angle at the top of 65.3° and rounding radius of 200 nm. Indentations were carried out in the checked load regime with preload of 0.001 mN.

For elasticity modulus calculation the obtained in the experiment by nanoindentation course dependences of load on indentation depth (strain) in ten points for each sample at loads of 0.01, 0.02, 0.03, 0.05, 0.10, 0.50, 1.0 and 2.0 mN were processed according to Oliver-Pharr method [47].

6.2.3 RESULTS AND DISCUSSION

In Fig. 1, the obtained according to the original methodics results of elasticity moduli calculation for nanocomposite butadiene-stirene rubber/nanoshungite components (matrix, nanofiller particle and interfacial layers), received in interpolation process of nanoindentation data, are presented. The processed in polymer nanocomposite SPIP image with shungite nanoparticles allows experimental determination of interfacial layer thickness l_{if}, which is presented in Fig. 1 as steps on elastomeric matrix-nanofiller boundary. The measurements of 34 such steps (interfacial layers) width on the processed in SPIP images of interfacial layer various section gave the mean experimental value l_{if}=8.7 nm. Besides, nanoindentation results (Fig. 1, figures on the right) showed, that interfacial layers elasticity modulus was only by 23–45% lower than nanofiller elasticity modulus, but it was higher than the corresponding parameter of polymer matrix in 6.0–8.5 times. These experimental data confirm, that for the studied nanocomposite interfacial layer is a reinforcing element to the same extent, as nanofiller actually [30, 32, 41].

Let us fulfill further the value l_{if} theoretical estimation according to the two methods and compare these results with the ones obtained experimentally. The first method simulates interfacial layer in polymer composites as a result of interaction of two fractals—polymer matrix and nanofiller surface [48, 49]. In this case there is a sole linear scale l, which defines these fractals interpenetration distance [50]. Since nanofiller elasticity modulus is essentially higher, than the corresponding parameter for rubber (in the considered case—in 11 times, see Fig. 1), then the indicated interaction reduces to nanofiller indentation in polymer matrix and then $l=l_{if}$. In this case it can be written [50]:

$$l_{if} \approx a\left(\frac{R_p}{a}\right)^{2\left(d-d_{surf}\right)/d}, \tag{1}$$

where a is a lower linear scale of fractal behavior, which is accepted for polymers as equal to statistical segment length l_{st} [51], R_p is a nanofiller particle (more precisely, particles aggregates) radius, which for nanoshungite is equal to ~84 nm [52], d is dimension of Euclidean space, in which fractal is considered (it is obvious, that in our case d=3), d_{surf} is fractal dimension of nanofiller particles aggregate surface.

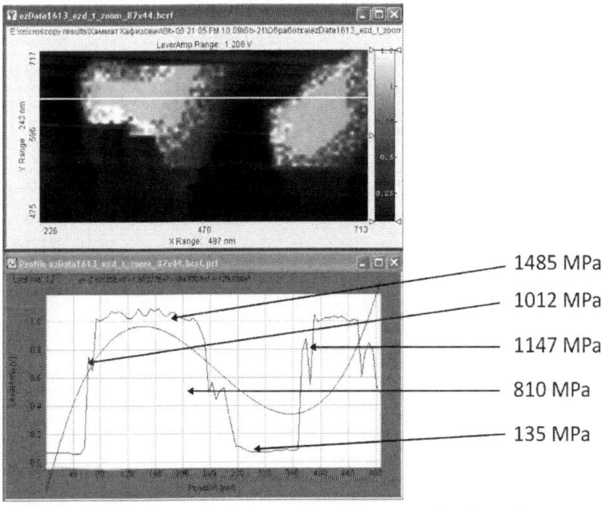

FIGURE 1 The processed in SPIP image of nanocomposite butadiene-styrene rubber/nanoshungite, obtained by force modulation method, and mechanical characteristics of structural components according to the data of nanoindentation (strain 150 nm).

The value l_{st} is determined as follows [53]:

$$l_{st} = l_0 C_\infty , \tag{2}$$

where l_0 is the main chain skeletal bond length, which is equal to 0.154 nm for both blocks of butadiene-stirene rubber [54], C_∞ is characteristic ratio, which is a polymer chain statistical flexibility indicator [55], and is determined with the help of the equation [51]:

$$T_g = 129\left(\frac{S}{C_\infty}\right)^{1/2} , \tag{3}$$

where T_g is glass transition temperature, equal to 217 K for butadiene-stirene rubber [32], S is macromolecule cross-sectional area, determined for the mentioned rubber according to the additivity rule from the following considerations. As it is known [56], the macromolecule diameter quadrate values are equal: for polybutadiene—20.7 Å^2 and for polystirene—69.8 Å^2. Having calculated cross-sectional area of macromolecule, simulated as a cylinder, for the indicated polymers according to the known geometrical formulas, let us obtain 16.2 and 54.8 Å^2, respectively. Further, accepting as S the average value of the adduced above areas, let us obtain for butadiene-stirene rubber S=35.5 Å^2. Then according to the Eq. (3) at the indicated values T_g and S let us obtain C_∞=12.5 and according to the Eq. (2)—l_{st}=1.932 nm.

The fractal dimension of nanofiller surface d_{surf} was determined with the help of the equation [3]:

$$S_u = 410 R_p^{d_{surf}-d} , \tag{4}$$

where S_u is nanoshungite particles specific surface, calculated as follows [28]:

$$S_u = \frac{3}{\rho_n R_p} , \tag{5}$$

where ρ_n is the nanofiller particles aggregate density, determined according to the formula [3]:

$$\rho_n = 0.188\left(R_p\right)^{1/3} \tag{6}$$

The calculation according to the Eqs. (4)–(6) gives d_{surf}=2.44. Further, using the calculated by the indicated mode parameters, let us obtain from the Eq. (1) the theoretical value of interfacial layer thickness l_{if}^{T}=7.8 nm. This value is close enough to the obtained one experimentally (their discrepancy makes up ~ 10%).

The second method of value l_{if}^{T} estimation consists in using of the two following equations [32, 58]:

$$\varphi_{if} = \varphi_n \left(d_{surf} - 2 \right) \tag{7}$$

and

$$\varphi_{if} = \varphi_n \left[\left(\frac{R_p + l_{if}^{T}}{R_p} \right)^3 - 1 \right], \tag{8}$$

where φ_{if} and φ_n are relative volume fractions of interfacial regions and nanofiller, accordingly.

The combination of the indicated equations allows to receive the following formula for l_{if}^{T} calculation:

$$l_{if}^{T} = R_p \left[\left(d_{surf} - 1 \right)^{1/3} - 1 \right] \tag{9}$$

The calculation according to the Eq. (9) gives for the considered nanocomposite l_{if}^{T}=10.8 nm, that also corresponds well enough to the experiment (in this case discrepancy between l_{if} and l_{if}^{T} makes up ~19%).

Let us note in conclusion the important experimental observation, which follows from the processed by program SPIP results of the studied nanocomposite surface scan (Fig. 1). As one can see, at one nanoshungite particle surface from one to three (in average—two) steps can be observed, structurally identified as interfacial layers. It is significant that these steps width (or l_{if}) is approximately equal to the first (the closest to nanoparticle surface) step width. Therefore, the indicated observation supposes, that in elastomeric nanocomposites at average two interfacial layers are formed: the first—at the expense of nanofiller particle surface with elastomeric matrix interaction, as a result of which molecular mobility in this layer is frozen and its state is glassy-like one, and the second—at the expense of glassy interfacial layer with elastomeric polymer matrix interaction. The most important question from the practical point of view, whether one interfacial layer or both serve as nanocomposite reinforcing element. Let us fulfill the following quantitative estimation for this question solution. The reinforcement degree (E_n/E_m) of polymer nanocomposites is given by the equation [32]:

$$\frac{E_n}{E_m} = 1 + 11\left(\varphi_n + \varphi_{if}\right)^{1.7}, \qquad (10)$$

where E_n and E_m are elasticity moduli of nanocomposite and matrix polymer, accordingly (E_m=1.82 MPa [32]).

According to the Eq. (7) the sum (φ_n+φ_{if}) is equal to:

$$\varphi_n + \varphi_{if} = \varphi_n\left(d_{surf} - 1\right), \qquad (11)$$

if one interfacial layer (the closest to nanoshungite surface) is a reinforcing element and

$$\varphi_n + 2\varphi_{if} = \varphi_n\left(2d_{surf} - 3\right), \qquad (12)$$

if both interfacial layers are a reinforcing element.

In its turn, the value φ_n is determined according to the equation [59]:

$$\varphi_n = \frac{W_n}{\rho_n}, \qquad (13)$$

where W_n is nanofiller mass content, ρ_n is its density, determined according to the Eq. (6).

The calculation according to the Eqs. (11) and (12) gave the following E_n/E_m values: 4.60 and 6.65, respectively. Since the experimental value E_n/E_m=6.10 is closer to the value, calculated according to the Eq. (12), then this means that both interfacial layers are a reinforcing element for the studied nanocomposites. Therefore the coefficient 2 should be introduced in the equations for value l_{if} determination (for example, in the Eq. (1)) in case of nanocomposites with elastomeric matrix. Let us remind, that the Eq. (1) in its initial form was obtained as a relationship with proportionality sign, that is, without fixed proportionality coefficient [50].

Thus, the used above nanoscopic methodics allow to estimate both interfacial layer structural special features in polymer nanocomposites and its sizes and properties. For the first time it has been shown, that in elastomeric particulate-filled nanocomposites two consecutive interfacial layers are formed, which are a reinforcing element for the indicated nanocomposites. The proposed theoretical methodics of interfacial layer thickness estimation, elaborated within the frameworks of fractal analysis, give well enough correspondence to the experiment.

For theoretical treatment of nanofiller particles aggregate growth processes and final sizes traditional irreversible aggregation models are inapplicable, since it is obvious, that in nanocomposites aggregates a large number of simultaneous growth takes place. Therefore the model of multiple growth, offered in Ref. [35], was used for nanofiller aggregation description.

In Fig. 2, the images of the studied nanocomposites, obtained in the force modulation regime, and corresponding to them nanoparticles aggregates fractal dimension d_f distributions are adduced. As it follows from the adduced values d_f^{ag} (d_f^{ag}=2.40–2.48), nanofiller particles aggregates in the studied nanocomposites are formed by a mechanism particle-cluster (P-Cl), that is, they are Witten-Sander clusters [61]. The variant A, was chosen which according to mobile particles are added to the lattice, consisting of a large number of "seeds" with density of c_0 at simulation beginning [60]. Such model generates the structures, which have fractal geometry on length short scales with value $d_f \approx 2.5$ (see Fig. 2) and homogeneous structure on length large scales. A relatively high particles concentration c is required in the model for uninterrupted network formation [60].

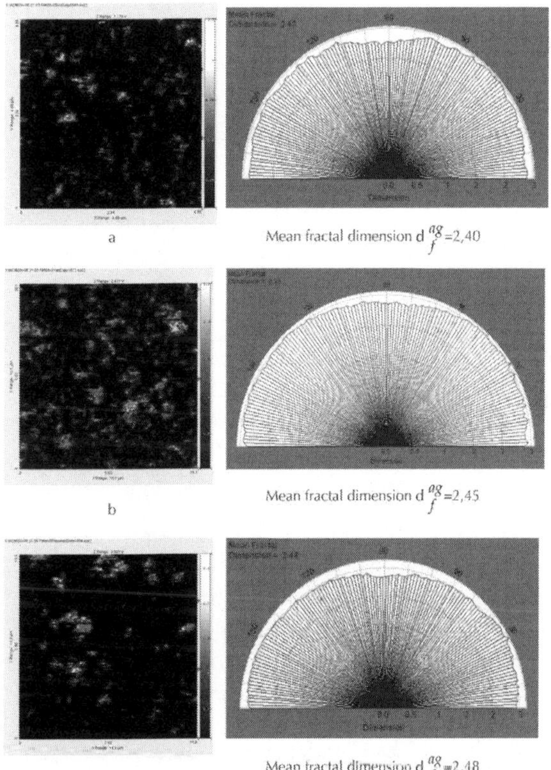

FIGURE 2 The images, obtained on atomic-force microscope in the force modulation regime, for nanocomposites, filled with technical carbon (a), nanoshungite (b), microshungite (c) and corresponding to them fractal dimensions d_f^{ag}.

In case of "seeds" high concentration c_0 for the variant A the following relationship was obtained [60]:

$$R_{max}^{d_f^{ag}} = N = c / c_0, \tag{14}$$

where R_{max} is nanoparticles cluster (aggregate) greatest radius, N is nanoparticles number per one aggregate, c is nanoparticles concentration, c_0 is "seeds" number, which is equal to nanoparticles clusters (aggregates) number.

The value N can be estimated according to the following equation [8]:

$$2R_{max} = \left(\frac{S_n N}{\pi \eta} \right)^{1/2} \tag{15}$$

where S_n is cross-sectional area of nanoparticles, of which an aggregate consists, η is a packing coefficient, equal to 0.74 [57].

The experimentally obtained nanoparticles aggregate diameter $2R_{ag}$ was accepted as $2R_{max}$ (Table 1) and the value S_n was also calculated according to the experimental values of nanoparticles radius r_n (Table 1). In Table 1 the values N for the studied nanofillers, obtained according to the indicated method, were adduced. It is significant that the value N is a maximum one for nanoshungite despite larger values r_n in comparison with technical carbon.

TABLE 1 The parameters of irreversible aggregation model of nanofiller particles aggregates growth.

Nanofiller	R_{ag}, nm	r_n, nm	N	R_{max}^T, nm	R_{ag}^T, nm	R_c, nm
Technical carbon	34.6	10	35.4	34.7	34.7	33.9
Nanoshungite	83.6	20	51.8	45.0	90.0	71.0
Microshungite	117.1	100	4.1	15.8	158.0	255.0

Further the Eq. (14) allows estimating the greatest radius R_{max}^T of nanoparticles aggregate within the frameworks of the aggregation model [60]. These values R_{max}^T are adduced in Table 1, from which their reduction in a sequence of technical carbon-nanoshungite-microshungite, that fully contradicts to the experimental data, that is, to R_{ag} change (Table 1). However, we must not neglect the fact that the Eq. (14) was obtained within the frameworks of computer simulation, where the initial aggregating particles sizes are the same in all cases [60]. For real nanocomposites the values r_n can be distinguished essentially (Table 1). It is expected, that the value R_{ag} or R^T_{max} will be the higher, the larger is the radius of nanoparticles, forming aggregate, is that is, r_n. Then theoretical value of nanofiller particles cluster (aggregate) radius R^T_{ag} can be determined as follows:

$$R^T_{ag} = k_n r_n N^{1/d_f^{ag}}, \tag{16}$$

where k_n is proportionality coefficient, in the present work accepted empirically equal to 0.9.

The comparison of experimental R_{ag} and calculated according to the Eq. (16) R^T_{ag} values of the studied nanofillers particles aggregates radius shows their good correspondence (the average discrepancy of R_{ag} and R^T_{ag} makes up 11.4%). Therefore, the theoretical model [60] gives a good correspondence to the experiment only in case of consideration of aggregating particles real characteristics and, in the first place, their size.

Let us consider two more important aspects of nanofiller particles aggregation within the frameworks of the model [60]. Some features of the indicated process are defined by nanoparticles diffusion at nanocomposites processing. Specifically, length scale, connected with diffusible nanoparticle, is correlation length ξ of diffusion. By definition, the growth phenomena in sites, remote more than ξ, are statistically independent. Such definition allows connecting the value ξ with the mean distance between nanofiller particles aggregates L_n. The value ξ can be calculated according to the equation [60]:

$$\xi^2 \approx \tilde{n}^{-1} R_{ag}^{d_f^{ag}-d+2}, \tag{17}$$

where c is nanoparticles concentration, which should be accepted equal to nanofiller volume contents φ_n, which is calculated according to the Eqs. (6) and (13).

The values r_n and R_{ag} were obtained experimentally (see histogram of Fig. 3). In Fig. 4 the relation between L_n and ξ is adduced, which, as it is expected, proves to be linear and passing through coordinates origin. This means, that the distance between nanofiller particles aggregates is limited by mean displacement of statistical walks, by which nanoparticles are simulated. The relationship between L_n and ξ can be expressed analytically as follows:

$$L_n \approx 9.6\xi, \text{ nm}. \tag{18}$$

The second important aspect of the model [60] in reference to nanofiller particles aggregation simulation is a finite nonzero initial particles concentration c or φ_n effect, which takes place in any real systems. This effect is realized at the condition $\xi \approx R_{ag}$, that occurs at the critical value $R_{ag}(R_c)$, determined according to the relationship [60]:

$$c \sim R_c^{d_f^{ag} - d} \tag{19}$$

The Eq. (19) right side represents cluster (particles aggregate) mean density. This equation establishes that fractal growth continues only, until cluster density reduces up to medium density, in which it grows. The calculated according to the Eq. (19) values R_c for the considered nanoparticles are adduced in Table 1, from which follows, that they give reasonable correspondence with this parameter experimental values R_{ag} (the average discrepancy of R_c and R_{ag} makes up 24%).

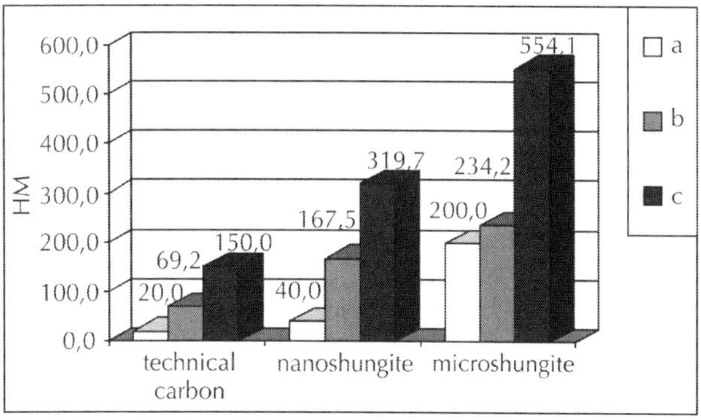

FIGURE 3 The initial particles diameter (a), their aggregates size in nanocomposite (b) and distance between nanoparticles aggregates (c) for nanocomposites, filled with technical carbon, nano- and microshungite.

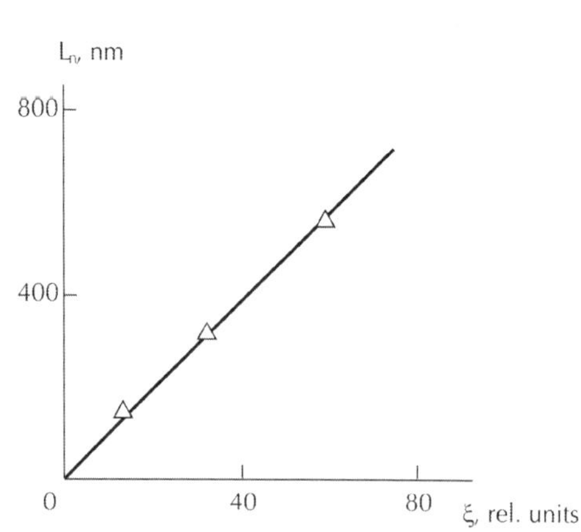

FIGURE 4 The relation between diffusion correlation length ξ and distance between nanoparticles aggregates L_n for considered nanocomposites.

Since the treatment [60] was obtained within the frameworks of a more general model of diffusion-limited aggregation, then its correspondence to the experimental data indicated unequivocally, that aggregation processes in these systems were controlled by diffusion. Therefore let us consider briefly nanofiller particles diffusion. Statistical walkers diffusion constant ζ can be determined with the aid of the relationship [60]:

$$\xi \approx (\zeta t)^{1/2}, \tag{20}$$

where t is walk duration.

The Eq. (20) supposes (at t=const) ζ increase in a number technical carbon-nanoshungite-microshungite as 196–1069–3434 relative units, that is, diffusion intensification at diffusible particles size growth. At the same time diffusivity D for these particles can be described by the well-known Einstein's relationship [62]:

$$D = \frac{kT}{6\pi\eta r_n \alpha}, \tag{21}$$

where k is Boltzmann constant, T is temperature, η is medium viscosity, α is numerical coefficient, which further is accepted equal to 1.

In its turn, the value η can be estimated according to the equation [63]:

$$\frac{\eta}{\eta_0} = 1 + \frac{2.5\varphi_n}{1-\varphi_n}, \tag{22}$$

where η_0 and η are initial polymer and its mixture with nanofiller viscosity, accordingly.

The calculation according to the Eqs. (21) and (22) shows, that within the indicated above nanofillers number the value D changes as 1.32–1.14–0.44 relative units, that is, reduces in three times, that was expected. This apparent contradiction is due to the choice of the condition t=const (where t is nanocomposite production duration) in the Eq. (20). In real conditions the value t is restricted by nanoparticle contact with growing aggregate and then instead of t the value t/c_0 should be used, where c_0 is the seeds concentration, determined according to the Eq. (14). In this case the value ζ for the indicated nanofillers changes as 0.288–0.118–0.086, that is, it reduces in 3.3 times, that corresponds fully to the calculation according to the Einstein's relationship (the Eq. (21)). This means, that nanoparticles diffusion in polymer matrix obeys classical laws of Newtonian rheology [62].

Thus, the disperse nanofiller particles aggregation in elastomeric matrix can be described theoretically within the frameworks of a modified model of irreversible aggregation particle-cluster. The obligatory consideration of nanofiller initial particles size is a feature of the indicated model application to real systems description. The indicated particles diffusion in polymer matrix obeys classical laws of Newtonian liquids hydrodynamics. The offered approach allows predicting nanoparticles aggregates final parameters as a function of the initial particles size, their contents and other factors number.

At present there are several methods of filler structure (distribution) determination in polymer matrix, both experimental [39, 64] and theoretical [33]. All the indicated methods describe this distribution by fractal dimension D_n of filler particles network. However, correct determination of any object fractal (Hausdorff) dimension includes three obligatory conditions. The first from them is the indicated above determination of fractal dimension numerical magnitude, which should not be equal to object topological dimension. As it is known [65], any real (physical) fractal possesses fractal properties within a certain scales range. Therefore the second condition is the evidence of object self-similarity in this scales range [66]. And at last, the third condition is the correct choice of measurement scales range itself. As it has been shown in Refs. [67, 68], the minimum range should exceed at any rate one self-similarity iteration.

The first method of dimension D_n experimental determination uses the following fractal relationship [69, 70]:

$$D_n = \frac{\ln N}{\ln \rho}, \tag{23}$$

where N is a number of particles with size ρ.

Particles sizes were established on the basis of atomic-power microscopy data (see Fig. 2). For each from the three studied nanocomposites no less than 200 particles were measured, the sizes of which were united into 10 groups and mean values N and ρ were obtained. The dependences $N(\rho)$ in double logarithmic coordinates was plotted, which proved to be linear and the values D_n were calculated according to their slope (see Fig. 5). It is obvious, that at such approach fractal dimension D_n is determined in two-dimensional Euclidean space, whereas real nanocomposite should be considered in three-dimensional Euclidean space. The following relationship can be used for D_n recalculation for the case of three-dimensional space [71]:

$$D3 = \frac{d + D2 \pm \left[(d - D2)^2 - 2 \right]^{1/2}}{2}, \tag{24}$$

where $D3$ and $D2$ are corresponding fractal dimensions in three- and two-dimensional Euclidean spaces, d=3.

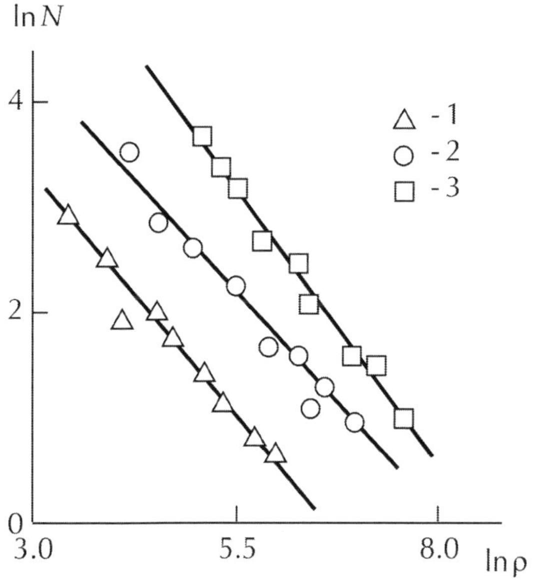

FIGURE 5 The dependences of nanofiller particles number N on their size ρ for nanocomposites BSR/TC (1), BSR/nanoshungite (2) and BSR/microshungite (3).

The calculated according to the indicated method dimensions D_n are adduced in Table 2. As it follows from the data of Table 2, the values D_n for the studied nanocomposites are varied within the range of 1.10–1.36, that is, they characterize more or less branched linear formations ("chains") of nanofiller particles (aggregates of particles) in elastomeric nanocomposite structure. Let us remind that for particulate-filled composites polyhydroxiether/graphite the value D_n changes within the range of ~2.30–2.80 [33, 39], that is, for these materials filler particles network is a bulk object, but not a linear one [65].

TABLE 2 The dimensions of nanofiller particles (aggregates of particles) structure in elastomeric nanocomposites.

Nanocomposite	D_n, the Eq. (23)	D_n, the Eq. (25)	d_0	d_{surf}	φ_n	D_n, the Eq. (29)
BSR/TC	1.19	1.17	2.86	2.64	0.48	1.11
BSR/nanoshungite	1.10	1.10	2.81	2.56	0.36	0.78
BSR/microshungite	1.36	1.39	2.41	2.39	0.32	1.47

Another method of D_n experimental determination uses the so-called "quadrates method" [72]. Its essence consists in the following. On the enlarged nanocomposite microphotograph (see Fig. 2) a net of quadrates with quadrate side size α_i, changing from 4.5 up to 24 mm with constant ratio α_{i+1}/α_i=1.5, is applied and then quadrates number N_i, in to which nanofiller particles hit (fully or partly), is counted up. Five arbitrary net positions concerning microphotograph were chosen for each measurement. If nanofiller particles network is a fractal, then the following relationship should be fulfilled [72]:

$$N_i \sim S_i^{-D_n/2}, \tag{25}$$

where S_i is quadrate area, which is equal to α_i^2.

In Fig. 6 the dependences of N_i on S_i in double logarithmic coordinates for the three studied nanocomposites, corresponding to the Eq. (25), is adduced. As one can see, these dependences are linear, that allows determining the value D_n from their slope. The determined according to the Eq. (25) values D_n are also adduced in Table 2, from which a good correspondence of dimensions D_n, obtained by the two described above methods, follows (their average discrepancy makes up 2.1% after these dimensions recalculation for three-dimensional space according to the Eq. (24)).

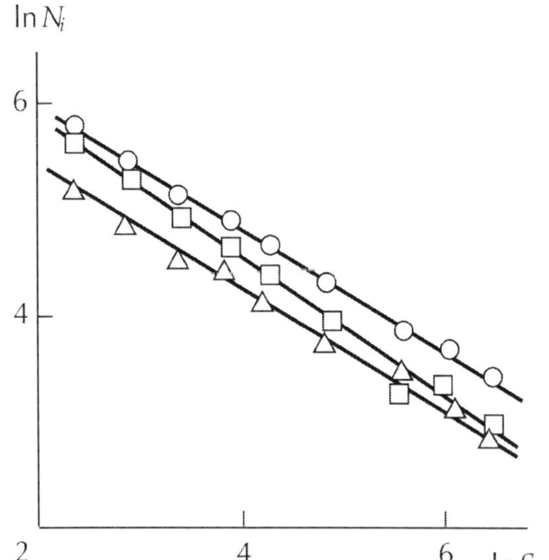

FIGURE 6 The dependences of covering quadrates number N_i on their area S_i, corresponding to the relationship (25), in double logarithmic coordinates for nanocomposites on the basis of BSR. The designations are the same, that in Fig. 5.

As it has been shown in Ref. [73], the usage for self-similar fractal objects at the Eq. (25) the condition should be fulfilled:

$$N_i - N_{i-1} \sim S_i^{-D_n} \tag{26}$$

In Fig. 7, the dependence, corresponding to the Eq. (26), for the three studied elastomeric nanocomposites is adduced. As one can see, this dependence is linear, passes through coordinates origin, that according to the Eq. (26) is confirmed by nanofiller particles (aggregates of particles) "chains" self-similarity within the selected α_i range. It is obvious, that this self-similarity will be a statistical one [73]. Let us note, that the points, corresponding to α_i=16 mm for nanocomposites butadiene-stirene rubber/technical carbon (BSR/TC) and butadiene-stirene rubber/microshungite (BSR/microshungite), do not correspond to a common straight line. Accounting for electron microphotographs of Fig. 2 enlargement this gives the self-similarity range for nanofiller "chains" of 464–1472 nm. For nanocomposite butadiene-stirene rubber/nanoshungite (BSR/nanoshungite), which has no points deviating from a straight line of Fig. 7, α_i range makes up 311–1510 nm, that corresponds well enough to the indicated above self-similarity range.

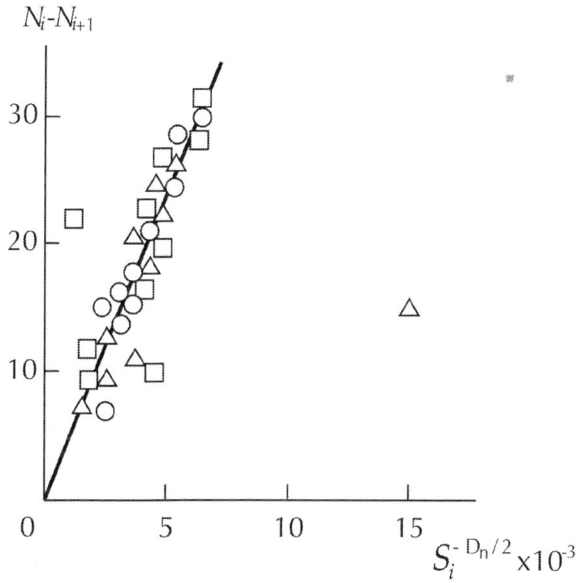

FIGURE 7 The dependences of (N_i-N_{i+1}) on the value $S_i^{D_n/2}$, corresponding to the relationship (26), for nanocomposites on the basis of BSR. The designations are the same, that in Fig. 5.

In Refs. [67, 68] it has been shown, that measurement scales S_i minimum range should contain at least one self-similarity iteration. In this case the condition for ratio of maximum S_{max} and minimum S_{min} areas of covering quadrates should be fulfilled [68]:

$$\frac{S_{max}}{S_{min}} > 2^{2/D_n} \tag{27}$$

Hence, accounting for the defined above restriction let us obtain S_{max}/S_{min}=121/20.25=5.975, that is larger than values $2^{2/D_n}$ for the studied nanocomposites, which are equal to 2.71–3.52. This means, that measurement scales range is chosen correctly.

The self-similarity iterations number μ can be estimated from the inequality [68, 69]:

$$\left(\frac{S_{max}}{S_{min}}\right)^{D_n/2} > 2^{\mu} \tag{28}$$

Using the above-indicated values of the included in the inequality Eq. (28) parameters, μ=1.42–1.75 is obtained for the studied nanocomposites, that is, in our experiment conditions self-similarity iterations number is larger than unity, that again confirms correctness of the value D_n estimation [64].

And let us consider in conclusion the physical grounds of smaller values D_n for elastomeric nanocomposites in comparison with polymer microcomposites, that is, the causes of nanofiller particles (aggregates of particles) "chains" formation in the first ones. The value D_n can be determined theoretically according to the equation [33]:

$$\varphi_{if} = \frac{D_n + 2.55d_0 - 7.10}{4.18}, \tag{29}$$

where φ_{if} is interfacial regions relative fraction, d_0 is nanofiller initial particles surface dimension.

The dimension d_0 estimation can be carried out with the help of the Eq. (4) and the value φ_{if} can be calculated according to the Eq. (7). The results of dimension D_n theoretical calculation according to the Eq. (29) are adduced in table 2, from which a theory and experiment good correspondence follows. The Eq. (29) indicates unequivocally to the cause of a filler in nano- and microcomposites different behavior. The high (close to 3, see table 2) values d_0 for nanoparticles and relatively small (d_0=2.17 for graphite [33]) values d_0 for microparticles at comparable values φ_{if} is such cause for composites of the indicated classes [33, 34].

Hence, the stated above results have shown, that nanofiller particles (aggregates of particles) "chains" in elastomeric nanocomposites are physical fractal within self-similarity (and, hence, factuality [70–73]) range of ~500–1450 nm. In this range their dimension D_n can be estimated according to the Eqs. (23), (25) and (29). The cited examples demonstrate the necessity of the measurement scales range correct choice. As it has been noted earlier [74], the linearity of the plots, corresponding to the Eqs. (23) and (25), and D_n nonintegral value do not guarantee object self-similarity (and, hence, factuality). The nanofiller particles (aggregates of particles) structure low dimensions are due to the initial nanofiller particles surface high fractal dimension.

In Fig. 8, the histogram is adduced, which shows elasticity modulus *E* change, obtained in nanoindentation tests, as a function of load on indenter P or nanoindentation depth *h*. Since for all the three considered nanocomposites the dependences *E*(*P*) or *E*(*h*) are identical qualitatively, then further the dependence *E*(*h*) for nanocomposite BSR/TC was chosen, which reflects the indicated scale effect quantitative aspect in the most clearest way.

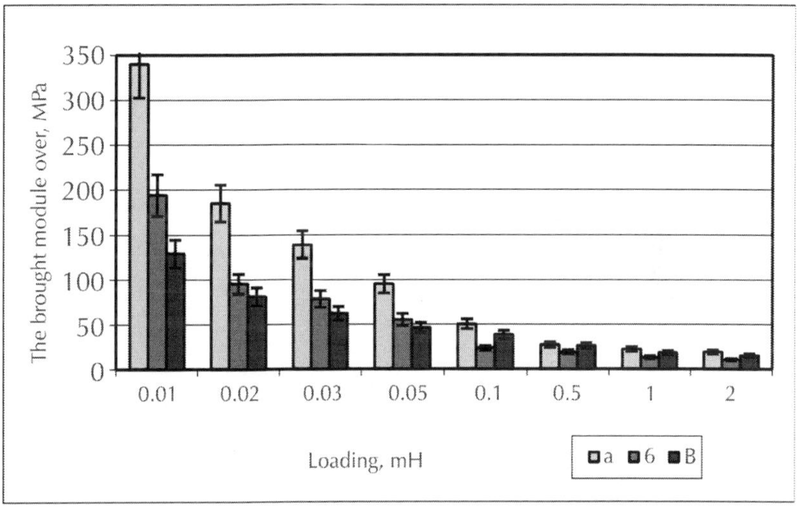

FIGURE 8 The dependences of reduced elasticity modulus on load on indentor for nanocomposites on the basis of butadiene-styrene rubber, filled with technical carbon (a), micro- (b) and nanoshungite (c).

In Fig. 9 the dependence of *E* on h_{pl} (see Fig. 10) is adduced, which breaks down into two linear parts. Such dependences elasticity modulus—strain is typical for polymer materials in general and is due to intermolecular bonds anharmonicity

[75]. In Ref. [76], it has been shown that the dependence $E(h_{pl})$ first part at $h_{pl} \leq 500$ nm is not connected with relaxation processes and has a purely elastic origin. The elasticity modulus E on this part changes in proportion to h_{pl} as:

$$E = E_0 + B_0 h_{pl}, \tag{30}$$

where E_0 is "initial" modulus, that is, modulus, extrapolated to $h_{pl}=0$, and the coefficient B_0 is a combination of the first and second kind elastic constants. In the considered case $B_0<0$. Further Grüneisen parameter γ_L, characterizing intermolecular bonds anharmonicity level, can be determined [76]:

$$\gamma_L \approx -\frac{1}{6} - \frac{1}{2}\frac{B_0}{E_0}\frac{1}{(1-2\nu)}, \tag{31}$$

where ν is Poisson ratio, accepted for elastomeric materials equal to ~0.475 [65].

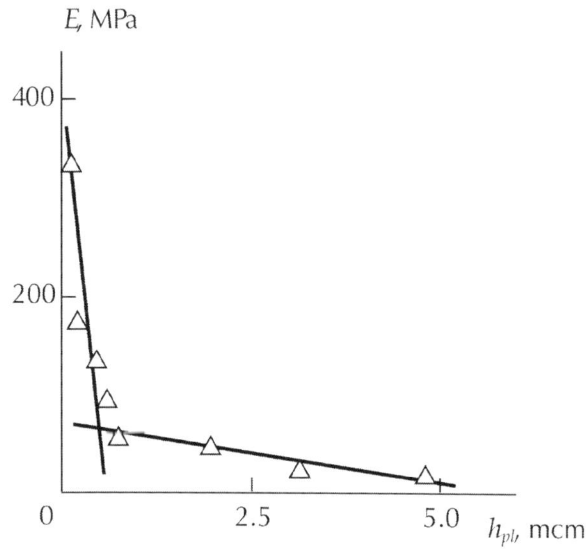

FIGURE 9 The dependence of reduced elasticity modulus E, obtained in nanoindentation experiment, on plastic strain h_{pl} for nanocomposites BSR/TC.

Calculation according to the Eq. (31) has given the following values γ_L: 13.6 for the first part and 1.50—for the second one. Let us note the first from γ_L adduced values is typical for intermolecular bonds, whereas the second value γ_L is much closer to the corresponding value of Grüneisen parameter G for intrachain modes [75].

Poisson's ratio ν can be estimated by γ_L (or G) known values according to the formula [75]:

$$\gamma_L = 0.7\left(\frac{1+\nu}{1-2\nu}\right) \tag{32}$$

The estimations according to the Eq. (32) gave: for the dependence $E(h_{pl})$ first part ν=0.462, for the second one—ν=0.216. If for the first part the value ν is close to Poisson's ratio magnitude for nonfilled rubber [65], then in the second part case the additional estimation is required. As it is known [76, 77], a polymer composites (nanocomposites) Poisson's ratio value ν_n can be estimated according to the equation:

$$\frac{1}{\nu_n} = \frac{\varphi_n}{\nu_{\text{TC}}} + \frac{1-\varphi_n}{\nu_m}, \tag{33}$$

where φ_n is nanofiller volume fraction, ν_{TC} and ν_m are nanofiller (technical carbon) and polymer matrix Poisson's ratio, respectively.

The value ν_m is accepted equal to 0.475 [65] and the magnitude ν_{TC} is estimated as follows [78]. As it is known [79], the nanoparticles TC aggregates fractal dimension d_f^{ag} value is equal to 2.40 and then the value ν_{TC} can be determined according to the equation [79]:

$$d_f^{ag} = (d-1)(1+\nu_{\text{TC}}) \tag{34}$$

According to the Eq. (34) ν_{TC}=0.20 and calculation ν_n according to the Eq. (33) gives the value 0.283, that is close enough to the value ν=0.216 according to the Eq. (32) estimation. The obtained by the indicated methods values ν and ν_n comparison demonstrates, that in the dependence $E(h_{pl})$ (h_{pl}<0.5 mcm) the first part in nanoindentation tests only rubber-like polymer matrix ($\nu=\nu_m\approx0.475$) is included and in this dependence the second part—the entire nanocomposite as homogeneous system [80]—$\nu=\nu_n\approx0.22$.

Let us consider further E reduction at h_{pl} growth (Fig. 9) within the frameworks of density fluctuation theory, which value ψ can be estimated as follows [51]:

$$\psi = \frac{\rho_n kT}{K_T}, \tag{35}$$

where ρ_n is nanocomposite density, k is Boltzmann constant, T is testing temperature, K_T is isothermal modulus of dilatation, connected with Young's modulus E by the relationship [75]:

$$K_T = \frac{E}{3(1-\nu)} \tag{36}$$

In Fig. 10, the scheme of volume of the deformed at nanoindentation material V_{def} calculation in case of Berkovich indentor using is adduced and in Fig. 11 the dependence $\psi(V_{def})$ in logarithmic coordinates was shown. As it follows from the data of this Figure, the density fluctuation growth is observed at the deformed material volume increase. The plot $\psi(\ln V_{def})$ extrapolation to $\psi=0$ gives $\ln V_{def} \approx 13$ or $V_{def}(V_{def}^{cr})=4.42\times10^5$ nm^3. Having determined the linear scale l_{cr} of transition to $\psi=0$ as $(V_{def}^{cr})^{1/3}$, let us obtain $l_{cr}=75.9$ nm, that is close to nanosystems dimensional range upper boundary (as it was noted above, conditional enough [35]), which is equal to 100 nm. Thus, the stated above results suppose, that nanosystems are such systems, in which density fluctuations are absent, always taking place in microsystems.

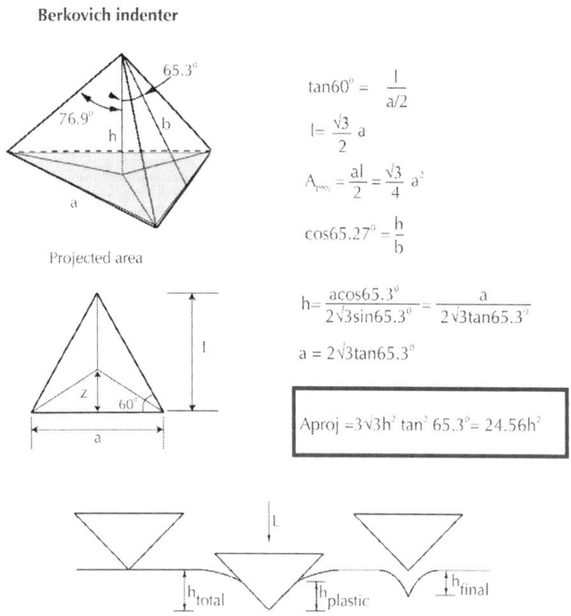

FIGURE 10 The schematic image of Berkovich indentor and nanoindentation process.

As it follows from the data of Fig. 9, the transition from nano- to microsystems occurs within the range h_{pl}=408–726 nm. Both the indicated above values h_{pl} and the corresponding to them values $(V_{def})^{1/3}\approx$814–1440 nm can be chosen as the linear length scale l_n, corresponding to this transition. From the comparison of these values l_n with the distance between nanofiller particles aggregates L_n (L_n=219.2–788.3 nm for the considered nanocomposites, see Fig. 3) it follows, that for transition from nano- to microsystems l_n should include at least two nanofiller particles aggregates and surrounding them layers of polymer matrix, that is the lowest linear scale of nanocomposite simulation as a homogeneous system. It is easy to see, that nanocomposite structure homogeneity condition is harder than the obtained above from the criterion ψ=0. Let us note, that such method, namely, a nanofiller particle and surrounding it polymer matrix layers separation, is widespread at a relationships derivation in microcomposite models.

It is obvious, that the Eq. (35) is inapplicable to nanosystems, since $\psi\to0$ assumes $K_T\to\infty$, which is physically incorrect. Therefore the value E_0, obtained by the dependence $E(h_{pl})$ extrapolation (see Fig. 9) to h_{pl}=0, should be accepted as E for nanosystems [78].

Hence, the stated above results have shown, that elasticity modulus change at nanoindentation for particulate-filled elastomeric nanocomposites is due to a number of causes, which can be elucidated within the frameworks of anharmonicity conception and density fluctuation theory. Application of the first from the indicated conceptions assumes, that in nanocomposites during nanoindentation process local strain is realized, affecting polymer matrix only, and the transition to macrosystems means nanocomposite deformation as homogeneous system. The second from the mentioned conceptions has shown, that nano- and microsystems differ by density fluctuation absence in the first and availability of ones in the second. The last circumstance assumes, that for the considered nanocomposites density fluctuations take into account nanofiller and polymer matrix density difference. The transition from nano- to microsystems is realized in the case, when the deformed material volume exceeds nanofiller particles aggregate and surrounding it layers of polymer matrix combined volume [78].

In Ref. [32] the following formula was offered for elastomeric nanocomposites reinforcement degree E_n/E_m description:

$$\frac{E_n}{E_m}=15.2\left[1-\left(d-d_{surf}\right)^{1/t}\right], \tag{37}$$

where t is index percolation, equal to 1.7 [57].

From the Eq. (37) it follows, that nanofiller particles (aggregates of particles) surface dimension d_{surf} is the parameter, controlling nanocomposites reinforcement degree [81, 82]. This postulate corresponds to the known principle about numerous division surfaces decisive role in nanomaterials as the basis of their properties change [83]. From the Eqs. (4)–(6) it follows unequivocally, that the value d_{surf} is defined by nanofiller particles (aggregates of particles) size R_p only. In its turn, from the Eq. (37) it follows, that elastomeric nanocomposites reinforcement degree E_n/E_m is defined by the dimension d_{surf} only, or, accounting for the said above, by the size R_p only. This means, that the reinforcement effect is controlled by nanofiller particles (aggregates of particles) sizes only and in virtue of this is the true nanoeffect.

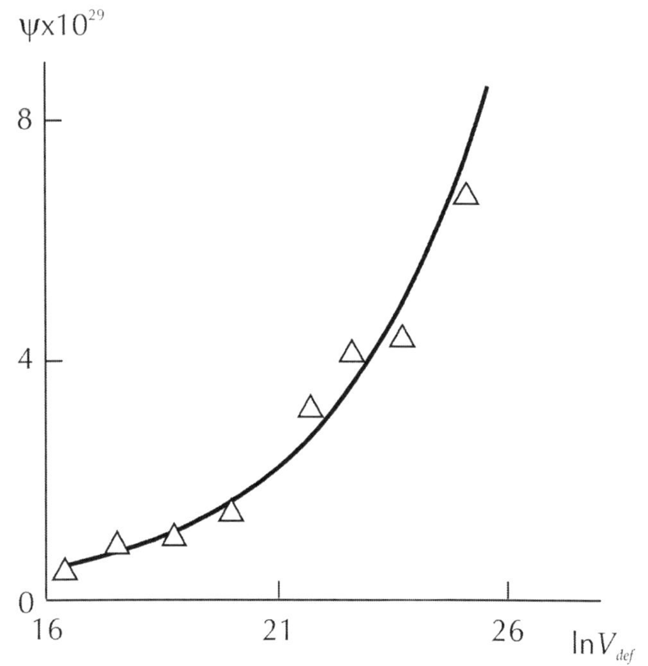

FIGURE 11 The dependence of density fluctuation ψ on volume of deformed in nanoindentation process material V_{def} in logarithmic coordinates for nanocomposites BSR/TC.

In Fig. 12, the dependence of E_n/E_m on $(d-d_{surf})^{1/1.7}$ is adduced, corresponding to the Eq. (37), for nanocomposites with different elastomeric matrices (natural and butadiene-stirene rubbers, NR and BSR, accordingly) and different nanofillers (technical carbon of different marks, nano- and microshungite). Despite the indicated distinctions in composition, all adduced data are described well by the Eq. (37).

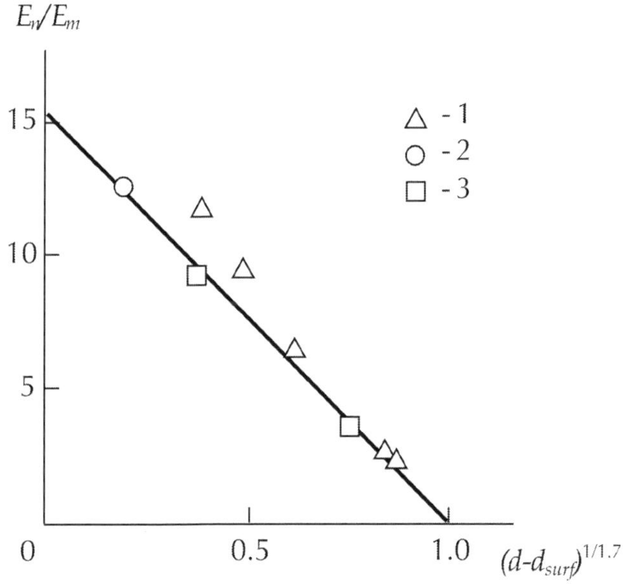

FIGURE 12 The dependence of reinforcement degree E_n/E_m on parameter $(d\text{-}d_{surf})^{1/1.7}$ value for nanocomposites NR/TC (1), BSR/TC (2) and BSR/shungite (3).

In Fig. 13, two theoretical dependences of E_n/E_m on nanofiller particles size (diameter D_p), calculated according to the Eqs. (4)–(6) and (37), are adduced. However, at the curve 1 calculation the value D_p for the initial nanofiller particles was used and at the curve 2 calculation—nanofiller particles aggregates size D_p^{ag} (see Fig. 3). As it was expected [34], the growth E_n/E_m at D_p or D_p^{ag} reduction, in addition the calculation with D_p (nonaggregated nanofiller) using gives higher E_n/E_m values in comparison with the aggregated one (D_p^{ag} using). At $D_p \leq 50$ nm faster growth E_n/E_m at D_p reduction is observed than at $D_p > 50$ nm, that was also expected. In Fig. 13 the critical theoretical value D_p^{cr} for this transition, calculated according to the indicated above general principles [83], is pointed out by a vertical shaded line. In conformity with these principles the nanoparticles size in nanocomposite is determined according to the condition, when division surface fraction in the entire nanomaterial volume makes up about 50% and more. This fraction is estimated approximately by the ratio $3l_{if}/D_p$, where l_{if} is interfacial layer thickness. As it was noted above, the data of Fig. 1 gave the average experimental value $l_{if} \approx 8.7$ nm. Further from the condition $3l_{if}/D_p \approx 0.5$ let us obtain $D_p \approx 52$ nm that is shown in Fig. 13 by a vertical shaded line. As it was expected, the value $D_p \approx 52$ nm is a boundary one for regions of slow ($D_p > 52$ nm) and fast ($D_p \leq 52$ nm) E_n/E_m growth at D_p reduction. In other words, the materials with nanofiller particles size $D_p \leq 52$ nm ("superreinforcing" filler according to the terminology of Ref. [34]) should be considered true nanocomposites.

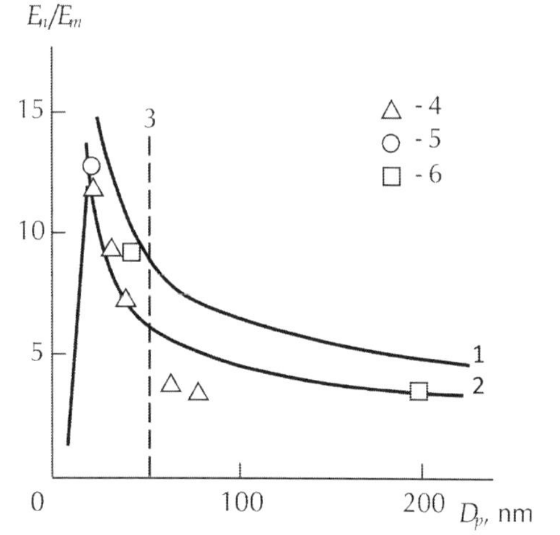

FIGURE 13 The theoretical dependences of reinforcement degree E_n/E_m on nanofiller particles size D_p, calculated according to the equations (4)-(6) and (37), at initial nanoparticles (1) and nanoparticles aggregates (2) size using. 3 – the boundary value D_p, corresponding to true nanocomposite. 4-6 – the experimental data for nanocomposites NR/TC (4), BSR/TC (5) and BSR/shungite (6).

Let us note in conclusion, that although the curves 1 and 2 of Fig. 13 are similar ones, nanofiller particles aggregation, which the curve 2 accounts for, reduces essentially enough nanocomposites reinforcement degree. At the same time the experimental data correspond exactly to the curve 2 that was to be expected in virtue of aggregation processes, which always took place in real composites [33] (nanocomposites [84]). The values d_{surf}, obtained according to the Eqs. (4)–(6), correspond well to the determined experimentally ones. So, for nanoshungite and two marks of technical carbon the calculation by the indicated method gives the following d_{surf} values: 2.81, 2.78 and 2.73, whereas experimental values of this parameter are equal to: 2.81, 2.77 and 2.73, that is, practically a full correspondence of theory and experiment was obtained.

6.2.4 CONCLUSIONS

Hence, the stated above results have shown, that the elastomeric reinforcement effect is the true nanoeffect, which is defined by the initial nanofiller particles size only. The indicated particles aggregation, always taking place in real materials, changes reinforcement degree quantitatively only, namely, reduces it. This effect theoretical treatment can be received within the frameworks of fractal analysis. For the considered nanocomposites the nanoparticle size upper limiting value makes up ~52 nm.

KEYWORDS

- **Berkovich indentor**
- **caveolae-mediated endocytosis**
- **clathrin-mediated endocytosis**
- **linear spatial structures**
- **nanocrystallite**
- **opsonins**

REFERENCES

1. Ozin, G. A.; Arsenault, A. C.; *Nanochemistry*; RSC Publishing: Cambridge, **2005**.
2. McFarland, A. D.; Haynes, C. L.; Mirkin, C. A.; "Color My Nanoworld," *J. Chem.* **2004,** *81,* 544A.
3. Mukherjee, S.; Ray, S.; Thakur, R. S.; *Ind. J. Pharm. Sci.* **2009**, 349–358.
4. Abdul Hasan A.; Sathali, M.; Priyanka, K.; Solid lipid nanoparticles: A Review, *Sci. Revs. Chem. Commun.* **2012**, *2(1),* 80–102.
5. Mozafari, M. R.; *Nano Carrier Technologies: Frontiers of Nano Therapy;* **2006**, 41–50.
6. Houli Li; Xiaobin Zhao; Yukun Ma; Guangxi Zhai; Ling Bing Li; Hong Xiang; Lou. J.; Cont. Release, **2009**, *133,* 238–244.
7. Melike Uner; Gulgun Yener; *Int. J. Nanomed.* **2007,** *2(3),* 289–300.
8. Kushwaha Anjali; Nanotherapeutics in Drug delivery: A Review**,** Pharmatutor ART 1291.
9. Mua, L.; Fenga, S. S.; A novel controlled release formulation for the anticancer drug paclitaxl (Taxol®): PLGA nanoparticles containing vitamin E TPGS; *J. Controlled Release*, **2003**, *86,* 33–48.
10. Lu, X. Y. ; Wu, D. C.; Li, Z. J.; Chen, G. Q.; *Polymer nanoparticles. Prog. Mol. Biol. Transl. Sci.* **2011,** *104,* 299–323.
11. Sinani, V. A.; Koktysh, D. S.; Yun, B. G.; Matts, R. L.; Pappas, T. C.; Motamedi, M.; Thomas, S. N.; Kotov, N. A.; Collagen coating promotes biocompatibility of semiconductor nanoparticles in stratified LBL films**.** *Nano Lett.*, **2003**, *3,* 1177–1182.

12. Zhang, Y.; Kohler, N.; Zhang, M.; Surface modification of superparamagnetic magnetite nanoparticles and their intracellular uptake. *Biomaterials*, **2002,** *23,* 1553–1561.
13. Gregoriadis, G.; Liposomes in the therapy of lysosomal storage diseases. *Nature* **1978,** *275,* 695–696.
14. Grislain, L.; Couvreur, P.; Lenaerts V, Roland, M.; Deprezdecampeneere, D.; Speiser, P.; Pharmacokinetics and distribution of a biodegradable drug-carrier. *Int. J. Pharm.* **1983,** *15,* 335–345.
15. Aderem, A.; Underhill, D.; Mechanisms of phagocytosis in macrophages. *Annu. Rev. Immunol.* **1999,** *17,* 593–623.
16. Rabinovitch, M.; Professional and nonprofessional phagocytes—an introduction. *Trends Cell Biol.* **1995,** *5,* 85–87.
17. Vonarborg, A.; Passirani, C.; Saulnier, P.; Benoit, J.; Parameters influencing the stealthiness of colloidal drug delivery systems. *Biomaterials* **2006**, *27,* 4356–4373.
18. Owens, D.; Peppas, N.; Opsonization, biodistribution, and pharmacokinetics of polymeric nanoparticles. *Int. J. Pharm.* **2006**, *307,* 93–102.
19. Groves, E.; Dart, A.; Covarelli, V.; Caron, E.; Molecular mechanisms of phagocytic uptake in mammalian cells. *Cell. Mol. Life. Sci.* **2008**, *65,* 1957–1976.
20. Vachon, E.; Martin, R.; Plumb, J.; Kwok, V.; Vandivier, R.; Glogauer, M.; Kapus, A.; Wang, X.; Chow, C.; Grinstein, S.; Downey, G.; CD44 is a phagocytic receptor. *Blood* **2006**, *107,* 4149–4158.
21. Caron, E.; Hall, A.; Identification of two distinct mechanisms of phagocytosis controlled by different Rho GTPases. *Science* **1998**, *282,* 1717–1721.
22. Swanson, J. A.; Baer, S. C.; Phagocytosis by zippers and triggers. *Trends Cell Biol.* **1995**, *5,* 89–93.
23. Claus, V.; Jahraus, A.; Tjelle, T.; Berg, T.; Kirschke, H.; Faulstich, H.; Griffiths, G.; Lysosomal enzyme trafficking between phagosomes, endosomes, and lysosomes in J774 macrophages. Enrichment of cathepsin H in early endosomes. *J. Biol. Chem.* **1998**, *273,* 9842–9851.
24. Mukherjee, S.; Ghosh, R. N.; Maxfield, F. R.; Endocytosis. *Physiol. Re***1997**, *77,* 759–803.
25. Bareford, L. M.; Swaan, P. W.; Endocytic mechanisms for targeted drug delivery. *Adv. Drug Deliv. Re***2007**, *59,* 748–758.
26. Conner, S. D.; Schmid, S. L.; Regulated portals of entry into the cell. *Nature 422,* 37–44. **2003,**
27. Mayor, S.; Pagano, R. E.; Pathways of clathrin-independent endocytosis. *Nat. Rev. Mol. Cell Biol.* **2007**, *8,* 603–612.
28. Swanson, J. A.; Watts, C.; Macropinocytosis. *Trends Cell Biol.* **1995**, *5,* 424–428.
29. Racoosin, E. L.; Swanson, J. A.; M-CSF-induced macropinocytosis increases solute endocytosis but not receptor-mediated endocytosis in mouse macrophages. *J. Cell Sci.* **1992**, *102,* 867–880.
30. Yanovskii, Yu. G.; Kozlov, G. V.; Karnet, Yu. N.; Mekhanika Kompozitsionnykh Materialov i Konstruktsii, **2011,** *17(2),* 203–208.
31. Malamatov, A. Kh.; Kozlov, G. V.; Mikitaev, M. A.; Reinforcement Mechanisms of Polymer Nanocomposites. Moscow, Publishers of the D. I.; Mendeleev RKhTU, **2006,** 240 p.

32. Mikitaev, A. K.; Kozlov, G. V.; Zaikov, G. E.; Polymer Nanocomposites: Variety of Structural Forms and Applications. Moscow, Nauka, **2009,** 278 p.
33. Kozlov, G. V.; Yanovskii, Yu. G.; Karnet, Yu. N.; Structure and Properties of Particulate-Filled Polymer Composites: the Fractal Analysis. Moscow, Al'yanstransatom, **2008,** 363 p.
34. Edwards, D. Mater, Sci.; C. J.; **1990,** *25(12),* 4175–4185.
35. Buchachenko, A. L.; Uspekhi Khimii, **2003,** *72(5),* 419–437.
36. Kozlov, G. V.; Yanovskii Yu.G.; Burya, A. I.; Aphashagova Z. Kh. Mekhanika Kompozitsionnykh Materialov i Konstruktsii, **2007,** *13(4),* 479–492.
37. Lipatov, Yu. S.; The Physical Chemistry of Filled Polymers. Moscow, Khimiya, **1977,** 304 p.
38. Bartenev, G. M.; Zelenev, Yu. V.; Physics and Mechanics of Polymers. Moscow, Vysshaya Shkola, **1983,** 391 p.
39. Kozlov, G. V.; Mikitaev, A. K.; Mekhanika Kompozitsionnykh Materialov i Konstruktsii, **1996,** *2(3–4*), 144–157.
40. Kozlov, G. V.; Yanovskii, Yu. G.; Zaikov, G. E.; Structure and Properties of Particulate-Filled Polymer Composites: the Fractal Analysis. New York, Nova Science Publishers, Inc.; **2010,** 282 p.
41. Mikitaev, A. K.; Kozlov, G. V.; Zaikov, G. E.; Polymer Nanocomposites: Variety of Structural Forms and Applications. New York, Nova Science Publishers, Inc.; **2008,** 319 p.
42. McClintok, F. A.; Argon, A. S.; Mechanical Behavior of Materials. Reading, Addison-Wesley Publishing Company, Inc.; **1966,** 440 p.
43. Kozlov, G. V.; Mikitaev, A. K.; Doklady AN SSSR, **1987,** *294(5),* 1129–1131.
44. Honeycombe, R. W. K.; The Plastic Deformation of Metals. Boston, Edward Arnold (Publishers), Ltd.; **1968,** 398 p.
45. Dickie, R. A. In: Polymer Blends. V. 1. Paul D. R.; Newman S. Ed.; New York, San-Francisco, London, Academic Press, **1980,** 386–431.
46. Kornev, Yu. V.; Yumashev, O. B.; Zhogin, V. A.; Karnet Yu.N.; Yanovskii Yu.G.; Kautschuk i Rezina, **2008,** *6,* 18–23.
47. Oliver, W. C.; Pharr, G. M. J.; *Mater. Res.* **1992,** *7(6),* 1564–1583.
48. Kozlov, G. V.; Yanovskii, Yu. G.; Lipatov, Yu. S.; Mekhanika Kompozitsionnykh Materialov i Konstruktsii, **2002,** *8(1),* 111–149.
49. Kozlov, G. V.; Burya, A. I.; Lipatov, Yu. S.; Mekhanika Kompozitnykh Materialov, **2006,** *42(6),* 797–802.
50. Hentschel, H. G. E.; Deutch, J. M.; *Phys. Rev. A*, **1984,** *29(3),* 1609–1611.
51. Kozlov, G. V.; Ovcharenko, E. N.; Mikitaev, A. K.; Structure of Polymers Amorphous State. Moscow, Publishers of the D. I.; Mendeleev RKhTU, **2009,** 392 p.
52. Yanovskii, Yu. G.; Kozlov Mater, G. V.; VII Intern. Sci.-Pract. Conf. "New Polymer Composite Materials." Nal'chik, KBSU, **2011,** 189–194.
53. Wu, S. J.; *Polymer Sci.: Part B: Polymer Phys.* **1989,** *27(4),* 723–741.
54. Aharoni, S. M.; *Macromolecules*, **1983,** *16(9),* 1722–1728.
55. Budtov V. P.; The Physical Chemistry of Polymer Solutions. Sankt-Peterburg, Khimiya, **1992,** 384 p.
56. Aharoni Macromolecules, S. M.; **1985,** *18(12),* 2624–2630.

57. Bobryshev, A. N.; Kozomazov, V. N.; Babin, L. O.; Solomatov, V. I.; Synergetics of Composite Materials. Lipetsk, NPO ORIUS, **1994,** 154 p.
58. Kozlov, G. V.; Yanovskii, Yu. G.; Karnet, Yu. N.; Mekhanika Kompozitsionnykh Materialov i Konstruktsii, **2005,** *11(3),* 446–456.
59. Sheng, N.; Boyce, M. C.; Parks, D. M.; Rutledge, G. C.; Abes, J. I.; Cohen, R. E.; *Polymer*, **2004,** *45(2),* 487–506.
60. Witten, T. A.; Meakin, P.; *Phys. Rev. B,* **1983,** *28(10),* 5632–5642.
61. Witten, T. A.; Sander, L. M.; *Phys. Rev. B,* **1983,** *27(9),* 5686–5697.
62. Happel J.; Brenner G.; Hydrodynamics at Small Reynolds Numbers. Moscow, Mir, **1976,** 418 p.
63. Mills N. J.; *J. Appl. Polymer Sci.* **1971,** *15(11),* 2791–2805.
64. Kozlov, G. V.; Yanovskii, Yu. G.; Mikitaev, A. K.; Mekhanika Kompozitnykh Materialov, **1998,** *34(4),* 539–544.
65. Balankin, A. S.; Synergetics of Deformable Body. Moscow, Publishers of Ministry Defence SSSR, **1991,** 404 p.
66. Hornbogen, E.; *Intern. Mater. Res.* **1989,** *34(6),* 277–296.
67. Pfeifer, P.; *Appl. Surf. Sci.* **1984,** *18(1),* 146–164.
68. Avnir D.; Farin D.; Pfeifer P.; *J. Colloid Interface Sci.* **1985,** *103(1),* 112–123.
69. Ishikawa Mater, K.; *J. Sci. Lett.* **1990,** *9(4),* 400–402.
70. Ivanova, V. S.; Balankin, A. S.; Bunin, I. Zh.; Oksogoev, A. A.; Synergetics and Fractals in Material Science. Moscow, Nauka, **1994,** 383 p.
71. Vstovskii, G. V.; Kolmakov, L. G.; Terent'ev, V. E.; *Metally*, **1993,** *4,* 164–178.
72. Hansen, J. P.; Skjeitorp, A. T.; *Phys. Rev. B,* **1988,** *38(4),* 2635–2638.
73. Pfeifer, P.; Avnir, D.; Farin D. J.; *Stat. Phys.* **1984,** *36(5/6)*, 699–716.
74. Farin, D.; Peleg, S.; Yavin, D.; Avnir, D.; *Langmuir*, **1985,** *1(4),* 399–407.
75. Kozlov, G. V.; Sanditov, D. S.; Anharmonical Effects and Physical-Mechanical Properties of Polymers. Novosibirsk, Nauka, **1994,** 261 p.
76. Bessonov, M. I.; Rudakov Vysokomolek, A. P.; *Soed. B,* **1971,** *13(7),* 509–511.
77. Kubat J.; Rigdahl M.; Welander, M. J.; *Appl. Polymer Sci.* **1990,** *39(5),* 1527–1539.
78. Yanovskii, Yu. G.; Kozlov, G. V.; Kornev, Yu. V.; Boiko, O. V.; Karnet, Yu. N.; Mekhanika Kompozitsionnykh Materialov i Konstruktsii, **2010,** *16(3),* 445–453.
79. Yanovskii, Yu. G.; Kozlov, G. V.; Aloev Mater, V. Z.; Intern. Sci.-Pract. Conf. "Modern Problems of APK Innovation Development Theory and Practice." Nal'chik, KBSSKhA, **2011,** 434–437.
80. Chow Polymer, T. S.; **1991,** *32(1),* 29–33.
81. Ahmed, S.; Jones, F. Mater, Sci.; R. J.; **1990,** *25(12),* 4933–4942.
82. Kozlov, G. V.; Yanovskii Yu.G.; Aloev Mater, V. Z.; Intern. Sci.-Pract. Conf.; dedicated to FMEP 50-th Anniversary. Nal'chik, KBSSKhA, **2011,** 83–89.
83. Andrievskii, R. A.; Rossiiskii Khimicheskii Zhurnal, **2002,** *46(5),* 50–56.
84. Kozlov, G. V.; Sultonov, N. Zh.; Shoranova, L. O.; Mikitaev, A. K.; Naukoemkie Tekhnologii, **2011,** *12(3),* 17–22.

CHAPTER 7

SOME NEW ASPECTS OF POLYMERIC NANOCOMPOSITES

A. K. MIKITAEV, A. Y. BEDANOKOV, and M. A. MIKITAEV

CONTENTS

7.1 Introduction ... 314
7.2 Structure of the Layered Silicates ... 314
7.3 Modification of the Layered Silicates ... 315
7.4 Structure of the Polymeric Nanocomposites on the Basis of the Montmorillonite ... 317
7.4.1 Determination of the Interlayer Space ... 318
7.4.2 The Degree of the Distribution of the Clay Particles in the Polymeric Matrix ... 320
7.5 Production of the Polymeric Nanocomposites on the Basis of the Aluminous Silicates ... 324
7.6 Properties of the Polymeric Nanocomposites ... 326
7.7 Conclusions ... 336
Keywords ... 337
References ... 337

7.1 INTRODUCTION

The polymeric nanocomposites are the polymers filled with nanoparticles, which interact with the polymeric matrix on the molecular level in contrary to the macrointeraction in composite materials. Mentioned nanointeraction results in high adhesion hardness of the polymeric matrix to the nanoparticles [1, 52].

Usual nanoparticle is less than 100 nm in any dimension, 1 nm being the billionth part of a meter [1, 2].

The analysis of the reported studies tells that the investigations in the field of the polymeric nanocomposite materials are very promizing.

The first notion of the polymeric nanocomposites was given in patent in 1950 [3]. Blumstain pointed in 1961 [4] that polymeric clay – based nanocomposites had increased thermal stability. It was demonstrated using the data of the thermogravimetric analysis that the polymethylmetacrylate intercalated into the Na^+ – methylmetacrylate possessed the temperature of destruction 40–50 °C higher than the initial sample.

This branch of the polymeric chemistry did not attract much attention until 1990 when the group of scientists from the Toyota Concern working on the polyamide – based nanocomposites [5–9] found two – times increase in the elasticity modulus using only 4.7 wt% of the inorganic compound and 100 °C increase in the temperature of destruction, both discoveries widely extending the area of application of the polyamide. The polymeric nanocomposites based on the layered silicates began being intensively studied in state, academic and industrial laboratories all over the world only after that.

7.2 STRUCTURE OF THE LAYERED SILICATES

The study of the polymeric nanocomposites on the basis of the modified layered silicates (broadly distributed and well – known as various clays) is of much interest. The natural layered inorganic structures used in producing the polymeric nanocomposites are the montmorillonite [10–12], hectorite [13], vermiculite [14], kaolin, saponine [15] and others. The sizes of inorganic layers are about 220 and 1 nm in length and width, respectively [16, 17].

The perspectives ones are the bentonite breeds of clays, which include at least 70% of the minerals from the montmorillonite group.

Montmorillonite (Na, K, Ca)(Al, Fe, Mg)$[(Si, Al)_4O_{10}](OH)_2{'}nH_2O$, named after the province Montmorillion in France, is the high – dispersed layered aluminous silicate of white or gray color in which appears the excess negative charge due to the nonstoichiometric replacements of the cations of the crystal lattice, charge being balanced by the exchange cations from the interlayer space. The main feature of the montmorillonite is its ability to adsorb ions, generally cations, and to exchange them. It produces plastic masses with water and may enlarge itself 10 times. Montmorillonite enters the bentonite clays (the term "bentonite" is given after the place Benton in USA).

The inorganic layers of clays arrange the complexes with the gaps called layers or galleries. The isomorphic replacement within the layers (such as Mg^{2+} replacing Al^{3+} in octahedral structure or Al^{3+} replacing Si^{4+} in tetrahedral one) generates the negative charges, which electrostatically are compensated by the cations of the alkali, or alkali-earth metals located in the galleries (Fig. 1) [18]. The bentonite is very hydrophilic because of this. The water penetrates the interlayer space of the montmorillonite, hydrates its surface and exchanges cations what results in the swelling of the bentonite. The further dilution of the bentonite in water results in the viscous suspension with bold tixotropic properties.

The more pronounced cation – exchange and adsorption properties are observed in the bentonites montmorillonite of which contains predominantly exchange cations of sodium.

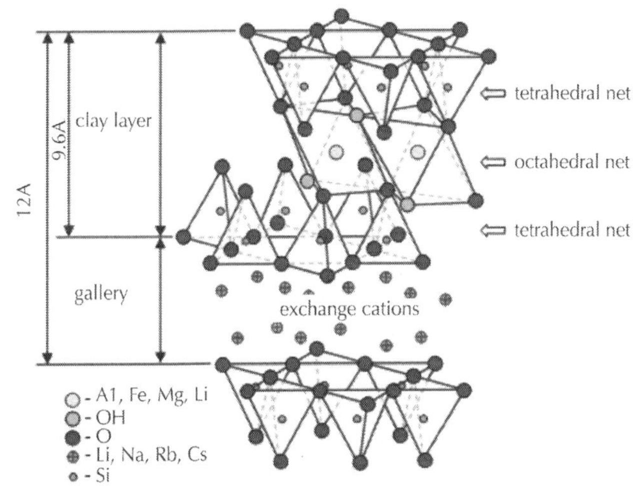

FIGURE 1 The structure of the layered silicate.

7.3 MODIFICATION OF THE LAYERED SILICATES

Layered silicates possess quite interesting properties – sharp drop of hardness at wetting, swelling at watering, dilution at dynamical influences and shrinking at drying.

The hydrophility of aluminous silicates is the reason of their incompatibility with the organic polymeric matrix and is the first hurdle needs to be overridden at producing the polymeric nanocomposites.

One way to solve this problem is to modify the clay by the organic substance. The modified clay (organoclay) has at least two advantages: (1) it can be well dispersed in polymeric matrix [19] and (2) it interacts with the polymeric chain [13].

The modification of the aluminous silicates can be done with the replacement of the inorganic cations inside the galleries by the organic ones. The replacing by the cationic surface – active agents like bulk ammonium and phosphonium ions increases the room between the layers, decreases the surface energy of clay and makes the surface of the clay hydrophobic. The clays modified such a way are more compatible with the polymers and form the layered polymeric nanocomposites [52]. One can use the nonionic modifiers besides the organic ones, which link themselves to the clay surface through the hydrogen bond. Organoclays produced with help of non – ionic modifiers in some cases become more chemically stable than the organoclays produced with help of cationic modifiers (Fig. 2a) [20].

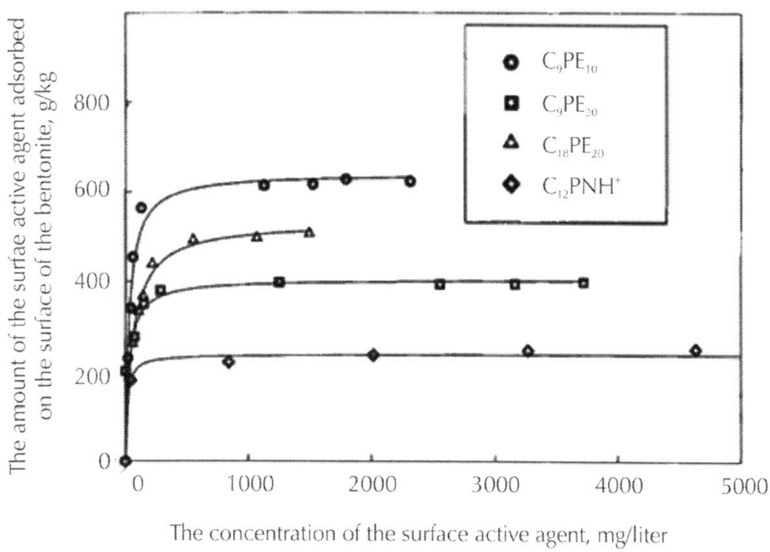

FIGURE 2A The adsorption of different modifiers on the clay surface.

The least degree of desorption is observed for non – ionic interaction between the clay surface and organic modifier (Fig. 2b). The hydrogen bonds between the ethylenoxide grouping and the surface of the clay apparently make these organoclays more chemically stable than organoclays produced with nonionic mechanism.

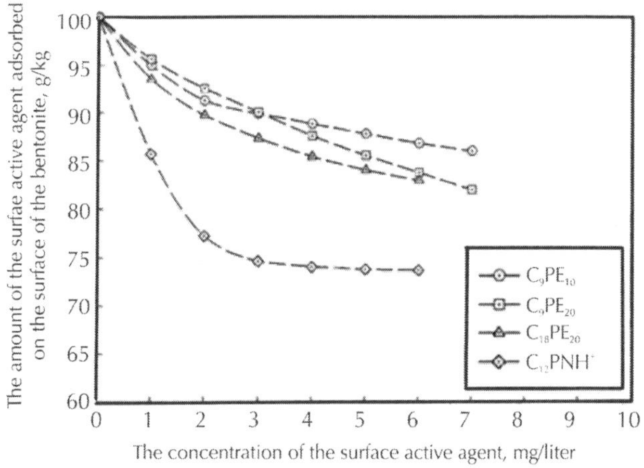

FIGURE 2B The desorption of different modifiers from the clay surface:
C_9PE_{10} – $C_9H_{19}C_6H_4(CH_2\ CH_2O)_{10}OH$;
C_9PE_{20} – $C_9H_{19}C_6H_4(CH_2\ CH_2O)_{20}OH$;
$C_{18}E_{20}$ – $C_{18}H_{37}(CH_2\ CH_2O)_{20}OH$;
$C_{12}PNH^+$ – $C_{12}H_{25}C_6H_4NH^+Cl$–.

7.4 STRUCTURE OF THE POLYMERIC NANOCOMPOSITES ON THE BASIS OF THE MONTMORILLONITE

The study of the distribution of the organoclay in the polymeric matrix is of great importance because the properties of composites obtained are in the direct relation from the degree of the distribution.

According Giannelis [21], the process of the formation of the nanocomposite goes in several intermediate stages (Fig. 3). The formation of the tactoid happens on the first stage – the polymer surrounds the agglomerations of the organoclay. The polymer penetrates the interlayer space of the organoclay on the second stage. Here the gap between the layers may reach 2–3 nm [22]. The further separation of the layers, third stage, results in partial dissolution and disorientation of the layers. Exfoliation is observed when polymer shifts the clay layers on more than 8–10 nm.

All mentioned structures may be present in real polymeric nanocomposites in dependence from the degree of distribution of the organoclay in the polymeric matrix. Exfoliated structure is the result of the extreme distribution of the organoclay. The excess of the organoclay or bad dispersing may born the agglomerates of the organoclays in the polymeric matrix what finds experimental confirmation in the X-ray analysis [11, 12, 21, 23].

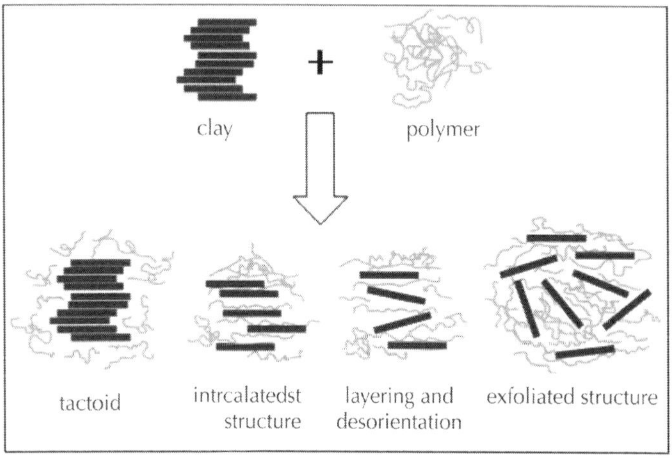

FIGURE 3 The schematic formation of the polymeric nanocomposite [24].

In the following subsections, we describe a number of specific methods used at studying the structure of the polymeric nanocomposites.

7.4.1 DETERMINATION OF THE INTERLAYER SPACE

The X-ray determination of the interlayer distance in the initial and modified layered silicates as well as in final polymeric nanocomposite is one of the main methods of studying the structure of the nanocomposite on the basis of the layered silicate. The peak in the small – angle diapason (2q = 6–8 °C) is characteristic for pure clays and responds to the order of the structure of the silicate. This peak drifts to the smaller values of the angle 2q in organomodified clays. If clay particles are uniformly distributed in the bulk of the polymeric matrix then this peak disappears, what witnesses on the disordering in the structure of the layered silicate. If the amount of the clay exceeds the certain limit of its distribution in the polymeric matrix, then the peak reappears again. This regularity was demonstrated on the instance of the polybutylenterephtalate (Fig. 4) [11].

The knowledge of the angle 2q helps to define the size of the pack of the aluminous silicate consisting of the clay layer and interlayer space. The size of such pack increases in a row from initial silicate to polymeric nanocomposite according to the increase in the interlayer space. The average size of that pack for montmorillonite is 1.2–1.5 nm but for organomodified one varies in the range of 1.8–3.5 nm.

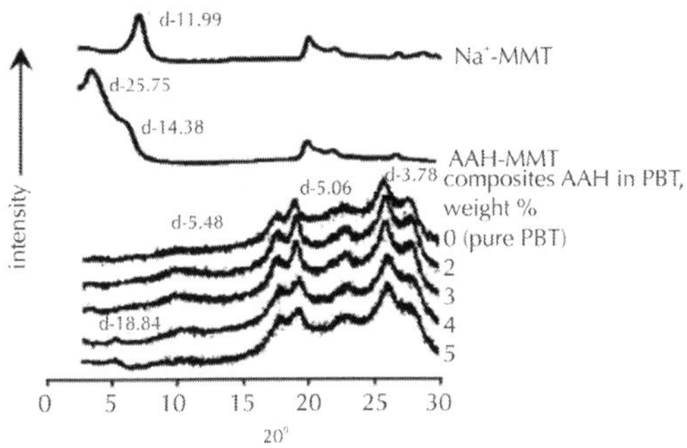

FIGURE 4 The data of the X-ray analysis for clay, organoclay and nanocomposite PBT/organoclay.

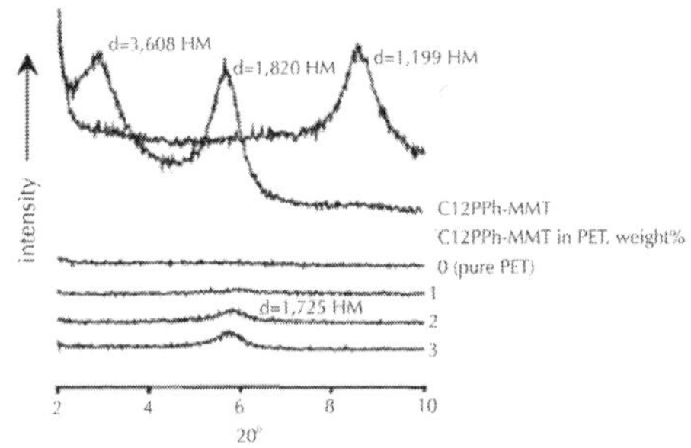

FIGURE 5 The data of the X-ray analysis for clay, organoclay and nanocomposite PET/organoclay.

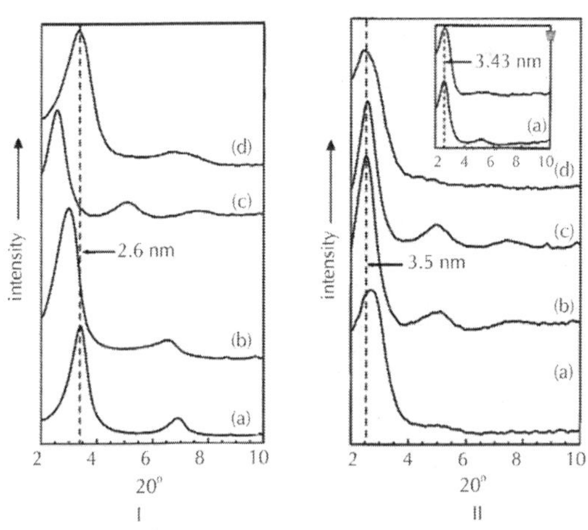

FIGURE 6 The data of the X-ray analysis for:
I. (a) – dimetyldioctadecylammonium (DMDODA) – hectorite;
(b) – 50% polystirene (PS)/50% DMDODA – hectorite;
(c) – 75% polyethylmetacrylate (PEM)/25% DMDODA;

(d) – 50% PS/ 50 % DMDODA – hectorite after 24 h of etching in cyclohexane.

II. mix of PS, PEM and ofganoclay:

(a) – 23.8% PS/71.2% PEM/5% DMDODA – hectorite;

(b) – 21.2% PS/63.8% PEM/15% DMDODA – hectorite;

(c) – 18.2% PS/54.8% PEM/27% DMDODA – hectorite;

(d) – 21.2% PS/63.8% PEM/15% DMDODA – hectorite after 24 h of etching in cyclohexane.

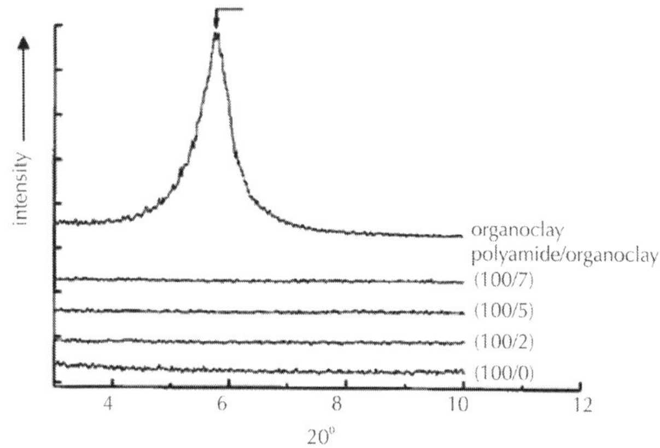

FIGURE 7 The data of the X-ray analysis for organoclay and nanocomposite polyamide acid/ organoclay.

Summing up we conclude that comparing the data of the X-ray analysis for the organoclay and nanocomposite allows for the determination of the optimal clay amount need be added to the composite. The data from the scanning tunneling (STM) and transmission electron (TEM) microscopes [27, 28] can be used as well.

7.4.2 THE DEGREE OF THE DISTRIBUTION OF THE CLAY PARTICLES IN THE POLYMERIC MATRIX

The two structures, namely the intercalated and exfoliated ones, could be distinguished with the respect to the degree of the distribution of the clay particles, Fig.8. One should note that clay layers are quite flexible though they are shown straight in the figure. The formation of the intercalated or exfoliated structures depends on many factors, for example the method of the production of the nanocomposite or the nature of the clay, and so forth [29].

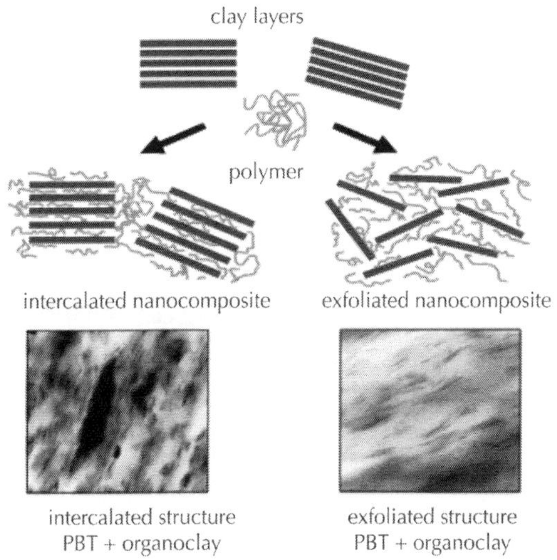

FIGURE 8 The formation of the intercalated and exfoliated structures of the nanocomposite.

The TEM images of the surface of the nanocomposites can help to find out the degree of the distribution of the nanosized clay particles, see plots (a) to (d) in the Fig. 9.

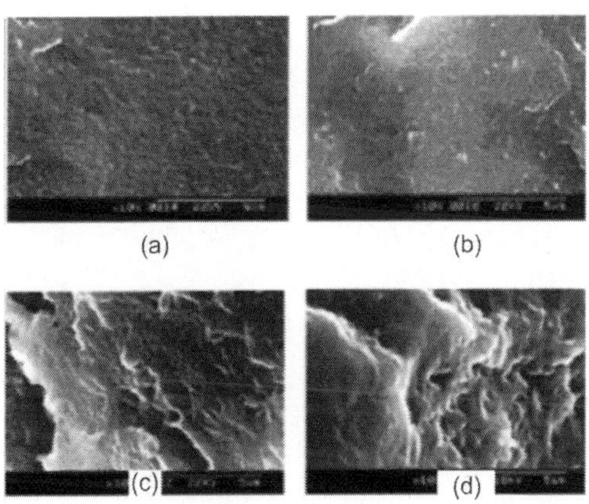

FIGURE 9 The images from scanning electron microscope for the nanocomposite surfaces:

(a) – pure PBT;

(b) – 3 wt% of organoclay in PBT;

(c) – 4 wt% of organoclay in PBT;

(d) – 5 wt% of organoclay in PBT.

The smooth surface tells about the uniform distribution of the organoclay particles. The surface of the nanocomposite becomes deformed with the increasing amount of the organoclay, see plots (a) to (d) in the Fig. 10. Probably, this is due to the influence of the clay agglomerates [30, 31].

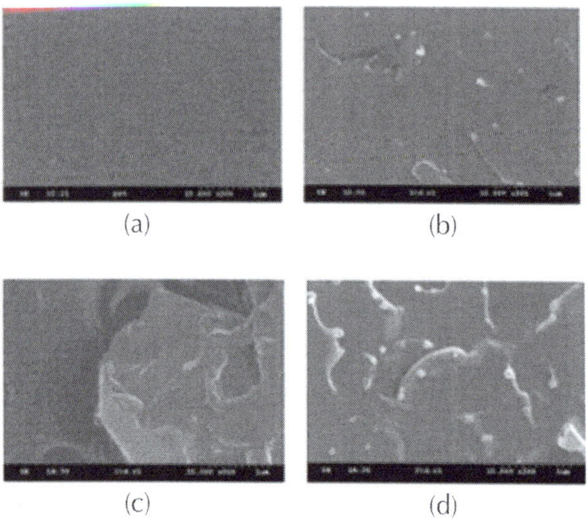

FIGURE 10 The images from scanning electron microscope for the nanocomposite surfaces:
(a) – pure PET;
(b) – 3 wt% of organoclay in PET;
(c) – 4 wt% of organoclay in PET;
(d) – 5 wt% of organoclay in PET.

Also one can use the STM images to judge on the degree of the distribution of the organoclay in the nanocomposite, Figs. 11 and 12. If the content of the organoclay is 2–3 wt% then the clay layers are separated by the polymeric layer of 4 to 10 nm width, Fig. 11. If the content of the organoclay reaches 4–5 wt % then the majority of the clay becomes well distributed, however, the agglomerates of 4 to 8 nm may appear.

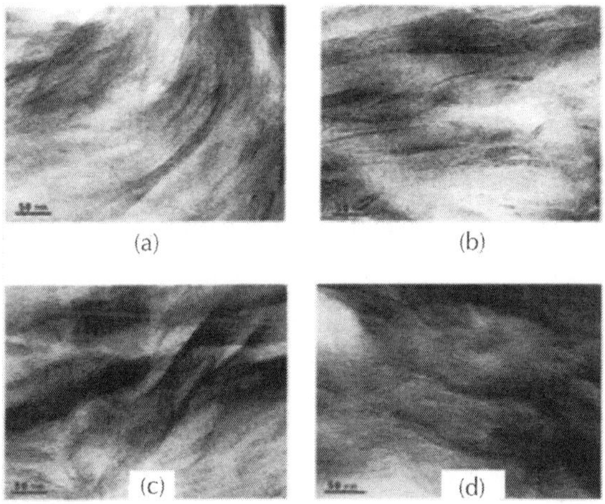

FIGURE 11 The images from tunneling electron microscope for the nanocomposite surfaces:
(a) – 2 wt% of organoclay in PBT;
(b) – 3 wt% of organoclay in PBT;
(c) – 4 wt% of organoclay in PBT;
(d) – 5 wt% of organoclay in PBT.

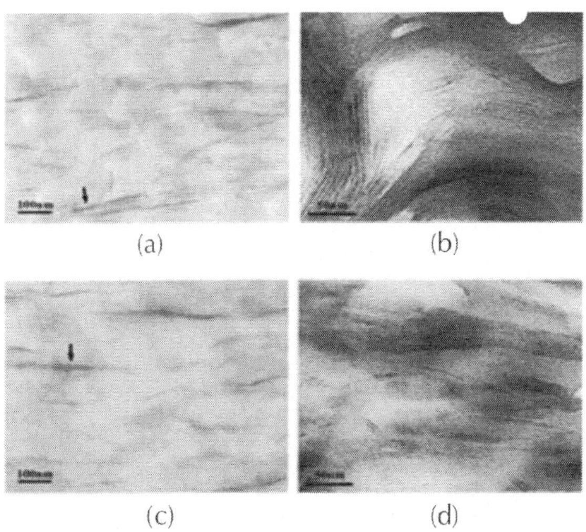

FIGURE 12 The images from tunneling electron microscope for the nanocomposite surfaces:
(a) – 1 wt% of organoclay in PET;
(b) – 2 wt% of organoclay in PET;
(c) – 3 wt% of organoclay in PET;
(d) – 4 wt% of organoclay in PET.

So the involvement of the X-ray analysis and the use of the microscopy data tell that the nanocomposite consists of the exfoliated clay at the low content (below 3 wt %) of the organoclay.

7.5 PRODUCTION OF THE POLYMERIC NANOCOMPOSITES ON THE BASIS OF THE ALUMINOUS SILICATES

Different groups of authors [32–35] offer following methods for obtaining nanocomposites on the basis of the organoclays: (1) in the process of the synthesis of the polymer [33, 36, 37], (2) in the melt [38, 39], (3) in the solution [40–46] and (4) in the sol-gel process [47–50].

The most popular ones are the methods of producing in melt and during the process of the synthesis of the polymer.

The producing of the polymeric nanocomposite in situ is the intercalation of the monomer into the clay layers. The monomer migrates through the organoclay galleries and the polymerization happens inside the layers [19, 51], Fig. 13.

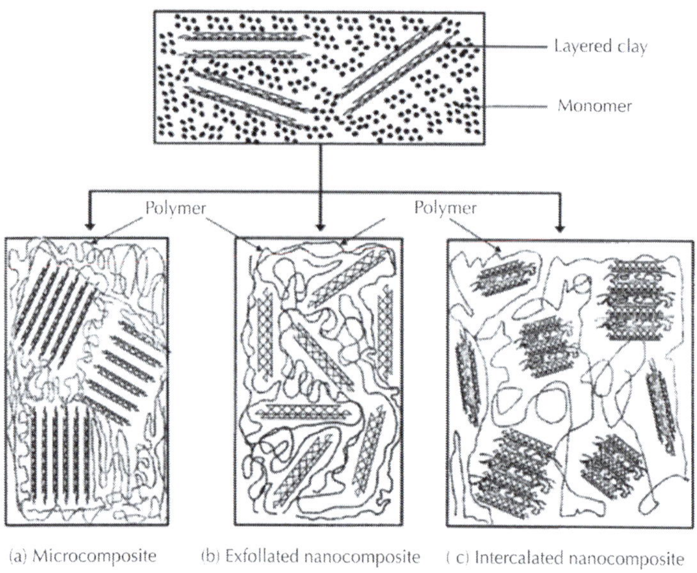

FIGURE 13 The production of the nanocomposite in situ:
(a) – microcomposite;
(b) – exfoliated nanocomposite;
(c) – intercalated nanocomposite [51].

The polymerization may be initiated by the heat, irradiation or other source. Obviously, the best results on the degree of the distribution of the clay particles in the polymeric matrix must emerge if using given method. This is associated with the fact that the separation of the clay layers happens in the very process of the inclusion of the monomer in the interlayer space. In other words, the force responsible for the separation of the clay layers is the growth of the polymeric chain whereas the main factor for reaching the necessary degree of the clay distribution in solution or melt is just satisfactory mixing. The most favorable condition for synthesizing the nanocomposites is the vacuuming or the flow of the inert gas. Besides, one has to use the fast speeds of mixing for satisfactory dispersing of the organoclay in the polymeric matrix.

The method of obtaining the polymeric nanocomposites in melt (or the method of extrusion) is the mixing of the polymer melted with the organoclay. The polymeric chains lose the considerable amount of the conformational entropy during the intercalation. The probable motive force for this process is the important contribution of the enthalpy of the interaction between the polymer and organoclay at mixing. One should add that polymeric nanocomposite on the basis of the organoclays could be successively produced by the extrusion [22]. The advantage of the extrusion method is the absence of any solvents what excludes the unnecessary leaks. Moreover the speed of the process is several times more and the technical side is simpler. The extrusion method is the best one in the industrial scales of production of the polymeric nanocomposites what acquires the lesser source expenses and easier technological scheme.

If one produces the polymer – silicate nanocomposite in solution then the organosilicate swells in the polar solvent such as toluene or N, N-dimethylformamide. Then the added is the solution of the polymer, which enters the interlayer space of the silicate. The removing of the solvent by means of evaporation in vacuum happens after that. The main advantage of the given method is that "polymer-layered silicate" might be produced from the polymer of low polarity or even nonpolar one. However, this method is not widely used in industry because of much solvent consumption [52].

The sol-gel technologies find application at producing nanocomposites on the basis of the various ceramics and polymers. The initial compounds in these technologies are the alcoholates of specific elements and organic oligomers.

The alcoholates are firstly hydrolyzed and obtained hydroxides being polycondensated then. The ceramics from the inorganic 3D net is formed as a result. Also the method of synthesis exists in which the polymerization and the formation of the inorganic glass happen simultaneously. The application of the nanocomposites on the basis of ceramics and polymers as special hard defensive coverages and like optic fibers [53] is possible.

7.6 PROPERTIES OF THE POLYMERIC NANOCOMPOSITES

Many investigations in physics, chemistry and biology have shown that the jump from macroobjects to the particles of 1–10 nm results in the qualitative transformations in both separate phases and systems from them [54].

One can improve the thermal stability and mechanical properties of the polymers by inserting the organoclay particles into the polymeric matrix. It can be done by means of joining the complexes of properties of both the organic and inorganic substances, that is combining the light weight, flexibility and plasticity of former and durability, heat stability and chemical resistance of latter.

Nanocomposites demonstrate essential change in properties if compared to the nonfilled polymers. So, if one introduces modified layered silicates in the range of 2 to 10 wt% into the polymeric matrix then he observes the change in mechanical (tensile, compression, bending and overall strength), barrier (penetrability and stability to the solvent impact), optical and other properties. The increased heat and flame resistance even at low filler content is among the interesting properties too. The formation of the thermal isolation and negligible penetrability of the charred polymer to the flame provide for the advantages of using these materials.

The organoclay as a nanoaddition to the polymers may change the temperature of the destruction, refractoriness, rigidity and rupture strength. The nanocomposites also possess the increased rigidity modulus, decreased coefficient of the heat expansion, low gas-penetrability, increased stability to the solvent impact and offer broad range of the barrier properties [54]. In Table 1, we gather the characteristics of the nylon-6 and its derivative containing 4.7 wt% of the organomodified montmorillonite.

TABLE 1 The properties of the nylon-6 and composite based on it [54].

	Rigidity modulus, GPa	Tensile strength, MPa	Temperature of the deformation, °C	Impact viscosity, kJ/m^2	Water consumption, weight %	Coefficient of the thermal expansion (x, y)
Nylon-6	1.11	68.6	65	6.21	0.87	13×10^{-5}
Nanocomposite	1.87	97.2	152	6.06	0.51	6.3×10^{-5}

It is important that the temperature of the deformation of the nanocomposite increases on 87 °C.

The thermal properties of the polymeric nanocomposites with the varying organoclay content are collected in Table 2.

TABLE 2 The main properties of the polymeric nanocomposites.

Property	Composition								
	Polybutyleneterephtalate + AAX-montmorillonite					Polyethyleneterephtalate + C_{12}PPh-montmorillonite			
	Organoclay content, %								
	0	2	3	4	5	0	1	2	3
Viscosity, dliter/g	0.84	1.16	0.77	0.88	0.86	1.02	1.26	0.98	1.23
T_g, °C	27	33	34	33	33	—	—	—	—
T_m, °C	222	230	230	229	231	245	247	245	246
T_d, °C	371	390	388	390	389	370	375	384	386
$W_{tR}^{600\,c}$, %	1	6	7	7	9	1	8	15	21
Strength limit, MPa	41	50	60	53	49	46	58	68	71
Rigidity modulus, GPa	1.37	1.66	1.76	1.80	1.86	2.21	2.88	3.31	4.10
Relative enlargement, %	5	7	6	7	7	3	3	3	3

The inclusion of the organoclay into the polybutyleneterephtalate leads to the increase in the glass transition temperature (T_g) from 27 to 33 degrees centigrade if the amount of the clay raises from zero to 2 wt%. That temperature does not change with the further increase of the organoclay content. The increase in T_g may be the result of two reasons [56–59]. The first is the dispersion of the small amount of the organoclay in the free volume of the polymer, and the second is the limiting of the mobility of the segments of the polymeric chain due to its interlocking between the layers of the organoclay.

The same as the T_g, the melting temperature T_m increases from the 222 to 230 °C if the organoclay content raises from 0 to 2 wt% and stays constant up to 5 wt%, see Table 2. This increase might be the consequence of both complex multilayer structure of the nanocomposite and interaction between the organoclay and polymeric chain [60, 61]. Similar regularities have been observed in other polymeric nanocomposites also.

The thermal stability of the nanocomposites polybutyleneterephtalate, briefly PBT, (or polyethyleleterephtalate)/organoclay determined by the thermogravimetric analysis is presented in Table 3 and in Figs. 14 and 15 [11, 12].

TABLE 3 The thermal properties of the fibers from PET with varying organoclay content.

Organoclay content, weight %	η_{inh}[a]	T_m (°C)	H_m[b] (J/g)	T^i_D[c] (°C)	Wt_R^{600}[d] (%)
0 (pure PET)	1.02	245	32	370	1
1	1.26	247	32	375	8
2	0.98	245	33	384	15
3	1.23	246	32	386	21

[a]viscosities were measured at 30 °C using 0.1 g of polymer on 100 mL of solution in mix phenol/tetrachlorineethane (50/50);
[b]change in enthalpy of melting;
[c]initial temperature of decomposition;
[d]weight percentage of the coke remnant at 600 °C.

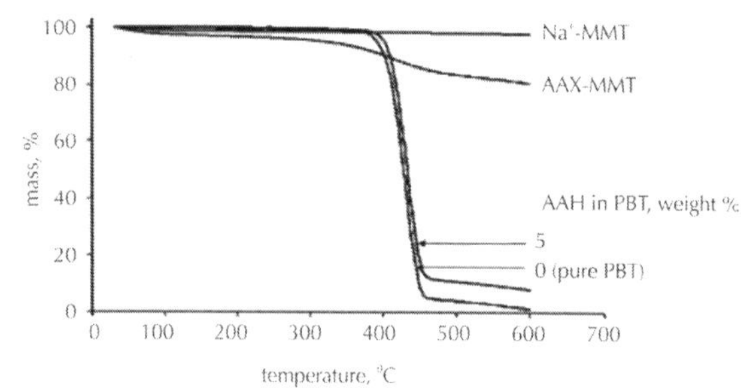

FIGURE 14 The thermogravimetrical curves for the montmorillonites, PBT and nanocomposites PBT/organoclay.

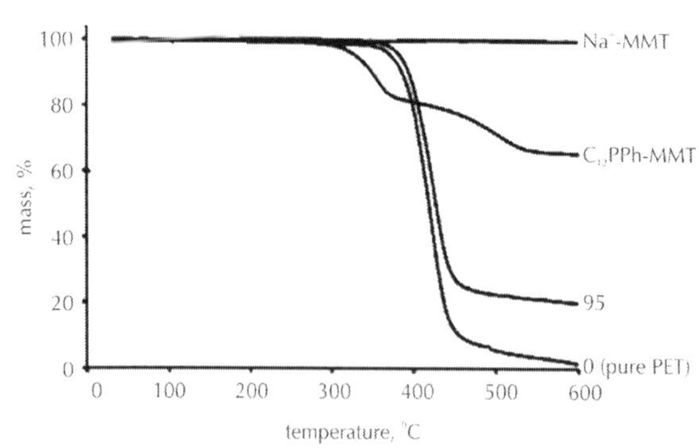

FIGURE 15 The thermogravimetrical curves for the montmorillonites, PET and nanocomposites PET/organoclay.

The temperature of the destruction, T_D, increases with the organoclay content up to 350 °C in case of the composite PBT/organoclay. The thermogravimetrical curves for pure and composite PBTs have similar shapes below 350 °C. The values of temperature T_D depends on the amount of organoclay above 350 °C. The organoclay added becomes a barrier for volatile products being formed during the destruction [61, 62]. Such example of the improvement of the thermal stability was studied in papers [63, 64]. The mass of the remnant at 600 °C increases with organoclay content.

Following obtained data authors draw the conclusion that the optimal results for thermal properties are being obtained if 2 wt% of the organoclay is added [11, 12, 19].

The great number of studies on the polymeric composite organoclay – based materials show [11, 12, 13, 19] that the inclusion of the inorganic component into the organic polymer improves the thermal stability of the latter, see Tables 3 and 4.

TABLE 4 The basic properties of the nanocomposite based on PBT with varying organoclay content.

Organoclay content, weight %	I.V.[a]	T_g	T_m (°C)	T_D^i [b] (°C)	Wt_R^{60} (°C) (%)
0 (pure PBT)	0.84	27	222	371	1
2	1.16	33	230	390	6
3	0.77	34	230	388	7
4	0.88	33	229	390	7
5	0.86	33	231	389	9

[a]viscosities were measured at 30 °C using 0.1 g of polymer on 100 mL of solution in mix phenol/tetrachlorineethane (50/50);

[b]initial temperature of the weight loss;

[c]weight percentage of the coke remnant at 600 °C.

The values of the melting temperature increase from 222 to 230 °C if the amount of the organoclay added reaches 2 wt% and then stay constant. This effect can be explained by both thermal isolation of the clay and interaction between the polymeric chain and organoclay [43, 64]. The increase in the glass transition temperature also occurs what can be a consequence of several reasons [55, 56, 58]. One of the main among them is the limited motion of the segments of the polymeric chain in the galleries within the organoclay.

If the organoclay content in the polymeric matrix of the PBT reaches 2 wt% then both the temperature of the destruction increases and the amount of the coke remnant increase at 660 °C and then both stay practically unchanged with the further increase of the organoclay content up to 5 wt%. The loss of the weight due to the destruction of the polymer in pure PBT and its composites looks familiar in all cases below 350 °C. The amount of the organoclay added becomes important above that temperature because the very clays possess good thermal stability and make thermal protection by their layers and form a barrier preventing the volatile products of the decomposition to fly off [43, 60]. Such instance of the improvement in thermal properties was observed in many polymeric composites [64–68]. The weight of the coke remnant increases with the rizing organoclay content up to 2 wt% and stays constant after that. The increase of the remnant may be linked with the high thermal stability of the organoclay itself. Also it is worth noticing that the polymeric chain closed in interlayer space of the organoclay has fewer degrees of oscillatory motion at heating due to the limited interlayer space and the formation of the abundant intermolecular bonds between the polymeric chain and the clay surface. And the best result is obtained at 2–3 wt% content of the organoclay added to the polymer.

If one considers the influence of the organoclay added to the polyethelenetere-phtalate, briefly PET, [69] then the temperature of the destruction increases on 16 °C at optimal amount of organoclay of 3 wt%. The coke remnant at 600 °C again increases with the rizing organoclay content, see Table 3.

Regarding the change in the temperature of the destruction in the cases of PET and PBT versus the organoclay content one can note that both trends look similar. However the coke remnant considerably increases would the tripheyldodecy-lphosphonium cation be present within the clay. The melting temperature does not increase in case of the organoclay added into the PET in contrary to the case of PBT. Apparently this may be explained by the more crystallinity of the PBT and the growth of the degree of crystallinity with the organoclay content.

It becomes obvious after analyzing the above results that the introduction of the organoclay into the polymer increases the thermal stability of the latter according the (1) thermal isolating effect from the clay layers and (2) barrier effect in relation to the volatile products of destruction.

The studying of the mechanical properties of the nanocomposites, see Table 2, have shown that the limit of the tensile strength increases with the organoclay added up to 3 wt% for the majority of the composites. Further addition of the organoclay, up to 5 wt%, results in the decreasing limit of the tensile strength. We explain this by the fact that agglomerates appear in the nanocomposite when the organoclay content exceeds the 3 wt% value [61, 70, 71]. The proof for the formation of the agglomerates have been obtained from the X-ray study and using the data from electron microscopes.

Nevertheless the rigidity modulus increases with the amount of the organoclay added into the polymeric matrix, the resistance of the clay itself being the explanation for that. The oriented polymeric chains in the clay layers also participate in the increase of the rigidity modulus [72]. The percentage of enlargement at breaking became 6–7 wt% for all mixes.

Using data of the Table 2 we explain the improvement in the mechanical properties of the nanocomposites with added organoclay up to 3 wt% by the good degree of distribution of the organoclay within the polymeric matrix. The degree of the improvement also depends on the interaction between the polymeric chain and clay layers.

The study of the influence of the degree of the extract of fibers on the mechanical properties has shown that the limit of strength and the rigidity modulus both increase in PBT whereas they decrease in nanocomposites, Table 5. This can be explained by the breaking of the bonds between the organoclay and PBT at greater degree of extract. Such phenomena have been observed in numerous polymeric composites [73–75].

TABLE 5 The ability to stretch of the nanocomposites PBT/organoclay at varying degrees of extract.

Organoclay content,	Limit of strength, MPa			Rigidity modulus, GPa		
weight %	DR=1	DR=3	DR=6	DR=1	DR=3	DR=6
0 (pure PBT)	41	50	52	1.37	1.49	1.52
3	60	35	29	1.76	1.46	1.39

The first notions on the lowered flammability of the polymeric nanocomposites on the organoclay basis appeared in 1976 in the patent on the composite based on the nylon-6 [5]. The serious papers in the field were absent till the 1995 [76].

The use of the calorimeter is very effective for studying the refractoriness of the polymers. It can help at measuring the heat release, the carbon monoxide depletion and others. The speed of the heat release is one of the most important parameters defining the refractoriness [77]. The data on the flame resistance in various polymer/organoclay systems such as layered nanocomposite nylon-6/organoclay, intercalated nanocomposites polystirene (or polypropylene)/organoclay were given in paper [78] in where the lowered flammability was reported, see Table 6. And the lowered flammability have been observed in systems with low organoclay content, namely in range from 2 to 5 wt%.

TABLE 6 Calorimetric data.

Sample	Remnant (%)±0.5	Peak of the HRR (Δ%) (kW/m^2)	Middle of the HRR (Δ%) (kW/m^2)	Average value H_c (MJ/kg)	Average value SEA (m^2/kg)	Average CO left (kg/kg)
Nylon-6	1	1010	603	27	197	0.01

TABLE 6 *(Continued)*

Nylon-6/organoclay, 2%, delaminated	3	686 (32%)	390 (35%)	27	271	0.01
Nylon-6/organoclay, 5%, delaminated	6	378 (63%)	304 (50%)	27	296	0.02
Polystirene	0	1120	703	29	1460	0.09
PS/organoclay, 3%, bad mixing	3	1080	715	29	1840	0.09
PS/organoclay, 3%, intercalated/delaminated	4	567 (48%)	444 (38%)	27	1730	0.08
PS w/DBDPO/Sb_2O_3, 30%	3	491 (56%)	318 (54%)	11	2580	0.14
Polypropylene	0	1525	536	39	704	0.02
PP/organoclay, 2%, intercalated	5	450 (70%)	322 (40%)	44	1028	0.02

H_c – heat of combustion;
SEA – specific extinguishing area;
DBDPO – dekabrominediphenyloxide;
HRR – speed of the heat release.

The curve of the heat release for the polypropylene and the nanocomposite on its basis (organoclay content varying from 2 to 4 wt%) is given in the Fig.16 from which one can see that the speed of the heat release for the nanocomposite enriched with the 4 wt% organoclay (the interlayer distance 3.5 nm) is 75% less than for pure polypropylene.

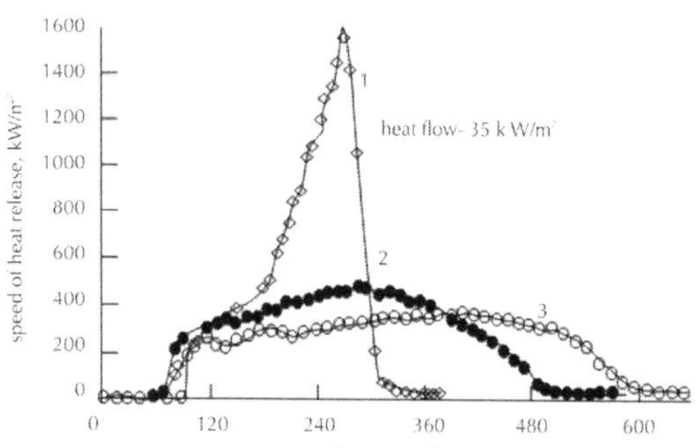

FIGURE 16 The speed of the heat release for:
1 – pure polypropylene;
2 – nanocomposite with 2 wt% of organoclay;
3 – nanocomposite with 4 wt% of organoclay.

The comparison of the experimental data for the nanocomposites on the basis of the nylon-6, polypropylene and polystirene gathered in Table 7 show that the heat of combustion, the smoke release and the amount of the carbon monoxide are almost constant at varying organoclay content. So we conclude that the source for the increased refractoriness of these materials is the stability of the solid phase and not the influence of the vapor phase. The data for the polystirene with the 30% of the dekabrominediphenyloxide and Sb_2O_3 are given in Table 6 as the proof of the influence of the vapor phase of bromine. The incomplete combustion of the polymeric material in the latter case results in low value of the heat of the combustion and high quantity of the carbon monoxide released [79].

One should note that the mechanism for the increased fire resistance of the polymeric nanocomposites on the basis of the organoclays is not, in fact, clear at all. The formation of the barrier from the clay layers during the combustion at their collapse is supposed to be the main mechanism. That barrier slows down the combustion [80]. In our paper we study the influence of the nanocomposite structure on the refractoriness. The layered structure of the nanocomposite expresses higher refractoriness comparing to that in intercalated nanocomposite, see Fig. 17.

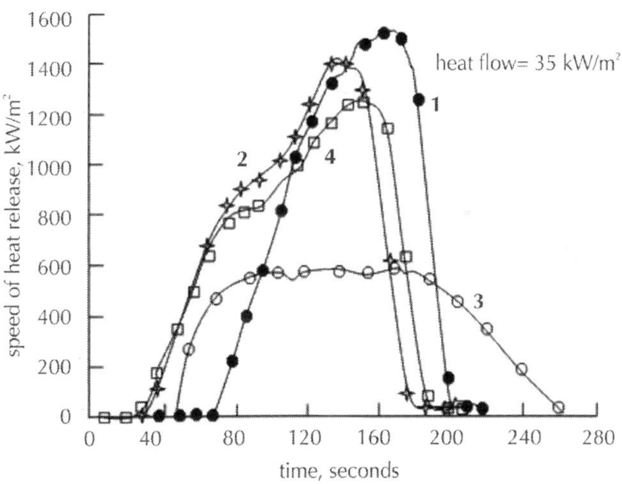

FIGURE 17 The speed of the heat release for:
1 – pure polystirene (PS);
2 – PS mixed with 3 wt% of Na^+ MMT;
3 – intercalated/delaminated PS (3 wt% 2C18-MMT) extruded at 170 °C;
4 – intercalated PS (3 wt% C14-FH) extruded at 170 °C.

The data on the polymeric polystirene – based nanocomposites presented in Fig.17 are for (1) initial ammoniumfluorine hectorite and (2) quaternary ammonium montmorillonite. The intercalated nanocomposite was produced in first case whereas the layered-intercalated nanocomposite was produced in the second one. But because the chemical nature and the morphology of the organoclay used was quite different it is very difficult to draw a unique conclusion about the flame resistance in polymeric nanocomposites produced. Nonetheless, one should point out that good results of the same quality were obtained for both layered and intercalated structures when studying the aliphatic groupings of the polyimide nanocomposites based on these clays. The better refractoriness is observed in case of polystirene embedded in layered nanocomposite while intercalated polystirene – based nanocomposite (with MMT) also exhibits increased refractoriness.

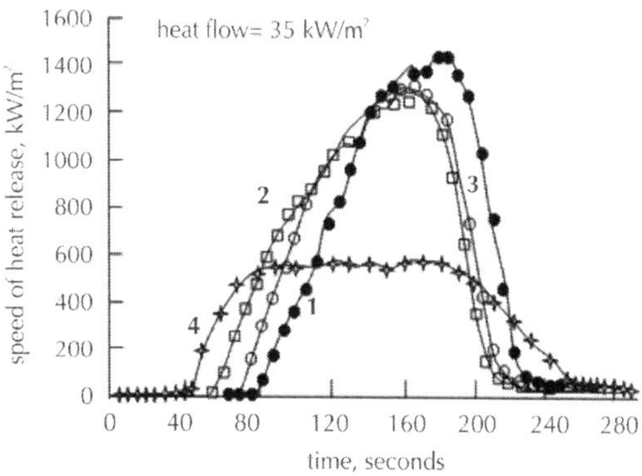

FIGURE 18 The speed of the heat release for:
1 – pure polysterene (PS);
2 – polystirene with Na-MMT;
3 – intercalated PS with organomontmorillonite obtained in extruder at 185 °C;
4 – intercalated/layered PS with organoclay obtained in extruder at 170 °C in nitrogen atmosphere or in vacuum.

As one can see the from the Fig.18, the speed of the heat release for the nanocomposite produced in nitrogen atmosphere at 170 °C is much lower than for other samples. Probably, the reason for the low refractoriness of the nanocomposite produced in extruder without the vacuuming at 180 °C is the influence of the high temperature and of the oxygen from the air what can lead to the destruction of the polymer in such conditions of the synthesis.

It is impossible to give an exact answer on the question about how the refractoriness of organoclay – based nanocomposites increases basing on only the upper experimental data but the obvious fact is that the increased thermal stability and refractoriness are due to the presence of the clays existing in the polymeric matrix as nanoparticles and playing the role of the heat isolators and elements preventing the flammable products of the decomposition to fly off.

There are still many problems unresolved in the field but indisputably polymeric nanocomposites will take the leading position in the chemistry of the advanced materials with high heat and flame resistance. Such materials can be used either as itself or in combination with other agents reducing the flammability of the substances.

The processes of the combustion are studied for the number of a polymeric nanocomposites based on the layered silicates such as nylon-6.6 with 5 wt% of Cloisite 15A – montmorillonite being modified with the dimethyldialkylammonium (alkyls studied C_{18}, C_{16}, C_{14}), maleinated polypropylene and polyethylene, both (1.5%) with 10 wt% Cloisite 15A. The general trend is two times reduction of the speed of the heat release. The decrease in the period of the flame induction is reported for all nanocomposites in comparison with the initial polymers [54].

The influence of the nanocomposite structure on its flammability is reflected in the Table 7. One can see that the least flammability is observed in delaminated nanocomposite based on the polystirene whereas the flammability of the intercalated composite is much higher [54].

TABLE 7 Flammability of several polymers and composites

Sample	Coke remnant, wt%	Max speed of heat release, kW/m²	Average value of heat release, kW/m²	Average heat of combustion, MJ/kg	Specific smoke release, m²/kg	CO release, kg/kg
Nylon-6	1	1010	603	27	197	0.01
Nylon-6 + 2% of silicate (delaminated)	3	686	390	27	271	0.01
Nylon-6 + 5% of silicate (delaminated)	6	378	304	27	296	0.02
Nylon-12	0	1710	846	40	387	0.02
Nylon-12 + 2% of silicate (delaminated)	2	1060	719	40	435	0.02
Polystirene	0	1562	803	29	1460	0.09
PS + 3% of silicate Na-MMT	3	1404	765	29	1840	0.09
PS + 3% of silicate C14-FH (intercalated)	4	1186	705	28	1790	0.09
PS + 3% of silicate 2C18-MMT (delaminated)	4	567	444	28	1730	0.08
Polypropylene	0	1525	536	39	704	0.02
PP + 2% of silicate (intercalated)	3	450	322	40	1028	0.02

The optical properties of the nanocomposites are of much interest too. The same materials could be either transparent or opaque depending on certain conditions. For example in Fig.19 we see transparency, plot (a), and turbidity, plot (c), of the material in dependence of the frequency of the current applied.

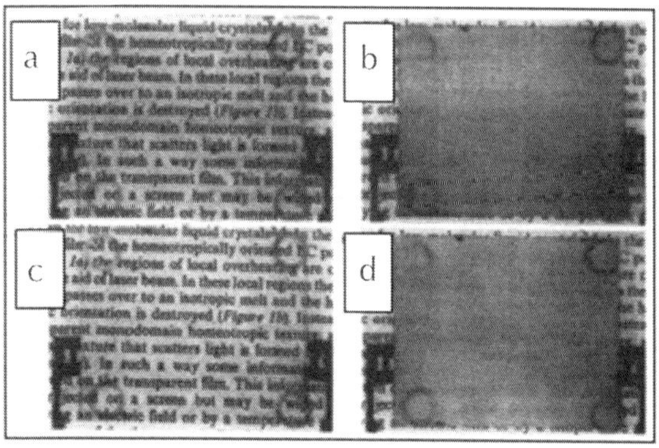

FIGURE 19 The optical properties of the clay – based nanocomposites in dependence of the applied electric current:
(a) – low frequency, switched on;
(b) – low frequency, switched off;
(c) – high frequency, switched on;
(d) – high frequency, switched off.

The effect in Fig. 19 is reversible and can be innumerately repeated. The transparent and opaque states exhibit the memory effect after the applied current switched off, plots (b) and (d) in Fig. 19. The study of the intercalated nanocomposites based on the smectite clays reveals that the optical and elecrooptical properties depend on the degree of intercalation [81].

7.7 CONCLUSIONS

The quantity of the papers in the field of the nanocomposite polymeric materials has grown multiple times in recent years. The possibility to use almost all polymeric and polycondensated materials as a matrix is shown. The nanocomposites from various organoclays and polymers have been synthesized. Here is just a small part of the compounds for being the matrix referenced in literature: polyacrylate [83], polyamides [82, 84, 85], polybenzoxasene [86], polybutyleneterephtalate [11, 82, 87], polyimides [88], polycarbonate [89], polymethylmetacrylate [90], polypropylene [91, 92], polystirene [90], polysulphones [93], polyurethane

[94], polybuthyleneterephtalate and polyethyleneterephtalate [10, 65, 68, 79, 95, 99–107], polyethylene [96], epoxies [97].

The organomodified montmorillonite is of the special interest because it can be an element of the nanotechnology and it can also be a carrier of the nanostructure and of asymmetry of length and width in layered structures. The organic modification is being usually performed using the ion-inducing surface-active agents. The nonionic hydrophobization of the surface of the layered structures have been reported either. The general knowledge about the methods of the study is being formed and the understanding of the structure of the nanocomposite polymeric materials is becoming clear. Also scientists come closer to the realizing of the relations between the deformational and strength properties and the specifics of the nanocomposite structure. The growth of researches and their direction into the nanoarea forecasts the fast broadening of the industrial involvement to the novel and attractive branch of the materials science.

KEYWORDS

- **bentonite**
- **methylmetacrylate**
- **organoclays**
- **polybutylenterephtalate**
- **polymethylmetacrylate**

REFERENCES

1. Romanovsky, B. V.; Makshina, E. V.; Sorosovskii obrazovatelniu zhurnal, **2004**, *8(2),* 50–55 (in Russian)
2. Golovin, Yu. I.; Priroda, 1–2004 (in Russian)
3. Carter, L. W.; Hendrics, J. G.; Bolley, D. S.; United States Patent **1950**, *2,* 531,396.
4. Blumstain A.; *Bull Chem. Soc.* **1961,** 899–905.
5. Fujiwara S.; Sakamoto T.; Japanese Application *(109),* 998; 1976.
6. Usuki A.; Kojima Y.; Kawasumi M.; Okada A.; Fukushima Y.; Kurauchi T.; Kamigatio O.; Journal of Appl polym science. **1995,** *55,* 119.
7. Usuki A.; Koiwai A.; Kojima Y.; Kawasumi M.; Okada A.; Kurauchi T.; Kamigaito O.; Journal of Appl polym science. **1995,** *55,* 119
8. Okada A.; Usuki A.; Mater Sci Engng. **1995,** *3,* 109.
9. Okada A.; Fukushima Y.; Kawasumi M.; Inagaki S.; Usuki A.; Sugiyama S.; Kurauchi T.; Kamigaito O.; United States Patent **1988**, *4,* 739,007.

10. Mikitaev, M. A.; Lednev, O. B.; Kaladjian, A. A.; Beshtoev, B. Z.; Bedanokov, A. Yu.; Mikitaev, A. K.; Second International Conference (Nalchik 2005).
11. Chang, J. -H.; An, Y. U.; Kim, S. J.; S. Im. Polymer, **2003,** *44,* 5655–5661.
12. Mikitaev, A. K.; Bedanokov, A. Y.; Lednev, O. B.; Mikitaev, M. A. Polymer/silicate nanocomposites based on organomodified clays/Polymers, Polymer Blends, Polymer Composites and Filled Polymers. Synthesis, Properties, Application. Nova Science Publishers: New York **2006**.
13. Delozier, D. M.; Orwoll, R. A.; Cahoon, J. F.; Johnston, N. J.; Smith, J. G.; Connell, J. W.; *Polymer,* **2002,** *43,* 813–822.
14. Kelly, P.; Akelah, A.; Moet, A. J.; *Mater. Sci.* **1994,** *29,* 2274–2280.
15. Chang, J. -H.; An, Y. U.; Cho, D.; Giannelis, E. P.; *Polymer*, **2003,** *44,* 3715–3720.
16. Yano, K.; Usuki, A.; Okada A. *J. Polym. Sci. Part A: Polym. Chem.* **1997,** *35,* 2289.
17. Garcia-Martinez, J. M.; Laguna, O.; Areso, S.; Collar EP. *J. Polym. Sci. Part B: Polym. Phys.* **2000,** *38,* 1564.
18. Giannelis, E. P.; Krishnamoorti R.; Manias E.; Advances in Polymer Science, 138 Springer-Verlag: Berlin Heidelberg, **1999**.
19. Delozier, D. M.; Orwoll, R. A.; Cahoon, J. F.; Ladislaw, J. S.; Smith, J. G.; Connell, Polymer, J. W.; **2003,** *44,* 2231–2241.
20. Shen, Y. -H.; *Chemosphere*, **2001,** *44,* 989–995.
21. Giannelis, E. P.; *Adv. Mater.* **1996,** *8,* 29–35.
22. Dennis, H. R.; Hunter, D. L.; Chang, D.; Kim, S.; White, J. L.; Cho, J. W.; Paul, Polymer, D. R.; **2001,** *42,* 9513–9522.
23. Kornmann, X.; Lindberg, H.; Berglund, L. A.; *Polymer* **2001,** *42,* 1303–1310.
24. Fornes, T. D.; Paul, D. R.; Formation and properties of nylon 6 nanocomposites. Polímeros 13 no.4 São Carlos Oct./Dec. 2003.
25. Voulgaris, D.; Petridis, D.; *Polymer*, **2002,** *43,* 2213–2218.
26. Tyan, H. -L.; Liu, Y. -C.; Wie, K. -H.; *Polymer*, **1999,** *40,* 4877–4886.
27. Davis, C. H.; Mathias, L. J.; Gilman, J. W.; Schiraldi, D. A.; Shields, J. R.; Trulove, P.; Sutto, T. E.; Delong, H. C. *J. Polym. Sci. Part B: Polym. Phys.* **2002,** *40,* 2661.
28. Morgan, A. B.; Gilman, J. W. *J. Appl. Polym. Sci.* **2003,** *87,* 1329.
29. John N.; Hay and Steve J.; Shaw. Organic-inorganic hybridssthe best of both worlds? Europhysics News **2003,** *34(3).*
30. Chang, J. H.; An YU, Sur GS. *J. Polym. Sci. Part B: Polym. Phys.* **2003,** *41,* 94.
31. Chang, J. H.; Park, D. K.; Ihn, K. J. *J. Appl. Polym. Sci.* **2002,** *84,* 2294.
32. Pinnavaia, T. J.; Science **1983,** *220,* 365.
33. Messersmith, P. B.; Giannelis, E. P.; *Chem. Mater.* **1993,** *5,* 1064.
34. Vaia, R. A.; Ishii, H.; Giannelis, E. P.; *Adv. Mater.* **1996,** *8,* 29.
35. Gilman, J. W.; *Appl. Clay. Sci.* **1999,** *15,* 31.
36. Fukushima, Y.; Okada, A.; Kawasumi, M.; Kurauchi, T.; Kamigaito O.; *Clay Miner.* **1988,** *23,* 27.
37. Akelah, A.; Moet A. J Mater Sci **1996,** *31,* 3589.
38. Vaia, R. A.; Ishii, H.; Giannelis, E. P.; *Adv. Mater.* **1996,** *8,* 29.
39. Vaia, R. A.; Jandt, K. D.; Kramer, E. J.; Giannelis, E. P.; Macromolecules **1995,** *28,* 8080.
40. Greenland DG. J Colloid Sci **1963,** *18,* 647.
41. Chang, J. H.; Park, K. M.; *Polym. Eng. Sci.* **2001,** *41,* 2226.

42. Greenland DG. J Colloid Sci **1963,** *18,* 647.
43. Chang, J. H.; Seo, B. S.; Hwang, D. H.; Polymer **2002,** *43,* 2969.
44. Vaia, R. A.; Jandt, K. D.; Kramer, E. J.; Giannelis, E. P.; Macromolecules **1995,** *28,* 8080.
45. Fukushima, Y.; Okada, A.; Kawasumi, M.; Kurauchi, T.; Kamigaito O.; *Clay Miner.* **1988,** *23,* 27.
46. Chvalun S. N.; Priroda **2000,** 7 (in Russian)
47. Brinker, C. J.; Scherer G. W.; Sol-Gel Science. Boston, 1990.
48. Mascia, L.; Tang, T.; Polymer **1998,** *39,* 3045.
49. Tamaki, R.; Chujo, Y.; *Chem. Mater.* **1999,** *11,* 1719.
50. Serge Bourbigot, E. A. Investigation of Nanodispersion in Polystirene–Montmorillonite Nanocomposites by Solid-State NMR.; Journal of Polymer Science: Part B: Polymer Physics, *41,*3188–3213 **2003,**
51. Lednev, O. B.; Kaladjian, A. A.; Mikitaev, M. A.; Tlenkopatchev, M. A.; New polybutylene terephtalate and organoclay nanocomposite materials. Abstracts of the International Conference on Polymer materials (México, 2005)
52. Tretiakov A. O.; Oborudovanie I instrument dlia professionalov (02(37) **2003,** (in Russian)
53. Sergeev Ros, G. B.; Chem. J. (Journal of D. I. Mendeleev Russian Chemical Society), **2002,** XLVI, 5 (in Russian)
54. Lomakin, S. M.; Zaikov Visokomol, G. E.; Soed. B. **2005,** *47(1),* 104–120 (in Russian)
55. Xu H, Kuo, S. W.; Lee, J. S.; Chang FC.; Macromolecules **2002,** *35,* 8788.
56. Haddad, T. S.; Lichtenhan, J. D.; Macromolecules **1996,** *29,* 7302.
57. Mather, P. T.; Jeon, H. G.; Romo-Uribe, A.; Haddad, T. S.; Lichtenhan, J. D.; Macromolecules **1996,** *29,* 7302.
58. Hsu, S. L. C.; Chang KC.; Polymer **2002,** *43,* 4097.
59. Chang, J. H.; Seo, B. S.; Hwang, D. H.; Polymer **2002,** *43,* 2969.
60. Fornes, T. D.; Yoon, P. J.; Hunter, D. L.; Keskkula, H.; Paul DR.; Polymer **2002,** *43,* 5915.
61. Chang, J. H.; Seo, B. S.; Hwang, D. H.; Polymer **2002,** *43,* 2969.
62. Fornes, T. D.; Yoon, P. J.; Hunter, D. L.; Keskkula, H.; Paul DR.; Polymer **2002,** *43,* 5915.
63. Wen, J.; Wikes, G. L.; *Chem. Mater.* **1996,** *8,* 1667.
64. Zhu, Z. K.; Yang, Y.; Yin, J.; Wang, X.; Ke Y, Qi Z. *J. Appl. Polym. Sci.* **1999,** *3,* 2063.
65. Mikitaev, M. A.; Lednev, O. B.; BeshtoevB. Z.; Bedanokov, A. Yu.; Mikitaev A. K.; Second International conference "Polymeric composite materials and covers" (Yaroslavl **2005,** may) (in Russian)
66. Fischer, H. R.; Gielgens, L. H.; Koster TPM.; Acta Polym **1999,** *50,* 122.
67. Petrovic, X. S.; Javni, L.; Waddong, A.; Banhegyi GJ. *J. Appl. Polym. Sci.* **2000,** *76,* 133.
68. Lednev, O. B.; Beshtoev, B. Z.; Bedanokov, A. Yu.; Alarhanova, Z. Z.; Mikitaev A. K.; Second International Conference (Nalchik 2005) (in Russian)
69. Chang, J. -H.; Kim, S. J.; Joo, Y. L.; S. Im. Polymer, **2004,** *45,* 919–926.
70. Lan, T.; Pinnavaia, T. J.; *Chem. Mater.* **1994,** *6,* 2216.

71. Masenelli-Varlot, K.; Reynaud, E.; Vigier, G.; Varlet, J. *J. Polym. Sci. Part B: Polym. Phys.* **2002,** *40,* 272.
72. Yano, K.; Usuki, A.; Okada A. J Polym Sci Part A: Polym Chem **1997,** *35,* 2289.
73. Shia, D.; Hui, Y.; Burnside, S. D.; Giannelis, E. P.; *Polym. Eng. Sci.* **1987,** *27,* 887.
74. Curtin WA. J Am Ceram Soc **1991,** *74,* 2837.
75. Chawla KK.; Composite materials science and engineering. New York: Springer; 1987.
76. Burnside, S.; D.; Giannelis, E. P.; *Chem. Mater.* **1995,** *7,* 4597.
77. Babrauskas, V.; Peacock, R. D.; *Fire Safety J.* **1992,** *18,* 225.
78. Gilman, J.; Kashiwagi, T.; Lomakin, S.; Giannelis, E.; Manias, E.; Lichtenhan, J.; Jones, P.; In: *Fire Retardancy of Polymers: the Use of Intumescence*. The Royal Society of Chemistry, Cambridge, **1998,** 203–221.
79. Mikitaev, A. K.; Kaladjian, A. A.; Lednev, O. B.; Mikitaev M. A.; Plastic masses **2004,** *12,* 45–50 (in Russian)
80. Gilman, J.; Morgan A. 10th Annual BCC Conference, May 24–26, 1999.
81. John N.; Hay and Steve, J.; Shaw. Organic-inorganic hybrids: the best of both worlds? Europhysics News **2003,** *34(3).*
82. Delozier, D. M.; Orwoll, R. A.; Cahoon, J. F.; Johnston, N. J.; Smith, J. G.; Connell, Polymer, J. W.; **2002,** *43,* 813–822.
83. Chen, Z.; Huang, C.; Liu, S.; Zhang, Y.; Gong K. J Apply Polym Sci **2000,** *75,* 796–801.
84. Okado, A.; Kawasumi, M.; Kojima, Y.; Kurauchi, T.; Kamigato O.; *Mater. Res. Soc. Symp. Proc.* **1990,** *171,* 45.
85. Leszek A.; Utracki, Jorgen Lyngaae-Jorgensen. Rheologica Acta, **2002,** *41,* 394–407.
86. Wagener, T. R.; Reizinger, J. G.; Polymer, **2003,** *44,* 7513–7518.
87. Li, X.; Kang, T.; Cho, W. J.; Lee, J. K.; Ha, C. S.; Macromol Rapid Commun.
88. Tyan, H. -L.; Liu, Y. -C.; Wie, K. -H.; *Polymer*, **1999,** *40,* 4877–4886.
89. Vaia, R.; Huang, X.; Lewis, S.; Brittain W.; Macromolecules **2000,** *33,* 2000–4.
90. Okamoto, M.; Morita, S.; Taguchi, H.; Kim, Y.; Kotaka, T.; Tateyama H.; Polymer **2000,** *41,* 3887–90.
91. Chow, W. S.; Mohd Ishak, Z. A.; Karger-Kocsis, J.; Apostolov, A. A.; Ishiaku, Polymer, U. S.; **2003,** *44,* 7427–7440.
92. Antipov, E. M.; Guseva, M. A.; Gerasin, V. A.; Korolev, Yu.M.; Rebrov, A. V.; Fisher, H. R.; Razumovskaya I. V.; Visokomol Soed. A. **2003,** *45(11),* 1885–1899 (in Russian).
93. Sur, G.; Sun, H.; Lyu, S.; Mark, J.; *Polymer* **2001,** *42,* 9783–9.
94. Wang, Z.; Pinnavaia T.; *Chem. Mater.* **1998,** *10,* 3769–71.
95. Bedanokov A. Yu.; Beshtoev B. Z. /Malij polimernij congress (Moscow 2005) (in Russian)
96. Antipov, E. M.; Guseva, M. A.; Gerasin, V. A.; Korolev, Yu.M.; Rebrov, A. V.; Fisher, H. R.; Razumovskaya Visokomol, I. V.; Soed, A. **2003,** *45(11),* 1874–1884 (in Russian).
97. Lan T.; Kaviartna P.; Pinnavaia T.; Proceedings of the ACS PMSE **1994,** *71,* 527–8.
98. Kawasumi, C., et al. Nematic liquid crystal/clay mineral composites. Science and Engineering, *6,* 135–143, 1998.

99. Lednev, O. B.; Kaladjian, A. A.; Mikitaev M. A.; Second International Conference (Nalchik 2005) (in Russian)
100. Mikitaev, A. K.; Kaladjian, A. A.; Lednev, O. B.; Mikitaev, M. A.; Davidov, E. M.; Plastic masses **2005,** *4,* 26–31(in Russian)
101. Eid, A.; Mikitaev, M. A.; Bedanokov, A. Y.; Mikitaev, A. K. Recycled Polyethylene Terephthalate/Organo-Montmorillanite Nanocomposites, Formation And Properties The first Afro-Asian Conference on Advanced Materials Science and Technology (AMSAT 06), Egypt, 2006.
102. Mikitaev, A. K.; Bedanokov, A. Y.; Lednev, O. B.; Mikitaev, M. A.; Polymer/silicate nanocomposites based on organomodified clays/ Polymers, Polymer Blends, Polymer Composites and Filled Polymers. Synthesis, Properties, Application. Nova Science Publishers. New York 2006.
103. Malamatov, A. H.; Kozlov, G. V.; Mikitaev M. A.; Mechanismi uprochnenenia polimernih nanokompositov, (Moscow, RUChT 2006) 240 p. (in Russian)
104. Eid ,A.; Doctor Thesis, (Moscow, RUChT 2006) 121 p. (in Russian)
105. Lednev, O. B.; Doctor Thesis (Moscow, RUChT 2006) 128 p. (in Russian)
106. Malamatov, A. H.; Professor Thesis (Nalchik, KBSU 2006) 296 p. (in Russian)
107. Borisov, V. A.; Bedanokov, A. Yu.; Karmokov, A. M.; Mikitaev, A. K.; Mikitaev, M. A.; Turaev, E. R.; *Plastic Mass.* 2007, 5 (in Russian).

CHAPTER 8

QUANTUM – CHEMICAL MODELING

M. A. CHASHKIN, V. I. KODOLOV, A. I. ZAKHAROV,
YU. M. VASILCHENKO, M. A. VAKHRUSHINA, V. V. TRINEEVA,
and G. E. ZAIKOV

CONTENTS

8.1 Introduction 344
8.2 Quantum – Chemical Investigation of Fragments Interaction in Epoxy Polymers Modified with Metal/Carbonic Nanocomposites 345
8.2 Experimental Part 349
8.2.1 Materials 349
8.2.2 Processing of Fine Suspensions 349
8.2.3 Processing of Epoxy Resins Modified with Cu/C Nanocomposite 350
8.2.4 Investigation Technique 351
8.3 Results and Discussion 352
8.3.1 Optimal Time Period of Ultrasound Processing of Fine Suspensions of Copper/Carbon Nanocomposite 352
8.3.2 Viscous Properties of Fine Suspensions of Copper/Carbon Nanocomposite 353
8.3.3 IR Spectroscopic Investigation 355
8.3.4 Adhesion 356
8.3.5 Thermogravimetric Investigations 356
8.4 Conclusion 357
Keywords 358
References 359

8.1 INTRODUCTION

This chapter is dedicated to the investigation of modification processes of cold hardened epoxy compositions with metal/carbon nanocomposites to improve their operational characteristics (adhesive strength and thermal stability). The work is theoretically substantiated with quantum-chemical modeling.

The chapter contains the investigation results of viscous properties of metal/carbon nanocomposite fine suspensions and thermal and physical properties of metal/carbon nanocomposite/epoxy compositions obtained.

The quantum-chemical modeling of the systems imitating the behavior of copper/carbon nanocomposite fine suspensions and epoxy resins cold hardened with them is given. IR spectra of polyethylene polyamine and metal/carbon nanocomposite fine suspensions based on it with concentrations 0.001–0.03% from polyethylene polyamine mass are analyzed and their comparative analysis is presented. The possible processes flowing in fine suspensions during the interaction of copper/carbon nanocomposite and polyethylene polyamine are described, as well as the processes influencing the increase in adhesive strength and thermal stability of metal/carbon nanocomposite/epoxy compositions.

This decade has been heralded by a large-scale replacement of conventional metal structures with structures from polymeric composite materials (PCM). Currently, we are facing the tendency of production growth of PCM with improved operational characteristics. In practice, PCM characteristics can be improved when applying modern manufacturing technologies, for example, the application of "binary" technologies of prepreg production [1], as well as the synthesis of new polymeric PCM matrixes or modification of the existing polymeric matrixes with different fillers.

The most cost-efficient way to improve operational characteristics is to modify the existing polymeric matrixes; therefore, currently the group of polymeric materials modified with nanostructures (NS) is of special interest. NS are able to influence the supermolecular structure, stimulate self-organization processes in polymeric matrixes in supersmall quantities, thus contributing to the efficient formation of a new phase "medium modified-nanocomposite" and qualitative improvement of the characteristics of final product – PCM. This effect is especially visible when NS activity increases which directly depends on the size of specific surface, shape of the particle and its ultimate composition [2]. Metal ions in the NS used in this chapter also contribute to the activity increase as they stimulate the formation of new bonds.

The increase in the attraction force of two particles is directly proportional to the growth of their elongated surfaces; therefore, the possibility of NS coagulation and decrease in their efficiency as modifiers increases together with their activity growth. This fact and the fact that the effective concentrations to modify polymeric matrixes are usually in the range below 0.01 mass % impose specific methods for introducing NS into the material modified. The most justified and widely applicable are the methods for introducing NS with the help of fine suspensions (FS). This introduction method allows most uniformly distributing particles in the volume of the medium modified, decreasing the possibility of their coagulation and preserving their activity during storage.

8.2 QUANTUM – CHEMICAL INVESTIGATION OF FRAGMENTS INTERACTION IN EPOXY POLYMERS MODIFIED WITH METAL/CARBONIC NANOCOMPOSITES

In the process of quantum-chemical modeling the fragments imitating the initial reagents are optimized: epoxy diane resin (EDR), polyethylene polyamine (PEPA), cobalt/carbon nanocomposite (Co/C NC), nickel/carbon nanocomposite (Ni/C NC), copper/carbon nanocomposite (Cu/C NC) with the inclusion of CO_2^+, Ni^{2+}, Cu^{2+} ions (Fig. 1). For each of these fragments the absolute value of binding energy is defined E_{EDR}, E_{PEPA}, $E_{NC(Me)}$, where Me – cobalt, nickel or copper ion.

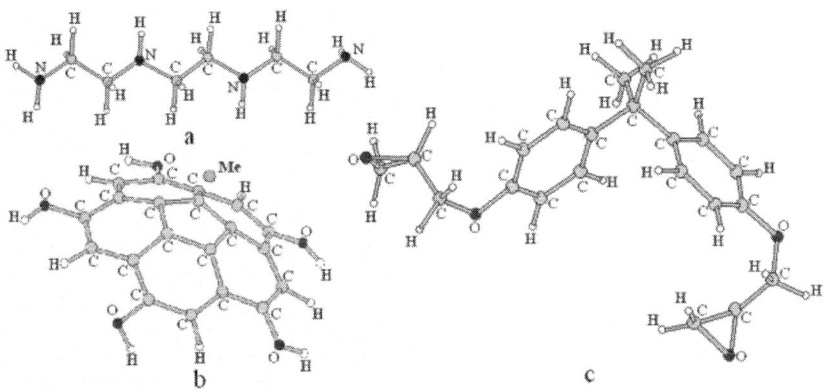

FIGURE 1 Fragments of initial substances: (a) PEPA fragment; (b) Me/C NC fragment; (c) EDR fragment.

As the modification process initially assumed the production of fine suspensions (FS) of NC on PEPA basis, the fragments imitating the behavior of the corresponding suspensions of Co/C, Ni/C, Cu/C nanocomposites were optimized (Fig. 2) and their absolute values of binding energy $E_{PEPA\text{-}NC(Co)}$, $E_{PEPA\text{-}NC(Ni)}$, $E_{PEPA-NC(Cu)}$ were defined.

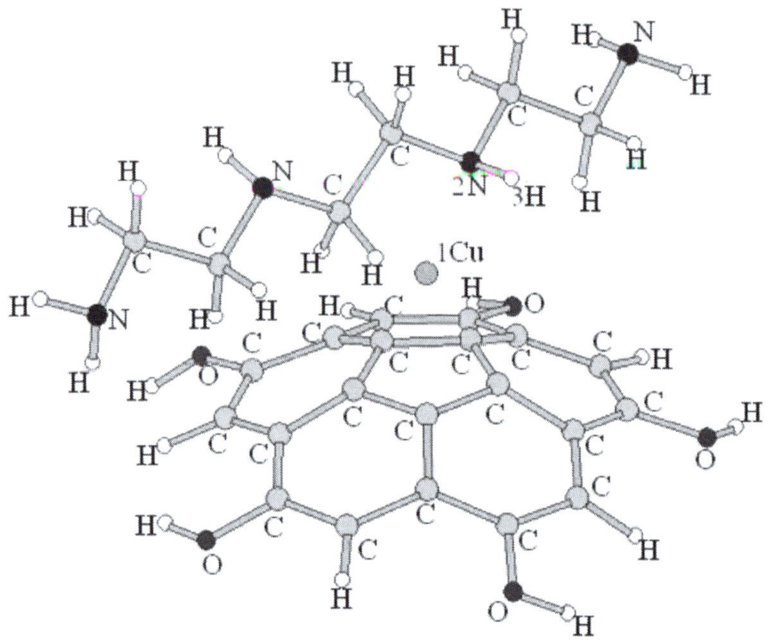

FIGURE 2 Fragment of Cu/C NC FS on PEPA basis.

The next step was to model the influence of fine suspensions of nanocomposites on epoxy resin. The complexes formed similarly with the previous ones were optimized, and the absolute values of binding energy were found for each of them $E_{PEPA\text{-}NC(Co)}$, $E_{PEPA\text{-}NC(Ni)}$, $E_{PEPA-NC(Cu)}$. The absolute values of binding energy are given in Table 1.

TABLE 1 Absolute binding energies of the fragments.

	$E_{NC(Me)}$, KJ/mol	$E_{PEPA\text{-}NC(Me)}$, KJ/mol	$E_{EDR\text{-}PEPA\text{-}NC(Me)}$, KJ/mol
CO_2^+	–19116.50	–29992.05	–51486.96
Ni^{2+}	–18562.38	–29098.04	–50621.15
Cu^{2+}	–18340.32	–28764.72	–50315.94
E_{EDR}, KJ/mol	–21424.25		
E_{PEPA}, KJ/mol	–10131.36		

Using the data from Table 1 by the following formula:

$$E_1 = E_{EDR-PEPA-NC(Me)} - E_2 - E_{EDR}$$

$$E_2 = E_{PEPA-NC(Me)} - E_{PEPA}$$

the relative interaction energies of molecular complexes E_1 are calculated and the diagram is arranged (Fig. 3).

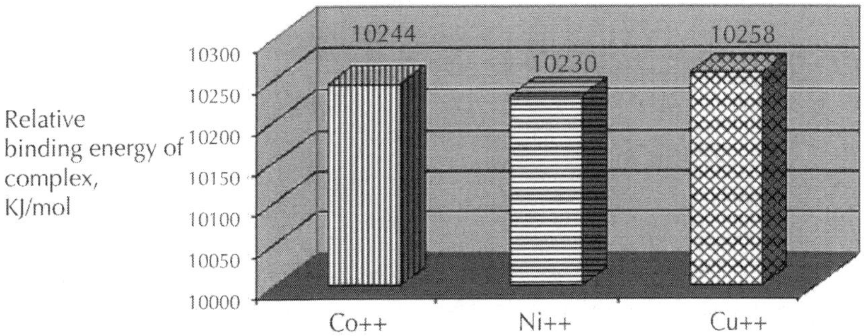

FIGURE 3 Diagram of relative interaction energies of molecular complexes.

From Fig. 3, it is seen that the relative interaction energy of molecular complexes with Cu/C NC is higher in comparison with the complexes with Co/C NC and Ni/C NC content. As the polymer forms the strongest complexes with Cu/C NC, therefore after the modification it will be the most effective. The detailed analysis of the lengths of the bonds formed and effective charges before and after the optimization of fragments imitating PEPA interaction with Cu/C NC (Tables 2 and 3) indicates that stable coordination bonds were formed between NC fragment and PEPA (between copper ion and nitrogen atom of amine group NH of PEPA). It was found that after the interaction of two fragments studied a part of electron density of N atom participating in the bond shifted to Cu atom, thus resulting in NH bond weakening.

TABLE 2 Bond lengths.

Bond designation	Bond length before optimization, Å	Bond length after optimization, Å
Cu-N	2.82	1.95
Cu-C	2.65	2.25

TABLE 3 Effective charges.

Atom number (see Fig. 2)	Atom designation	Effective charge before optimization	Effective charge after optimization
1	Cu (copper)	0.138	–0.250
2	N (nitrogen)	–0.066	0.420
3	H (hydrogen)	0.045	0.027

The bond weakening is indirectly confirmed by the increase in the effective charge of H atom and slight change in the wave number in oscillatory spectra calculated. The wave number of NH bond before the optimization was 3360 cm^{-1}, and after 3354 cm^{-1}, which correlates with the data of IR spectra obtained with the help of IR Fourier spectrometer. For instance, in the spectrum of PEPA and NC suspension on PEPA basis with nanocomposite concentration 0.03% the shift of wave numbers of peaks of amine groups is observed from 3280 cm^{-1} to 3276 cm^{-1}.

The modeling of hardening process with the participation of epoxy resin, polyethylene polyamine and Cu/C nanocomposite is given in Figs. 4a and 4b, respectively. From the geometry of optimized molecular systems it is seen that the introduction of Cu/C NC into the system leads to its self-organization (Fig. 4b) and formation of presumably coordination polymer. The formation of coordination polymer due to the introduction of nanocomposite active particles can result in increasing the adhesive strength and thermal stability as the total number of bonds in polymer grid grows and more energy will be required for its thermal destruction. At the same time, the formation of nanocomposite metal coordination bond with PEPA nitrogen can increase the stability of fine suspension formed. Such coordination results in polyethylene polyamine activity growth during the hardening of epoxy diane resin.

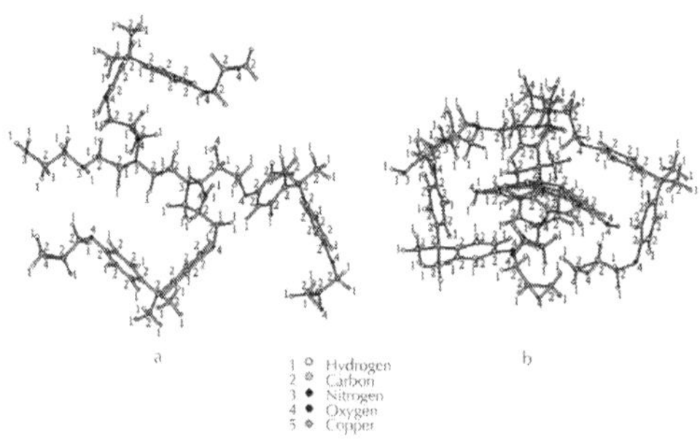

FIGURE 4 Influence of Cu/C NC on epoxy resin (ER) hardening: (a) ER hardening without Cu/C NC; (b) ER hardening in the presence of Cu/C NC.

Thus, quantum-chemical modeling allows predicting the interaction processes of components when hardening the epoxy resin with polyethylene polyamine and active participation of copper/carbon nanocomposite.

8.2 EXPERIMENTAL PART

8.2.1 MATERIALS

Currently there is a huge need in modern epoxy systems with improved operational characteristics [3] that can be reached, as previously mentioned, when modifying them with nanostructures. Therefore in this chapter the modification processes of cold-hardened model epoxy composition containing epoxy diane resin ED-20 State Standard (GOST) 10587-84 in the amount of 100 weight fractions and polyethylene polyamine (PEPA) grade A Technical Condition (TU) 2413-357-00203447-99 in the amount of 10 weight fractions were considered as the research object. The modification was carried out when introducing Metal/C nanocomposite into the epoxy resin.

PEPA represents the mixture of linear branched ethylene polyamines with average molecular mass $200-250$ and very wide molecular-mass distribution. The general structural formula of PEPA is as follows:

$$NH_2[-CH_2CH_2NH-]_{n-}H, \text{ where } n = 2-8.$$

Synthesized Co/C, Ni/C, Cu/C nanocomposites were studied with the help of transmission electron microscopy (TEM) and electron microdiffraction (EMD) in the shared centers of research institutions in Ekaterinburg and Moscow.

The investigations of Cu/C and Ni/C nanocomposites revealed that their average sizes (r) differ approximately in two times: r of Cu/C NC is 25 nm, r of Ni/C NC is about 11 nm. The average size of Co/C nanocomposite practically equals the average size of Ni/C NC. Such correlation of average sizes of nanocomposites is apparently connected with the ability to form metal containing clusters. Atom magnetic moments of nanocomposites, as demonstrated in [], are greater than the corresponding moments of microparticles of the same metals. The availability of magnetic properties of nanoproducts widens their application possibilities.

8.2.2 PROCESSING OF FINE SUSPENSIONS

The fine suspension was prepared in a number of stages:

1. Preliminary grinding of Cu/C NC in mechanical mortar.

2. Mechanical and chemical activation of suspension components when combining PEPA and Cu/C nanocomposite.
3. Ultrasound processing for complete and uniform dispergation of Cu/C NC particles in PEPA volume.

8.2.3 PROCESSING OF EPOXY RESINS MODIFIED WITH CU/C NANOCOMPOSITE

The samples of epoxy polymer modified with Cu/C NC were produced mixing the epoxy resin heated up to 60 °C and fine suspension of Cu/C NC on PEPA basis in proportion 10:1. The mixing took 5–10 min. A part of the polymer modified with NC was poured into the mold with copper wire to test the adhesive strength (Fig. 5) and was further hardened in the mold. Another part was hardened in the form of plates to grind the samples for thermogravimetric investigation.

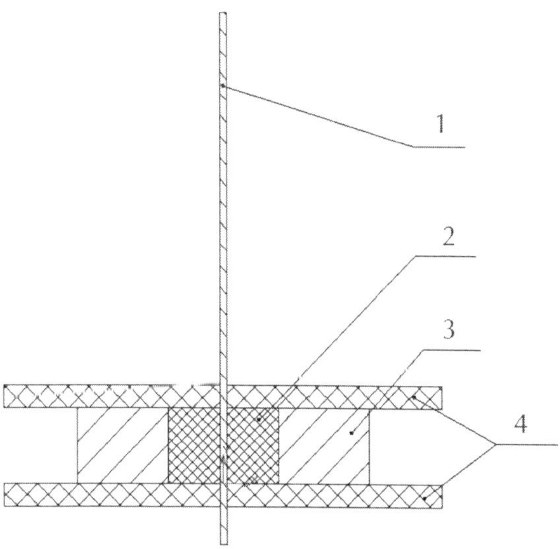

FIGURE 5 Sample for determining the adhesive strength of cold-hardened epoxy composition (CHEC): (1) copper wire; (2) hardened composition; (3) metal mold; (4) antiadhesive liner.

Two sets of samples were produced to determine the adhesive strength, including the reference and modified FS of Cu/C NC with concentrations 0.001%,

0.01% and 0.03%. For oxide film removal and degreasing the copper wire was treated with 0.1 M solution of hydrochloric acid and acetone.

For thermogravimetric investigation also two sets of samples were produced, including the reference and modified FS of Cu/C NC with concentrations 0.001%, 0.01%, 0.03%, 0.05%.

8.2.4 INVESTIGATION TECHNIQUE

1. *Quantum-chemical modeling.* In the frameworks of this method the fragments imitating the behavior of Co/C, Ni/C, Cu/C nanocomposites, fine suspensions of nanocomposites and systems being epoxy resins modified with nanocomposites were constructed and optimized with the help of software HyperChem v. 6.03 and semiempirical methods. At the same time, the absolute binding energies of certain components of the molecular systems formed were found and their relative interaction energies were calculated. The oscillatory spectra of fine suspensions of Cu/C nanocomposite were calculated with the help of semiempirical method PM3.

2. *IR spectroscopy.* For IR spectra IR Fourier-spectrometer FSM 1201 was used. The IR spectra of liquid films of PEPA, NC FS with concentrations 0.001%, 0.01% and 0.03% were taken. The spectra were taken on KBr glasses in wave number range 399–4500 cm^{-1}.

3. *Optical spectroscopy.* Spectral photometer KFK-03-01 was used to define the optimal time period of NC FS ultrasound processing. The optimal time period value was the interval when the optical density (D) was at the maximum, that is, corresponded to the maximum saturation of FS with NC. The optical density of NC FS samples with the concentration 0.01% processed in ultrasound bath Sapfir UZV 28 within 0, 3, 7, 10, 15, 20, 30 min, respectively, at US power 0.5 kW and frequency 35 kHz was measured in 5-mL quartz cuvettes. The work wavelength was found when defining the optical density in the range λ=320–920 nm.

4. *Determination of relative viscosity.* The relative viscosity of PEPA and FS on its basis was found with the help of viscometer VZ-246 in accordance with State Standard (GOST) 8420-74. The relative viscosity values were translated into the kinematic following GOST 8420-74. The viscosity was measured for PEPA processed and not processed with ultrasound, and also for fine suspensions of NC with the concentrations 0.001%, 0.01% and 0.03% processed and not processed with ultrasound.

5. *Thermogravimetric technique.* Thermal stability was found by the destruction temperatures of modified epoxy resins. The temperatures of destruction beginning were determined by thermogravimetric (TG) curves. Derivatographer DIAMOND TG/DTA was applied to obtain TG curves. TG curves of modified

epoxy resins with NC concentrations 0.001%, 0.03%, 0.05% from PEPA weight was taken. The sample heating rate was 5 °C/min.

6. *Determination of adhesive strength.* The adhesive strength of NC/EC was found with the technique described (Fig. 5). The tests were carried out on the tensile testing machine. The strength was found comparing breaking stresses of CHEC and NC/EC samples.

8.3 RESULTS AND DISCUSSION

8.3.1 OPTIMAL TIME PERIOD OF ULTRASOUND PROCESSING OF FINE SUSPENSIONS OF COPPER/CARBON NANOCOMPOSITE

In the process of selecting the work wavelength (λ) on spectral photometer KFK-03-01 to define the optimal time period of ultrasound processing of FS on PEPA basis the optimal wavelength 413 nm was found. This wavelength was selected as the optical density of FS with the concentration of Cu/C NC 0.01% not processed with ultrasound was 0.800 that corresponded to the middle of the operational range of this instrument (operational range D = 0.001–1.5). At the wavelength λ = 413 nm the optical densities of all FS investigated were defined. This is reflected in the graphic dependence of optical density D upon the time period of ultrasound processing τ_{us} (Fig. 6).

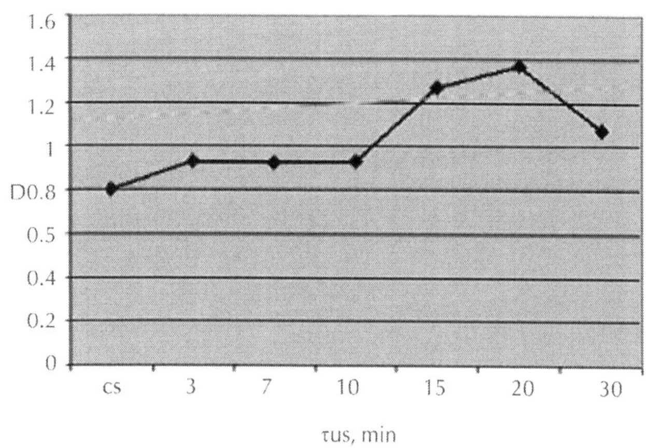

FIGURE 6 Dependence of the change in optical density (D) of Cu/C NC FS upon the time period of ultrasound processing τ_{us}.

The analysis of optical density dependence upon the time period of ultrasound processing (Fig. 6) demonstrates that the optimal processing time of Cu/C NC fine suspension on PEPA basis is 20 min. Further, ultrasound processing is useless as in 20 min the maximum optical density 1.37 is reached.

Taking into account the above data for defining the optimal time period of ultrasound processing, the preparation of fine suspension for IR investigations, for finding the viscosity and for producing the modified epoxy resin was carried out with preliminary ultrasound processing within 20 min.

8.3.2 VISCOUS PROPERTIES OF FINE SUSPENSIONS OF COPPER/CARBON NANOCOMPOSITE

The following diagram was prepared based on the results of viscosity measurements of PEPA and fine suspension (FS) with different Cu/C NC content before and after US processing (Fig. 7).

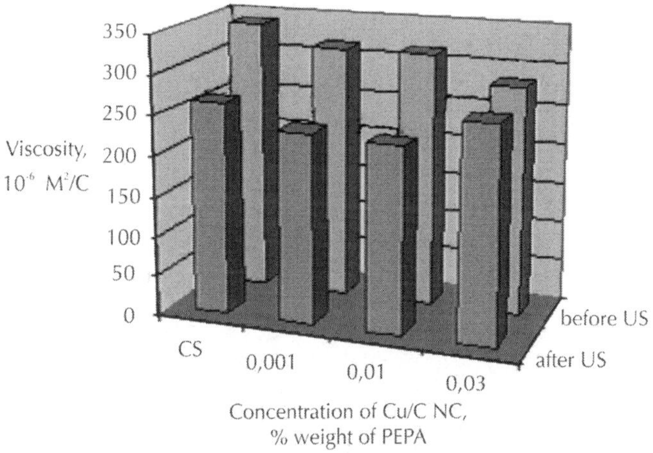

FIGURE 7 Diagram of the dependence of FS kinematic viscosity on Cu/C NC concentration.

The decrease in FS viscosity with the increase in Cu/C NC concentration on the diagram (Fig. 7) is explained by the fact that the system consisting of two phases, in our case (PEPA – dispersion medium and Cu/C NC – disperse phase) tends to surface energy decrease. This tendency is expressed by self-decrease in interface surface due to sorption. PEPA molecules start sorbing on Cu/C NC surface, probably producing the interface between Cu/C NC particles being the obstacle for coagulation. Due to such localization the interaction between medium particles diminishes, resulting in intermolecular friction decrease, and, consequently, decrease in system viscosity (Fig. 8a). When FS are processed with ultrasound, the size of disperse phase decreases, thus leading to the growth of surface energy and increase in sorption ability, the action region on disperse medium increases (Fig. 8b). The action regions of Cu/C nanocomposite in fine suspension with concentrations 0.001% and 0.01% do not overlap that is confirmed by viscosity decrease, and for FS with concentration 0.03% the action regions overlap and, consequently, the increase in intermolecular friction and viscosity are observed.

Taking into account that the action degree on the medium mainly depends on the action time of Cu/C NC on the medium modified, FS were studied for 48 h. After this time interval FS layered into floccular structures with no sediment on the bottom observed (Fig. 8c).

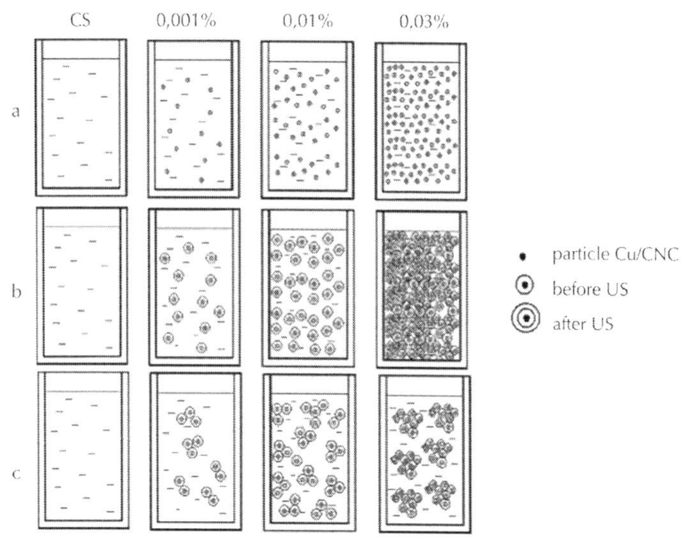

FIGURE 8 Distribution of Cu/C NC in PEPA: (a) before ultrasound processing; (b) after ultrasound processing; (c) after 48 h.

Mechanical action resulted in FS recovery, which is indirectly confirmed by the availability of PEPA stable layer between Cu/C nanocomposite particles preventing the coagulation and possibility of suspension production with the ability to recover the distribution of nanocomposite particle distribution.

8.3.3 IR SPECTROSCOPIC INVESTIGATION

The comparative IR spectroscopic investigation of *bis*(2-aminoethyl)amine – one of PEPA olygomers demonstrates the consistency of wave numbers appropriate for these compounds.

In both spectra the wave numbers appropriate for the oscillations of amine groups $\nu_s(NH_2)$ 3352 cm^{-1} and asymmetric $\nu_{as}(NH_2)$ 3280 cm^{-1} are available, there are wave numbers that refer to symmetric $\nu_s(CH_2)$ 2933 cm^{-1} and asymmetric valence $\nu_{as}(CH_2)$ 2803 cm^{-1}, deformation wagging oscillations $\nu_d(CH_2)$ 1349 cm^{-1} of methylene groups, deformation oscillations ν_d (NH) 1596 cm^{-1} and $\nu_d(NH_2)$ 1456 cm^{-1} of amine groups, and also the oscillations of skeleton bonds are vivid ν(CN) 1059–1273 cm^{-1} and ν(CC) 837 cm^{-1}.

IR spectra of PEPA and Cu/C NC FS were taken (Fig. 9).

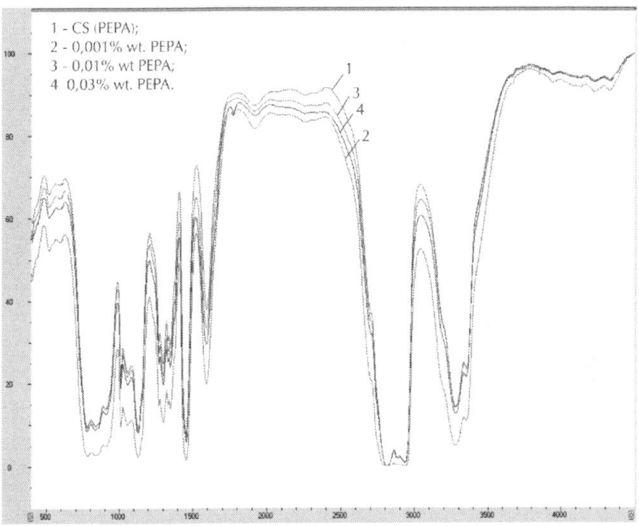

FIGURE 9 IR spectra of PEPA and FS of Cu/C nanocomposite.

The comparison of IR spectra of polyethylene polyamine and fine suspension of Cu/C NC on PEPA basis (Fig. 9) indicates that practically all changes of wave numbers in the spectra are within the error ± 2 cm^{-1}. However, in FS spectra the vivid increase in peak intensity corresponding to deformation oscillations of NH bonds is observed. These changes can spread onto the vast areas arranging a certain supermolecular structure, apparently involving the adjoining amine groups into the process, which is demonstrated by the intensity change of these peaks.

8.3.4 ADHESION

The tests for defining the adhesion of modified epoxy resin to copper wire were carried out on tensile testing machine, the values of destruction load were found. The adhesive strength was calculated by the following formula:

$$\sigma = F/A \text{ [MPa]}$$

where F – average load values at which the breaking-off took place, (kgs); A – area of wire interaction with the hardened composition, (cm^2).

$$A = 2{\cdot}\pi{\cdot}r{\cdot}h = 2{\cdot}3,\ 14{\cdot}0,\ 05{\cdot}1 = 0.314\ (cm^2),$$

where *r* – wire radius 0.05 (cm); h – height of metal mold 1.00 (cm).

The following diagram was prepared based on tests (Fig. 10)

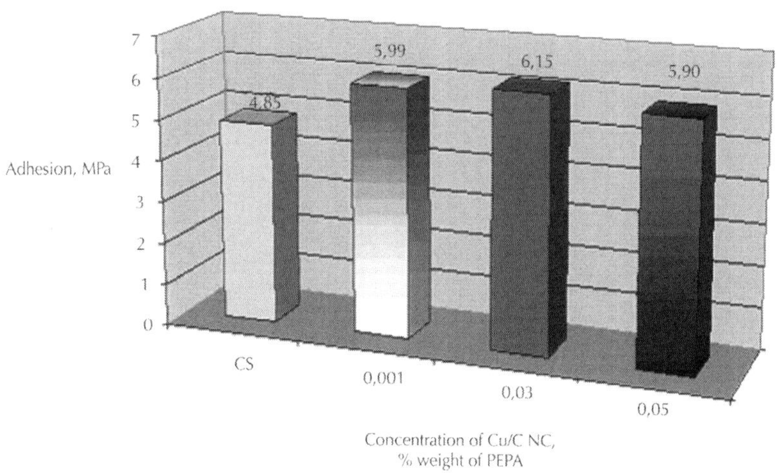

FIGURE 10 Dependence of adhesive strength of NC/EC on Cu/C NC composition.

8.3.5 THERMOGRAVIMETRIC INVESTIGATIONS

To define the influence of nanocomposite on thermal stability of epoxy composition, a number of thermogravimetric investigations were carried out on reference and modified samples. The concentrations 0.001%, 0.03% and 0.05% from PEPA weight were used. Based on the results of thermogravimetric investigations the following diagram was prepared (Fig. 11).

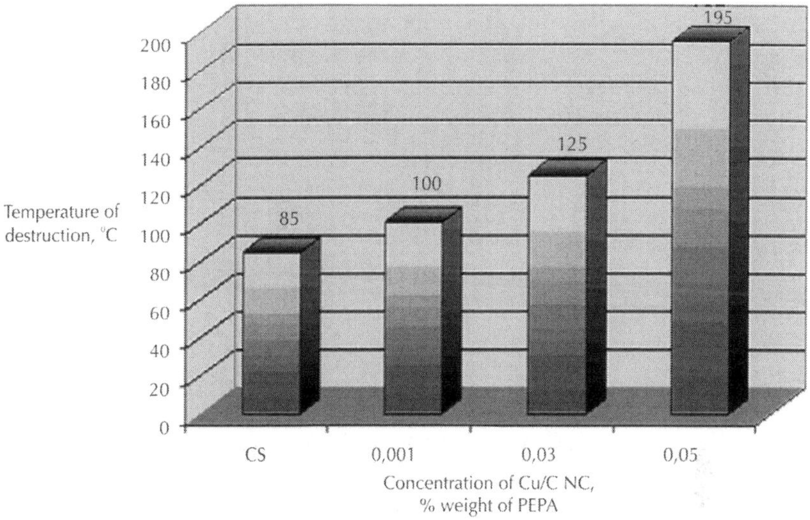

FIGURE 11 Dependence of thermal stability of cold-hardened epoxy composition on the concentration of copper/carbon nanocomposite

The data from the diagram of adhesive strength (Fig. 10) indicate that the strength maximum of the modified epoxy resin is reached when Cu/C nanocomposite concentration is 0.03%. Probably the maximum strength of the modified epoxy resin at this concentration is conditioned by the optimal number of a new phase growth centers. The adhesion decrease with the concentration increase indicates that the number of Cu/C nanocomposite particles exceeded the critical value, which depends on their activity [2]. Therefore probably the number of cross-links in the polymer grid increased – the material became brittle. At the same time, when the concentration of nanocomposite elevates, the growth of polymer thermal stability is observed (Fig. 11). The thermal stability growth is apparently connected with the increase in the number of coordination bonds in epoxy polymer.

8.4 CONCLUSION

Quantum-chemical modeling methods allow quite precisely defining the typical reaction of interaction between the system components, predict the properties of molecular systems, decrease the number of experiments due to the imitation of technological processes. The computational results are completely comparable with the experimental modeling results.

The optimal composition of nanocomposite was found with the help of software HyperChem v. 6.03. Cu/C nanocomposite is the most effective for modifying the epoxy resin. Nanosystems formed with this NC have higher interaction energy in comparison with nanosystems produced with Ni/C and Co/C nanocomposites. The effective charges and geometries of nanosystems were found with semiempirical methods. The fact of producing stable fine suspensions of nanocomposite on PEPA basis and increasing the operational characteristics of epoxy polymer was ascertained. It was demonstrated that the introduction of Cu/C nanocomposite into PEPA facilitates the formation of coordination bonds with nitrogen of amine groups, thus resulting in PEPA activity increase in epoxy resin hardening reactions.

It was found out that the optimal time period of ultrasound processing of copper/carbon nanocomposite fine suspension is 20 min.

The dependence of Cu/C nanocomposite influence on PEPA viscosity in the concentration range 0.001–0.03% was found. The growth of specific surface of NC particles contributes to partial decrease in PEPA kinematic viscosity at concentrations 0.001%, 0.01% with its further elevation at the concentration 0.03%.

IR investigation of Cu/C nanocomposite fine suspension confirms the quantum-chemical computational experiment regarding the availability of NC interactions with PEPA amine groups. The intensity of these groups increased in several times when Cu/C nanocomposite was introduced.

The test for defining the adhesive strength and thermal stability correlate with the data of quantum-chemical calculations and indicate the formation of a new phase facilitating the growth of cross-links number in polymer grid when the concentration of Cu/C nanocomposite goes up. The optimal concentration for elevating the modified epoxy resin (ER) adhesion equals 0.003% from ER weight. At this concentration the strength growth is 26.8%. From the concentration range studied, the concentration 0.05% from ER weight is optimal to reach a high thermal stability. At this concentration the temperature of thermal destruction beginning increases up to 195 °C.

Thus, in this chapter the stable fine suspensions of Cu/C nanocomposite were obtained. The modified polymers with increased adhesive strength (by 26.8%) and thermal stability (by 110 °C) were produced based on epoxy resins and fine suspensions.

KEYWORDS

- **epoxy compositions**
- **metal/carbon nanocomposites**
- **polymeric composite materials**
- **quantum-chemical modeling**

REFERENCES

1. Panfilov, B. F. Composite materials: production, application, market tendencies. *Polymeric Materials*, **2010,** *2–3,* 40–43.
2. Kodolov, V. I.; Khokhriakov, N. V.; Trineeva, V. V.; Blagodatskikh, I. I. Activity of nanostructures and its expression in nanoreactors of polymeric matrixes and active media. *Chemical Physics and Mesoscopy*, **2008,** *10(4),* 448–460.
3. Bobylev, V. A.; Ivanov, A. V. New epoxy systems for glues and sealers produced by CJSC "Chimex Ltd." *Glues. Sealers. Paints*, **2008,** *2*, 162–166.
4. Kodolov, V. I.; Kovyazina, O. A.; Trineeva, V. V.; Vasilchenko, Yu. M.; Vakhrushina, M. A.; Chmutin I. A. On the production of metal/carbon nanocomposites, water and organic suspensions on their basis. VII International Scientific-Technical Conference "Nanotechnologies to the production, 2010." *Proceedings Fryazino*, **2010,** 52–53.

CHAPTER 9

NOVEL PHASE OF ELEMENTAL SILVER NANO-PARTICLES FORMATION

N. I. NAUMKINA, O. V. MIKHAILOV, and T. Z. LYGINA

CONTENTS

9.1 Introduction 362
9.2 Experimental Part 363
9.3 Results and Discussion 364
Acknowledgement 375
Keywords 375
References 375

9.1 INTRODUCTION

The process of "reprecipitation" of elemental silver according to scheme Ag→$Ag_4[Fe(CN)_6]$→Ag into gelatin-immobilized matrices, on the first stage of which these matrices are processed with water-alkaline solutions containing potassium hexacyanoferrate(II) and potassium hexacyanoferrate(III), on the second stage, with water-alkaline solutions containing tin(II) dichloride and organic or inorganic substance forming rather stable soluble coordination compounds with Ag(I). **D** = f(D^{Ag}) and **D** = f(C_{Ag}^{V}) dependences where **D** is an optical density of "reprecipitated" silver in the matrix corresponding to initial density D^{Ag} and volume concentrations of elemental silver C_{Ag}^{V}. It has been found that a degree of influence of complex-formed substance (**D**/D^{Ag}) is determined with its nature as well as its concentration in reducing solution; besides, the most considerable this influence is in the case of ethanediamine-1,2, the least considerable one, in the case of ammonia. It has been found that silver formed in gelatin matrix, contains two various phases distinguished between themselves by optical and XRD parameters; in addition, one of these phases consists of nano-particles of elemental silver.

Even more 40 years ago in Ref. [1] it was mentioned about colloid element silver with rather small size of the particles, formed in gelatin layers at development of silver halide photographic emulsions. A number of works, in which there are indications on existence of a separate phase of the element silver consisting from nano-particles and received as a result of photochemical reduction of Ag(I) salts, was appeared lately in the literature, in particular Refs. [2–11]. Before in Refs. [10, 11], it has been noted that at development of gelatin layers of silver halide photographic emulsions by water-alkaline solutions containing tin(II) dichloride and some inorganic or organic substance forming rather stable coordination compounds with Ag(I), formation only of element silver occurs, too, however, besides the gelatin layer is tinged brown or red but not black color as it takes place at standard development by using hydroquinone developers. It is significant that with increase of optical densities of the gelatin layer containing such elemental silver, red tone in coloring of gelatin layer becomes more and more clearly expressed. The similar phenomenon takes place, too, when instead of silver halide AgHal in a gelatin matrix there is such silver(I) compound as silver(I) hexacyanoferrate(II) $Ag_4[Fe(CN)_6]$. Whether is this totality of particles a novel phase of element silver? Or it is only a variety of known phases of the given simple substance? This question remains till now opened and deserves special consideration.

As it is known through Refs. [12–15], physical-chemical processes in the so-called gelatin-immobilized matrix systems, in particular the reaction of nucleophilic substitution, ion exchange, and "self-assembly" of metal chelates are often accompanied by the formation of such compounds that can not be obtained by carrying out similar processes in solution or solid phase. The reason for this is, on the one hand, the fact that these processes occur in the intermolecular cavities of gelatin; on the other hand, the fact that in this particular reaction system, there is a preliminary reduction of entropy, which are made possible through such processes that normally thermodynamic forbidden. In this connection, it may be expected that in the gelatin-immobilized matrix systems, specific redox processes, in particular, with the formation of particles of elemental metals, can occur. In this connection, Chapter 9 is devoted to consideration of redox-reactions connected with obtaining element silver.

9.2 EXPERIMENTAL PART

As initial material to obtain silver-containing gelatin-immobilized matrix implants (hereafter **GIM**), X-ray film *Structurix D-10 (Agfa-Gevaert, Belges)* was. Samples of the given film (which actually is nothing but AgHal-**GIM**) having format 20×30 cm^2 were exposed to X-ray radiation with an irradiation dose at a range 0.05–0.50 Röntgen. These exposed sample were further subjected to processing according to the following technology [10, 11]:

- Development in D-19 standard developer as it was indicated in Ref. [10, 11], for 6 min at 20–25 °C;
- Washing with running water for 2 min at 20–25 °C;
- Fixing in 25% water solution of sodium trioxosulphidosulfate(VI) ($Na_2S_2O_3$) for 10 min at 20–25 °C;
- Washing with running water for 15 min at 18–25 °C.

Three first stages of standard processing indicated (development, washing and fixing) were carried out at nonactinic green-yellow light, final washing – at natural light. The samples of **GIM** obtained which contained elemental silver (Ag-**GIM**), then were processed according to next technology:

1. Oxidation in water solution containing (g×L^{-1})

Potassium hexacyanoferrate(III)	50.0
Potassium hexacyanoferrate(II)	20.0
Potassium hydroxide	10.0
Sodium trioxocarbonate(IV)	5.0
(Na_2CO_3) Water	up to 1000 ml for 6 min at 20–25 °C;

2. Washing with running water for 2 min at 20–25 °C;
3. Reduction in water solution containing (g×L^{-1})

Tin(II) chloride	50.0
Sodium N, N'-ethylenediaminetetraacetate	35.0
Potassium hydroxide	50.0
Reagent formed water-soluble complex with Ag(I)	1.0–100.0
Water	up to 1000 ml
	for 1 min at 20–25 °C;

4. Washing with running water for 15 min at 18–25 °C;
5. Drying for 2–3 hours at 20–25 °C.

As complex-forming reagents that form water-soluble complexes with Ag(I), ammonia NH_3, potassium thiocyanate KSCN, sodium trioxosulphidosulfate(VI) $Na_2S_2O_3$, ethanediamine-1,2 $H_2N–CH_2–CH_2–NH_2$, 2-aminoethanol $H_2N–CH_2–CH_2–OH$ and 3-(2-hydroxyethyl)-3-azapenthanediol-1,5 $N(CH_2–CH_2–OH)_3$, were used. At the first stage of given processing of Ag-**GIM** obtained, conversion of Ag-**GIM** into $Ag_4[Fe(CN)_6]$-**GIM** occurred, on the second stage, reduction of $Ag_4[Fe(CN)_6]$-**GIM** with Sn(II) to elemental silver took place. And so, peculiar "reprecipitation" of elemental silver into gelatin matrix occurred incidentally.

An isolation substances from Ag-**GIM** was carried out by means of influence on them of water solutions of some proteolytic enzymes (e.g., trypsin or *Bacillus mesentericus*) destroying the polymeric carrier of a **GIM** (gelatin), and the subsequent separation of a solid phase from mother solution according to a technique described in Ref. [16]. The substances isolated thus from **GIM** further analyzed by X-ray diffraction method with using of spectrometer D8 Advance (Bruker, Germany). A scanning was carried out in an interval from 3 up to 65° 2θ, a step was 0.05–2θ.

Calculation of intensities of reflexes (*I*) and interplane distances (*d*) carried out with application of standard software package EVA. Theoretical XRD spectra (X-ray patterns) were calculated under *PowderCell* program described in Refs. [17, 18]. Optical density Ag-**GIM** was measured by means of Macbeth TD504 photometer (Kodak, USA) in a range 0.1–5.0 units with accuracy of ±2 % (rel.).

9.3 RESULTS AND DISCUSSION

Already at visual observation over a course of process of transformation $Ag_4[Fe(CN)_6]$-**GIM** into Ag-**GIM,** following circumstance attracts its attention. Ag-**GIM** received as a result of standard processing of exposed AgHal-**GIM**, at rather small optical density (D^{Ag}) have gray color, at big D^{Ag}, black color. Coloring

Ag-**GIM** containing the "reprecipitated" element silver, varies from black-brown to red depending on the nature and quantity of complex-forming reagent contained in solution.

It is significant, however, that absorption spectra of both initial and the "reprecipitated" element silver in visible area do not contain any accurately expressed maxima. Besides, optical density Ag-**GIM** with the "reprecipitated" silver (**D**^{Ag}), at the same volume concentration of element silver (C_{Ag}^{V}) in **GIM**, as a rule, is essentially more than D^{Ag} values and, also, depends on nature and quantity of complex-forming reagent in solution contacting with **GIM**. Examples of **D** = f(D^{Ag}) и **D** = f(C_{Ag}^{V}) dependences for inorganic and organic reagents are presented in Figs. 1–6. It is significant that (**D**/D^{Ag}) value, as a rule, greater than 1.0, and in some cases, it reaches very high values (as in the case of potassium thiocyanate – nearly 5.0). Attention is drawn to the fact that the stronger the color of the gelatin layer with the "reprecipitated" elemental silver is different from the gray-black tones of the gelatinous layer initially Ag-gelatin-immobilized matrix, the greater is the (**D**/D^{Ag}) value. The maximal degree of amplification (**D**/D^{Ag})$_{max}$ is also very much depends on the nature of the complexing agent (Table 1); the most profound effect on this parameter has etandiamin-1,2 [(**D**/D^{Ag})$_{max}$ = 5.80], the least severe – ammonia, the degree of possibility of which, nevertheless, is also quite high [(**D**/D^{Ag})$_{max}$ = 3.40]. For the ammonia, the growth (**D**/D^{Ag}) values is typical with increasing concentration of NH_3 in reducing solution to a relatively small its value (~0.30 mol×L^{-1}), after which the optical density **D** begin to fall (Fig. 1). It is noteworthy that red-brown color of gelatin layer attained at the indicated concentration, with a further increase in the concentration of ammonia does not change. Analogous situation occurs in the case of the other two we studied inorganic complexing agents – trioxosulfidosulfate(VI) and the thiocyanate anion (Figs. 2–3), with the only difference being that in the case of $S_2O_3^{2-}$ maximum degree of amplification is achieved with less high in comparison with NH_3 concentration (0.15 mol×L^{-1}), in the case of SCN^- – at a higher concentration (~ 0.70 mol×L^{-1}). In this regard, it was quite natural to try the difference marked with different stability of 1:2 complexes formed by Ag(I) with NH_3, $S_2O_3^{2-}$ and SCN^- (pK = 7.25, 13.32 and 8.39, respectively). However, in the presence of such a correlation, the value of concentration indicated for SCN^- should be lower than for NH_3, which in fact is not observed. In reality, for the three organic complex-forming substances studied by us, molar concentrations at which the maximum value (**D**/D^{Ag}) is reached, significantly greater than those for inorganic complex-forming substances (Table 1). Stability of the complexes of silver(I) with each of these ligands is lower than with NH_3, and the correlation function between these concentrations and the stability of coordination compounds of Ag(I) with given ligands in varying degrees, still visible. But any explosion-term relationship that very stability of the (**D**/D^{Ag}) values do not see: how it may be easily noticed when comparing the data of Table 1, the maximum degree of amplification decreases in the direction of ethanediamine-1,2 > 2-aminoethanol > SCN^- > $S_2O_3^{2-}$> NH_3> 3-(2-hydroxyethyl)-3-azapenthane-diol-1,5, while the resistance formed by these ligands complexes with silver(I) – in the direction $S_2O_3^{2-}$ > SCN^- > ethanediamine-1,2 > NH_3> 2-aminoethanol > 3-(2-hydroxyethyl)-3-azapenthanediol-1,5. Thus, the complexing is though important, but not the sole determinant of the degree of influence of complex-forming agents on the redox process considered.

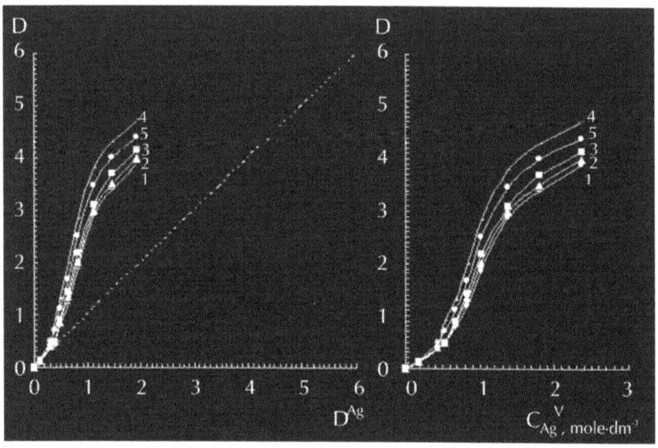

FIGURE 1 Dependences of **D** = f(D^{Ag}) and **D** = f(C_{Ag}^{V}) of reduction process $Ag_4[Fe(CN)_6]$→Ag with using of NH_3 at concentration 1.5 g×L^{-1} (curve 1), 3.0 g×L^{-1} (2), 4.5 g×L^{-1} (3), 6.0 g×L^{-1} (4) and 7.5 g×L^{-1} (5). Optical densities D^{Ag} and **D** were measured with blue light-filter with a transmission maximum at 450 nm.

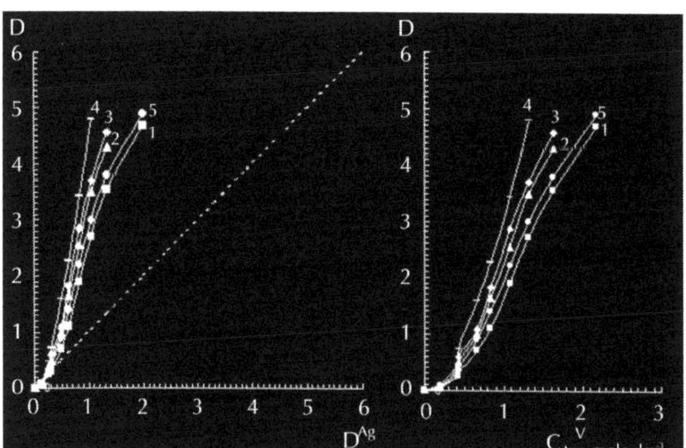

FIGURE 2 Dependences of **D** = f(D^{Ag}) and **D** = f(C_{Ag}^{V}) of reduction process $Ag_4[Fe(CN)_6]$→Ag with using of $Na_2S_2O_3$ at concentration 2.0 g×L^{-1} (curve 1), 4.0 g×L^{-1} (2), 8.0 g×L^{-1} (3), 24.0 g×L^{-1} (4) and 40.0 g×L^{-1} (5). Optical densities D^{Ag} and **D** were measured with blue light-filter with a transmission maximum at 450 nm.

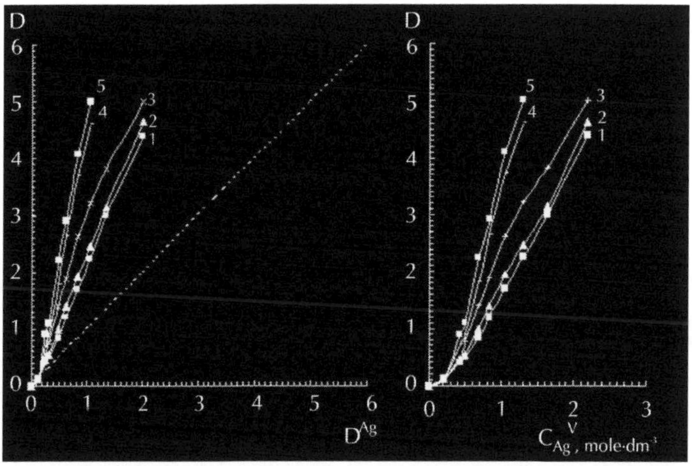

FIGURE 3 Dependences of **D** = f(D^{Ag}) and **D** = f(C_{Ag}^{V}) of reduction process $Ag_4[Fe(CN)_6]\rightarrow Ag$ with using of KSCN at concentration 2.0 g×L^{-1} (curve 1), 4.0 g×L^{-1} (2), 8.0 g×L^{-1} (3), 24.0 g×L^{-1} (4) and 60.0 g×L^{-1} (5). Optical densities D^{Ag} and **D** were measured with blue light-filter with a transmission maximum at 450 nm.

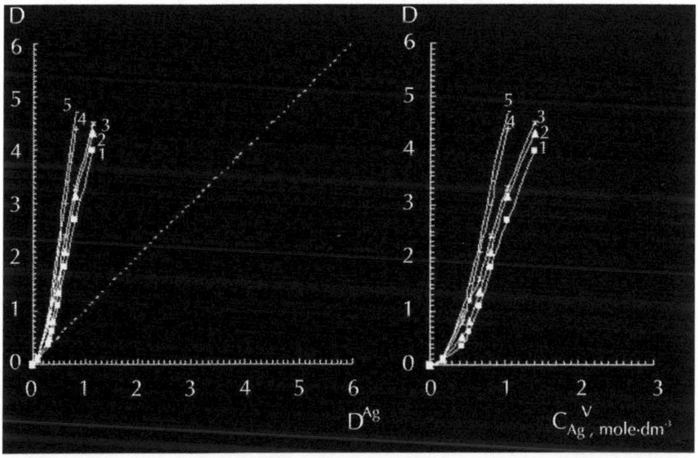

FIGURE 4 Dependences of **D** = f(D^{Ag}) and **D** = f(C_{Ag}^{V}) of reduction process $Ag_4[Fe(CN)_6]\rightarrow Ag$ with using of 2-aminoethanol $H_2N\text{–}(CH_2)_2\text{–}OH$ at concentration 7.5 g×L^{-1} (curve 1), 15.0 g×L^{-1} (2), 55.0 g×L^{-1} (3), 110.0 g×L^{-1} (4) and 150.0 g×L^{-1} (5). Optical densities D^{Ag} and **D** were measured with blue light-filter with a transmission maximum at 450 nm.

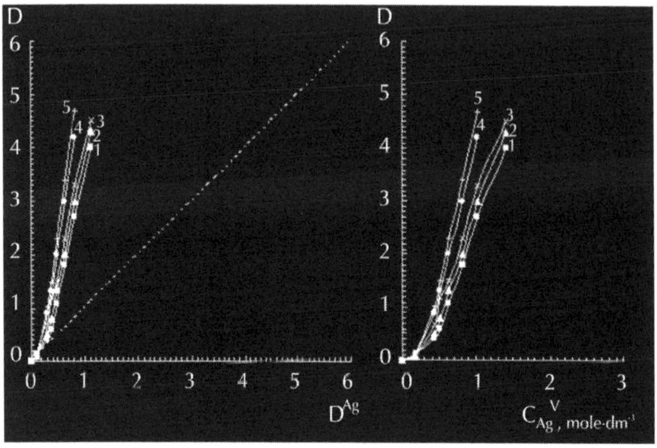

FIGURE 5 Dependences of **D** = f(D^{Ag}) and **D** = f(C_{Ag}^{V}) of reduction process $Ag_4[Fe(CN)_6]$→Ag with using of ethanediamine-1,2 $H_2N–(CH_2)_2–NH_2$ at concentration 5.0 g×L^{-1} (curve 1), 10.0 g×L^{-1} (2), 20.0 g×L^{-1} (3), 40.0 g×L^{-1} (4) and 80.0 g×L^{-1} (5). The optical densities D^{Ag} and **D** were measured with blue light-filter with a transmission maximum at 450 nm.

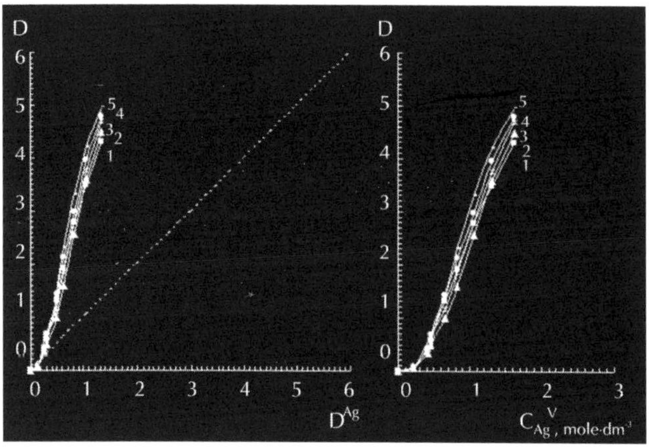

FIGURE 6 Dependences of **D** = f(D^{Ag}) and **D** = f(C_{Ag}^{V}) of reduction process $Ag_4[Fe(CN)_6]$→Ag with using of 3-(2-hydroxyethyl)-3-azapenthanediol-1,5 $N(CH_2–CH_2–NH_2)_3$ at concentration 10.0 g×L^{-1} (curve 1), 20.0 g×L^{-1} (2), 35.0 g×L^{-1} (3), 50.0 g×L^{-1} (4) and 100.0 g×L^{-1} (5). The optical densities D^{Ag} and **D** were measured with blue light-filter with a transmission maximum at 450 nm.

TABLE 1 The maximal (D/D^{Ag}) and pK_s values of Ag(I) complexes for various complex-forming reagents.

Complex-forming reagent	$(D/D^{Ag})_{max}$	Concentration of complex-forming reagent in solution at which reaches $(D/D^{Ag})_{max}$, $g \times L^{-1}$ ($mol \times L^{-1}$)	pK_s of Ag(I) complex having 1:2 composition
NH_3	3.40	4.5 (0.27)	7.25
$Na_2S_2O_3$	4.13	23.7 (0.15)	13.32
KSCN	4.92	69.7 (0.70)	8.39
$HO–(CH_2)_2–NH_2$	5.40	109.8 (1.80)	6.62
$H_2N–(CH_2)_2–NH_2$	5.80	77.0 (1.28)	7.84
$N(CH_2–CH_2–OH)_3$	3.93	99.8 (0.67)	3.64

At the first stage of the given process, reaction, which may be described by Eq. (1), takes place (in the braces **{....}**, formulas of gelatin-immobilized chemical compounds have been indicated):

$$4\{\mathbf{Ag}\} + 4[Fe(CN)_6]^{3-} \rightarrow \{\mathbf{Ag_4[Fe(CN)_6]}\} + 3[Fe(CN)_6]^{4-} \quad (1)$$

Each of complex-forming reagents under examination forms with Ag(I) *soluble* complex having a metal ion: ligand ratio of 1:2. That is why, formation of silver(I) complex with corresponding ***CR*** will occur to some extent when $Ag_4[Fe(CN)_6]$-**GIM** is at the contact with the solution containing any of complexation reagent indicated. Gelatin-immobilized silver(I) hexacyanoferrate(II) as well as any of these soluble complexes, can participate in process of reduction with Sn(II). In this connection, proceeding two parallel processes Ag(I)→Ag(0) will take place at contact of $Ag_4[Fe(CN)_6]$-**GIM** with the above indicated solution, containing Sn(II) and complexation reagent:

- gelatin-immobilized silver(I) hexacyanoferrate(II) reduction proceeding in *a polymer layer*;
- Ag(I) complex with complexation reagent reduction proceeding *on interface of phases* a **GIM**/solution.

In water solutions at pH=12–13, Sn(II) is mainly in a form of hydroxo-complex $[Sn(OH)_3]^-$. In this connection, Eq. (2) may be offered for the first of these processes in the case of negative charged "acid"-ligands may be ascribed (**L** – symbol of ligand, *z* – its charge). The particles of element silver formed as a result of Eqs. (3) and (4), theoretically should have smaller sizes than the particles of element silver arising in polymer layer of **GIM**. To be a part of substance immobilized in **GIM**, these particles should place freely in intermolecular cavities of a gelatin layer. Only in this case, they may diffuse in **GIM** and may be immobilized in gelatin mass.

$$\{\mathbf{Ag_4[Fe(CN)_6]}\} + 2[Sn(OH)_3]^- + 6OH^- \rightarrow 4\{\mathbf{Ag}\} + 2[Sn(OH)_6]^{2-} + [Fe(CN)_6]^{4-} \quad (2)$$

For the second of these processes, Eq. (3)

$$2[Ag\mathbf{L}_2]^+ + [Sn(OH)_3]^- + 3OH^- \rightarrow 2Ag + 4\mathbf{L} + [Sn(OH)_6]^{2-} \quad (3)$$

in the case of noncharged ligands and Eq. (4)

$$2[Ag\mathbf{L}_2]^{(2z-1)-} + [Sn(OH)_3]^- + 3OH^- \rightarrow 2Ag + 4\mathbf{L}^{z-} + [Sn(OH)_6]^{2-} \quad (4)$$

Gelatin has an extremely high surface area and an extensive system of micropores. The fragment of its structure has been shown in Fig. 7; as may be seen, it contains many intermolecular cavities. It may be valued the average size of intermolecular cavity in the gelatin structure [12–15].

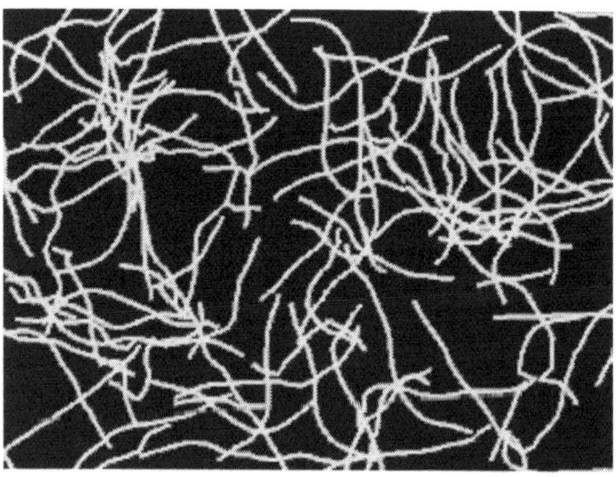

FIGURE 7 The fragment of gelatin structure containing intermolecular cavities.

For example, the volume of polymer layer of ***GIM*** (V_{gl}) having area 1 cm^2 and thickness 20 μm is $(1.0 \times 1.0 \times 20 \times 10^{-4})$ cm^3 = 2.0×10^{-3} cm^3, so that the mass of gelatin containing in such a layer, at average value of its density 0.5 g·cm^{-3}, is $(0.5 \times 2.0 \times 10^{-3})$ g = 1.0×10^{-3} g. Molecular mass of gelatin (M_{Gel}) is known to be ~$(2.0–3.0) \times 10^5$ c.u. [12, 13], the number of its molecules in given mass will be $(1.0 \times 10^{-3}/M_{Gel}) \times (6.02 \times 10^{23}) = (2.0–3.0) \times 10^{15}$. As discussed above, gelatin molecule in average has a length ~285,000 pm and diameter ~1400 pm, and if it may be considered as narrow cylinder, total volume of gelatin molecules V_M will be equal to $(1/4)\pi D^2h = (1/4) \times 3.14 \times (285{,}000 \times 10^{-10}$ cm$) \times (1400 \times 10^{-10}$ cm$)^2 = 4.38 \times 10^{-19}$ cm^3. In the case of maximal compact arrangement, these molecules occupy total volume equal to 4.38×10^{-19} cm^3 $(2.0–3.0) \times 10^{15} = (8.76–13.15) \times 10^{-4}$ cm^3. It may be postulated that the volume of cavities indicated, is equal to total volume of polymer massif minus the volume occupied by gelatin molecules, namely $(2.0 \times 10^{-3} – (8.76–13.15) \times 10^{-4})$ cm^3 that will be in the end $(0.69–1.12) \times 10^{-3}$ cm^3. Then, the average volume of one intermolecular cavity may be found as a quotient from division of their total volume into number of gelatin molecules and, as it may be easily noted, will be $(3.4–5.6) \times 10^{-19}$ cm$^3 = (3.4–5.6) \times 10^{11}$ pm^3. The linear size of such an "average" cavity in the case when it has *spherical* form, will be equal to $d = (6\ V/\pi)^{1/3} = [6 \times (3.7–5.6) \times 10^{11}\ \text{pm}^3/3.14]^{1/3} = (89.1–102.2) \times 10^2$ pm; when it has *cubic* form, equal to $a = V^{1/3} = [(3.7–5.6) \times 10^{11}\ \text{pm}^3)^{1/3} = (71.8–82.4) \times 10^2$ pm. As one can see from these values, these cavities are nano-sized. Therefore, only *nano-particles* of substance can entry into these cavities. By entering into such cavities, nano-particles of element silver are isolated from each other. In consequence of thereof, their aggregation with each other becomes rather difficult.

With growth of concentration of any of complexation reagent mentioned above, concentration of coordination compounds formed given complexation reagent with Ag(I) must increase. Correspondingly, the quantity of nano-particles of the element silver formed as a result of reduction of these coordination compounds by $[Sn(OH)_3]^-$ complex, should increase, too. In this connection, it may be expected that when concentration of these complexation reagent in a solution increases, the share of nano-particles contained in the "reprecipitated" elemental silver, should accrue gradually. Thus, at the same concentration nano-particles of element silver owing to their higher dispersion degree in comparison with microparticles should provide higher degree of absorption of visible light (and, accordingly, higher optical density) polymeric layer **GIM**. The experimental data presented in Figs. 1–6, are in full conformity with the given prediction.

The particles of element silver formed as a result of Eqs. (3) and (4), are one- or two-nuclear. While it is not enough of them (it occurs, when concentration of Ag(I) complexes on interface of phases a **GIM**/solution is low enough), these particles owing to their remoteness from each other have not time to be aggregated. They diffuses in polymeric layer of **GIM**, where are immobilized without change of their sizes. With increase of complex-forming reagent concentration (and, accordingly, of concentration of Ag(I) complex with given reagent), quantity of nano-particles indicated on interface of phases the **GIM**/solution accrues. It leads to increase of a number of such particles, diffused into **GIM**. However, at some rather high concentration complex-forming reagent in a solution, the effect of aggregation of nano-particles of element silver starts to affect. One- and two-nuclear particles of element silver formed at reduction of corresponding Ag(I) complex, begin to unite to some extent with each other in larger particles. Polynuclear particles of element silver resulting such an association are not so mobile and, consequently, will be not diffuse into polymeric layer of **GIM**. They will be precipitated in it near to interface **GIM**/solution (or even to escape as a solid phase in the solution contacting with **GIM**). As a result, rates of an increment of number of one-and two-nuclear particles of elemental silver with further growth of concentration complex-forming reagent begin to be slowed down. Thus, inevitably there should come the moment when the number of similar particles will reach some limiting value. That is why, since certain "threshold" concentration complex-forming reagent in a solution, growth of $\mathbf{D}^{Ag}$ values must stop. Moreover, at excess of this "threshold" concentration, certain decrease $\mathbf{D}^{Ag}$ should begin. The point is that an alignment between number of the aggregated particles and number one- and two-nuclear ones with growth of concentration of Ag(I) complex continuously grows and has no restrictions. These polynuclear particles are precipitated in frontier zone **GIM** on rather small depth and form, as a matter of fact, the same kind of elemental silver, as the microparticles of elemental silver formed as a result of reduction of gelatin-immobilized $Ag_4[Fe(CN)_6]$ according to Eq. (2). That is why, $\mathbf{D}^{Ag}$ values with increase of complex-forming reagent concentration at first increase, reach a maximum and then decrease.

By taking into consideration the foregoing, it may be assumed that "reprecipitated" gelatin-immobilized silver should contain, as a minimum, two phases of the silver particles, one of which is formed by nano-particles, and another, by microparticles. In order to corroborate the given conclusion, we carried out the analysis of elemental silver isolated from initial Ag-**GIM** and elemental silver isolated from Ag-**GIM** after end of "reprecipitation" process by X-ray powder diffraction method. X-ray powder diffraction patterns (XRD-patterns) of samples obtained are presented in Figs. 8–10. As may be seen from them, XRD-pattern of initial elemental silver with gray-black color of gelatin layer (Fig. 8) and XRD-patterns of "reprecipitated" elemental silver (Figs. 9 and 10), rather essentially differ from each other. So, in XRD-patterns of "reprecipitated" elemental silver obtained at an availability of any of studied ***CR*** in a solution contacting with **GIM**, there are accurate reflexes having $d = 333.6$, 288.5, 166.7 and 129.1 pm that are absent in XRD-pattern of initial elemental silver. At the same time, reflexes with $d = 235.7$, 204.1, 144.4, 123.1 and 117.9 pm are observed on them. These reflexes are characteristic for the known phase of elemental silver isolated from initial Ag-**GIM**. In this connection, there are all reasons to believe that the "reprecipitated" elemental silver obtained with using of a solution containing any of complex-forming reagent indicated above, contains at least two structural modifications of elemental silver.

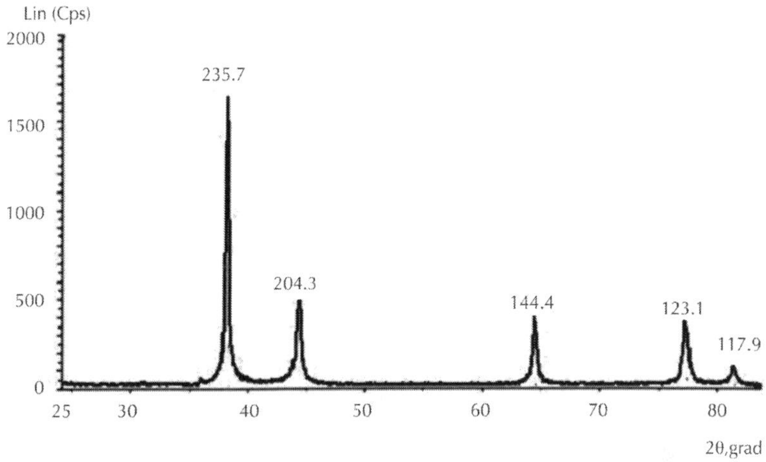

FIGURE 8 XRD-pattern of elemental silver isolated from initial Ag-**GIM**.

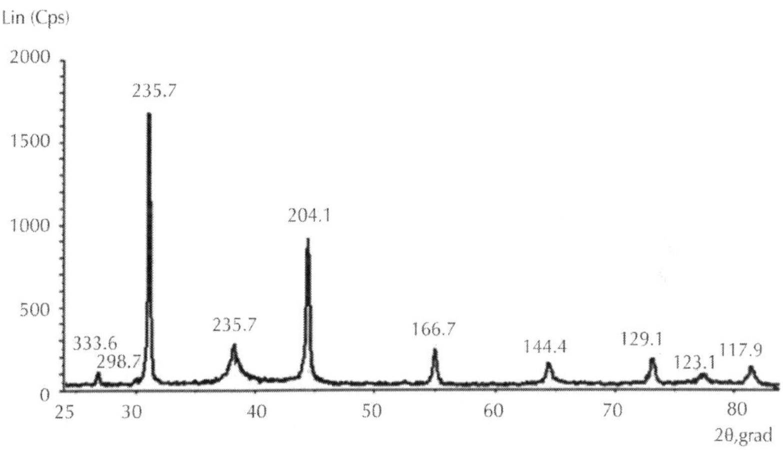

FIGURE 9 XRD-pattern of substance isolated from Ag-**GIM** containing "reprecipitated" elemental silver and obtained with using of solution containing $Na_2S_2O_3$ in concentration 20.0 g×L^{-1}.

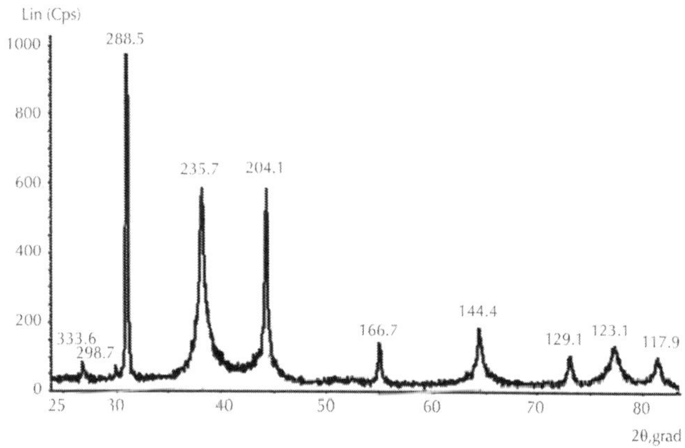

FIGURE 10 XRD-pattern of substance isolated from Ag-**GIM** containing "reprecipitated" elemental silver and obtained with using of solution containing ethanediamine-1,2 in concentration 20.0 g×L^{-1}.

The next curious circumstance attracts its attention: reflexes with d = 333.6, 288.5, 204.2, 166.7 and 129.1 pm are rather close to d values of reflexes of silver(I) bromide AgBr (number of card PDF 06–0438, parameter of an elementary cell a_0 = 577.45 pm, face-centered lattice, cubic syngonia, *Fm3 m* group of symmetry according to the international classification [14, 15]). In this connection, it may be assumed that the structure of the novel phase contained in "reprecipitated" elemental silver, at least in outline, resembles structure AgBr and its crystal lattice is similar to a lattice of silver(I) bromide where positions of atoms Br occupy atoms of silver.

To answer on the question, whether the reflexes indicated can belong to elemental silver with such a space structure in principle, theoretical XRD-patterns of assumed structure of element silver with use of program *PowderCell* described in Ref. [18], have been constructed by us. These XRD-patterns are presented in Fig. 11. As may be seen from them, the theoretical d values calculated by us (333.6, 288.7, 204.2, 174.1, 166.7, 144.4, 132.5, 129.1, 117.9 pm) for specified above structure with an elementary cell parameter a = 288.72 pm and *Pm3 m* symmetry group, practically coincide with d values experimentally observed in XRD-pattern of the "reprecipitated" elemental silver (d = 333.6, 288.5, 166.7 and 129.1 pm). It should be noted in this connection that d values calculated theoretically for the elemental silver isolated from initial Ag-**GIM** (235.4, 204.3, 144.5, 123.2 and 118.0 pm), correspond to compact-packed crystal structure having an elementary cell parameter a = 408.62 pm and *Fm3 m* symmetry group, interplane distances in which are 235.7, 204.1, 144.4, 123.1 and 117.9 pm.

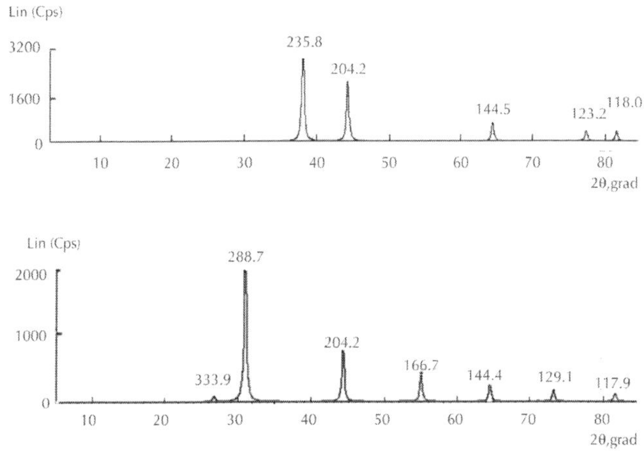

FIGURE 11 Theoretical XRD-patterns of elemental silver contained in initial Ag-**GIM** (*a*) and of elemental silver formed in **GIM** as a result of "reprecipitation" process (*b*).

Thus, formation of a novel phase of elemental silver, which probably was not described in the literature up to now, takes place here indeed.

ACKNOWLEDGEMENT

The Russian Foundation of Basic Researches (RFBR) is acknowledged for the financial support (Grant No. 09-03-97001).

KEYWORDS

- ***Bacillus mesentericus***
- **gelatin matrix**
- ***Powder Cell* program**
- **reprecipitation**

REFERENCES

1. Skillman, D. G.; Berry, C. R. Effect of particle shape on the spectral absorption of colloid silver in gelatin. *J. Chem. Phys.* **1968,** *48(7),* 3297–3304.

2. Linnert, T.; Mulvaney, P.; Henglein, A.; Weller, H. Long-lived nonmetallic silver clusters in aqueous solution: preparation and photolysis. *J. Am. Chem. Soc.* **1990,** *112(12),* 4657–4664.
3. Fedrigo, S.; Harbich, W.; Butter, J. Collective dipole oscillations in small silver clusters embedded in raregas matrices. *Phys. Rev.* **1993,** *47(23)*, 10706–10715.
4. Satoh, N.; Hasegawa, H.; Tsujii, K.; Kimura, K. Photoinduced coagulation of Ag nanocolloides. *J. Phys. Chem.* **1994,** *98(7)*, 2143–2147.
5. Sato, T.; Ishikawa, T.; Ito, T.; Yonazawa, Y.; Kodono, K.; Sakaguchi, T.; Miya, M.; *Chem. Phys. Lett.* **1995,** *242(3),* 310–314.
6. Al-Obaidi, A. H. R.; Rigbi, S. J.; McGarvey, J. J.; Wamsley, D. G.; Smith, K. W.; Hellemans, I.; Snauwaert, J. Microstructural and spectroscopies studies of metal liquidlike films of silver and gold. *J. Phys. Chem.* **1994,** *98(24)*, 11163–11168.
7. Ershov, B. G.; Henglein, A. Reduction of Ag^+ on polyacrilate chains in aqueous solutions. *J. Phys. Chem. B* **1998**, *102(24),* 10663–10666.
8. Kapoor, S. Surface modification of silver particles. *Langmuir*, **1998,** *14(5)*, 1021–1025.
9. Sergeev, B. M.; Kiryukhin, M. V.; Prusov, A. N. Effect of light on the disperse composition of silver hydrosols stabilized by partially decarboxylated polyacrylate. *Mendeleev Commun.* **2001,** *11(2)*, 68–69.
10. Mikhailov, O. V.; Guseva, M. V.; Krikunenko, R. I. An amplification of silver photographic images by using of processes changing disperse of image carrier. *Zh. Nauchn. Prikl. Foto-Kinematogr*. **2003,** *48(4),* 52–57 (in rus.)
11. Mikhailov, O. V.; Kondakov, A. V.; Krikunenko, R. I. Image intensification in silver halide photographic materials for detection of high-energy radiation by reprecipitation of elemental silver. *High Energy Chem.*; **2005,** *39(5)*, 324–329.
12. Mikhailov, O. V. Reactions of nucleophilic, electrophilic substitution and template synthesis in the metalhexacyanoferrate(II) gelatin-immobilized matrix. *Rev. Inorg. Chem.*; **2003,** *23(1)*, 31–74.
13. Mikhailov, O. V. Gelatin-Immobilized Metalcomplexes: Synthesis and Employment. *J. Coord. Chem.* **2008,** *61(7)*, 1333–1384.
14. Mikhailov, O. V. Self-Assembly of Molecules of Metal Macrocyclic Compounds in Nanoreactors on the Basis of Biopolymer-Immobilized Matrix Systems. *Nanotechnol. Russ.* **2010,** *5(1–2)*, 18–25.
15. Mikhailov, O. V. Soft template synthesis of Fe(II, III), Co(II, III), Ni(II) and Cu(II) metalmacrocyclic compounds into gelatin-immobilized matrix implants. *Rev. Inorg. Chem.*; **2010,** *30(4)*, 199–273.
16. Mikhailov, O. V. Enzyme-assisted matrix isolation of novel dithiooxamide complexes of nickel(II). *Indian J. Chem.* **1991,** *30A(2),* 252–254.
17. Kraus, W.; Nolze, G. Powder cell – a program for the representation and manipulation of crystal structures and calculation of the resulting X-ray powder patterns. *J. Appl. Cryst.* **1996,** *29(3)*, 301–303.
18. Powder Diffract File. Search Manual Fink Method. Inorganic. USA, Pennsylvania: JCPDS – International Centre for Diffraction Data, **1995** (release 2000).

CHAPTER 10

A STUDY ON POLYCARBONATE MODIFIED WITH CU/C NANOCOMPOSITE

YU. V. PERSHIN and V. I. KODOLOV

CONTENTS

10.1 Introduction ... 378
10.2 Experimental Part ... 378
10.3 Results and Discussion ... 379
10.4 Conclusion ... 383
Keywords ... 383
References ... 383

10.1 INTRODUCTION

Modification of polycarbonate with copper/carbon nanocomposite was investigated. The changes in spectral, optical and thermal-physical characteristics of compositions based on polycarbonate modified with copper/carbon nanocomposites were observed.

Recently, the materials based on polycarbonate have been modified to improve their thermal-physical and optical characteristics and to apply new properties to use them for special purposes. The introduction of nanostructures into materials facilitates self-organizing processes in them. These processes depend on surface energy of nanostructures, which is connected with energy of their interaction with the surroundings. It is known [1] that the surface energy and activity of nanoparticles increase when their sizes decrease.

For nanoparticles, the surface and volume are defined by the defectiveness and form of conformation changes of film nanostructures depending on their crystallinity degree. However, the possibilities of changes in nanofilm shapes with the changes in medium activity are greater in comparison with nanostructures already formed. At the same time, sizes of nanofilms formed and their defectiveness, i.e., tears and cracks on the surface of nanofilms, play an important role [2].

When studying the influence of super-small quantities of substances introduced into polymers and considerably changing their properties, apparently we should consider the role, which these substances play in polymers possessing highly organized super-molecular regularity, both in crystalline and amorphous states. It can be assumed that the mechanism of this phenomenon is in the nanostructure energy transfer to polymer structural formations through the interface resulting in the changes in their surface energy and mobility of structural elements of the polymeric body. Such mechanism is quite realistic as polymers are structural-heterogenic (highly dispersed) systems.

Polycarbonate modification with super-small quantities of Cu/C nanocomposite is possible using fine suspensions of this nanocomposite, which contributes to uniform distribution of nanoparticles in polycarbonate solution.

10.2 EXPERIMENTAL PART

Polycarbonate "Actual" was used as the modified polycarbonate.

Fine suspensions of copper/carbon nanocomposite were prepared combining 1.0; 0.1; 0.01; 0.001% of nanocomposite in polycarbonate solution in ethylene dichloride. The suspensions underwent ultrasonic processing.

To compare the optical density of nanocomposite suspension in polycarbonate solution in ethylene dichloride, as well as polycarbonate and polycarbonate samples modified with nanocomposites, spectrophotometer KFK-3-01a was used.

Samples in the form of modified and nonmodified films for studying IR spectra were prepared precipitating them from suspension or solution under vacuum. The obtained films about 100 mcm thick were examined on Fourier-spectrometer FSM 1201.

To investigate the crystallization and structures formed the high-resolution microscope (up to 10 mcm) was used.

To examine thermal-physical characteristics the lamellar material on polycarbonate and polymethyl methacrylate basis about 10 mm high was prepared. Three layers of polymethyl methacrylate and two layers of polycarbonate were used. Thermal-physical characteristics (specific thermal capacity and thermal conductivity) were investigated on calorimeters IT-c-400 and IT-λ-400.

10.3 RESULTS AND DISCUSSION

During the investigation the nanocomposite fine suspension in polycarbonate solution in ethylene dichloride was studied, polycarbonate films modified with different concentrations of nanocomposite were compared with the help of optical spectroscopy, microscopy, IR spectroscopy and thermal-physical methods of investigation.

The results of investigation of optical density of Cu/C nanocomposite fine suspension (0.001%) based on polycarbonate solution in ethylene dichloride are given in Fig. 1. As seen in Fig. 1, the introduction of nanocomposite and polycarbonate into ethylene dichloride resulted in transmission increase in the range 640–690 nm (approximately in three times). At the same time, at 790 nm and 890 nm the significant increase in optical density was observed. The comparison of optical densities of films of polycarbonate and modified materials after the introduction of different quantities of nanocomposite (1.0; 0.1; 0.01%) into polycarbonate is interesting (Fig. 2).

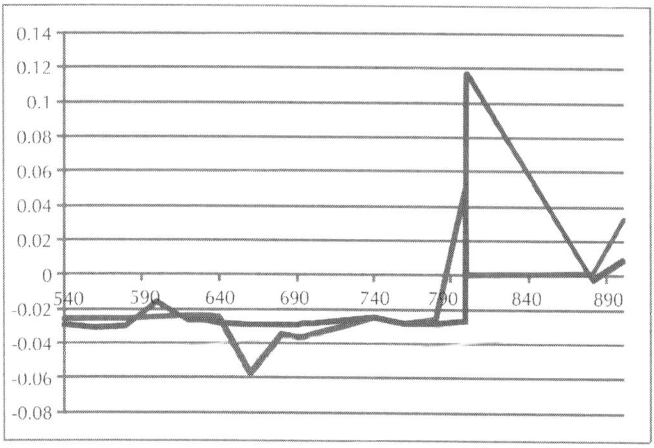

FIGURE 1 Curves of optical density of suspension based on ethylene dichloride diluted with polycarbonate and Cu/C nanocomposite in the concentration to polycarbonate 0.001% (1) and ethylene dichloride (2).

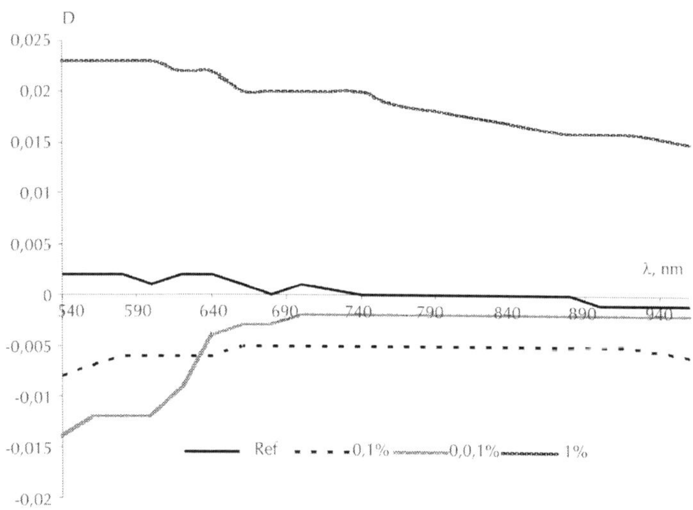

FIGURE 2 Curves of optical density of reference sample modified with Cu/C nanocomposites in concentration 1%, 0.1%, and 0.01%.

Comparison of optical density of suspension containing 0.001% of nanocomposite and optical density polycarbonate sample modified with 0.01% of nanocomposite indicates the proximity of curves character. Thus the correlation of optical properties of suspensions of nanocomposites and film materials modified with the same nanocomposites is quite possible.

Examination of curves of samples optical density demonstrated that when the nanostructure concentration was 1% from polycarbonate mass, the visible-light spectrum was absorbed by about 4.2% more if compared with the reference sample. When the nanocomposite concentration was 0.1%, the absorption decreased by 0.7%. When the concentration was 0.01%, the absorption decreased in the region 540–600 nm by 2.3%, and in the region 640–960 nm by 0.5%.

During the microscopic investigation of the samples the schematic picture of the structures formed was obtained at 20-mcm magnification. The results are given in Fig. 3. From the schematic pictures it is seen that volumetric structures of regular shape surrounded by micellae were formed in polycarbonate modified with 0.01% Cu/C nanocomposite. When 0.1% of nanocomposite was introduced, the linear structures distorted in space and surrounded by micellae were formed. When Cu/C nanocomposite concentration was 1%, large aggregates were not observed in polycarbonate.

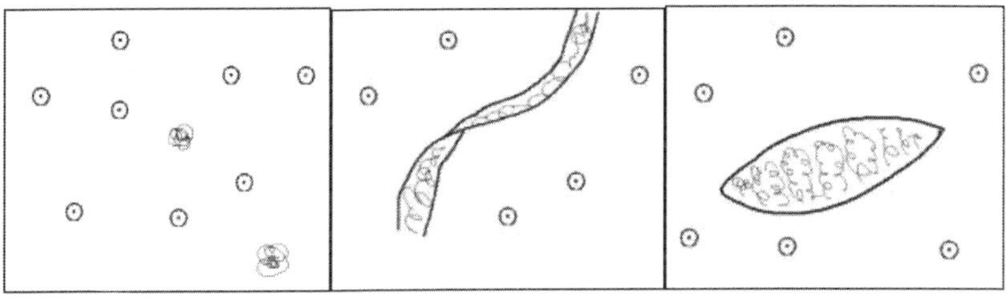

FIGURE 3 Schematic picture of structure formation in polycarbonate modified with Cu/C nanocomposites in concentrations 1%, 0.1%, and 0.01% (from left to the right), 20-mcm magnification

Apparently, the decrease in nanocomposite concentration in polycarbonate can result in the formation of self-organizing structures of bigger size. In Ref. [3], there is the hypothesis on the transfer of nanocomposite oscillations onto the molecules of polymeric composition, the intensity of bands in IR spectra of which sharply increases even after the introduction of super-small quantities of nanocomposites. In our case, this hypothesis was checked on modified and non-modified samples of polycarbonate films.

In Fig. 4, you can see the IR spectra of polycarbonate and polycarbonate modified with 0.001% of Cu/C nanocomposite.

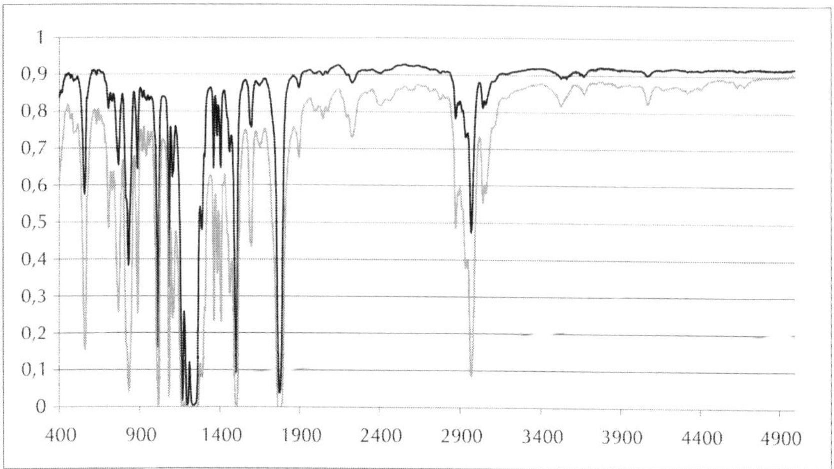

FIGURE 4 IR spectra of reference sample (upper) and polycarbonate modified with Cu/C nanocomposites in concentration 0.001% (lower).

As seen in IR spectra, the intensity increases in practically all regions, indicating the influence of oscillations of Cu/C nanocomposite on all the system. The most vivid changes in the band intensity are observed at 557 cm^{-1}, in the region 760–930 cm^{-1}, at 1400 cm^{-1}, 1600 cm^{-1}, 2970 cm^{-1}.

Thus, polycarbonate self-organization under the influence of Cu/C nanocomposite takes place with the participation of certain bonds, for which the intensity of absorption bands goes up. At the same time, the formation of new phases is possible, which usually results in thermal capacity increase. But thermal conductivity can decrease due to the formation of defective regions between the aggregates formed.

The Table 1 contains the results of thermal-physical characteristics studied.

TABLE 1 Thermal-physical characteristics of polycarbonate and its modified analogs.

	Nanocomposite content in polycarbonate, %			
	0%, ref.	**1%**	**0.1%**	**0.01%**
m 10^{-3}, kg	1.9	1.955	1.982	1.913
h 10^{-3}, m	9.285	9.563	9.76	9.432
Csp, J/kg K	1440	1028	1400	1510
Λ	0.517	0.503	0.487	0.448

It is demonstrated that when the nanostructure concentration in the material decreases, thermal capacity goes up that is confirmed by the results of previous investigations. Thermal conductivity decline, when the nanostructure concentration decreases, is apparently caused by the material defectiveness.

10.4 CONCLUSION

When Cu/C nanocomposites are introduced into the material modified, the nanostructures can be considered as the generator of molecules excitation, which results in wave process in the material.

It is found that polycarbonate modification with metal/carbon containing nanocomposites results in the changes in polycarbonate structure influencing its optical and thermal-physical properties.

KEYWORDS

- **copper/carbon nanocomposite**
- **ethylene dichloride**
- **IR spectra**
- **polycarbonate**

REFERENCES

1. Kodolov, V. I.; Khokhriakov, N. V. Chemical physics of the processes of formation and transformation of nanostructures and nanosystems. Izhevsk, Publishing office of Izhevsk State Agricultural Academy, **2009,** *1–2*, 728 p.
2. Kodolov, V. I.; Khokhriakov, V. N. V.; Trineeva, V.; Blagodatskikh, I. I. Activity of nanostructures and its demonstration in nanoreactors of polymeric matrixes and active media. *Chem. Phys. Mesoscopy*, **2008,** *10(4)*, p. 448.
3. Kodolov, V, V. I.; Trineeva, V. Perspectives of idea development about nanosystems self-organization in polymeric matrixes; Publishing office of Izhevsk State Agricultural Academy, **2012,** *1(4)*, 650 p.

CHAPTER 11

NANOFIBERS AND SOLAR CELLS

V. MOTTAGHITALAB, M. SAJEDI, M. S. MOTLAGH, M. ABBASI, and A. K. HAGHI

CONTENTS

11.1 Introduction 386
11.2 The Basic Concept of Solar Cell 387
11.2.1 Electrical Measurement Background 389
11.2.2 Inorganic Solar Cell 389
11.2.3 Organic Solar Cells (OSCs) 391
11.2.4 Dye-Sensitized Solar Cells 392
11.3 Textile Solar Cells 395
11.3.1 Recent Progress of PV Fibers 396
11.3.2 PV Integration in Textile 400
11.4 Nanofibers as a Potential Key Element in Textile Solar Cells 401
11.4.1 Electrospun Nanofiber 401
11.4.2 Nanofibers in DSSC Solar Cells 404
11.4.3 Electrospun Nonwoven Organic Solar Cell 407
11.4.4 Future Prospects of Organic Nanofiber Textile Solar 408
11.5 Conclusion 409
Keywords 409
References 410

11.1 INTRODUCTION

Nowadays, the wearable wireless electronic device and system extensively needs a sustainable energy supply as potential alternative in order to overcome the urgent needs to regular battery with finite power. The basic concept of photovoltaic effect reviewed and the structure of a range of energy conversion devices including inorganic either homojunction or multijunction, DSSC and polymer based solar cells briefly introduced to consolidate the idea regarding to the development of textile solar cells. This chapter attempts to presents an overview on photovoltaic devices by focusing on flexible solar cells composed of nanofibers as key element. A detailed review on cell material selection and their effect on energy conversion are considered to elucidate the role of nanofiber in energy conversion. The chapter also presents promizing potentials of electrospun nanofibers web for being a key element in photovoltaic devices. As a new alternative, the transparent ITO glass can be replaced by the transparent conductive fabric. The sprayed anatase TiO_2 layer introduced on transparent conductive fabric as blocking layer, electrospun anatase TiO_2 nanofibers web laminated on TiO_2 coated conductive fabric as the photoanode, and a transparent PET film employed as the seal film to protect the gel electrolyte. The preparation of core–shell nanofibers by coaxial electrospinning process of two components such as poly(3-hexyl thiophene) (P3HT) (a conducting polymer) or P3HT/PCBM as the core and poly(vinyl pyrrolidone) (PVP) as the shell material open a new horizon for integration of nanofibers in textile solar cells. Latest experience regarding to polymer solar cell and low band gap material needs to be considered for ambitious plans with nanofiber morphology.

In recent years, renewable energies attract considerable attention due to the inevitable end of fossil fuels and due to global warming and other environmental problems. Photovoltaic solar energy is being widely studied as one of the renewable energy sources with key significance potentials and a real alternate to fossil fuels. Solar cells are in general packed between w80, brittle and rigid glass plates. Therefore, increasing attention is being paid to the construction of lighter, portable, robust, multipurpose and flexible substrates for solar cells. Textiles substrates are fabricated by a wide variety of processes, such as weaving, knitting, braiding and felting. These fabrication techniques offer enormous versatility for allowing a fabric to conform to even complex shapes. Textile fabrics not only can be rolled up for storage and then unrolled on site but also they can also be readily installed into structures with complex geometries.

Textiles are engaging as flexible substrates in that they have an enormous variety of uses, ranging from clothing and household articles to highly sophisticated technical applications. Last innovations on photovoltaic technology have allowed obtaining flexible solar cells, which offer a wide range of possibilities, mainly in wearable applications that need independent systems. Nowadays, entertainment, voice and data communication, health monitoring, emergency, and surveillance functions, all of which rely on wireless protocols and services and sustainable energy supply in order to overcome the urgent needs to regular battery with finite power. Because of their steadily decreasing power demand, many portable devices can harvest enough energy from clothing-integrated solar modules with a maximum installed power of 1–5 W. [1]

Increasingly textile architecture is becoming progressively of a feature as permanent or semipermanent constructions. Tents, such as those used by the military and campers, are the best known textile constructions, as are sun shelter, but currently big textile constructions are used extensively for exhibition halls, sports complexes and leisure and recreation centers. Although all these structures provide protection from the weather, including exposure to the sun, but solar concept offers an additional precious use for providing power. Many of these large textile architectural constructions cover huge areas, sufficient to supply several kilowatts of power. Even the fabric used to construct a small tent is enough to provide a few hundred watts. In addition to textile architecture, panels made from robust solar textile fabrics could be positioned on the roofs of existing buildings. Compared to conventional and improper solar panels for roof structures lightweight and flexible solar textile panels is able to tolerate load-bearing weight without shattering.

Moreover, natural disaster extensively introduces the huge potential needs the formulation of unusual energy package based on natural source. Over the past five years, more than 13 million *people have lost their home* and possessions because of earthquake, bush fire, flooding or other natural disaster. The victims of these disasters are commonly housed in tents until they are able to rebuild their homes. Whether they stay in tented accommodation for a short or long time, tents constructed from solar textile fabrics could provide a source of much needed power. This power could be stored in daytime and used at night, when the outdoor temperature can often fall. There are also a number of other important potential applications. The military would benefit from tents and field hospitals, especially those in remote areas, where electricity could be generated as soon as the structure is assembled.

11.2 THE BASIC CONCEPT OF SOLAR CELL

In 1839 French scientist, Edmond Becquerel found out photovoltaic effect when he observed increasing of electricity generation while light exposure to the two metal electrodes immersed in electrolytic solution [2]. Light is composed of energy packages known as photons. Typically, when a matter exposed to the light, electrons are excited to a higher level within material, but they return to their initial state quickly. When electrons take sufficient energy more than a certain threshold (band gap), move from the valance band to the conduction band holes with positive charge will be created. In the photovoltaic effect electron-hole pairs are separated and excited electrons are pulled and fed to an external circuit to buildup electricity [3] (Fig. 1).

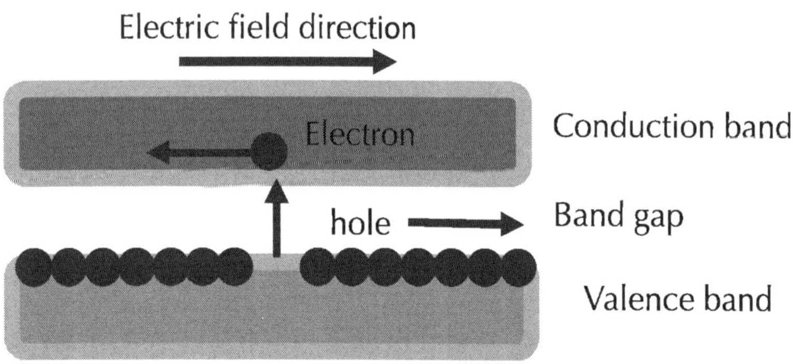

FIGURE 1 Electron excitation from valence band to conduction band.

An effective solar cell generally comprises an opaque material that absorbs the incoming light, an electric field that arises from the difference in composition between the semiconducting layers comprising the absorber, and two electrodes to carry the positive and negative charges to the electrical load. Designs of solar cells differ in detail but all must include the above features (Fig. 2).

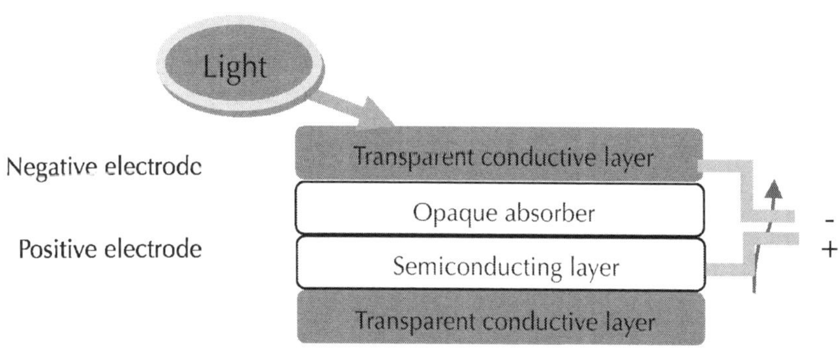

FIGURE 2 The general structure of a Solar cell.

Generally, Solar cells categorized into three main groups consisting of inorganic, organic and hybrid solar cells (Dye sensitized solar cells). Inorganic solar cells cover more than 95% of commercial products in solar cells industry.

11.2.1 ELECTRICAL MEASUREMENT BACKGROUND

All current–voltage characteristics of the photovoltaic devices were measured with a source measure unit in the dark and under simulated solar simulator source was calibrated using a standard crystalline silicon diode. The current-voltage characteristics of Photovoltaic devices are generally characterized by the short-circuit current (*I*sc), the open-circuit voltage (*V*oc), and the fill factor (*FF*). The photovoltaic power conversion efficiency (η) of a solar cell is defined as the ratio between the maximum electrical power (*P*max) and the incident optical power and is determined by Eq. (1). [4]

$$\eta = \frac{I_{sc} \times V_{oc} \times FF}{P_{in}} \tag{1}$$

In Eq. (1), The short circuit current (I_{sc}) is the maximum current that can run through the cell. The open circuit voltage (V_{oc}) depends on the highest occupied molecular orbital (homo) level of the donor (p-type semiconductor quasi Fermi level) and the lowest unoccupied molecular orbital (lumo) level of the acceptor (*n*-type semiconductor quasi Fermi level), linearly. *P*in is the incident light power density. FF, the fill-factor, is calculated by dividing *P*max by the multiplication of *I*sc and *V*oc and this can be explained by the following equation (2):

$$FF = \frac{\mathrm{I}_{mpp} \times V_{mpp}}{I_{sc} \times V_{oc}} \tag{2}$$

In the Eq. (2), Vmpp and Impp represent, respectively, the voltage and the current at the maximum power point (MPP), where the product of the voltage and current is maximized [4].

11.2.2 INORGANIC SOLAR CELL

Inorganic solar cells based on semiconducting layer architecture can be divided into four main categories including P–N homo junction, hetrojunction either P–I–N or N–I–P and multi junction.

The P–N homojunction is the basis of inorganic solar cells in which two different doped semiconductors (n-type and p-type) are in contact to make solar cells (Fig. 3a). P-type semiconductors are atoms and compounds with fewer electrons in their outer shell, which could create holes for the electrons within the lattice of p-type semiconductor. Unlike p-types semiconductors, the n-type have more electrons in their outer shell and sometimes there are exceed amount of electron on n-type lattice result lots of negative charges [5].

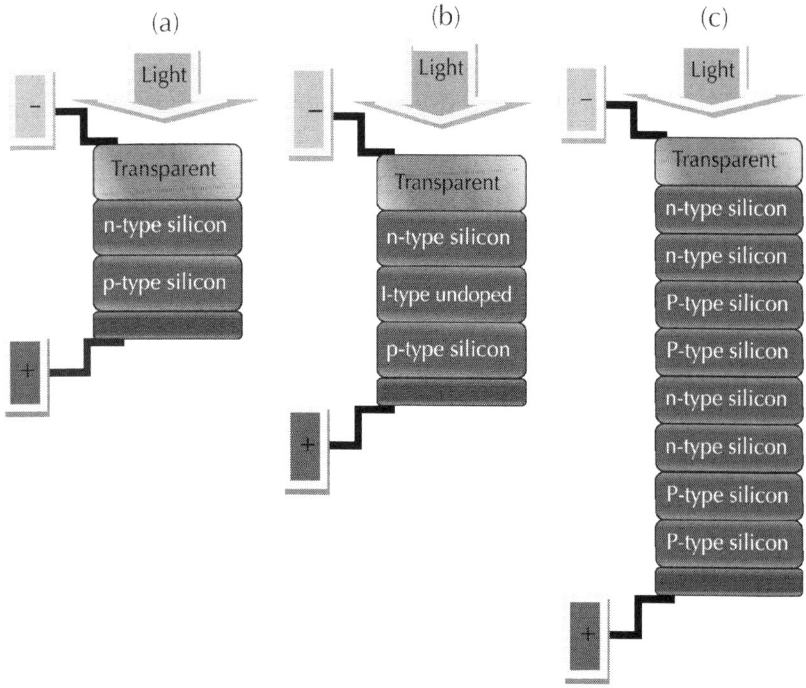

FIGURE 3 A general scheme of inorganic solar cell (a) P–N homojunction (b) P–I–N hetrojunction (c) multijunction

Compared to homojunction structure amorphous silicon thin-film cells use a P–I–N hetrojunction structure, whereas cadmium telluride (CdTe) cells use a N–I–P arrangement. The overall picture embraces a three-layer sandwich with a middle intrinsic (i-type or undoped) layer between an N-type layer and a P-type layer (Fig. 3b). Multiple junction cells have several different semiconductor layers stacked together to absorb different wavebands in a range of spectrum, producing a greater voltage per cell than from a single junction cell which most of the solar spectrum to electricity lies in the red (Fig. 3c).

Variety of semiconducting material such as single and poly crystal silicon, amorphous silicon, Cadmium-Telluride (CdTe), Copper Indium/Gallium Di Selenide (CIGS) have been employed to form inorganic solar cell based on layers configuration to enhance absorption efficiency, conversion efficiency, production and maintenance cost.

11.2.3 ORGANIC SOLAR CELLS (OSCS)

Photoconversion mechanism in organic or excitonic solar cells is differing from conventional inorganic solar cells in which exited mobile state are made by light absorption in electron donor. While, light absorption creates free electron-hole pairs in inorganic solar cells [5]. It is due to law dielectric constant of organic materials and weak noncovalent interaction between organic molecules. Consequently, exciton dissociation of electron-hole pairs occurs at the interface between electron donor and electron acceptor components [6] . Electron donor and acceptor act as semiconductor p-n junction in inorganic solar cells and should be blended together to prevent electron-hole recombination (Fig. 4).

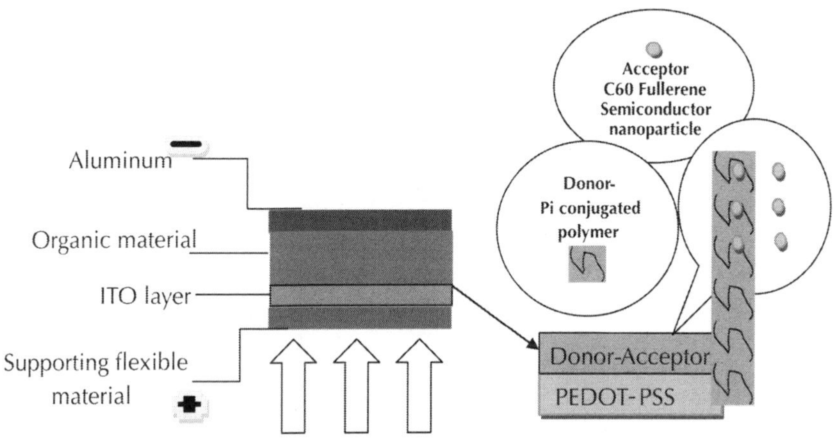

FIGURE 4 Schematic of organic solar cell.

There are two main types of PSCs including: bilayer heterojunction and bulk-heterojunction [7]. Bulk-heterojunction PSCs are more attractive due to their high surface area junction that increases conversion efficiency. This type of polymer solar cell consists of Glass, ITO, PEDOT-PSS, active layer, calcium and aluminum in which conjugated polymer are used as active layer [8]. The organic solar cells with maximum conversion efficiency about 6% still are at the beginning of development and have a long way to go to compete with inorganic solar cells. Indeed, the advantages of polymers including low-cost deposition in large areas, low weight on flexible substrates and sufficient efficiency are promizing advent of new type of solar cells [9]. Conjugated small molecule attracted as an alternative approach of organic solar cells. Development of small molecule for OSCs interested because of their properties such as well-defined molecular structure, definite molecular weight, high purity, easy purification, easy mass-scale production, and good batch-to-batch reproducibility [10–12].

11.2.4 DYE-SENSITIZED SOLAR CELLS

Dye-sensitized solar cells (DSSC) use a variety of photosensitive dyes and common, flexible materials that can be incorporated into architectural elements such as windowpanes, building paints, or textiles. DSSC technology mimic photosynthesis process whereby the leaf structure is replaced by a porous titania nanostructure, and the chlorophyll is replaced by a long-life dye. The general scheme of DSSC process is shown in Fig. 5. Although traditional silicon-based photovoltaic solar cells currently have higher solar energy conversion ratios, dye-sensitive solar cells have higher overall power collection potential due to low-cost operability under a wider range of light and temperature conditions, and flexible application [13].

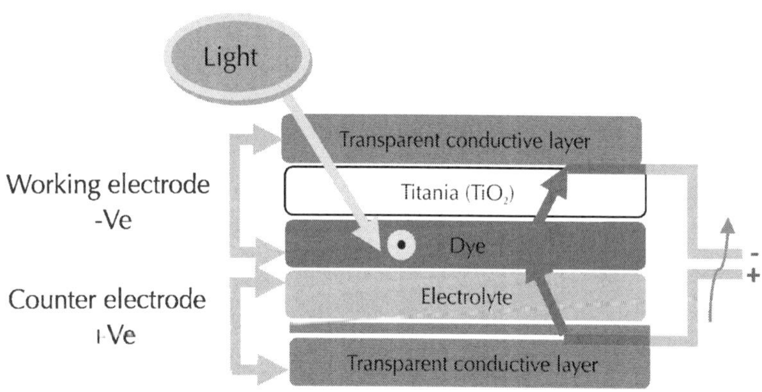

FIGURE 5 Schematic of Dye Senetesized Solar Cell (DSSC).

Oxide semiconductors materials such as TiO_2, ZnO_2 and SnO_2 have a relatively wide band gap and cannot absorb sunlight in visible region and create electron. Nevertheless, in sensitization process, visible light could be absorbed by photosensitizer organic dye results creation of electron. Consequently, excited electrons are penetrated into the semiconductor conduction band. Generally, DSSC structures consist of a photoelectrode, photosensitizer dye, a redox electrolyte, and a counter electrode. Photoelectrodes could be made of materials such as metal oxide semiconductors. Indeed, oxide semiconductor materials, particularly TiO_2, are choosing due to their good chemical stability under visible irradiation, nontoxicity and cheapness. Typically, TiO_2 thin film photoelectrode prepared via coating the colloidal solution or paste of TiO_2 and then sintering at 450 °C to 500 °C on the surface of substrate, which led to increase of dye absorption drastically by TiO_2 [14]. The substrate must have high transparency and low ohmic resistance to high performance of cell could be achieved. Recently, many researches focused on the both organic and inorganic dyes as sensitizer regarding to their extinction coefficients and performance. Among them, B4 (N3): RuL2(NCS)2:L=(2,2′-bipyridyl-4,4′-dicarboxylic acid) and B2(N719): {*cis–bis* (thiocyanato)-*bis*(2,20-bipyridyl-4,40-dicarboxylato)-ruthenium(II) *bis*-tetrabutylammonium} due to its outstanding performance was interested (Scheme 1).

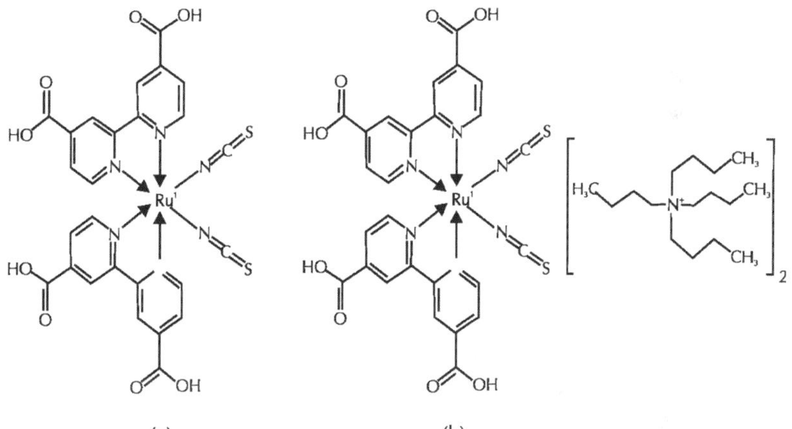

SCHEME 1 The chemical structure of (a) B4 (N3): RuL2(NCS)2 L(2,2′-bipyridyl-4,4′-dicarboxylic acid), (b) B2(N719) {*cis–bis* (thiocyanato)-*bis*(2,20-bipyridyl-4,40-dicarboxylato)-ruthenium(II) *bis*-tetrabutylammonium

B2 is the most common high performance dye and a modified form of B4 to increase cell voltage. Up to now different methods have been performed to develop of new dyes with high molar extinction coefficients in the visible and near-IR in order to outperform N719 as sensitizers in a DSSC [15]. DNH2 is a hydrophobic dye, which very efficiently sensitizes wide band-gap oxide semiconductors, like titanium dioxide. DBL (otherwise known as "black dye") is designed for the widest range spectral sensitization of wide band-gap oxide semiconductors, like titanium dioxide up to wavelengths beyond 800 nm (Scheme 2).

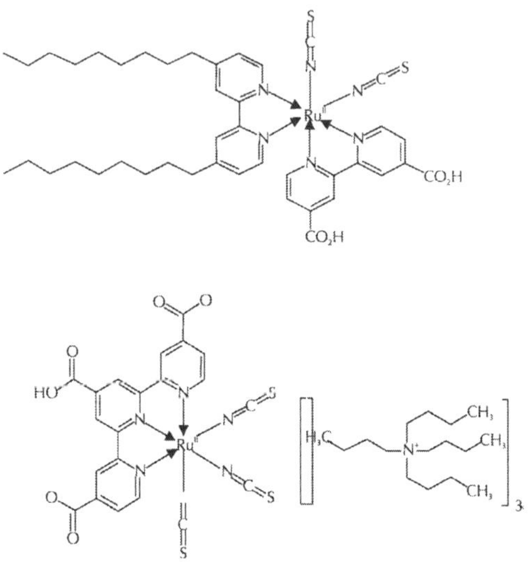

SCHEME 2 The chemical structure of (a) DNH2 (Z907) RuLL'(NCS)2, L=2,2'-bipyridyl-4,4'-dicarboxylic acid, L'= 4,4'-dinonyl-2,2'-bipyridine (b) DBL (N749) [RuL(NCS)3]: 3 TBA
L= 2,2':6,'2''-terpyridyl-4,4,'4''-tricarboxylic acid TBA=tetra-n-butylammonium.

In order to continuous electron movement through the cell, the oxidized dye should be reduced by electron replacement. The role of redox electrolyte in the DSSCs is to mediate electrons between the photoelectrode and the counter electrode. Common electrolyte used in the DSSC is based on I^-/I^-_3 redox ions [16]. The mechanism of photon to current has been summarized in the following equations:

$$\text{Dye} + \text{light} \rightarrow \text{Dye}^* + e^- \qquad (a)$$

$$\text{Dye}^* + e^- + TiO_2 \rightarrow e^-(\text{TiO2}) + \text{oxidized Dye} \qquad (b)$$

$$\text{Oxidized Dye} + 3/2I^- \rightarrow \text{Dye} + 1/2I^{-3} \qquad (c)$$

$$1/2I^-_3 + e^-(\text{counter electrode}) \rightarrow 3/2I^- \qquad (d)$$

Despite of many advantages of DSSCs, still lower efficiency compared to commercialized inorganic solar cells is a challenging area. Recently, one-dimensional nanomaterials, such as nanorods, nanotubes and nanofibers, have been proposed to replace the nanoparticles in DSSCs because of their ability to improve the electron transport leading to enhanced electron collection efficiencies in DSSCs.

Subsequent sections attempts to provide fundamental knowledge to general concept of textile solar cells and their recent progress based on. Of particular interest are electrospun TiO_2 nanofibers playing the role as a key material in DSSCs and other organic solar cell, which have been shown to improve the electron transport efficiency and to enhance the light harvesting efficiency by scattering more light in the red part of the solar spectrum. A detailed review on cell material selection and their effect on energy conversion are considered to elucidate the potential role of nanofiber in energy conversion for textile solar cell applications.

11.3 TEXTILE SOLAR CELLS

Clothing materials either for general or specific use are passive and the ability to integrate electronics into textiles provides great opportunity as smart textiles to achieve revolutionary improvements in performance and the realization of capabilities never before imagined on daily life or special circumstances such as battlefield. In general, smart textiles address diverse function to withstand an interactive wearable system. Development, incorporation and interconnection of flexible electronic devices including sensor, actuator, data processing, communication, internal network and energy supply beside basic garment specifications sketch the road map toward smart textile architecture. Regardless of the subsystem functions, energy supply and storage play a critical role to propel the individual functions in overall smart textile systems.

The integration of photovoltaic (PVs) into garments emerges new prospect of having a strictly mobile and versatile source of energy in communications equipment, monitoring, sensing and actuating systems. Despite of extremely good power efficiency, most conventional crystalline silicon based semiconductor PVs is intrinsically stiff and incompatible with the function of textiles where flexibility is essential. Extensive research has been conducted to introduce the novel potential candidates for shaping the textile solar cell (TSC) puzzle. In particular, polymer-based organic solar cell materials have the advantages of low price and ease of operation in comparison with silicon-based solar cells. Organic semiconductors, such as conductive polymers, dyes, pigments, and liquid crystals, can be manufactured cheaply and used in organic solar cell constructions easily. In the manufacturing process of organic solar cells, thin films are prepared using specific techniques, such as vacuum evaporation, solution processing, printing [17, 18], or nanofiber formation [19] and electrospinning [20] at room temperatures. Dipping, spin coating, doctor blading, and printing techniques are mostly used for manufacturing organic solar cells based on conjugated polymers [17].Recent TSC studies revealed two distinctive strategies for developing flexible textile solar cell and its sophisticated integration.

1. The first strategy involved the simple incorporation of a polymer PV on a flexible substrate Such as poly thyleneterphthalate (PET) directly into the clothing as a structural power source element.

2. The second strategy was more complicated and involved the lamination of a thin anti reflective layer onto a suitably transparent textile material followed by plasma, thermal or chemical treatment. The next successive step focuses on application of a photoanode electrode onto the textile material. Subsequent procedure led to the deposition of the active material and finally evaporation of the cathode electrode complete the device as a textile PV composed of organic, inorganic and also their composites

Regardless of many gaps need to be bridged before large-scale application of this technology, the TSC fabrication based on second strategy may be envisaged through two routes to solve pertinent issues of efficiency and stability. The solar cell architecture in first approach is founded based on knitted or woven textile substrate, however second alternative follows a roadmap to develop a wholly PV fiber for further knitting or weaving process that may form energy-harvesting textile structures in any shape and structure.

Irrespective to fabric or fiber shaped of the photovoltaic unit, the light penetration and scattering in photo anode layer needs a waveguide layer. This basic requirement naturally mimicked by polar bear hair. The optical functions of the polar bear hair are scattering of incident light into the hair, luminescence wave shift and wave-guide properties due to total reflection. The hair has an opaque, rough-surfaced core, called the medulla, which scatters incident light. The simulated synthetic coreshell fiber can be manufactured through spinning of a core fiber or with sufficient wave-guiding properties followed by finishing with an optically active, that is, fluorescing, coating as a shell to achieve a polar bear hair' effect [21]. As described in Tributsch's original work [22], a high-energy conversion can be expected from high frequency shifts as difference between the frequencies of absorbed and emitted light. According to Tributsch et al. [22] for the polar bear hair, the frequency shift is of the order of 2×10^{14} Hz.

In principle, the refractive index varies over the fiber diameter and this would provide a certain wave-guiding property of the fiber. In charge of wave-guiding property most specifically With regard to manufacturing on a larger scale, fiber morphology, crystalinity, alignment, diameter and geometry can be altered for preferred optical performance as solar energy transducer. The focus on both approached of second strategy and illuminates their opportunities and challenges are main area of interest in next parts.

11.3.1 RECENT PROGRESS OF PV FIBERS

Fiber shaped organic solar cells has been subject of a few patent, project and research papers. Polymers, small molecules and their combinations were used as light-absorbing layers in previous studies. A range of synthetic substrate with various level of flexibility including optical [23], polyimide [24] and poly propylene (PP) [25] subjected to processes in which functional electrodes and light absorbing layer continuously forms on fiber scaffold. In recent PV fiber studies, conducting polymers such as P3HT in combination to small molecules nanostructure materials such as branched fullerene plays main role as photoactive material. For instance, Ref. [23] introduces a light-absorbing layer on optical fiber composed of poly (3-hexylthiophene) (P3HT): phenyl-C61-butyric acid methyl ester (P3HT:PCBM) (Fig. 6). While the light was traveling through the optical fiber and generating hole–electron pairs, the 100-nm top metal electrode (which does not let the light transmit from outside) was used to collect the electrons [23].

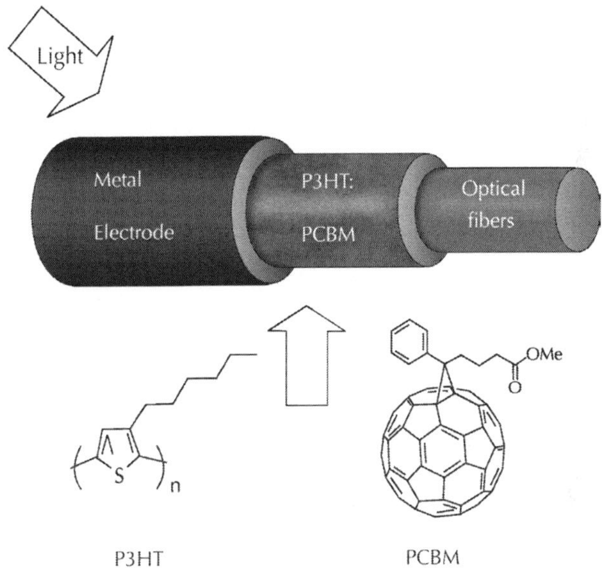

FIGURE 6 Simplified pattern of hetrojunction photoactive layer in fiber solar cell.

One of the important challenges of flexible solar cell concentrated in hole collecting electrode, which is most widely used ITO as a transparent conducting material. However, the inclusion of ITO layer in flexible solar cell could not be applicable. The restrictions are mostly due to the low availability and expense of indium, employment of expensive vacuum deposition techniques and providing high temperatures to guarantee highly conductive transparent layers. Accordingly, there are some ITO-free alternative approaches, such as using carbon nanotube (CNT) layers or different kinds of poly(3,4-ethylenedioxythiophene):poly(stirene sulfonate) (PEDOT:PSS) and its mixtures [26–28], or using a metallic layer [29] to perform as a hole-collecting electrode (Scheme 3).

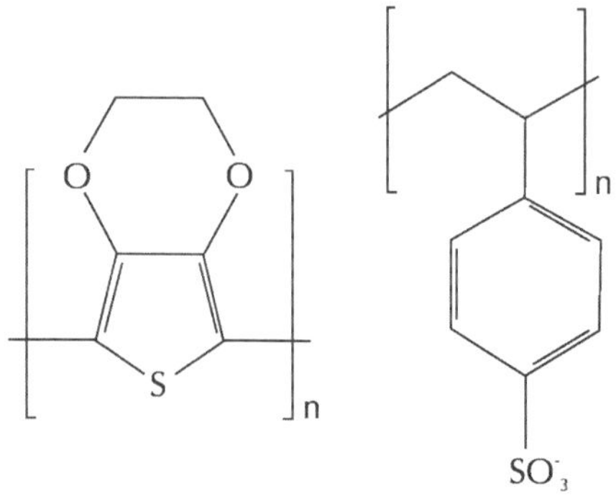

SCHEME 3 The chemical structure of poly (3,4-ethylenedioxythiophene):poly(stirenes ulfonate).

The ITO free hole-collecting layer was realized using highly conductive solution of PEDOT:PSS as a polymer anode that is more convenient for textile substrates in terms of flexibility, material cost, and fabrication processes compared with ITO material. Based on procedure described in Ref. [25] a sophisticated and simple design was presented to show how thin and flexible could be a solar cell panel (Fig. 7).

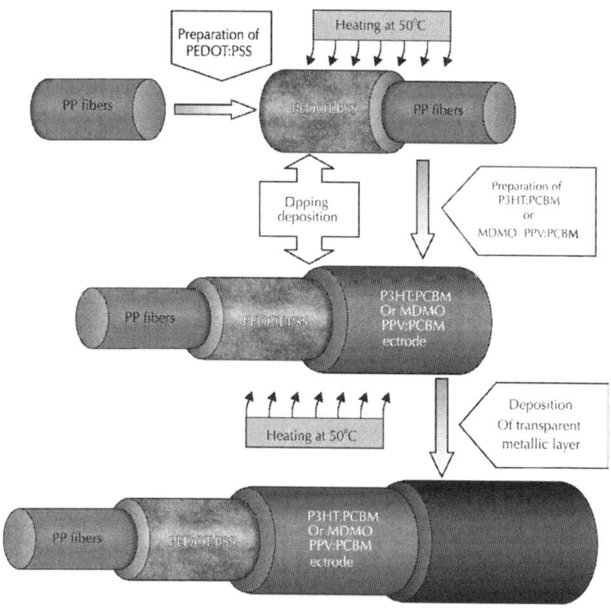

FIGURE 7 The Schematic of procedure for preparation of PV fiber.

TABLE 1 Photoelectrical characteristics of photovoltaic fibers having different having different photoactive layers (P3HT: PCBM and MDMO-PPV:PCBM) [25].

Solar cell pattern	Voc (mV)	Isc (mA/cm^2)	FF (%)	η (%)
PP|PEDOT:PSS| P3HT:PCBM|LiF/Al	360	0.11	24.5	0.010
PP|PEDOT:PSS| MDMO-PPV:PCBM| LiF/Al	300	0.27	26	0.021
ITO|PEDOT:PSS| MDMO-PPV:PCBM| LiF/Al	740	4.56	43.4	1.46

Based on implemented pattern, the sunbeams entered into the photoactive layer with 4–10 mm^2 active area by passing through a 10 nm of lithium fluoride/aluminum (LiF/Al) layer as semitransparent cathode outer electrode.

Table 1 gives the current density data versus voltage characteristics of the photovoltaic fibers consisting of P3HT: PCBM and MDMO-PPV: PCBM blends. Based on given results of open-circuit voltage, short-circuit current density and fill factor for two types of photoactive material, and also Eqs. (1) and (2) in Section 11.2.1, the power conversion efficiency of the MDMO-PPV: PCBM based photovoltaic fiber was higher than the P3HT: PCBM based photovoltaic fiber.

Comparing the solar cell characteristics of second and third pattern shows greater performance of ITO|PEDOT:PSS| MDMO-PPV:PCBM| LiF/Al rigid organic solar cell compared to PP|PEDOT:PSS| MDMO-PPV:PCBM| LiF/Al fiber solar cell. Since a same cathode, anode and photoactive material used in both pattern, the higher power conversion efficiency of ITO solar cell can be attributed to different wave-guide property and transparency of cell pattern in sun's ray entrance angle (i.e., LiF/AL versus ITO glass).

Since using ITO is strictly restricted for PV fiber, enhancing the power conversion efficiency of needs to improving existing materials and techniques. In particular, the optical band gap of the polymers used as the active layer in organic solar cells is very important. Generally, the best bulk heterojunction devices based on widely studied P3HT: PCBM materials are active for wavelengths between 350 and 650 nm. Polymers with narrow band gaps can absorb more light at longer wavelengths, such as infrared or near-infra-red, and consequently enhance the device efficiency. Low band gap polymers (<1.8 eV) can be an alternative for better power efficiency in the future, if they are sufficiently flexible and efficient for textile applications [30, 31].The variety of factors influence on polymer band gaps which can be categorized as intrachain charge transfer, substituent effect, π-conjugation length.

Systematically the fused ring low band gap copolymer composes of a low energy level electron acceptor unit coupled with a high energy level electron donor unit. The band gap of the donor/acceptor copolymer is determined by the HOMO of the donor and LUMO of the acceptor, and therefore a high energy level of the HOMO of the donor and a low energy level of the LUMO of the acceptor results in a low band gap [32].

The substituent on the donor and acceptor units can affect the band gap. The energy level of the HOMO of the donor can be enhanced by attaching electron-donating groups (EDG), such as thiophene and pyrrole. Similarly, the energy level of the LUMO of the acceptor is lowered, when electron-withdrawing groups (EWG), such as nitrile, thiadiazole and pyrazine, are attached. This will result in improved donor and acceptor units, and hence, the band gap of the polymer is decreased [33].

11.3.2 PV INTEGRATION IN TEXTILE

Having a complete functional textile solar cell motivates the researchers to attempt an approach for direct incorporation of photovoltaic cell elements onto the textile. The textile substrates inherently scatter most part of the incident light outward. Therefore, it was found necessary to apply a layer of the very flexible polymer PE onto the textile substrate to have a surface compatible with a layered device. The textile-PE substrate was plasma treated before application of the transparent PEDOT electrode in order to obtain good adhesion of the PEDOT layer to the PE carrier. Then screen-printing was employed for the application of the active polymer poly[2-methoxy-5-(2'-ethylhexyloxy)-p-phenylene vinylene] (MEH-PPV) [33].

The traditional solar cell geometry was reinvented in fractal forms that allow the building of structured modules by sewing the 25–40 cm cells realized. Figure 8 shows an step-by-step approach for fabrication textile solar cell pattern based on polymer photo absorbing layer.

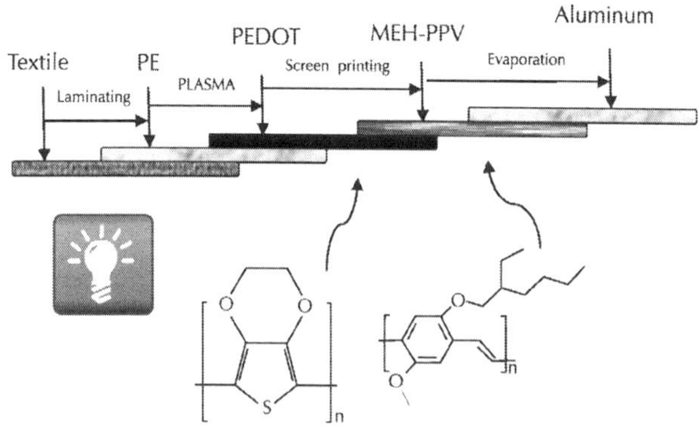

FIGURE 8 A typical fabrication procedure and key elements of textile solar cell.

The pattern designed was particularly challenging for application in solar cells and reduced the active area to 190 cm^2 (19% of the real area). The best module output power was found to be 0.27 mW with a Isc = 3:8 μA, Voc = 275 mV and a FF% of 25.7%. The pattern designed allows connections in different site of the cloth cell with reproducible performances within 5–10%.

11.4 NANOFIBERS AS A POTENTIAL KEY ELEMENT IN TEXTILE SOLAR CELLS

Previous sections present variety of solar cell structure and their corresponding elements and power conversion performance to indicate opportunities and challenges of producing of solar energy harvesting module based on a wholly flexible textile based photovoltaic unit. Current state of Textile Solar Cells is extremely far from commercial inorganic hetrojunction solar cells that showing around 45% conversion efficiency. Current section addresses promizing potential of nanofiber 1D morphology to be used as solar cell elements. Of particular, enhancement of photovoltaic unit demanding properties is a great of importance. Two different strategies can be presumed including integration of functional photoanode, photo cathode, scattering layer, photoactive or acceptor—donor materials in the form of nanofiber on to textile substrate or developing fully integrated multilayer nonwoven solar cloth.

11.4.1 ELECTROSPUN NANOFIBER

Fibers with a diameter of around 100 nm are generally classified as nanofibers. What makes nanofibers of great interest is their extremely small size. Nanofibers compared to conventional fibers, with higher surface area to volume ratios and smaller pore size, offer an opportunity for use in a wide variety of applications. To date, the most successful method of producing nanofibers is through the process of electrospinning. The electrospinning process uses high voltage to create an electric field between a droplet of polymer solution at the tip of a needle and a collector plate. When the electrostatic force overcomes the surface tension of the drop, a charged, continuous jet of polymer solution is ejected. As the solution moves away from the needle and toward the collector, the solvent evaporates and jet rapidly thins and dries. On the surface of the collector, a nonwoven web of randomly oriented solid nanofibers is deposited. Material properties such as melting temperature and glass transition temperature as well as structural characteristics of nanofiber webs such as fiber diameter distribution, pore size distribution and fiber orientation distribution determine the physical and mechanical properties of the webs. The surface of electrospun fibers is important when considering end-use applications. For example, the ability to introduce porous surface features of a known size is required if nanoparticles need to be deposited on the surface of the fiber.

The conventional setup for producing a nonwoven layer can be manipulated to fabricate diverse profile and morphology including oriented [34], Core shell [35] and hollow [36] nanofiber. Figure 9 shows the latest nanofiber profiles and its corresponding electrospinning production instrument. The variety and propagation of nanofiber products opens new horizon for development of functional profile respect to demanding application. Amongst developed techniques, coaxial electrospinning forms coreshell and/or hollow nanofiber through combination of different materials in the core or shell side, novel properties and functionalities for nanoscale devices can be found.

Increasing demands for the manufacturing of bi-component structures, in which one is surrounded by the other or the particles of one are encapsulated in the matrix of the other, at the micro or nano level, show potential for a wide range of uses.

Application includes minimizing chances of decomposition of an unstable material, control releasing a substance to a particular receptor and improving mechanical properties of a core polymer by its reinforcing with another material. The electro-spinneret consists of concentric inner and outer syringe by witch two fluids are introduced to the spinneret, one in the core of the inner syringe and the other in the space between in the inner syringe and outer syringe.

(a)

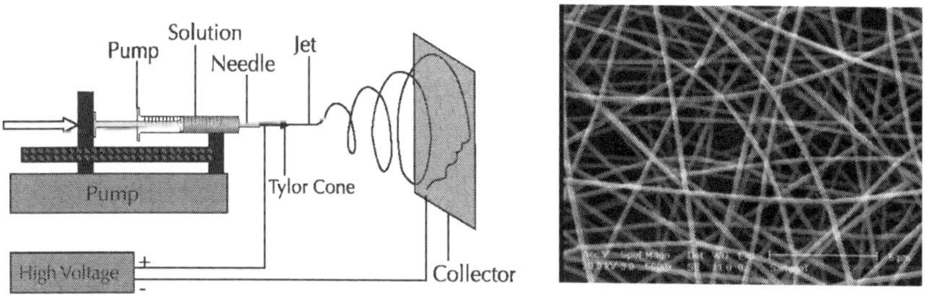

FIGURE 9 *(Continued)*

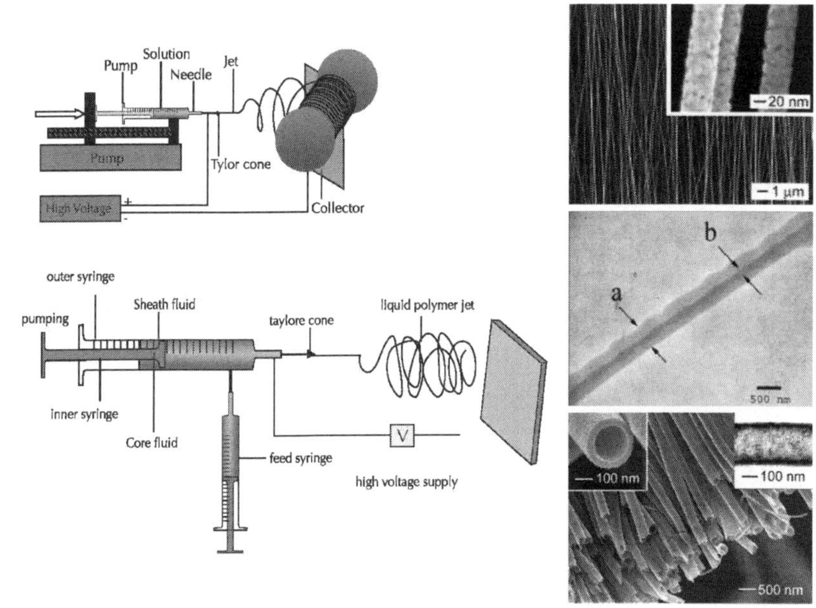

FIGURE 9 Electrospinning setup and its corresponding nanofiber profile (a) conventional nanofiber] (b) Oriented nanofiber [34] (c) Core shell nanofiber [35], (d) Hollow nanofiber [36].

The droplet of the sheath solution elongates and stretches due to the repulsing between charges and form a conical shape. When the applied voltage increases, the charge accumulation reaches a certain value so a thin jet extends from the cone. The stresses are generated in the sheath solution cause to the core liquid to deform into the conical shape and a compound coaxial jet develops at the tip of the cones (Fig. 10).

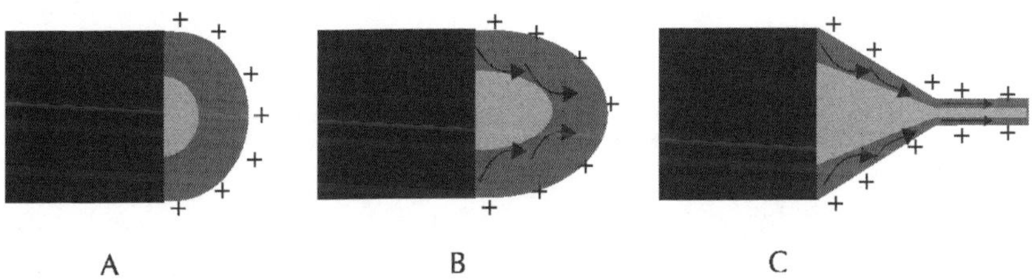

FIGURE 10 Schematic illustration of compound Taylor cone formation (a) Surface charges on the sheath solution, (b) viscous drag exerted on the core by the deformed sheath droplet, © Sheath-core compound Taylor cone formed due to continuous viscous drag).

11.4.2 NANOFIBERS IN DSSC SOLAR CELLS

It can be presumed that the electrospun nanofiber offers high specific surface areas (ranging from hundreds to thousands of square meters per gram) and bigger pore sizes than nanoparticle or film. Meanwhile, referring to Section 2.3, the particle-based titanium dioxide layers have low efficiencies due to the high density of grain boundaries, which exist between nanoparticles. The 1D morphology of metal oxide fibers attracts more interest because of the lower density of grain boundaries compared to those of sintered nanoparticles [37].

11.4.2.1 ELECTROSPUN SCATTERING LAYER

Reference [38] compared the effect of TiO_2 nanofiber and nanoparticle as scattering layer and indicated the significant enhancement of all photovoltaic specifications. In another attempt ZNO nanofiber used instead of TiO2 nanofiber to form photo anode [39].

TABLE 2 The effect of TiO_2 nanofiber on cell performance for DSSC solar cell [38–40].

Solar cell pattern	Voc (mV)	Isc (mA/cm^2)	FF (%)	η (%)	Ref.
FTO|TiO_2 nanoparticle|N719|LiI/I2/TBT|Pt|FTO	630	11.3	54	3.85	[38]
FTO|TiO_2 nanofiber|N719|LiI/I2/TBT|Pt|FTO	660	14.3	53	4.9	[38]
FTO|ZNO nanofiber|N719|LiI/I2/TBT|Pt|FTO	690	2.87	44	0.88	[39]
FTO|TiO_2 nanofiber: Ag nano P|N719|LiI/I2/TBT|Pt|FTO	800	7.57	55	3.3	[40]

However, the measured photocurrent density–voltage shows poorer results compared to TiO_2. The influence of Ag nanoparticle was also studied showed nearly same fill factor but lower power conversion efficiency compared to neat TiO_2 nanofiber scattering layer [40]. Recently, a composite anatase TiO_2 nanofibers/nanoparticle electrode was fabricated through electrospinning [41]. This method avoided the mechanical grinding process, and offered a higher surface area, so conversion efficiencies of 8.14% and 10.3% for areas of 0.25 and 0.052 cm^2, respectively, were reported Hybrid TiO_2 nanofibers with moderate multi walled carbon nanotubes (MWCNTs) content also can prolong electron recombination lifetimes [41]. Since MWCNTs can quickly transport charges generated during photocatalysis, the opportunity for charge recombination is reduced. Furthermore, MWCNTs decrease the agglomeration of TiO_2 nanoparticles and increase the surface area of TiO_2. These advantages make this hybrid electrode a promizing candidate for DSSCs.

11.4.2.2 NANOFIBER ENCAPSULATED ELECTROLYTE

Nanofiber can be considered as promizing candidate for preparation of solid or semi solid electrolyte. This is mostly because of the inherent long-term instability of electrolyte used in DSSCs usually consists of triiodide/iodide redox coupled in organic solvents [42]. Many solid or semisolid viscous electrolytes with low level of penetration to TiO_2 layer such as ionic liquids [43], and gel electrolytes [44] used to triumph over these problems. However, nanofiber with may increase the penetration of viscous polymer gel electrolytes through large and controllable pore sizes.

A few research conducted to fabricate the Electrospun PVDF-HFP membrane by electrospinning process from a solution of poly(vinylidenefluoride-cohexafluoropropylene) in a mixture of acetone/N, N-dimethylacetamide to encapsulate electrolyte solution [45, 46]. Although the solar energy-to electricity conversion efficiency of the quasi-solid-state solar cells with the electrospun PVDF-HFP membrane was slightly lower than the value obtained from the conventional liquid electrolyte solar cells, this cell exhibited better long-term durability because of the prevention of electrolyte solution leakage.

11.4.2.3 FLEXIBLE NANOFIBER AS COUNTER ELECTRODE

Efficient charge transfer from a counter electrode to an electrolyte is a key process during the operation of dye-sensitized solar cells. One of the greatest flexible counter electrode could be polyaniline (PAni) nanofibers on graphitized polyimide (GPi) carbon films for use in a tri-iodide reduction. These results are due to the high electrocatalytic activity of the PAni nanofibers and the high conductivity of the flexible GPi film. In combination with a dye-sensitized TiO_2 photoelectrode and electrolyte, the photovoltaic device with the PAni counter electrode shown in Table 3 energy conversion efficiency of 6.85%. Short-term stability tests indicate that the photovoltaic device with the PAni counter electrode approximately preserves its initial performance [47].

TABLE 3 Current–voltage characteristics of the dye-synthesized solar cells with various electrodes [47].

Solar cell pattern	Voc (mV)	Isc (mA/cm^2)	FF (%)	η (%)
FTO|TiO_2 nanoparticle|N719|PMII/I2/TBP |Pt|FTO	820	12.61	62.3	6.44
FTO|TiO_2 nanoparticle|N719|PMII/I2/TBP |PAni|FTO	831	12.22	62.1	6.31
FTO|TiO_2 nanoparticle|N719|PMII/I2/TBP |PAni	856	11.59	58.7	5.82
FTO|TiO_2 nanoparticle|N719|PMII/I2/TBP |PAni|GPi	901	9.68	28.3	2.49

The major concern for the application of alternative counter electrodes to conventional platinized TCOs in DSSCs is long-term stability. Many publications indicate that during prolonged exposure in corrosive electrolyte, catalysts will detach from the substrate and deposit onto the surface of the semiconductor photoelectrode.

11.4.2.4 A PROPOSED MODEL FOR DSSC TEXTILE SOLAR CELL USING NANOFIBERS

Choosing proper material and structure for DSSC textile solar cell using previously mentioned nanofiber propose potential candidates for designing of an integrated photovoltaic unit. As can be seen in Fig. 11 a multilayer textile DSSC solar cell composed of a complicated pattern while nanofiber is dominant in step-by-step fabrication process.

The major concern regarding the DSSC textile solar is TiO_2 nanofiber that needs to subjected to high temperature for being scattering layer. This strategy is not compatible with other textile element and it is believed that the usage of Anatase TiO_2 spinning solution provide the possibility to avoid high temperature treatment. The proposed strategy is under intensive investigation in our laboratory and future results probably mostly illuminate the opportunities and challenges.

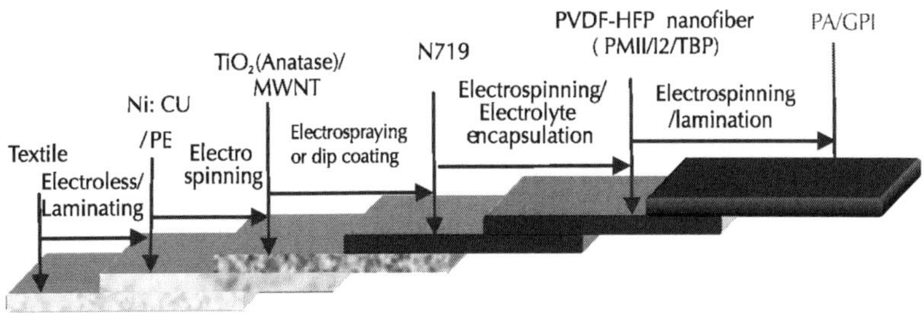

FIGURE 11 A proposed model for DSSC textile solar cell using nanofibers in successive layer.

11.4.3 ELECTROSPUN NONWOVEN ORGANIC SOLAR CELL

The idea of generating nonwoven photovoltaic (PV) cloths using organic conducting polymers by electrospinning is quite new and has not been intensively investigated. Based on previously reported PV fiber composed of conjugated hole conducting polymer (see Section 11.3.3), a novel methodology was reported to generate a nonwoven organic solar cloth. The fabrication of core–shell nanofibers has been achieved by coelectrospinning of two components such as poly(3-hexyl thiophene) (P3HT) (a conducting polymer) or P3HT/PCBM as the core and poly(vinyl pyrrolidone) (PVP) as the shell using a coaxial electrospinning set up [see section 4.1].[48]

Initial measurements of the current density vs. voltage of the P3HT/PCBM solar cloth were carried out and showed current density (Isc), open circuit voltage (Voc) and fill factor (FF) of the fiber cloth around 3.2×10^{-6} mA/cm^2, –0.12 V and 22.1%, respectively. In addition, a six order of magnitude lower photo conversion efficiency of the fiber cloth around 8.7×10^{-8} was observed that might sound disappointing. The low photovoltaic (PV) parameters of the fiber cloth could be attributed to the following factors:

(a) The fiber cloth processing steps including electrospinning as well as ethanol washing were carried out under ambient conditions.
(b) The thickness of the fiber cloth was ~5 μm compared to the diffusion length of the charge carriers in organic solar cells, which is only several nm.

Therefore, most of the charge carriers were lost in the fiber matrix itself. Drop casted films of the same thickness also showed similar PV parameters.

11.4.4 FUTURE PROSPECTS OF ORGANIC NANOFIBER TEXTILE SOLAR

Nanofiber revolutionizes the future trend of material selection to enhance the characteristics of textile solar cell. Latest experience regarding to polymer solar cell and low band gap material needs to be considered for ambitious plans with nanofiber morphology. Figure 12 shows a schematic for layer-by-layer hetrojunction textile solar cell according to previously mentioned concerns and promizing potential solar absorbing and photoactive material. The usage of anti reflective and protective materials is extremely crucial for industrial scale production. A protective layer will save the organic material from moisture and oxygen. The antireflective layer in solar cells can obstruct the reflection of light and also contribute to the device performance. This is an important point that should be overcome in the case of large-scale production of textile solar cell. The PEDOT:PSS combination with CNTs forms first P–N junction in the form of bi-layer nanofiber. The second layer composed of core shell or general bi-layer nanofiber of MDMO/PPV:PCBM, which absorb visible spectrum. The proposed plan can be realized in large-scale production through a needles electrospinning setup. The continuity of process also is not beyond expectation and a range of materials as nanofiber has been already provided on given substrates.

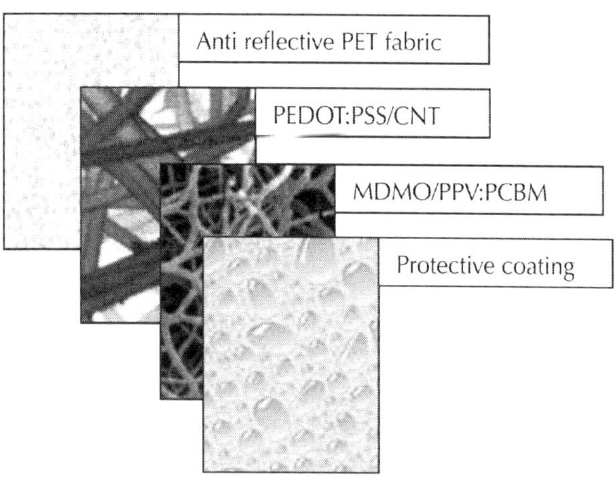

FIGURE 12 A general scheme of hetrojunction Organic nanofiber textile solar.

11.5 CONCLUSION

The multilayer solar cell energy conversion unit although obey simple theory of light scattering, absorption, electron excitation, charge transfer and its compensation but each layer for specific prescribed functions needs to be intensively investigated for being applicable in diverse circumstances. The commercial solar cell products including silicon either homo or heterojunctions, dye synthesized solar cell and organic solar cell subjected to demanding research and reached to high level of maturity. The huge experience in silicon solar cell and other photovoltaic system should be reviewed for developing new generation of flexible solar cell. Therefore, current work in first part has a quick glance on variety of solar cells including inorganic, organic and hybrid structures. In overall, regardless of type of solar cell and its corresponding elements, following points needs to be addressed to find fascinating performance:

- The wider absorption wavebands in a range of spectrum;
- The thinner the solar cell;
- The lowering the band gap;
- The higher surface area per unit mass;
- The lowering the cathode thicknesses;
- Using antireflective coating;
- Using protective coating;
- The multijunction cell to cover range of spectrum;
- Using semitransparent wave guide material;
- The lifetime enhancement.

Incorporation of organic solar cells into textiles has been realized reaching encouraging performances. Stability issues need to be solved before future commercialization can be envisaged. The mechanical stability of the devices was not limiting the function of the devices prepared. It would seem that low power conversion efficiency much more pertinent than the mechanical stability on the timescale of commercial.

KEYWORDS

- **antireflective coating**
- **photovoltaic system**
- **protective coating**
- **silicon solar cell**

REFERENCES

1. Schubert, M. B.; In: Conf. Rec. 31st IEEE Photovolt. Specialists Conf.; IEEE, New York, **2005,** 1488.
2. Nelson, J.; The Physics Of Solar Cells: Imperial College Press, **2003,** 1–16.
3. Miles, R. W.; Forbes. H.; I.; Photovoltaic solar cells: An overview of state-of-the-art cell development and environmental issues. Progress in Crystal Growth and Characterization of Materials, **2005,** *51,* 1–42.
4. Günes, S.; Beugebauer, H.; and Sariciftci, N. S.; Conjugated Polymer-based Organic Solar Cells, *Chem. Rev.;* **2007,** *107,* 1324–1338.
5. Castañer, T. M.; Solar Cells. In: Castañer, T. M. L. (Ed.), Practical Handbook of Photovoltaics: Fundamentals and Applications: Elsevier, **2003,** 71–95.
6. Thompson, B. C.; Polymer–Fullerene Composite Solar Cells. Angew. *Chem. Int. Ed.* **2008,** *47,* 58–77.
7. Shrotriya, V.; Yao, Y, Moriarty, T, Emery, K.; Yang.Y.; Accurate Measurement and Characterization of Organic Solar Cells. *Adv. Funct. Mater.* **2006,** *16,* 2016–2023.
8. Krebs, F. C. In F. C.; Krebs (Ed.), Polymer Photovoltaics A Practical Approach: SPIE, 1–10. **2008,**
9. Cai, W.; Cao, Y.; Polymersolarcells: Recent development and possible routes for improvement in the performance. *Solar Energy Materials and Solar Cells* **2010,** *94,* 114–127.
10. Dutta, P.; Eom, S. H.; Lee, S. H.; Synthesis and characterization of triphenylamine flanked thiazole-based small molecules for high performance solution processed organic solar cells. Organic Electronics, **2012,** *13(2),* 273–282.
11. Soa, S.; Koa, H. M.; Kima, C.; Paeka, S.; Choa, N.; Songb, K.; Leec, J. K.; Koa, J.; Novel unsymmetrical push–pull squaraine chromophores for solution processed small molecule bulk heterojunction solar cells. *Solar Energy Materials and Solar Cells,* **2012,** *98,* 224–232.
12. Lina, Y.; Liua, Y.; Shia, Q.; Hua, W.; Y.; Lia, Zhan, X.; Small molecules based on bithiazole for solution-processed organic solar cells. Organic Electronics, **2012,** *13(4),* 673–680.
13. Gratzel, M.; *Nature* **2001,** *414,* 338–344.
14. Ginger, D. S.; N. C. G. **2004,** Electrical Properties of Semiconductor Nanocrystals. In V. I.; Klimov (Ed.), Semiconductor and Metal Nanocrystals (pp. 236–285): Marcel Dekker, Inc. Ferrazza, F. **2002,** Large size multicrystalline silicon ingots. Proc. E-MRS **2001,** Spring Meeting, Symposium E on Crystalline Silicon Solar Cells. *Sol. Energy Mater. Sol. Cells, 72,* 77–81.
15. Kisserwan, H.; Enhancement of photovoltaic performance of a novel dye, "T18," with ketene thioacetal groups as electron donors for high efficiency dye-sensitized solar cells. *Inorganica Chimica Acta,* **2010,** *363,* 2409–2415.
16. Wei, D. **2010,** Dye Sensitized Solar Cells, *Int. J. Mol. Sci. 11,*1103–1113.
17. Günes, S.; Beugebauer, H.; and Sariciftci, N. S.; Conjugated Polymer-based Organic Solar Cells, *Chem. Rev.; 107,* 1324–1338 **2007,**
18. Brabec, C. J.; Dyakonov, V.; Parisi, J.; and Sariciftci, N. S.; "Organic Photovoltaics Concepts and Realization," 1st Ed.; Springer: New York, **2003.**

19. Berson, S.; de Bettignies, R.; Bailly, S.; and Guillerez S.; Poly(3-hexylthiophene) Fibers for Photovoltaic Applications, *Adv. Funct. Mater.* **2007,** *17,* 1377–1384.
20. Gonzalez, R.; and Pinto, N. J.; Electrospun poly(3-hexylthiophene-2,5-diyl) Fiber Field Effect Transistor, *Synthetic Metals*, **2005,** *151*, 275–278.
21. Bahners, T.; Schlosser, U.; Gutmann, R.; Schollmeyer, E.; Textile solar light collectors based on models for polar bear hair, *Solar Energy Materials and Solar Cells* **2008,** *92,* 1661–1667.
22. Tributsch, H.; Goslowski, H.; Ku, U.; Wetzel, H.; Light collection and solar sensing through the polar bear pelt, *Sol. Energy Mater.* **1990,** *21,* 219–236.
23. Liu, J.; Namboothiry, M. A. G.; Carroll, D. L.; Fiber based Architectures for Organic Photovoltaics, *Appl. Phys. Lett.* **2007,** *90,* 063501.
24. O'Connor, B.; Pipe, K. P.; Shtein, M.; Fiber Based Organic Photovoltaic Devices, *Appl. Phys. Lett.* **2008,** *92,* 193306.
25. Bedeloglu, A.; Demir, A.; Bozkurt, Y.; Sariciftci, N. S.; A Photovoltaic Fiber Design for Smart Textiles, *Textile Res. J.* **2007,** *80(11),* 1065–1074.
26. Ouyang, J.; Chu, C. W.; Chen, F. -C.; Xu, Q.; Yang, Y.; High-conductivity Poly 3, 4-ethylenedioxythiophene): Poly(stirene sulfonate) Film and its Application in Polymer Optoelectronic Devices, *Adv. Funct. Mater.* **2005,** *15,* 203–208.
27. Kushto, G. P.; Kim, W.; Kafafi, Z. H.; Flexible Organic Photovoltaics using Conducting Polymer Electrodes, *Appl. Phys. Lett.* **2005,** *86,* 093502.
28. Huang, J.; Wang, X.; Kim, Y.; deMello, A. J.; Bradley, D. D. C.; deMello, J. C.; High Efficiency Flexible ITO-free Polymer/fullerene Photodiodes, *Phys. Chem. Chem. Phys.* **2006,** *8,* 3904–3908
29. Tvingstedt, K.; Inganäs, O.; Electrode Grids for ITO-free Organic Photovoltaic Devices, *Adv. Mater.* **2007,** *19,* 2893–2897.
30. Perzon, E.; Wang, X.; Admassie, S.; Inganäs, O.; Andersson, M. R.; An Alternating Low Band-gap Polyfluorene for Optoelectronic Devices, *Polymer*, **2006,** *47,* 4261–4268.
31. Campos, L. M.; Tontcheva, A.; Günes, S.; Sonmez, G.; Neugebauer, H.; Sariciftci, N. S.; Wudl, F.; Extended Photocurrent Spectrum of a Low Band Gap Polymer in a Bulk eterojunction Solar Cell, *Chem. Mater.* **2005,** *17,* 4031–4033.
32. Shaheen, S. E.; Brabec, C. J.; Sariciftci, N.; Padinger, F.; Fromherz, T, Hummelen, J. C.; *Appl. Phys. Lett.* **2001,** *78,* 841–843.
33. Scharber, M. S.; Mühlbacher, D.; Koppe, M, Denk, P, C.; Waldauf, Heeger, A. J.; Brabec, C. J.; *Adv. Mater.* **2006,** *18,* 789–794.
33. Krebs, F. C.; Biancardo, M.; Jensen, B. W.; Spanggard, H.; Alstrup, J, Strategies for incorporation of polymer photovoltaics into garments and textiles, *Solar Energy Materials and Solar Cells*, **2006,** *90,* 1058–1067.
34. Li, D.; Wang, Y.; Y.; Xia, Electrospinning Nanofibers as Uniaxially Aligned Arrays and Layer-by-Layer Stacked Films, *Adv. Mat.*, **2004,** *16(14)*, 361–366.
35. Yu, J. H.; Fridrikh, S. V.; Rutledge, G. C.; Production of Submicrometer Diameter Fibers by Two-Fluid Electrospinning, *Adv. Mat.*, **2004,** *16(17)*, 1562–1566.
36. Li, D.; Xia, Y.; Direct Fabrication of Composite and Ceramic Hollow Nanofibers by Electrospinning, *Nano Lett.*, **2004,** *4(5)*, 933–938.
37. Chuangchote, S.; Sagawa, T.; Yoshikawa, S.; *Appl. Phys. Lett.* **2008,** *93*, 033310.

38. Zhao, X.; Lin, H.; Li, X.; Li, J.; The application of freestanding titanate nanofiber paper for scattering layers in dye-sensitized solar cells, *Materials Lett.* **2011,** *65*, 1157–1160.
39. Li, S.; Zhang, X, Jiao, X.; Lin, H.; One-step large-scale synthesis of porous ZnO nanofibers and their application in dye-sensitized solar cells, *Materials Lett.* **2011,** *65*, 2975–2978.
40. Li, J.; Chen, X, Ai, N.; Hao, J.; Chen, Q.; Strauf, S.; Shi, Y, Silver nanoparticle doped TiO2 nanofiber dye sensitized solar cells, *Chem. Phys. Lett.* **2011,** *514*, 141–145.
41. Hu, G. J.; Meng, X. F.; Feng, X. Y.; Ding, Y. F.; Zhang, S. M.; Yang, M. S.; *J. Mat. Sci.* **2007,** *42*, 7162–7170.
42. Kubo, W.; Kitamura, T.; Hanabusa, K.; Wada, Y.; Yanagida, S.; Chemical Communication. **2002,** 374–375.
43. Wang, P.; Zakeeruddin, S. M.; Comte, P.; Exnar, I.; Gratzel, M.; *J. Am. Chem. Soc.*, **2003,** *125*, 1166–1167.
44. Wang, P.; Zakeeruddin, S. M.; Moser, J. E.; Nazeeruddin, M. E.; Sekiguchi, T.; Gratzel, M.; Natural Material. **2003,** *2,* 402–407.
45. Park, S. H.; Kim, J. U.; Lee, S. Y.; Lee, W. K.; Lee, J. K.; Kim, M. R.; *J. Nanosci. Nanotech.* **2008,** *8*, 4889–4894.
46. Kim, J. U.; Park, S. H.; Choi, H. J.; Lee, W. K.; Lee, J. K.; Kim, M. R.; Sol. Energy Mater. Sol. Cells **2009,** *93*, 803–807.
47. Chen, J.; Lia, B.; Zheng, J.; Zhao, J.; Jing, Zhu, Z.; Polyaniline nanofiber/carbon film as flexible counter electrodes in platinum-free dye-sensitized solar cells, Electrochimica Acta **2011,** *56,* 4624–4630.
48. Sundarrajan, S.; Murugan, R.; Nair, A. S.; Ramakrishna, S.; Fabrication of P3HT/PCBM solar cloth by electrospinning technique, *Materials Lett.* **2010,** *64,* 2369–2372.

CHAPTER 12

NEW GENERATION OF NANOMOLECULAR STRUCTURES

A. L. IORDANSKII, S. Z. ROGOVINA, I. AFANASOV, and A. A. BERLIN

CONTENTS

12.1 Introduction 414
12.2 Modified Cardiavascular Stents as Implants Providing Drug Delivery Function 417
12.3 Modern Orthopedic Implants and Their Further Development 419
12.4 Modern Genitourinary Implants 425
12.5 Nanoimplants with Encapsulated Drugs for Ophthalmology 428
12.6 Prospects of Neurological Implants 431
12.7 Miniature Electromechanical Implants for Programmed Delivery of Biologically Active Substances 432
12.8 Conclusions 434
Keywords 435
References 435

12.1 INTRODUCTION

Challenges and development perspectives on nanopatterned implants loaded with drugs intended to replace or improve human organs and tissues are analyzed. An innovative approach to polymer and composite design combines the surface modification on the molecular and nanosized levels, formation of the implant matrix as hybrid nanocomposite, and drug encapsulation aimed at ensuring their targeted and programed delivery. The economic and scientometric situations in the world development of composite and hybrid implant systems are briefly described. The major fields in nanoimplantology for the nearest decade are represented, including cardiology, ophthalmology, genitourology and orthopedy. The prospects for biodegradable polymers, such as poly(α-hydroxy acids) (PLA PGA, PLGA) and poly(β-hydroxyalkanoates) (PHB, PHVB), are considered, as well as nanoscale biochips and sensors, miniature electromechanical systems (MEMS and NEMS), and neurological conduits.

When designing the new generation of medical implants, it should be considered many factors and processes that guarantee biomedical efficiency, the retention of the necessary properties during the service life, patient's comfort, as well as other parameters characterizing implant behavior in biological media [1–5]. In addition to the above requirements, the implants should also sustain drug delivery [6–10] and be arranged with nanoscale structures [11, 12]. The latter requirement is determined by implant—cell interactions at the receptor and immune levels, which are most effective if composite implants, have a nanostructured surface [13, 14].

The implanted medical device (IMD) is an intracorporeal system (device or material) intended for (a) the replacement of lost organs, tissues, or cell assembly; (b) an enhance in the efficiency of the body's biological function; and (c) a patient postsurgery rehabilitation and mechanical support at arrangement in vivo. Implants differ from transplants in their artificial origin, that is, implants are manufactured under plant, pilot, or laboratory conditions. Traditionally, implants are produced from metals, inorganic compounds, and polymers. Recently, biocomposites, multilayered and hybrid constructs, and nanoscale modifications of implants have been elaborated. IMD specific features are their biocompatibility and, when necessary, controlled biodegradability. State-of-the-art implants include electronic or computer devices (pacemakers and auricular prostheses), trigger regulators, and feedback systems, preferably wireless. A separate group of implants is the combined systems for dosed and targeted drug delivery in the body. Unlike traditional drug formulation, the active substance in this case is encapsulated directly into the matrix or reservoir of an implanted device as, for example, in the combined cardiac stents with an active polymer coating. This group of implants is appropriate for therapeutic procedures, surgery (injectors), and diagnosing (in vivo electronic chips).

Note that, if both passive and active transports may be implemented in the case of the nanoparticles assisted drug delivery, the drug encapsulated in the implant enters the adjacent tissues exclusively via a combination of the hydrolytic (enzymatic) process [15] and diffusion [16]. This suggests that, in order to provide the appropriate drug flux from the implant's matrix that is necessary for therapy, the optimally high sorption capacity of implant is required, that is, the critical concentration of the functional groups in polymer for drug immobilization.

To ensure the efficient function and safety of the implants and coordination of the efforts on their design and commercial promotion, European Union has issued a set of directives (for example, [17] that strictly defines what is an IMD. Over the last three years, Russian metrology has also formulated a number of requirements to IMDs. Analogous directives have also been elaborated by the corresponding authorities in the United States [18] and Canada [19].

In 2010, the manufacture of biomedical devices and functional implants worldwide reached $272.2 billion at market price. According to the expert forecasts, by the year 2015, this value will increase by 65%, despite the economic crisis and will almost double in Eastern European countries (from $4.1 billion in 2010 to $7.8 billion in 2015). The United States, which occupies the widest sector on the market of medical products and devices (over 40%), is first in the list of manufacturers of implanted systems, followed by EU countries (Western European countries, 27%) [20].

This development of the market reflects a high demand for medical devices and implants in clinics and diagnostic centers worldwide. Only in the United States, the number of patients using various IMDs exceeds 25 million [20]. Over 2 million stents are annually implanted in cardiac surgery and urology. An even higher number of patients use temporary medical devices, such as catheters (400 million) and kidney dialyzers (25 million) [21]. First on the list of the manufactured IMDs are cardiovascular implants (stents, defibrillators, and pacemakers), followed by orthopedic implants. The next group includes neurological stimulators and signal transducers, as well as gastrointestinal bands and laparoscopy systems. Auditory prostheses are the most dynamically developing group of polymer implants with the production prospect of $27.9 billion by the year of 2016. Similar to most high technologies and taking into account the regulations developed by state control institutions, the period from laboratory research to commercial production for the nanomedical innovations is 12–15 years. The first successful results in the field of nanomedicine were obtained in the early 2000 s; correspondingly, a considerable increase in the number of nanoscale drugs and devices can be expected as soon as the end of the current decade. By the end of 2010, FDA approved 22 ready-to-use nanopharmaceutical implanted systems, and the number of systems that undergo different phase of clinical trials is even higher. Concurrently, about 25 items of IMDs in Europe are in the final stage of clinical trials.

Note that the commercialization of nanotechnology products is accompanied by exponential growth in the corresponding research and development activities; moreover, as is evident from Fig. 1, the rate of increase in the number of publications (patents and research papers) in the field of nanobiomedicine exceeds that rate of the corresponding publications in the remaining fields of nanotechnology. Note also that the improvement of regulations and legislation for design and use of implants and prostheses, which took place in 1989, had a significant effect on the intensity of the corresponding research, as is reflected by the breakpoint in curve *2*, which is denoted for clarity by the corresponding symbol. According to the location within or outside the body, all the currently known implants fall into intracorporeal and intercorporeal. In addition, the devices intended for analysis, diagnosis, and drug delivery are traditionally used under in vivo, ex vivo, and in vitro conditions. Excluding all intercorporeal implants, such as ocular contact lenses, dental prostheses and fillings, periodontal membranes, transdermal systems, wound coverings, and so on from consideration, this review will consider intracorporeal miniature systems except for the traditional orthopedic devices, cartilage implants, surgical meshes, pericardium, and other macroscale medical constructs.

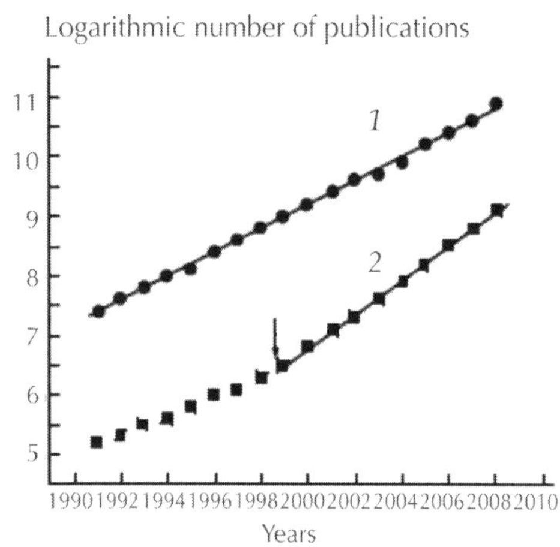

FIGURE 1 Exponential growth in the number of publications (research papers and patents) in (*1*) all nanotechnology areas and (*2*) nanobiomedicine area. Arrow denotes approval datum of the first regulation documents for implants in the United States and one year later in Europe.

12.2 MODIFIED CARDIAVASCULAR STENTS AS IMPLANTS PROVIDING DRUG DELIVERY FUNCTION

Cardiovascular diseases are first on the list of mortality and morbidity, exceeding the cancer diseases and casualties. Cardiovascular pathologies are among the basic causes of short lifespan and deteriorating demographic situation. In 2008, the World Health Organization (WHO) acknowledged Romania as the most unfavorable country, with cardiovascular diseases accounting for 52% of fatal cardiac cases. In 2005, Russian Federation headed this list in Europe, with 57% of fatal cases, which is twofold higher than the most trouble-free developed countries [22]. According to the World Health Organization (WHO), 17.3 million people died from cardiovascular related diseases in 2008, and this number is projected to rise to an estimated 23.6 million by 2030. Therefore, it is quite natural that the prevention and therapy of cardiovascular diseases is considered to be a most important priority in the field of demography, health promotion, and increase in the lifespan of population [22, 23].

The latest advance in the field of nanotechnology most pronouncedly appeared in construction of cardiovascular stents. First and foremost, this concerns the main concept in design of the stents that provide a prolonged and local drug delivery. In addition, the obligatory requirement to cardiovascular stents is effective hemocompatibility. The combination of controlled drug delivery and biocompatibility may be attained by nanostructuring the surfaces of cardiac implants, for example, constructing meso- and nanoporous layers responsible for controllable release, protein adsorption, and adhesion of blood cells (platelets, erythrocytes, leukocytes, etc.).

A typical complication when implanting a vascular stent is restenosis as a result of neointimal hyperplasia, on the one hand, and thrombosis activated by platelets, on the other hand. After the active clinical and technical search for efficient macro and micro constructs (the search for an optimal geometry of bare metal spiral stents) have been depleted, it has become clear that the classic paradigm for introduction of metal devices should be replaced with elaboration of surface modified devices able to dose drugs and provide their prolonged elution. The pioneers in modern technologies for producing next generation cardiac implants with a modified surface (drug eluting stents) are several companies, that is, Boston Scientific, Johnson & Johnson, Medtronic, Guidant and the others. Currently, designs of the newest stents with biodegradable or inert coatings provide for the targeted delivery of various cardiac drugs, such as Paclitaxl, Sirolimus, Zotrolimus, and so forth, which are responsible for the inhibition of cell proliferation and, thereby, decrease the incidence rate of restenosis compared to helical metal stents of the first generation.

However, the antiproliferation effect is most pronounced only soon after implanting (the first six months); however, as recently shown [24, 25], these constructs can initiate thrombosis because of a decrease in cell proliferation and partial degeneration of the neointima [26]. Construction of the nanoporous surface structures with aluminum oxide or titanium oxide for release of Tacrolimus

[27], carbon nanoparticles with chromium additions that form the surface layer of the implants for dispensing Paclitaxl [28], and gold [29] is aimed at solving a polyfunctional problem, namely, concurrent provision of (a) biocompatibility of implant and (b) the necessary kinetic profile of drug release. When improving biocompatibility, it is necessary to take into account the topography characteristic of the stent surface (e.g., roughness), which should simulate the natural vascular surface in a nanoscale range. Achieving this effect should lead to improved cell adhesion and stent endothelization and, as a consequence, decrease the probability of thrombosis due to the tight contact between the stent surface and endothelium [30]. The results of constructing the surface structures with nickel titanate [31] or hydroxyapatite [32], which are analogous to the specific features of natural vascular surface, suggest with a certain degree of reserve that one of the promizing directions in nanotechnology stent constructs is designing of their nanopatterned surfaces. These surfaces should concurrently provide the following two functions: targeted drug delivery and regulated cell adhesion as the factor that limits restenosis in short-terms and the factor that prevents long-term thrombosis as well as promote endotheliazation [33] see the scheme in Fig. 2. In the case of specific therapy of vascular tissues and concurrent diagnostics, it is proposed to use magnetized metals as a stent core structure [34]; these metals attract superparamagnetic nanoparticles (e.g., iron oxides) from the bloodstream, thereby allowing not only targeted drug delivery in the absence of external magnetic field, but also to contrast abnormal regions on the vascular surface [35].

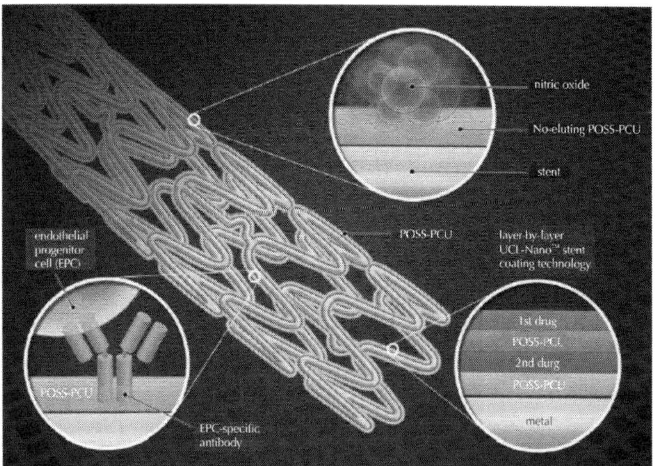

FIGURE 2 The novel generation cardiovascular stents coating with special nanocomposite polymers like POSS-PCU (polyhedral oligomeric silsesquioxane and copolymer carbonate-urea urethane: trade named UCL-NanoTM) has been developed specifically to capture oxide nitrogen and EPC specific antibodies (factor of endothelialization) simultaneously.

Another promizing direction in the development of implanted stents is a complete rejecting of metal alloys as basic coiled structures. It is proposed to use for these purposes biodegradable polymeric materials, namely, polylactide-coglycolides, PLGAs, or poly(3-hydroxybutyrate-co3-hydroxyvalerate), PHBV. Controlled biodegradation allows, first, for a prolonged drug release to local regions of the cardiovascular system and, second, for a decrease in the risk of repeated thrombosis and hyperplasia with biodegradation of stent [36]. When designing this type of implants, two important conditions should be met, that is, (1) the rate of stent biodegradation should be approximately equal to the cell proliferation rate; and (2) the biodegradation products should not be toxic or induce undesirable immune responses. The mentioned biopolymers almost completely satisfy these criteria, since their intermediate degradation products are members of the physiological Krebs cycle and the final products are molecules of CO_2 and H_2O [37].

12.3 MODERN ORTHOPEDIC IMPLANTS AND THEIR FURTHER DEVELOPMENT

The modern orthopedic implants (OIs) belong to the class of nanocomposites with rather intricate structural organization. They should not only effectively withstand the stresses, but also should be biocompatible and provide a prolonged targeted drug delivery of biologically active substances. In the general case, OI biocompatibility is determined by the cell response of bone tissues to the implanted material, which characterizes the degree of OI integration into the implanted region. The innovative approach to OI construction as nanobiocomposites is in that the material used to replace bone tissue is represented by a composite with a successive hierarchical structure of spatial elements of organic and inorganic origins. In the nanoscale range, the used structure forming elements are collagen fibers, hydroapatite, and proteoglycans, which together provide for the structural stiffness, stability, and concurrent plasticity of a natural bone.

The effective use of a bone implant is based on its biocompatibility at the cell level. Most researchers who work in this field believe that it is possible to influence the cell behavior via biochemical mediators (with involvement of biologically active substances), surface topography of implant (creation of modifying surface layer or coating), and induction of excitation signals [38]. All of these processes provide a stable cell adhesion and, consequently, form a perfect contact between nonphysiological material and bone tissue. In this situation, the implants maximally biologically mimic the natural bone material. In the absence of such correspondence, the cells in contact with the implant surface change at both the morphological and functional levels, as well as alter their behavior during adhesion, proliferation, endocytosis, and gene regulation. The OI surface is modified using various state-of-the-art technologies that form a nanoscale structure, including nanolithography, phase separation, chemical etching, molecular self-assembly, and electrospinning of micro and nanofibers.

Nanostructuring of implants creates highly developed surfaces, while the implants acquire fundamentally new mechanical, electric, optical, superparamagnetic, and other characteristics. The set of these properties will allow for the successful application of OI in targeted delivery of low molecular weight drugs, active macromolecules, and cells in the nearest future, as well as their use as diagnostic and biosensor elements. The current concepts of the interactions between the bone tissue and composite materials include adsorption of specific proteins, such as fibronectin and vitronectin, and subsequent adhesion of osteoblasts and other bone cells [39, 40]. In this case, both the composition and structure of the protein layer determine the quality and intensity of cell adhesion and, consequently, the integration ability of bone tissue which determines the differentiation, growth, and death of bone cells on the surface of implant [39, 41].

Currently, intensive studies in this field aim to clarify how the surface energy, chemical composition, charge, morphology, and nanoscale topography of the phase boundary influence the adsorption of specific proteins [42, 43]. In particular, fibronectin predominantly adsorbs on the calcium–phosphate surface of composite [44]. Then, the adhesion intensity of osteoblasts increases in the presence of the amino acid motif of the absorbed protein on the surface, such as arginine–glycine–asparagine (RGD sequence in Fig. 3), as well as when osteoprosthesis is heparinized [45]. In an ideal situation, the protein adsorption layer should stimulate adhesion of the cells that positively influence biocompatibility of the implant, for example, osteoblasts, and prevent adhesion of fibroblasts, which negatively influence the biocompatibility of implants [39]. As we already mentioned, one of the most important factors in OI biocompatibility, along with adsorption processes, is the surface topography of the implant. This particular factor has a considerable effect on the proliferation and differentiation of osteoblasts, the young cells responsible for the formation of new bone tissue [46]. It is not completely clear how the nanoscale surface topological elements act on the behavior of these cells [47]. However, there are serious reasons to believe that surface structuring in a nanoscale range modulates the cell membrane receptors, thereby modifying the communication between cells [48]. In addition, a nanoscale landscape of the implant surface influences the adsorption and conformational changes of integrin-binding proteins (see Fig. 3), which also activates the cell signaling receptors [49].

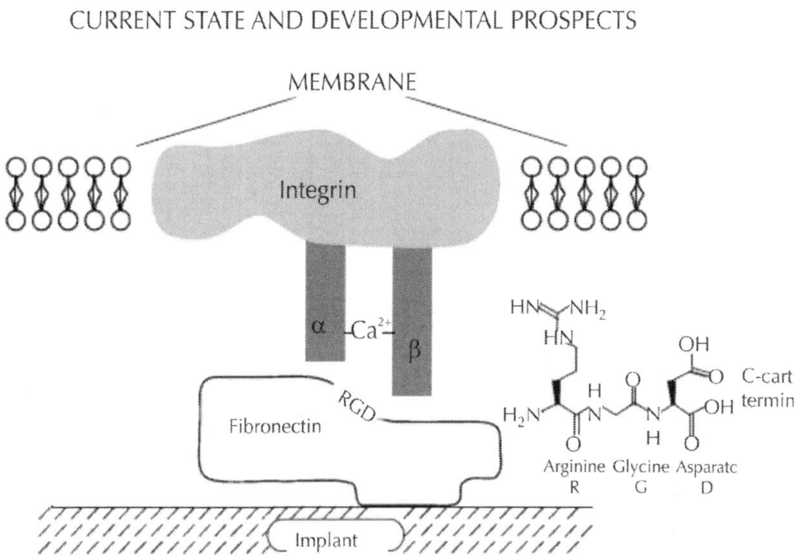

FIGURE 3 Protein adsorption on ostheoimplant surface and following interactions adsorbed proteins with cells of bone tissue.

Nanoscale implant surface topography comprises structural elements, such as pores, corners, slot-shaped hollows, nanofiber structures, nodular intertangling, and their combinations. The specific topographic features of the surface considerably influence its interaction with the cells of surrounding tissue, and this effect on cell adhesion and proliferation is frequently more significant compared to the effect of implant's chemical structure [50]. The situation is somewhat complicated by that individual cell types respond differently to the surface topography. For example, the adhesion of osteoblasts increases with the surface energy of carbon nanofibers, whereas the adhesion of fibroblasts and smooth muscle cells decreases [51]. This is why the detailed mechanism underlying the cell interactions with implant's surface elements is still vague and, in most cases, designers form the topographic landscape in a rather empirical manner.

The surface chemistry of implants plays an important role in their service. During implantation, proteins and other biomolecules adsorb on the implant's surface. The changes in native conformations of biomolecules on a non-physiological surface initiate a trigger cascade of inflammatory (immune) responses, which interfere with the integration of the implant into the living tissue or enhance the formation of fiber capsule. A certain success in the prevention of the negative processes is achieved via a controlled administration of anti-inflammatory drugs. The formation of the surface layers, for example, of gold modified with thiol groups and carrying spatially structured functional groups (–CH3, –OH, –COOH, and –NH2) able to bind integrin or regulating cell adhesion, is no less promizing [52]. Hydrophilic–hydrophobic balance of the implant surface influences the behavior of an adhered cell.

In addition to modern methods (atomic force microscopy, transmission electron microscopy, etc.), surface wettability (method of contact angle) is used as a simplified rapid assay for hydrophilicity. Amazing as it may seem, a macroscopic characteristic, such as wettability, reflects the OI structure on the nanoscale level. In particular, the surface wetting of aluminum can be increased by decreasing the particle size from 167 to 24 nm [53]. In another work [54], ultrathin titanium crystals displayed a high wettability, which stimulates osteoblasts to proliferation and adhesion and, as a consequence, accelerates the implant integration with the bone tissue. Regarding the bone tissue as a nanocomposite, the current strategy of bone engineering is based on creating a combination of inorganic compounds (calcium triphosphate, hydroxyapatite, organic silicon glass, as well as carbon tubes [55, 56] and organic polymers, such as copolymer of lactic-coglycolic acid, PLGA; poly(3-hydroxybutyrate), PHB; polycaprolactone; polyphosphazene; chitin; and chitosan. Polymeric matrices form 3D structures for the repair or regeneration of the bone tissue. An original method for OI construction consists of the incorporation of a nanoceramic-reinforcing agent that ensures an optimal surface topography and increases the strength of material [55]. The presence of mineral nanoparticles on the OI surface stabilizes its contacts with the cell tissue (integration into tissue) and strengthens the signal transduction of growing cells, thereby enhancing osteogenic activity.

The matrices of the nanofibers impregnated with drugs display the characteristics necessary for growing organs and tissues in cell engineering. Characteristic of the polymer nanofibers produced by electrospinning are high surface-to-volume ratios, large specific porosity, and adjustable size of nanopores. This set of characteristics makes the nanofiber matrices the most promizing material for growing bone and other biological tissues [57, 58]. The ultrathin fiber sizes and their branched surface create optimal conditions for immobilization of growing cells, while the system of pores enhances the metabolic processes associated with cell growth and vital activities [59]. Stem cells are of considerable interest in this connection [60]. Another nanotechnology approach to this problem consists of improving the biocompatibility of the surface. This is attained by coating the implant surface with nanofibers and increasing the flow of drug encapsulated in them [61].

Biocompatibility can be improved by introducing biologically active substances into the modified surface layers of a bone nanoimplant. Controlled release of antiseptics, peptides, extracellular proteins, osteoinductive growth factor, and osteogenic cells is used to stimulate osteogenesis (formation of new bone tissue) and the replacement of biodegradable regions of the implant with bone tissue [62, 63].

Thus, state-of-the-art bone implants should simultaneously contain nanostructured biocompatible surface and the active compounds in their surface layer, which creates an antiseptic effect, as well as controls the state and growth of bone tissue at the boundary with the implant. The adhesion and subsequent colonization of bacteria on the OI surface lead to subsequent infectious complications after implantation. Despite preliminary aseptic procedures, bacterial infections present a serious problem during post-surgery period, when IO is in the body. In particular, the rate of complications in hip arthroplasty reaches 3% and the rate of reinfection, even 14%. Antibiotics, the immobilization of antimicrobial agents within the matrix or on its surface, the addition of antiseptic metals (such as silver, copper, or titanium), and the encapsulation of nitrogen oxide are used to prevent the propagation of pathogenic microorganisms [62,64]. Pioletti et al. [65] have reviewed the latest data on the controlled drug release from OIs. A considerable segment of the innovative research and developments includes nanotechnology modifications of traditional implants, first and foremost, involving titanium and its alloys, in particular with nickel (TiNi). The specific feature of new generation titanium implants is their nanostructured surface. For this purpose, titanium nanotubes vertically oriented relative to the OI surface are currently synthesized to provide the encapsulation of the drug with its subsequent controlled release into the bone tissue. The rate of drug elution from titanium nanotubes can be regulated by the thickness and composition of biodegradable polymer layer used to coat the tube surface and represent an additional barrier for drug diffusion (see Fig. 4) [66].

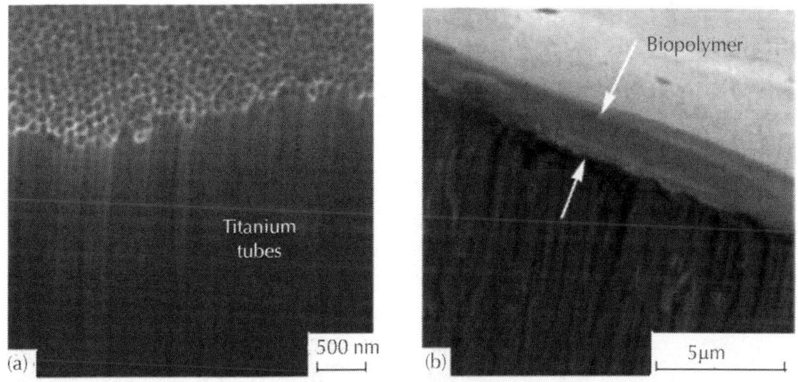

FIGURE 4 Nanostructured surface of a titanium osteoimplant (a) formed as tubes and (b) covered with a layer of poly(lactide-coglycolide).

Researchers from Brown University (United States) have demonstrated that the quality of bone titanium implants can be considerably improved by coating them with a film of polypyrrole (a conductive polymer). This makes it possible to control and program the elution rate of antibiotics and anti-inflammatory drugs by exciting an electric pulse in this electroconductive polymer [67]. Bisphosphonates, correctors of metabolism in the bone and cartilage tissues, prevent the lysis of the bone tissue near the implant and, consequently, enhance its long-term fixation at the implantation site. Several representatives of this class of drugs, for example, Zoledronate® (both salt and acid variants), pamidronate, and ibandronate, encapsulated in hydroxyapatite nanoparticles display a local prolongation effect for inhibiting the activity of osteoclasts (giant cells) and block calcium desorption from bones as a manifestation of osteoporosis. According to the definition by the FDA [66], in this case, OIs are used as a combined system that comprises a support function and the role of the matrix for controllable drug release.

A factor that complicates the use of nanostructured composite implants and scaffolds is the destruction of their elements in the body. Because of the potential cytotoxicity of the products of implant enzymatic hydrolysis, it is necessary to use polymeric materials that are degradable to nontoxic products, in particular CO_2 and H_2O, as is the situation with the biodegradable polymers of poly(3-hydroxyalkanoates) (PHAs and their derivatives) or poly(α-hydroxyacids) (PGA, PLA, and their copolymers PLGA). The use of carbon in implants carries a certain toxicological risk; however, clearance may be increased via hydrophilization of their walls with the functional groups, such as –COOH, –OH, and –NH2 [69]. Application of biodegradable materials in the current OIs or scaffolds requires serious studies aimed at regulating the rate of biodegradation [70], since, in an ideal situation, the rate of implant resorption should very precisely coincide with the growth rate of bone tissue in (or its penetration into) OI pores.

In summary, it can be expected that new commercial OIs will appear in clinics in the next years that will differ from the currently available implants in (a) nanopatterned surface and/or matrix; (b) combination of orthopedic and therapeutic functions; and (c) high biocompatibility and antibacterial resistance. Currently, the well-known manufacturers like Johnson and Johnson, Biomet, Apex Surgical LLC, Encore Orthopedics Inc., Optetrack Inc., Osteoimplant Technology Inc., Smith and Nephew Inc., Stryker Howmedica Osteonics, Zimmer Inc., Sulzer Orthopedics, and Depuy Co., have brought to the market the composite osteoimplants for replacing bone tissue in knees and hips, in spinal cord surgery, maxillofacial surgery, and so on. In 2009, Medtronic Inc. (Minneapolis, United States) announced the production of Mastergraft ceramic scaffolds for growing spinal implants that elute the corresponding drugs. SurModics Inc. (Minnesota, United States) has reached certain success in surface modification and subsequent drug immobilization in the surface layer. However, OIs that meet all three above characteristics are practically still absent and will be the focus of innovative developments in this decade.

12.4 MODERN GENITOURINARY IMPLANTS

Gynitourinary implants (GIs) are widely used for contraception and prolonged therapy (1 year and more) of female pelvic organs, for example, in endometriosis. Specific construction features of such implants are their elasticity, compatibility with adjacent tissues, and, which is most important, drug encapsulation in their matrix. However, in some cases, specialized laparoscopically introduced micro-implants are used to induce a cell rejection response and subsequent growth of fibrous tissue for blocking fallopian tubes, thereby preventing pregnancy.

A polymer-metal device, Essure (Conceptus Inc., San Carlo, California, United States), registered and approved by the FDA, is an example [71]. In 2006, the contraceptive implants Implanon (Organon USA Inc.) and Nexplanon (Merck and Co. Inc., USA) were registered by the FDA and appeared on the market; these microimplants are multilayered copolymer constructs carrying up to 70 mg of etonogestrel. Unlike the widely known contraceptive rings and their analogs (Norplant and Jadelle), these microimplants provide a daily dose of ~30 μg and more of long-term drug elution (for 3 years) without hurting patient tissues [72, 73]. To visualize their location, these implants are provided with contrasting (X-ray or NMR) characteristics, for example, by copper doping [74]. Among the latest devices of this type, one should note the Mirena (Bayer, Germany) polysiloxane implant, produced in 2007, which is capable of the controlled release of hormonal compounds for 5 years. The disadvantages of all the currently available commercial GIs are, on one hand, a limited contribution of innovative technologies to their production and, on the other, the lack of the possibility to modify the rate of drug elution according to circadian and monthly body cycles. The current chronobiology recommends synchronizing a targeted drug delivery to the activity of the corresponding organ. A programed pattern of targeted drug delivery can soon be achieved by implanting microchips into the problem areas of the body. These feedback devices are currently produced by innovative methods (nanolithography, etching, electron micrography, imprinting, etc.). Depending on the changes in biochemical composition of the ambience, these miniature-combined chips provide for regulated drug release, similar to the transdermal and insulin-containing electromechanical systems [75].

Nanotechnology is applied to designing new methods of treating urological diseases in two main directions. When dealing with functional abnormalities, such as bladder dyskinesia, prostatitis, and hyperplasia, various types of nanoparticles are used as diagnostic and therapeutic agents. The second direction of application of nanobiomedicine in urology is the reconstruction of lost or congenitally abnormal elements of the genitourinary system using cell-engineering methods. This involves the restoration of ureters, bladder repair, cosmetic surgery of genitals, and so on. In this review, the latter aspect associated with the implantation of reconstituted urological elements and tissues is of interest. In this case, a key role of cell engineering is evident, namely, the growth of cells, tissues, and elements of organs under special conditions and with the help of specialized 2D and 3D matrix systems (scaffolds). Simplified sketch of tissue growth on matrices is displayed in Fig. 5.

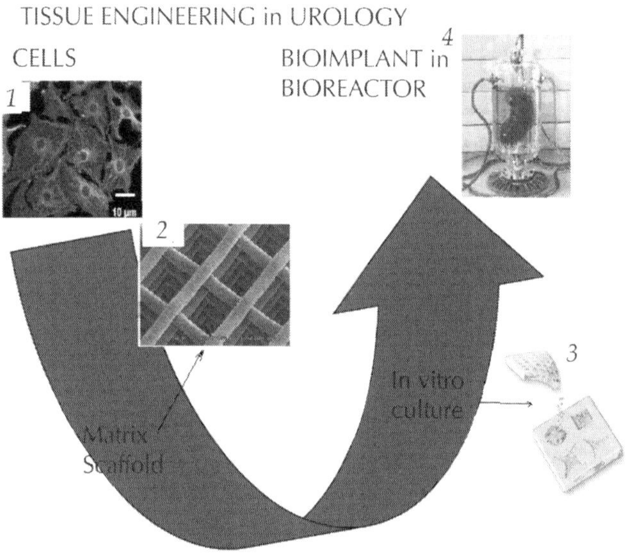

FIGURE 5 (a) The principal stages of tissue engineering for urology. (b) Compilation elements adopted from Ref. [76]. Perspectives on the role of nanotechnology in bone tissue Engineering. (c) Ref. [77] (d) Cit. David Mack/Science Photo Library [78]. Recent advances in the field of tissue bioengineering have taken scientists closer to growing organs in the laboratory.

Cell technology today has started to compensate for the deficiencies in tissues acquired due to traumas or surgery. In particular, Pattison et al. [79] pioneered the construction of a nanoscale biodegradable 3D matrix for the successful growth of bladder smooth muscle cells. The fibrinogen nanofibers produced by electrospinning were used to design a new type of porous matrices to produce cell material in urology [80]. Bioartificial (hybrid) implants as analogs of the kidney [81] are developed at the University of Michigan (United States). Their working element is a nanoporous silicon membrane that concurrently works as a filter and a matrix for immobilization of renal tubular cells. These devices are intended for the cases of acute kidney failure. As is mentioned above, a drug with antiseptic and cell stimulating functions can be immobilized in a polymeric material, for example, biodegradable poly(lactide-coglycolide).

Trauma, toxic or inflammatory circumstances as well as congenital diseases of a genitourinary tract may result in serious damages and even the loss of function. In this situation novel relevant biomaterials are crucially demanded to substitute damaged organs. For example the search for an alternative reservoir to replace the native bladder has proved to be crucially intricate. Synthetic, nonbiodegradable materials, such as polyolefines, silicone, rubber, polytetrafluoroethylene and others were initially used in urological surgery. However, these genitourinary implants quickly covered with a crust, were prone to infection and subject to host–foreign body reactions [82]. Currently the tissue engineering is playing an increasing role in the management of genitourinary implant variety. Scaffolds for cell cultivation may use not only biopolymers but also artificial tissues or combination of the two. In particular, a combined scaffold of collagen and polyglycolic acid even enhances even tissue regeneration [83]. Sharma et al. [84] performed bladder augmentation in rats with a special polymeric scaffold (elastomeric poly(1,8-octanediol-cocitrate)). The polymer matrix was seeded on opposing surfaces with human mesenchymal stem cells (MSC) and urothelial cells (UC). At 10 weeks morphometry of MSC-polymer-UC sandwiches revealed muscle/collagen ratios approximately 2 times more than controls. Additionally the data of work [83] displayed that MSC colony support partial regeneration of bladder tissue in vivo, A new approach in the area of biomaterials also for the lower urinary tract could be the use of electrospun nanofibers which reveal a superior viability in comparison to other materials, most probably due to their high surface/volume ratio imitating an extracellular matrix structure. Growth of stem cells on the nanofiber material and differentiation into the different cell types has been confirmed in histological analysis [85].

To make the picture complete, the innovative works in the field of urological surgery should be mentioned. In particular, an EnSeal™ implant (SurgRx, Palo Alto, California, United States) used in laparoscopic resection of the prostate and comprising millions of nanoparticles, which enhances healing injured small veins, thereby minimizing the tissue damage in the involved organ [86]. Submicron tweezers, a potential urological surgery tool, is now used in surgery of spermatic ducts or, for example, in vasectomies [87]. High-molecular-weight implants, such as hydrogels or elastic polymeric constructs, are used when treating and replacing genitourinary elements. Hydrogel membranes are formed for a number of implanted devices to separate the ambience and drug container. The therapy of prostate tumors, as well as the regulation of adolescent sexual maturation, involve the administration of hormones; for this purpose, the Hydron, Supprelin, and Vantas gel implants with about 1-year-long drug elution have been developed [88]. The reservoir systems that use an osmotic pump principle, that is, ALZET (Alza) [89] and DUROS [90], have been designed; their separation membranes are formed of cellulose ester and polyurethane, respectively, and they elute at a constant rate a protein, LHRH agonist, administered to treat prostate cancer.

Biodegradable polymers represent a separate group of polymer implants in urology. By analogy with bioresorbable suture materials and nanoparticles, the following groups of natural polymers have been used for their production: poly(α-hydroxyacids) (polylactides and their copolymers) and poly(β-hydroxyalkanoates) (PHAs) and the copolymers with hydroxyvalerate, hydrooxyhexanoate, and so forth. An example is the multifunctional implanted system Zoladex™ (Astra Zeneca, Canada), which was designed over 20 years ago and is currently widely used in clinical practice for the long-term therapy of endometriosis, breast tumors, and prostate tumors [91]. Note that polyhydroxyacids (PLA and PGLA) and poly(ortho esters) were certified by the FDA US in 2009. Catheters of various types and diameters are now widely used as transient implants. They are temporarily introduced into various body parts, such as the bladder, to replace ureters, to enhance the therapy of renal tissues, and to prevent gall bladder dyskinesia. The main requirements to polymeric catheters are nontoxicity; biocompatibility; and, in particular, their antiseptic effect. For this purpose, similar to vascular catheters, their surfaces are modified and enriched for encapsulated drugs, most frequently, a bactericidal agent. The controlled delivery of the drug desorbed from the walls of an implant in this situation allows for the avoidance of numerous complications caused by urine-containing, bacterially active media.

12.5 NANOIMPLANTS WITH ENCAPSULATED DRUGS FOR OPHTHALMOLOGY

The eye diseases present a serious problem, first and foremost, for the population of industrially developed countries. These diseases include age related retinal changes, diabetic retinopathy, structural alterations in the lens, and so on. A systemic drug delivery to the anatomically intricate organ, such as the eye, and especially to the intraocular elements is not a simple task, which interferes with the therapy of the uvea, vitreous body, and optic nerve. Because of multiple tissue barriers and lacrimal drainage, only 1–3% of the active substance of traditionally administered drugs (e.g., as eye drops) reaches the problem areas within the eye [92=83]. Therefore, it is planned in the nearest future to widely use nanoparticles and polymer implants with encapsulated drugs for prolonged and targeted drug delivery. The eye lens spatially separates the eyeball into the anterior segment, which comprises a cornea, sclera, and conjunctiva, and the posterior segment, with all of its elements localized beyond the lens deep in the eye (retina, vascular membrane, vitreous body, and optic nerve). This is why the ocular implants are classified according to the eye anatomical structure into anterior ocular and (e.g., contact lenses) and intraocular devices.

The installation of ocular micro and macroscale implants is associated with a number of issues that should be briefly mentioned in this section. First, the eye is a paired organ; correspondingly, this provides for an objective comparison of the outcome of innovative treatment, be it a surgery or therapy, of one eye with the intact state of the other eye, using it as a control [94]. Second, the eyeball structure is rather transparent, which allows for an efficient use of optical and photosensitive methods in diagnosis and control. Third, the anatomical structure of the eye includes miniature elements in both the anterior (in front of the lens) and posterior (beyond the lens) chambers. Correspondingly, the ocular implants are limited by small sizes, including their nanoscale geometry. Fourth, the loss of sight is not necessarily associated with the loss of life, but considerably complicates it; therefore, the efforts for expanding the market of ocular preparations are very active, and the ophthalmological market by 2011 had reached several billions of dollars (Table 1).

TABLE 1 Dynamics and forecast for the world market of implanted medical devices and prostheses (billions of $$) [18].

Years	2004	2006	2008	2010	2015 (forecast)
United States and Canada	83.14	93.97	105.88	118.23	164.63
Western Europe	52.50	59.04	65.65	73.11	103.00
Japan	24.14	26.64	29.31	32.65	44.76
Asian countries (without Japan)	15.72	18.36	23.13	30.20	54.97
South America	3.48	3.94	4.83	6.11	11.18
Eastern Europe	3.98	4.50	5.96	7.73	14.90
Africa	2.49	2.82	3.41	4.13	7.79
Total production	185.45	209.27	238.17	272.16	401.23

The current state and nearest future of implants in ophthalmology are to a certain degree reflected by Tables 2 and 3, which lists the IMDs for local drug delivery both commercially produced and at various stages of clinical trials. All the tools are partitioned according to the target of their prolonged delivery, namely, the anterior eye chamber (cornea, conjunctiva, sclera, etc.) and the posterior chamber (retina, optic nerve, vitreous body, macula, etc.). According to the current trends, the analogous drug forms already tested in ophthalmology but designed for gene therapy will be further developed. In particular, it has been shown at the initial stage of clinical trials in the United States that some RNAs (such as low molecular weight interfering RNAs) implanted into biodegradable polymer complexes display a high efficiency in treatment of the congenital retinal degeneration and macular degeneration [95]. In addition, it is purposeful to focus on the design and production of the drainage systems for treatment of glaucoma [95], one of the most widespread diseases. In an ideal situation, these systems (ANDIs) should contain the polymer lines for drainage of lacrimal fluid and exudate and provide for a targeted delivery of the drugs decreasing intraocular pressure and preventing destruction of the optic nerve. Advance in designing the new generation implants for the anterior eye chamber will allow the miniature sensors for biochemical eye surface monitoring [96] and concurrent drug elution [97] to be installed directly into contact lenses or into posterior compartment (see Fig. 6).

TABLE 2 Promising implanted devices for local drug delivery to the anterior eye chamber [65].

Active agent	Trademark	Type	Matrix	For treating of	Stage of development
Betaxolol	Betoptic S®	Eye drops	Amberlite IRP-69	Glaucoma	In market
Azithromycin	AzaSite®	Eye drops	Polycarbophil	Bacterial conjunctivitis	In market
Ketotifen		Soft contact lenses		Allergic conjunctivitis	Phase 3
Latanoprost		Tampon cylinder	PVA and cellulose	Glaucoma	Phase 2
Latanoprost		Implant of eye connective tissue (subconjuctival)	PVA–PLGA composite	Glaucoma	Phase 2
				Keratoconjunctivitis	Phase 3
Cyclosporine (LKh-201)		Implant in the eye episclera	Silicon	Uveitis	Phase 2
Dexamethasone phosphate EGP-437	EyeGate II®	Device for microionophoresis	–	Difficulties in lacrimation (dry eye symptom)	Phase 3

TABLE 3 Promising implanted devices for local drug delivery to the posterior eye chamber [65].

Active agent	Trademark	Type	Matrix	For treating of	Stage of development
Ganciclovir	Vitrasert®	Intravitreal implant	Poly(ethylene-co-vinyl alcohol)	Viral retinitis (inflammation of retina)	In market
Fluocinolone acetonide	Retisert®	Intravitreal implant	PVA–silicone	Uveitis	In market
Fluocinolone acetonide	Iluvien®	Intravitreal implant	Polyimide–PVA	Diabetic macular edema	Phase 3
Dexamethasone	Ozurdex®	Intravitreal implant	Biodegradable PLGA	Renal venous congestion	In production
Brimonidine	–	Intravitreal implant	Biodegradable PLGA	Age-related macular degeneration	Phase 2
Triamcinolone acetonide	I-vation™ TA	Intravitreal implant	PMMA/ethylvinyl acetate	Diabetic macular edema	Phase 2
Triamcinolone acetonide	IBI-20089	Intravitreal implant	Oil medium	Renal venous congestion	Phase 1
Triamcinolone acetonide	RETAAC	Intravitreal implant	Biodegradable PLGA	Diabetic macular edema	Phase 1

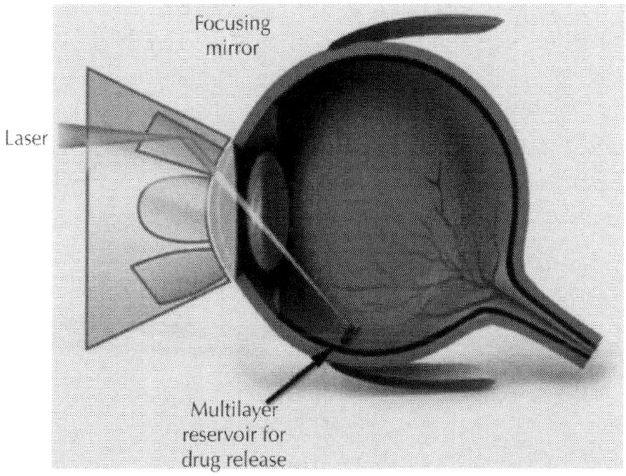

FIGURE 6 Photodynamically activated system for controlled drug delivery within the intraocular region of eye.

The current nanomedicine provides the possibilities for replacement of several eye elements with artificial implants. The replacement therapy in ophthalmology includes the repair of the retinal nerve (formation of polymer conduit) and optical nerve [95]. A separate field in cell engineering uses 2D and 3D biodegradable scaffolds that enhance cell proliferation and function for the growth of eye tissues and, in the nearest future, also elements of the eye, for example, the cornea [98]. In this case, the surface topography of the polymer should meet special requirements, since the nanostructured surface of the scaffold provides controllable cell adhesion and migration. Thus, we again meet the problem of controllable drug elusion from polymer matrix. For example, the delivery of the peptides capable of spatial self-organization and self-assembly will be used in the nearest future for the regeneration of the retinal axon [99].

12.6 PROSPECTS OF NEUROLOGICAL IMPLANTS

Neurology is an interesting and most promizing field for implant application. Typical implants used for this purpose are polymer conduits intended for the restoration of the integrity and conductivity of nervous conductors. It is known that the nerve cells have only limited ability to repair and the nerve tissue has a limited ability to regenerate after injuries, for example, craniocerebral traumas. Moreover, an increase in the rate of diseases, such as Alzheimer's or Parkinson's disease, as well as of cerebrovascular impairments and subsequent dystrophy of nerve fibers, requires an ever-increasing number of micro and nanoimplants providing for transduction of the signals in the central and peripheral nervous systems. The major component for these implants is biodegradable polymers of polylactide or polyoxyalkanoate type. The prospects for nanoimplants in this field, third in the list of most widely used devices (after cardiac implants and bone prostheses), have been detailed in several reviews [100, 101]. Here, organic silicon (polymer) systems are the most promizing. Useful properties of structured silicon dioxide have been recently discovered; its nanoporous surface provides a more stable contact with neurons compared to smooth surfaces and prevents tissue overproliferation (gliosis) [102]. The pioneer experiments have demonstrated that the activation of semiconductor nanoparticles with light pulses stimulates rat brain neurons, which allows for the contactless (electrode-free) excitation of a neuronal signal [103]. In this field of regeneration medicine, the implanted materials should be polyfunctional, namely, should work as flexible nanoscale backbone constructs, enhance excitation of signal transduction, and concurrently elute a drug that activate development of a nerve impulse.

12.7 MINIATURE ELECTROMECHANICAL IMPLANTS FOR PROGRAMMED DELIVERY OF BIOLOGICALLY ACTIVE SUBSTANCES

Innovative developments in the chemistry of polymers and supramolecular chemistry, composite engineering, and design of micro and nanoscale electromechanical systems (MEMS and NEMS) are directed towards obtaining of fundamentally new therapeutic systems for controllable release of low-molecular-weight drugs, macromolecules, and cells [104, 105]. Polymer and inorganic implants equipped with electronic chips create kinetic profiles with different complexities to release the biologically active component. Several implanted systems additionally contain mechanical manipulators, drives, and nanocantilevers. The combination of computer, electronic, and mechanical elements in an implant determines that it is a principally new electromechanical device (MEMS/NEMS) [106–108]. The miniaturization of these devices, which use state-of-the-art technologies (nanolithography and imprinting), allows these implants to be considered an independent universal class of the systems for targeted drug delivery [109–111]. The introduction of MEMS/NEMS as drug-delivery-programed systems into clinical practice is determined to a considerable degree by advances in a new field of biomedicine, personalized medicine, which is a custom iced therapy of each patient that takes into account his/her individual genetic features and specificity of biochemical processes [112, 113]. The information about symptomatology of disease and data on the metabolites assayed with biomarkers are input into a chip to provide a programed release of a set of drugs from a system of containers according to these integrate data. Here, the most impressive results are attained when using an integrated treatment of complex pathologies, such as tumor diseases, liver cirrhosis, and blood disorders.

Using a feedback principle, an MEMS is used to design a vitally important organ, the artificial pancreas, which monitors the sugar level in the blood and releases the necessary dose of insulin according to the sugar concentration [114–116]. Currently, there are three hindrances to the commercialization of this system for diabetes control, namely, (a) a reliable programed device; (b) glucose sensors stable in their characteristics and able to stably function for a long time in the specific aggressive medium, such as blood; and (c) still high price of the final product [117]. Nonetheless, taking into account the high rate of diabetic diseases, in the next 10–12 years, these implanted systems should become a mass therapeutic tool.

Once the first two problems are associated with advances in nanotechnology, the third issue has a rather moral and ethical aspect. No matter how large the expenditures for treating a patient using precision and expensive electromechanical platforms, an improvement in his/her quality of life cannot be limited by material and financial parameters. The last statement may be illustrated by one rare disease with a very limited number of cases. It is known that metabolic disorders require constant therapy to sustain a patient's comfort. In this situation, implanted NEMS devices for the controllable release of deficient enzymes find their own niche in the world pharmaceutical market. For example, Genzyme Therapeutics (United States) [117] specializes in the production of a Cerezyme implant, which elutes recombinant proteins for the long-term therapy of the Gaucher disease. Despite that this disease is very rare (1 per 200,000 in the population), in 2008, this company acquired $1.2 billion from selling this device, which accounts for 27% of its annual profit. As in the case with the NEMS implants, large expenditures for implantation and therapy ($200000 USD annually per patient) [112] are completely justified, since any alternative method that alleviates suffering and prevents a preliminary fatal outcome is absent.

The specific construction features of the MEMS-and NEMS-type implants include not only their miniature size and stable operation, but also the use of bioinert and biocompatible materials. In their operation principle, these devices are miniature reservoirs that contain a drug, which is eluted in response to electrical signal or movement of piston/micropump walls. The common characteristics for the main types of electromechanical implants are listed in Table 4.

At the micro and nanoscales, several electromechanical devices for controllable drug release act as pumps with either a drive or by the effect of osmotic forces [118]. In the latter case, there is no need for any electromechanical drive, and strictly speaking, such devices cannot be regarded as MEMS/NEMS. Only a very limited number of implanted systems that use a micropump principle are manufactured in batches. The examples include a MiniMed Medtronic Pump for controlled insulin release and SynchroMed II Programmable Pump (commercially available since 1988) for critical therapy (a high level of pain and neurological indications). A series of the Debiotech pumps, which use the piezoelectric effect in their design, allows the drug solution to be eluted at a rate of 100 μL/h [119]. Another operating principle of a micropump is to use materials with a shape memory effect [119]; they allow the therapeutic rate of drug delivery to be reduced to 5 μL/h.

TABLE 4 Comparative characteristics of nanoimplants utilizing different drug release mechanisms

Principle of drug release	Main characteristics	Therapeutic potential
Implanted electronic chip	Limitations on volume of device Only for highly active compounds High probability of tissue capsule formation Stricter requirements to bioinert and nontoxic materials in long-term operation Invasive implantation procedure High price	Local and more economical drug delivery Precise and time-controllable drug delivery More preferable than parenteral drug administration Possibility of pulse and more complex kinetic profile of drug release Has advantages as compared with aerosols and oral administration Combined therapy and diagnosing (monitoring)
Implanted micropump	Limitations on volume of device Only for highly active compounds Exclusively for release in liquid form (difficulties in releasing polypeptides due to their low stability in aqueous medium) Possible structural changes in active component because of shear stress Precipitation and subsequent plugging of pump piston with the drug precipitated from solution Invasive implantation procedure Considerable expenditures for implantation and production	Local and more economical drug delivery More preferable than parenteral drug administration Possibility of pulse and more complex kinetic profile of drug release; Has advantages as compared with aerosol, oral, and parenteral drug administration routes
Implanted polymer chip	Low invasive Limitations on volume of device Only for highly active compounds Stricter requirements to drug stability (absence of hydrolysis) Broad therapeutic window Difficulties in precise dosing	Local delivery provides advantages compared to traditional drug injection and aerosol administrations Perfect alternative to parenteral drug administration Tolerance to delivery via systemic or local administration routes Biodegradable elements do not require repeated invasion for extracting worked out system
Implanted selective membranes	Limitations on volume of device Only for highly active compounds Requires surgical intervention Risk for stoppage of membrane pores with loss of production capacity and selectivity Appropriate only for drug solutions Favorable conditions for cell material	Due to local delivery, has advantages as compared with traditional drug injection and aerosol administration Perfect alternative to parenteral drug administration Depending on implantation site, allows for controlled delivery via bloodstream or to local body area Does not require repeated invasion for extracting worked out system

12.8 CONCLUSIONS

The transition from a macroscopic description of implanted systems to molecular and submolecular consideration is a necessary condition for the development of new generation bioimplants. Nowadays, there are good grounds to believe that the structuring of the implant surface in a nanoscale range modulates cell membrane receptors that modify the communication between cells and has a positive effect on adsorption and conformational changes in the proteins responsible for immune and toxic responses. The successive consideration of the cardiac implants based on the example of new generation stents and bone implants, as well as ophthalmological, urological, and others, demonstrates that the combination of the following three factors determines the nearest prospects in implantology: (1) topography and chemical composition of the surface corresponding to the natural organ; (2) the ability to perform mechanical, optical, ion conducting, and other functions characteristic of the replaced element; and (3) the ability to encapsulate drugs with the possibility of their controllable (programed) and prolonged targeted delivery to a local area of the body.

KEYWORDS

- **biodegradable polymers**
- **Gynitourinary implants**
- **hybrid nanocomposite**
- **implanted medical device**
- **modern orthopedic implants**
- **neurological implants**

REFERENCES

1. Logothetidis, S.; *NanoScience and Technology.* **2012,** *61*, 1–26.
2. Huebsch, N.; Mooney, D. J.; *Nature*, **2009,** *462(7272),* 426–432.
3. Hasirci, V.; Vrana, E.; Zorlutuna, P.; Ndreu, A.; Yilgor, P.; Basmanav, F. B.; Aydin, E.; *Journal of Biomaterials Science, Polymer* **2006,** *17 (11),* 1241–1268.
4. Bazhenov, S. L.; Berlin, A. A.; Kul'kov, A. A.; and Oshmyan, V. G.; *Polimernye kompozitsionnye materialy.Prochnost' i tekhnologiya* (Polymer Composite Materials. Durability and Technology), Rus. Edition. Dolgoprudnyi: Intellekt, 2010.
5. Jorfi, M.; Roberts, M. N.; Foster, E. J.; Weder, C. *ACS Applied Materials and Interfaces.* **2013,** *5 (4),* 1517–1526.
6. Frima, H. J.; Gabellieri, C.; Nilsson, M. -I.; *Journal of Controlled Release.* **2012,** *161 (2),* 409–415. Drug delivery research in the European Union's Seventh Framework Program for Research

7. Drug Delivery Systems US Industry Study with Forecasts for **2012,** and **2017,** Study #2294 | March **2008,** 338 The Freedonia Group. Cleveland, OH. USA. www.freedoniagroup.com
8. Staples, M.; *Wiley Interdisciplinary Reviews: Nanomedicine and Nanobiotechnology* **2010,** *2 (4),* 400–417.
9. Leucuta, S. E.; *Current Clinical Pharmacology* **2012,** *7 (4),* 282–317.
10. *Stevenson, C. L.; Santini Jr.; J. T.; Langer R.;* Advanced Drug Delivery Reviews.2012. In press doi: 10.1016/j.addr.2012.02.005. Reservoir-Based Drug Delivery Systems Utilizing Microtechnology.
11. Sirivisoot, S.; Pareta, R. A.; Webster, T. J.; *Recent Patents on Biomedical Engineering* **2012,** *5 (1),* 63–73.
12. Global Industry Analysts Inc.; Nanomedicine: a global strategic business report, 2009.
13. Murday, J. S., et al. Nanomedicine: Nanotechnology, Biology, and Medicine, **2009,** *5,* 251–273.
14. Ekdahl, K. N.; Lambris, J. D.; H.; Elwing, D.; Ricklin, Nilsson, P. H.; Y.; Teramura, I. A.; Nicholls Bo Nilsson. *Advanced Drug Delivery Reviews* **2011,** *63,* 1042–1050.
15. Bonartsev, A. P.; Boskhomedgiev, A. P.; Iordanski, A. L. et al. In: Kinetics, Catalysis and Mechanism of Chemical Reactions. From pure to applied science. V. 2 -Tomorrow and perspectives. (eds. Islamova, R. M.; Kolesov, S. V.; Zaikov, G. E.) Nova Science Publishers: New York **2012,** Ch.27. 335–350. Degradation of poly(3-hydroxybytyrare) and its derivatives: Characterization and kinetic behavior.
16. Iordanskii, A. L.; Rudakova, T. E.; Zaikov, G. E.; *Interaction of Polymers with Bioactive and Corrosive Media. Ser. New Concepts in Polymer Science.* VSP Science Press. Utrecht –Tokyo Japan. **1994**, 298p.
17. Directive 2007–47-EC of the European Parliament and of the council.pdf
18. http://www.fda.gov/MedicalDevices/DeviceRegulationandGuidance/Overview/ClassifyYourDevice/ucm051512.htm
19. http://laws-lois.justice.gc.ca/eng/regulations/CRC,c.870/index.html
20. Market Report: World Medical Devices Market September **2007,** Acmite Market Intellegence. http://www.acmite.com/market-reports/medicals/world-medical-devices-market.html.
21. Report of Ministry of Healthcare and social development RF Review [online]. http://www.mzsrrf.ru/press_smi/388.html.
22. (a) The World Health Organization Report, **2006,** [online]. http://www.who.int/entity/whr/2006/whr06_en.pdf. (b) The World Health Organization Report, **2011,** Cardiovascular disease. http://www.who.int/cardiovascular diseases/en/
23. Document: Conception of demography development in Russia until **2015,** http://www.demoscope.ru/weekly/knigi/koncepciya/koncepciya.html
24. Kastrati, A. et al. *N. Engl. J. Med.* **2007,** *356*, 1030.
25. Stone, G. W. et al. *N. Engl. J. Med.* **2007,** *356*, 998.
26. Lagerqvist, B. et al. *N. Engl. J. Med.* **2007,** *356*, 1009.
27. Wieneke, H. et al. *Catheter Cardiovasc. Interv.* **2003,** *60*, 399.
28. Bhargava, B. et al. *Catheter Cardiovasc. Interv.* **2006,** *67*, 698.
29. Erlebacher, J. et al. *Nature* **2001,** *410*, 450.
30. Caves, J. M.; Chaikof, E. Vasc, L. J.; Surg. **2006,** *44*, 1363.

31. Samaroo, H. D. et al. Int. J.; Nanomedicineю **2008,** *3*, 75.
32. Tana, Y.; Farhatnia, Achala de Mel, J.; Rajadas, Alavijeh, M. S.; Seifalian, A. M.; Journal of Biotechnology **2013,** *164,* 151–170. Inception to actualization: Next generation coronary stent coatings incorporating nanotechnology.
33. Godin, B.; Sakamoto, J. H.; Serda ,R. E.; Grattoni, A.; Bouamrani ,A.; Ferrari, M. Trends in Pharmacological Sciences. **2010,** *31 (5)*, 199.
34. Rosengart, A. J.; Kaminski, M. D.; Chen H.; Caviness, P. L.; Ebner, A. D.; Ritter J. Magn, A. *J. Magn. Mater.* **2005,** *293*, 633.
35. Cregg, P. J.; Murphy K.; Mardinoglu A.; Prina-Mello A.; Journal of Magnetism and Magnetic Materials. **2010,** 322, 2087.
36. Zilberman M.; Nelson, K. D.; Eberhart R. C.; Mechanical Properties and In Vitro Degradation of Bioresorbable Fibers and Expandable Fiber-Based Stents. Published online 30 June **2005,** in Wiley InterScience (www.interscience.wiley.com). DOI: 10.1002/jbm.b.30319
37. Yun-Xuan Weng, Xiu-Li Wang, Yu-Zhong Wang. Polymer Testing. **2011,** *30*, 372.
38. Kumbar, S. G.; Kofron, M. D.; Nair, L. S.; Laurencin CT.; Cell behavior toward nanostructured surfaces. In: Gon-salves, K. E.; Laurencin, C. T.; Halberstadt, C.; Nair L. S., eds. Biomedical Nanostructures. New York: John Wiley & Sons; **2008,** 261–295.
39. Balasundarama G.; Webster T. Mater, J. J.; Chem. **2006,** 16, 3737.
40. Schakenraad J. M.; Biomaterial Science / Eds. Ratner, B. D.; Hoffman, A. S.; Schoen, F. J.; Lemons J. E. (San Diego: Academic Press, Inc.; 1996), 140.
41. Kennedy, S. B.; Washburn N. R, Simon, C. G.; Amis Biomaterials, *E. J.;* **2006,** 20, 3817.
42. Hersel U.; Dahmen C.; Kessler H.; Biomaterials. **2003,** *24*, 4385.
43. Feng Y.; Mrksich M.; Biochemistry, **2004,** *43*, 15811.
44. El-Ghannam A.; Ducheyne P.; Shapiro I. *M.* J.*;* Orthop. Res. **1999,** *17*, 340.
45. Sagnella S.; Anderson E.; Sanabria N.; Marchant, R. E.; Kottke-Marchant K.; Tissue Eng. **2005,** *11*, 226.
46. Anselme K.; Bigerell M. *J. Mater. Sci. Mater. Med.* **2006,** *17*, 471.
47. Curtis A. *IEEE Trans. Nanobiosci.* **2004,** *3,*P. 293.
48. Price, R. L.; Ellison K.; Haberstroh, K. M.; Webster T. *J. Biomed, J. Mater. Res.* **2004,** V. *70*, 129.
49. Webster, T. J.; Schadler, L. S.; Siegel, R. W.; Bizios R.; *Tissue Eng.* **2001,** *7,* 291.
50. Curtis, A.; Britland S.; Surface modification of biomaterials by topographic and chemical patterning. In: Advanced Biomaterials in Biomedical Engineering and Drug Delivery Systems. Ogata, N.; Kim, S. W.; Feijen, J.; Okano, T.; eds. Tokyo: Springer-Verlag, **1996**, 158–167.
51. Woo, K. M.; Chen V. J.; Ma P. X. *J. Biomed. Mater. Res..* **2003,** *67A (2),* 531.
52. Keselowsky, B. G.; Collard, D. M.; Garcia A. J. J Biomed. Mater. Res. **2003,** *66A (2),* 247.
53. Webster, T. J.; Ergun C.; Doremus, R. H.; Siegel, R. W.; Bizios R. *J. Biomed. Mater. Res..* **2000,** 51. 3, 475.
54. Faghihi, S.; Azari, F.; Zhilyaev, A. P.; et al. Biomaterials **2007,** V. 28. (27 P. 3887.
55. Khan, Y.; El-Amin S. F, Laurencin C. T. Conference Proceedings of the IEEE Engineering in Medicine and Biology Society.; New York, **2006,** *1,* 529.
56. Harrison B. S, Atala A.; Biomaterials **2007,** *28(2)*, 344.

57. Li, W. J.; Laurencin, C. T.; Caterson, E. J.; Tuan, R. S.; Ko F. Biomed, K. J.; Mater. Res. **2002,** 60, 613.
58. Nukavarapu, S. P.; Kumbar, S. G.; Nair, L. S.; Laurencin C. T.; Biomedical Nanostructures / Eds. Gonsalves, K. E.; Laurencin, C. T.; Halberstadt C.; Nair L. S. (NewYork: Wiley, 2008) P. 377.
59. Kumbar, S. G.; Nukavarapu, S. P.; Roshan R.; Nair, L. S.; Laurencin C. T. . Biomed Mater **2008,** 3, 1.
60. Pelled G.; Tai K.; Sheyn D.; Zilberman Y.; Kumbar, S. G.; et al. J.; Biomech **2007,** 40, 399.
61. Kumbar, S. G.; Nair, L. S.; Bhattacharyya S.; Laurencin C. T. J Nanosci. Nanotechnol. **2006,** 6, 2591–2607.
62. Nablo, B. J.; Rothrock, A. R.; Schoenfisch M. H.; Biomaterials **2005,** 26, 917.
63. Simchi A.; Tamji E.; Pishbin F.; Boccaccini Nanomedicine, Nanotechnology, A. R.; Biology, and Medicine. **2011,** 7, 22.
64. Popat K. C, Eltgroth, M.; LaTempa T. J, Grimes C. A, Desai T. A.; Small **2007,** 3, 1878.
65. Pioletti, D. P.; Gauthier O.; Stadelmann V. A. et al. Orthopedic implant used as drug delivery system: Clinical situation and state of the research. Current Drug Delivery, **2008,** 5, 59–63.
66. Gulati K.; Ramakrishnan S.; Aw, M. S.; Atkins, G. J.; Findlay, D. M.; Losic D.; Acta Biomaterialia **2012,** 8, 449.
67. Sirivisoot S.; Pareta R.; and Webster, T. J.; Electrically controlled drug release from nanostructured polypyrrole coated on titanium. **2011,** *Nanotechnology* 22–085101 doi: 10.1088/0957–4484/22/8/085101
68. http: //www.fda.gov/oc/combination/
69. Laurencin, C. T.; Kumbar, S. G.; Nukavarapu Interdiscipl, S. P.; Rev. Nanomed. Nanobiotechnol. **2009,** 1, 6.
70. Ivantsova, E. L.; Kosenko R. Yu.; Iordanskii, A. L.; et al. Polymer Science, Ser. A, **2012,** 54,No. 2,pp. 87–93
71. Shavell V. I.; Abdallah, M. E.; Shade Jr, G. H.; Diamond, M. P.; Berman J. M. J Min. Invasive Gynecol. **2000,** *16(1),* 22.
72. Shulman L. P, Gabriel H.; Contraception **2006,** 73, 325.
73. http: //www.merck.com/newsroom/news-release-archive/prescription-medicine-news/2011_1109.html
74. Correia L.; Ramos A. B, Machado, A. I.; Rosa D.; Marques C.; Contraception **2011,** in press doi: 10.1016/j.contraception.2011.10.011
75. Staples M.; Microchips and controlled release drug reservoirs. WIREs Nanomed Nanobiotechnol 2010–2 400–417.
76. E.; Saiz, Zimmermann, E. A.; Lee, J. S.; U. Wegst, G. K.; Tomsi, A. P.; Dental Materials **2013,** 29, 103–115.
77. Fennema E.; Rivron N.; Rouwkema J.; van Blitterswijk C.; de Boer J.Trends in Biotechnology, **2013,** *31(2),* 271–282. http: //dx.doi.org/10.1016/j.tibtech.2012.12.003.
78. Chopa N.; Kayes O. Trends in Urology and men's health. **2013,** No 1–2 14–16. Recent advances in the field of tissue bioengineering have taken scientists closer to growing organs in the laboratory. Cit. David Mack/Science Photo Library.
79. Pattison M. A, Wurster, S.; Webster, T. J.; Haberstroh K. M.; Biomaterials **2005,** 26, 2491.

80. McManus, M.; Boland, E.; Bowlin, G.; Simpson, D.; Espy, P.; Koo H.; American Urological Association Annual Meeting, May 8–13, **2004,** San Francisco, CA, USA.
81. Humes H. D.; *Semin. Nephrol.* **2000,** *20*, 71.
82. Lee, D.; Lee, J. T.; Sheperd, D.; Abrahams H.; Lee, D.; Preliminary use of the Enseal™ system for sealing of the dorsal venous complex during robotic assisted laparoscopic prostatectomy. American Urological Association Annual Meeting, May 21–26, **2005,** San Antonio, TX, USA.
83. Eberli D.; Yoo, J. J.; Atala A.; Urologe. **2007,** A*46*, 32.
84. Sharma, A. K.; Hota, P. V.; Matoka, D. J., et al. Biomaterials. **2010,** *31 (24),* 6207–62178.
85. Feil G.; Daum L.; Amend B.; et.al. Advanced Drug Delivery Reviews. **2011,** 63 375–378. From tissue engineering to regenerative medicine in urology — The potential and the pitfalls.
86. Murphy, D. G.; Costello Nanotechnology, A. J.; In: New Technology in Urology. (P.; Dasgupta et al., eds) Springer Verlag. London 201 268.
87. Cruz C. G. M. PHD Thesis's. pH-Triggered Dynamic Molecular Tweezers for Drug Delivery Applications. Queen's University. Kingston, Ontario, Canada September **2011,** 180.
88. Wright J. C.; Hoffman, A. S.; Historical Overview of Long Acting Injections and Implants in Long Acting Injections and Implants (J. C.; Wright and Burgess D. J.; (eds.)), Advances in Delivery Science and Technology, Chapter 2. DOI 10.1007/978-1-4614-0554-2_2, © Controlled Release Society, Springer. 2012.
89. Theeuwes, F.; Yum S. I. Ann Biomed. Eng. **1976,** 4343–353. Principles of the design and operation of generic osmotic pumps for the delivery of semisolid or liquid drug formulations
90. Wright, J. C.; Leonard, S. T.; Stevenson, C. L.; Beck, J. C.; Chen, G.; Jao, R. M.; Johnson, P. A.; Leonard, J.; Skowronski, R. An in vivo/in vitro comparison with a leuprolide osmotic implant for the treatment of prostate cancer. J Control Release. **2001,** 75, 1–10.
91. Duncan R.; Nature Review. **2006,** *6,* 688–701. Polymer conjugates as anticancer nanomedicines.
92. Maurice, D. M.; Mishima, S.; Ocular pharmacokinetics, in Handbook of Experimental Pharmacology, Sears, M. L.; Ed.; Berlin Heidelberg: Springer, **1984,** 16.
93. Kuno, N.; Fujii, S.; Polymers, **2011,** *3*, 193.
94. Thomson, H.; Lotery, A.; Nanomedicine, **2009,** *4,* 599.
95. Pan, T.; Brown, J. D.; and Ziaie, B.; IEEE Eng. Med. Biol. Soc.; **2006,** *1,* 3174.
96. Parviz, B. A.; Shen, T. T.; and Ho, H.; Invest. Ophtalmol. Vis. Sci.; **2008,** *49,* Abstr. 4783, Functional contact lens with integrated inorganic microstructures. www.arvo.org
97. Kapoor, Y.; Thomas, J. C.; Tan, G.; John, V. T.; Chauhan, A.; Biomaterials, **2009,** *30,* p. 867.
98. Liu, B. S.; Yao, C. H.; Hsu, S. H.; et al.; *J. Biomater. Appl.* **2004,** *19*, 21.
99. Elder J. B, Liu, C. Y.; Apuzzo MLJ.; Neurosurgery in the realm of 10–9, Part 2, Applications of nanotechnology to neurosurgery—present and future. Neurosurgery. **2008,** 62 269–84.
100. Ellis-Behnke R. Med Clin of North America. **2007,** 91, 937–962.

101. Lebedev, M. A.; Nicolelis MAL. Brain-machine interfaces past, present, future. Trends Neurosci **2006,** 29, 536–46.
102. Moxon, K. A.; Hallman, S.; Aslani, A.; Kalkhoran, N. M.; Lelkes P. I.; Bioactive properties of nano-structured porous silicon for enhancing electrode to neuron interfaces. *J Biomater Sci Polym.* **2007,**18, 1263–1281.
103. Zhao, Y.; Larimer, P.; Pressler, R. T.; Strowbridge, B. W.; Burda C.; Wireless activation of neurons in brain slices using nanostructured semiconductor photoelectrodes. Angew Chem Int Edn. **2009,** 48, 2407–2410.
104. Madou M. J.; Fundamentals of Microfabrication: The Science of Miniaturization, 2nd ed.; CRC Press: Boca Raton, **2002,** 752.
105. Borzenko A. G.; Scientifical instrument engineering. **2005,** *15(3)*, 8–24. Analytical devices: Mili- micro and nanohorizons. (In Rus: Борзенко А. Г. Аналитические приборы: мили—, микро, и нано - горизонты // Научное приборостроение. 15 **2005,** *3,* 8–24.)
106. Kim, K.; Lee J-B. MEMS for drug delivery, Chapter 12. In: WangW, Soper, S. A.; eds. Bio-MEMS: Technologies and Applications. Boca Raton, FL: CRC Press; **2006,** 325–348.
107. Santini J. T. Jr., Richards A. C, Scheidt R. A, Cima M. J, Langer R. S.; Microchip technology in drug delivery. Ann Med, **2000,** *32*, 377–379.
108. Santini Jr, J. T.; Cima, M. J. Langer, R. A controlled-release microchip. *Nature*, **1999,** *397,* 335–338.
109. 99. Gardner, D. P.; Microfabricated nanochannel implantable drug delivery devices: trends, limitations and possibilities. Exp Opin Drug Deliv, **2006,** *3*, 479–487.
110. Shrivistava, S. D.; Dash Applying nanotechnology to human health. J Nanotech **2009,** 1–14.
111. US Food and Drug Administration. Nanotechnology page. Available at: http://www.fda. gov/nanotechnology/. (Accessed February 1, 2010).
112. The Personalized Medicine Coalition. Home page. Available at: http://www.personalized medicinecoalitition.org/sciencepolicy/personalmed-101 overview.php. (Accessed January 2, 2010).
113. Amgen J. Future of biotechnology in healthcare, Chapter 9. In: An Introduction to Biotechnology. **2009,** 31–35. Available at: http://www.amgenscholars.eu/web/guest/futureof-biotechnology-in-healthcare (accessed January 2, 2010).
114. Heller. Integrated medical feedback systems for drug delivery. Am Inst Chem Eng, J.; **2005,** 51, 1054–1066.
115. Brunetti, P.; Federici, M. O.; Benedetti, M. M.; The artificial pancreas. Artif Cells Blood Substit Immobil Biotechnol. **2003,** *31*, 127–138.
116. M.; Staples. Microchips and controlled release drug reservoirs. WIREs Nanomed Nanobiotechnol**2010,** 2, 400–417.
117. Genzyme website. The cost of enzyme replacement therapy. Available at: http: // www.genzyme.com/commitment/patients/costof treatment.asp. (Accessed January 2, 2010).
118. Shoji, S.; Esashi M.; Microflow devices and systems. J.; Micromech Microeng. **1994,** *4,* 157–171.
119. Amirouche, F.; Zhou, Y.; Johnson T.; Current micropump technologies and their biomedical applications. Microsys Tech, **2009,** *15,* 647–667.

CHAPTER 13

INFLUENCE OF VARIOUS METAL/ CARBON NANOCOMPOSITES ON CHANGES IN THE PROPERTIES OF COMPOSITIONS ON LIQUID GLASS BASIS

L. F. AKHMETSHINA, V. I. KODOLOV, and G. E. ZAIKOV

CONTENTS

13.1 Introduction 442
13.2 Experimental Part 442
 13.2.1 Nanocomposites Structure 442
 13.2.2 Equipment and Investigation Technique 444
13.3 Results and Discussion 446
 13.3.1 Preparation of Fine Suspension 446
 13.3.2 Analysis of Compositions on Liquid Glass Basis 447
13.4 Conclusion 454
Keywords 454
References 455

13.1 INTRODUCTION

This chapter presents the investigation results of the influence of metal/carbon nanocomposites on heat-physical and viscous properties of silicate compositions. The introduction of nanocomposite facilitates the decrease in material temperature conductivity by nearly 50%. The article discusses the investigation results of liquid glass suspensions with IR spectroscopy. The changes in liquid glass properties depend on nanocomposite. It is found that relative viscosity of iron/carbon nanocomposites decreases, if compared with nickel/carbon.

Nanotechnology is based on self-organization of nanosize particles resulting in the improvement of material properties and obtaining of products and items with unique characteristics. The modification of silicates and liquid glass with nanocomposites containing metal clusters is perspective. The interest to liquid glass is conditioned, first of all, by its ecological friendliness and production and application simplicity, inflammability and nontoxicity, biological stability, and raw material availability. Nanostructures can be used to improve such characteristics of paints as elasticity, adhesion to the base, hydrophobic behavior and also helps solving the problems connected with coating flaking-off from the base, discoloration, limited color range, and so forth. Due to the unique properties of nanostructures we can apply new properties to silicate paints, for example, to produce the coating protecting from electromagnetic action.

The application of nanostructures in silicate materials to improve their stability to external action is known [1].

The introduction of nanosize structures into the liquid glass allows qualitatively changing the material behavior in a positive way. This possibly occurs in the process of binder supermolecular structure change.

Based on the aforesaid, it is important to improve operational properties of compositions on liquid glass basis modifying them with metal/carbon nanocomposites, providing a wider field of their application.

The aim of this chapter is the investigation and analysis of the influence of various metal/carbon nanocomposites on changes in the properties of compositions on liquid glass basis.

13.2 EXPERIMENTAL PART

13.2.1 NANOCOMPOSITES STRUCTURE

To modify the materials nickel and iron containing nanocomposites were obtained with low-temperature synthesis in nanoreactors of polymeric matrixes of polyvinyl chloride [2] and polyvinyl alcohol [3]. Iron oxide (3) was selected as metal containing phase, which in the process of nanostructures obtaining is reduced to magnetite, and nickel oxide, which is also reduced.

To define the sizes, shape and structure the nanoproduct obtained was investigated with transmission electron microscopy (TEM) and electron diffraction (ED). Iron/carbon nanocomposite (Fe/C NC) is mainly represented as film nanostructures with metal inclusions (Fig. 1).

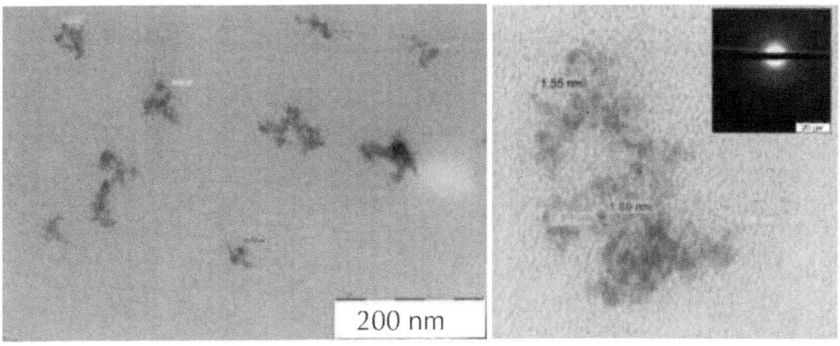

FIGURE 1 TEM picture of Fe/C nanocomposite.

The particles of metal containing phase have a globular shape and are located between the layers of carbon films connected with them. There are also nanostructures close to a spherical shape. In TEM pictures of Fe/C NC dark spots characterize metal, light ones—carbon films. Electron-diffraction patterns are presented as ring reflexes. This indicates the availability of amorphous phase in the sample. The average size of particles constituting the aggregates is 17 nm.

Fig. 2 demonstrates large particles and aggregations of nickel clusters. A significant number of particles have the shape close to spherical, however, there are elongated particles as well. Globular film nanostructures with the inclusion of nickel nanoparticles and its compounds are present in the samples. The electron-diffraction pattern is presented as ring reflexes, however dot reflexes are also present. This indicates the availability of both crystalline and amorphous phases in the sample. The average size of particles is 11 nm.

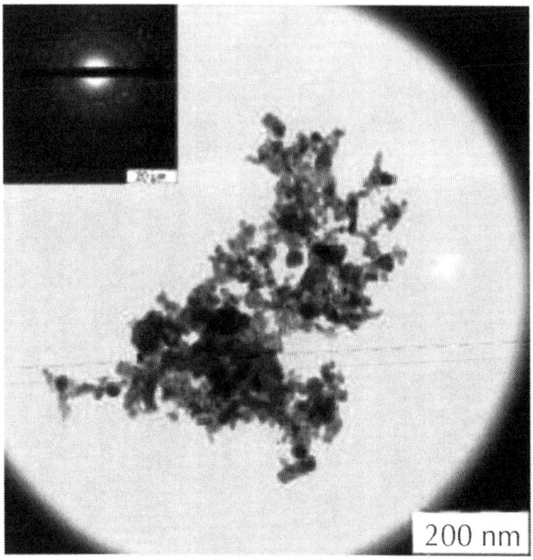

FIGURE 2 TEM picture of Ni/C nanocomposite.

13.2.2 EQUIPMENT AND INVESTIGATION TECHNIQUE

13.2.2.1 SPECTROPHOTOMETRY

To define the interaction between liquid glass and nanocomposite the optical density of liquid glass sample films with nanostructures was investigated. The films were produced applying a thin layer of silicate modified with nanostructures onto transparent film with further drying to remove the moisture. Films without nanocomposites were used as the reference. The investigations were carried out with spectrophotometer KFK-3-01 in the wave range 950–350 nm.

13.2.2.2 ULTRASOUND PROCESSING OF SUSPENSIONS

Spectrophotometer KFK-3-01 was also used to define the optimal action time interval on nanocomposite fine suspension processed with ultrasound. In the process of selecting the work wavelength (λ) on spectrophotometer KFK-3-01 to define the optimal time interval of ultrasound processing of fine suspension on liquid glass basis, the optimal wavelength of 430 nm was found, as with this wavelength the optical density of liquid glass with NC before the US processing was 0.818, thus corresponding to the midpoint of the work range of this instrument (work range: optical density (D) = 0.001–1.5). At λ = 430 nm the optical densities of all fine suspensions investigated were found. The optical density was measured for the samples of nanocomposite fine suspension and liquid glass with the concentration 0.003% processed in ultrasound bath Sapfir UZV 28 with US power 0.5 kW and frequency 35 kHz. Liquid glass not modified with nanostructures was used as a reference sample. The processing time with maximum D was selected as optimal.

13.2.2.3 INVESTIGATION OF HEAT-PHYSICAL CHARACTERISTICS

Heat capacity of the samples was investigated with dynamic technique measuring the specific heat capacity of solids with c-calorimeter IT-c-400 [4]. The measuring device was based on comparative technique of c-calorimeter with heat meter and adiabatic shell. Temperature delay time on heat meter was measured with stopwatch at room temperature and the specific heat capacity value was found by the following formula:

$$C = K_T/m_0(\tau_T - \tau_T^{\circ}),$$

where C is the specific heat capacity, J/kg*K; m_0 is the sample mass, g; K_T is the instrument constant which depends on the temperature and is found by the instrument calibration; τ_T° is the temperature delay time with empty ampoule, s; τ_T is the temperature delay time on heat meter, s.

λ-calorimeter IT-λ-400 based on monotonous heating mode was used to measure heat conductivity. In the experiment, the readings of n_0 is the temperature difference by plate thickness was measured with a stopwatch by the following formula to define the heat resistance:

$$R_s = ((1 + \sigma_c)\, S_{n0}/K_T \times n_T) - R_K,$$

where R_s is the sample heat resistance, $m^2 \times K/W$; S is the sample area, m^2; $\sigma_c = C_0/2(C_0 + C_c)$ is the allowance taking into account the sample heat capacity C_0 (C_0 is the total heat capacity of the sample tested, J/(kg*K); found before with the instrument IT-s-400; C_c, K_T, and R_K are the constants of the measuring instrument. The time was measured with a stopwatch with the accuracy 0.01 s.

13.2.2.4 DETERMINATION OF RELATIVE AND ABSOLUTE VISCOSITY OF SUSPENSIONS

Relative viscosity of the modified liquid glass was found with viscometer VZ-246 [5]. The viscometer was placed in the rack and fixed horizontally with the balance. The vessel was placed under the viscometer nozzle. The nozzle hole was closed with a finger, the material to be tested was poured into the viscometer to excess in order to have convex meniscus above the upper edge of the viscometer. The viscometer was filled slowly to avoid air bubbles. Then the nozzle hole was opened and as soon as the material being tested is appeared, the stopwatch was started, then stopped and the discharge time was calculated. The values of relative viscosity were translated into the kinematic based on Russian Standard 8420-74.

The absolute (dynamic) viscosity of liquid glass was measured with a falling sphere viscometer (Geppler viscometer) intended for precise measurement of transparent Newton liquids and gasses. In accordance with Geppler principle, the liquid viscosity is proportional to the falling time being measured. The falling time of a sphere rolling down inside the inclined pipe filled with the liquid being tested was measured. The time required for the sphere to cover the distance between the upper and lower ring marks on the pipe with the sample was measured with a stopwatch with the resolution 0.01 s. Further the values of dynamic viscosity were calculated.

13.2.2.5 IR SPECTROMETRY TECHNIQUE

IR Fourier-spectrometer FSM 1201 was used to obtain IR spectra of suspensions based on liquid glass and nanocomposites. The spectra were taken in the range of wave numbers 399–4500 cm^{-1}.

13.3 RESULTS AND DISCUSSION

13.3.1 PREPARATION OF FINE SUSPENSION

According to reference sources, the most advantageous and urgent technique to introduce nanostructures into material is the application of fine suspensions of nanoproducts. They provide the fine modifier uniform distribution through the volume of the material modified.

The necessary concentration of nanostructures in suspension was chosen based on the basic binder mass and varied from 0.003 up to 0.3%. The suspensions were prepared in mechanical mortar mixing nanocomposites with liquid glass following the preselected mode with further processing in ultrasound bath

(Fig. 3) to prevent the nanostructure coagulation. The metal in nanostructures will possibly interact with liquid glass forming insoluble silicates. Thus, the hydrosilicic acid is isolated from the liquid glass as an insoluble gel that results in compressing the liquid glass structure and increasing its moisture stability.

13.3.2 ANALYSIS OF COMPOSITIONS ON LIQUID GLASS BASIS

13.3.2.1 ULTRASOUND PROCESSING

According to reference sources, the ultrasound technique is an effective dispergation method. Such techniques are widely applied today. However, the lengthy soaking in ultrasound baths can result in partial or complete decomposition of nanostructures and their recoagulation. The decomposition of nanostructures due to lengthy ultrasound action is conditioned by high temperature and pressure during the cavitation. The coagulation can be caused with the decomposition of solvate shell on disperse phase particles. It is necessary to define the processing time interval during which the optical density will be at the maximum, i.e., it will correspond to the maximal saturation of nanocomposite suspension.

To define the time interval of ultrasound processing four suspensions were prepared for each soaking period—3, 5, 10, 15 min, respectively, and also the reference solution without US processing—for comparison. The concentration in all solutions was 0.003%.

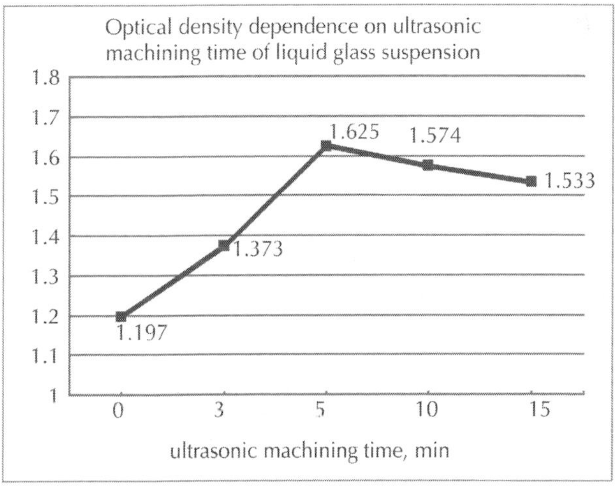

FIGURE 3 Diagram of suspension optical density on liquid glass.

Thus the optimal time interval for ultrasound processing is 5 min. Further processing of suspensions for investigation will be carried out within this interval.

13.3.2.2 SPECTROPHOTOMETRIC INVESTIGATIONS

In accordance with the results of spectrophotometric investigations of modified films in the glue sample with Fe and glue with Ni the shift of optical density is observed at some wavelengths, indicating the changes taking place when nanostructures are introduced (Fig. 4).

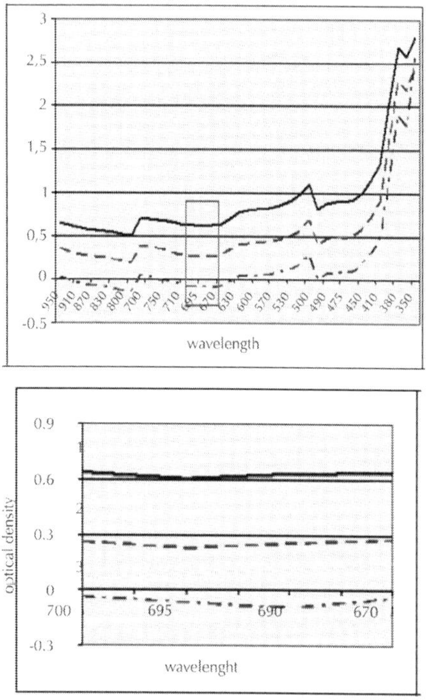

FIGURE 4 Shift of optical density for Fe in relation to Ni and liquid glass (1—Fe/C NC and liquid glass, 2—Ni/C NC and liquid glass, 3—liquid glass).

At other wavelength values the optical density only increases indicating the possibility to apply nanostructures as coloring pigments in paints. At the same time, the increase in the optical density of films with nanostructures in comparison with liquid glass films possibly indicates the increase in the density of compositions and formation of new structural elements in them.

13.3.2.3 INVESTIGATION OF HEAT-PHYSICAL PROPERTIES

To obtain the details of changes in heat-physical characteristics, the heat capacity and thermal conductivity of the samples based on cardboard and modified liquid glass were investigated. The samples were prepared gluing several layers of cardboard with liquid glass modified with nanostructures. The sample dimensions were found by the technique for measuring heat capacity and thermal conductivity. The sample being tested was 15 mm in diameter and 10 mm in height. The sample dimensions were measured with micrometer with 0.01 mm accuracy. The sample mass was measured with the allowance not exceeding 0.001 g. Heat capacity of the samples was investigated on c-calorimeter IT-s-400. To find thermal conductivity the calorimeter IT-λ-400 was used. Further the specific thermal conductivity of the sample was calculated as follows:

$$\lambda = h/R_s,$$

where λ is the specific thermal conductivity, W/m*K; h is the sample thickness, m.

The results of heat-physical investigations are given in Table 1.

TABLE 1 Heat-physical characteristics of the samples.

Samples	Cardboard / Glue	Cardboard / Glue with Fe (change in %)	Cardboard / Glue with Ni (change in %)
Density, kg/m^3	624.5	744 (↑19%)	669 (↑7%)
Heat capacity C_{spec}, J/kg*K	1790	2156 (↑20%)	2972 (↑66%)
Thermal conductivity λ, W/m*K	0.083	0.061 (↓27%)	0.064 (↓23%)

Heat capacity of the sample with Fe increased by 20% in comparison with the reference, sample with Ni by 66%. At the same time, thermal conductivity decreased by 27% and 23% for the samples with Fe and Ni, respectively. So when nanostructures are introduced, in the average the characteristics change as follows: density increases by 13%, heat capacity by 40%, thermal conductivity decreases by 25%. Further, using the experimental results, the temperature conductivity is calculated by the following formula:

$$a = \lambda / c\rho,$$

where a is the temperature conductivity coefficient, λ is the heat conductivity coefficient, c is the heat capacity, ρ is the density.

Inserting the experimental data into the formula, we can define the temperature conductivity values (in percent):

$$a = \lambda/c\rho = 0.75\lambda_0/1.4c_0 \times 1.13\rho_0 = 0.47a_0.$$

Or calculate separately:

$$a_1/a_0 = \lambda_1\rho_0C_0/\lambda_0\rho_1C_1 = 0.51$$
$$a_2/a_0 = \lambda_2\rho_0C_0/\lambda_0\rho_2C_2 = 0.43$$

where a_1, λ_1, ρ_1, C_1 is the characteristics of the sample with Fe/C NC; a_2, λ_2, ρ_2, C_2 is the characteristics of the sample with Ni/C NC; a_0, λ_0, ρ_0, C_0 is the characteristics of nonmodified sample.

Thus, the temperature conductivity decreased by nearly 50% in comparison with the initial values (a_0).

When nanostructures are introduced, self-organization takes place. Nanoparticles structure the silicate matrix leading to the formation of new elements in the structure, thus increasing the material density and influencing its heat-physical characteristics. When additional structural elements and new bonds are formed, the system internal energy increases leading to heat capacity elevation and, consequently, temperature conductivity decrease. Thermal conductivity decrease of silicate paints when applied as a coating allows improving heat-physical characteristics of the whole protective structure of a building. In turn, temperature conductivity decrease results in decreasing the amount of heat passing through the coating, thus preserving adhesive characteristics of the coating for a long time.

13.3.2.4 IR SPECTROSCOPY

To determine the interactions between nanostructures and liquid glass the IR spectroscopy investigations were carried out as well. The spectra were taken in relation to water as the suspensions contained nanocomposite solution with liquid glass and water (Fig. 5). The spectra were read with the reference tables. Water spectra demonstrate the region 2750–3750 cm^{-1} connected with O–H bond oscillations. They were observed in other regions as indicated by the values of wave numbers: 2380, 3850, and 3840 cm^{-1}. At the same time, the values 3200, and 3500 cm^{-1} appropriate for valence bound OH appeared in the spectra of nanostructures on Fe basis. In the region 1600–1650 cm^{-1} the bands of deformation oscillations of OH-groups were observed. Bands appropriate to the oscillations of Si-O-Me and Si-O-bonds (1100–400 cm^{-1}) [5] were seen in low frequency region.

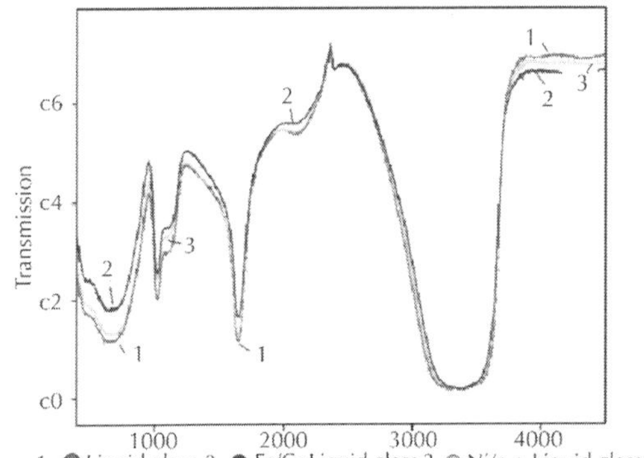

FIGURE 5 IR spectrum of suspensions on the basis of liquid glass and various types of nanocomposites (1—liquid glass, 2—Fe/C NC and liquid glass, 3—Ni/C NC and liquid glass).

IR spectra contained wave numbers reflecting the interaction between metal/carbon nanocomposites and liquid glass, for example, peaks with wave numbers 407–420 cm^{-1} (Si-Me). The spectra of suspensions vividly demonstrated peaks at 1020 and 1100 cm^{-1}, appropriate to the oscillations of Si-O-Si and Si-O-C. However, in the spectrum of liquid glass containing Fe/C nanocomposite those values were shifted from 1104 to 1122 cm^{-1}, there was also the shift from 600 to 660 cm^{-1} caused by the change in oscillations of some bonds under the nanocomposite action. The spectrum obtained resembles the liquid glass spectrum from reference sources [5].

The spectra demonstrate the absence of strong interactions, mostly the intensity changes thus indicating the formation of a large number of definite bonds, for example, Si-O showing the structuring of liquid medium.

For suspensions on liquid glass basis the relative and absolute viscosity were found. Taking into account the aforementioned data for defining the optimal ultrasound processing time, the suspensions for viscosity test were prepared with ultrasound preprocessing within 5 min.

13.3.2.5 DETERMINATION OF RELATIVE AND ABSOLUTE VISCOSITY OF SUSPENSIONS

Dynamic viscosity is the ration between force unit required to shift the liquid layer for distance unit and layer area unit. In metric system it is given as dyne-second per square centimeter called "poise."

$$\eta = \tau \cdot (\rho_1 - \rho_2) \cdot \kappa \cdot F,$$

where η is the dynamic viscosity [mPa·s], τ is the sphere movement time [s], ρ_1 is the sphere density according to the test certificate [g/cm^3], ρ_2 is the sample density, [g/cm^3], κ is the sphere constant according to the test certificate [mPa·cm^3/g], F is the work angle constant.

The experiment was carried out with pipe inclination for the sample 80° => F = 1.0

ρ_1 = 15.2 g/cm^3

ρ_2 = 1.45 g/cm^3

κ = 0.7

Liquid glass viscosity: τ_{cp} = 9.60 s

$$\eta = 9.6(15.2\text{–}1.45) \times 0.7 \times 1 = 92.4 \text{ Pa·s}$$

Viscosity of liquid glass modified with iron containing nanocomposite:

$$\eta = 6.97(15.2\text{–}1.45) \times 0.7 \times 1 = 67.09 \text{ Pa·s}$$

Viscosity of liquid glass modified with nickel containing nanocomposite:

$$\eta = 8.76(15.2\text{–}1.45) \times 0.7 \times 1 = 84.32 \text{ Pa·s}$$

Kinematic viscosity is the ratio between dynamic viscosity and liquid density.

Based on the data obtained when measuring the viscosity of liquid glass and fine suspension with Fe/C and Ni/C NC, the diagrams were constructed (Figs. 6 and 7).

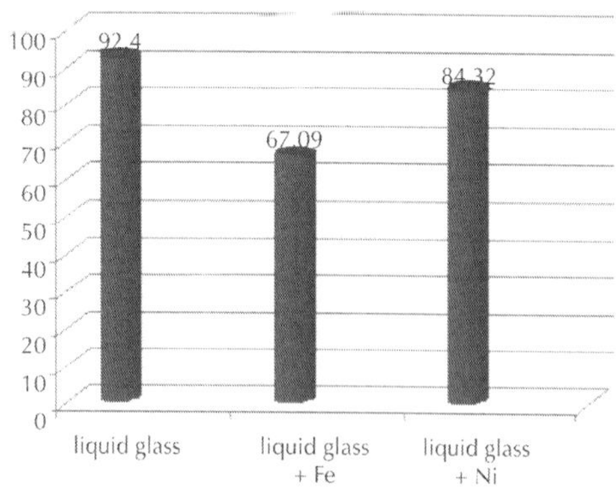

FIGURE 6 Dynamic viscosity.

Having analyzed this diagram, we can conclude that when a nanocomposite is introduced the viscosity decreases. Such phenomenon will have a positive effect in the process of silicate material production and application. For example, better application—in silicate paint, much easier foaming process—in foam glass.

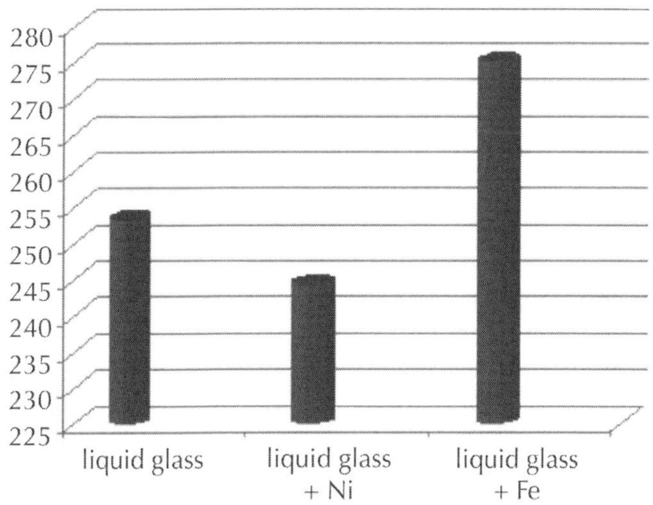

FIGURE 7 Kinematic viscosity

In Fig. 7, we can see that nanostructures work differently. The viscosity of fine suspension with Ni/C NC is less than kinematic viscosity with Fe/C NC. This can be caused by coagulation, aggregation of Fe/C NC (Table 2).

TABLE 2 Characteristics of kinematic and dynamic viscosity of liquid glass and its analogs modified with Ni/C and Fe/C nanocomposites.

Sample	Kinematic viscosity, $mm^2 \cdot s^{-1}$	Dynamic viscosity, Pa·s (change in %)
Liquid glass	253.3	92.4
Liquid glass with Ni/C nanocomposite	244.3 (↓4%)	84.32 (↓9%)
Liquid glass with Fe/C nanocomposite	275.1 (↑9%)	67.09 (↓27%)

The viscosity of compositions on liquid glass basis depends on nanocomposite composition: in some cases the viscosity increase is observed, in other cases—the decrease. The viscosity increase is connected with the elevation of intermolecular friction, which can be caused by the influence of nanoparticles on liquid glass structuring and density increase. The decrease in suspension viscosity can be explained by the fact that NC is surrounded by liquid glass molecules resulting in the decrease in their interactions and intermolecular friction.

13.4 CONCLUSION

Based on the results obtained it can be concluded that nanostructures have a positive effect on liquid glass properties.

The optimal time interval of suspension ultrasound processing was found experimentally. It equals 5 min and corresponds to the maximum solution saturation with nanocomposites.

The data of optical density investigation of liquid glass films and IR spectra of liquid glass solutions indicate the changes in material structure when introducing metal/carbon nanocomposites.

The changes in heat-physical and viscous properties of suspensions on liquid glass indicate the self-organization processes when introducing nanostructures. At the same time, the results differ when introducing nanocomposites of different compositions.

The application of nanocomposites will allow increasing the life of paints and, consequently, their storage and covering ability. Also not only potassium but sodium liquid glass can be used as well, since potassium liquid glass is significantly more expensive than sodium one and its manufacturing is limited. To improve the characteristics only small concentrations of nanoproduct are required, which will positively affect the material production cost.

KEYWORDS

- **Geppler principle**
- **IR spectroscopy**
- **metal/carbon nanocomposites**
- **poise**
- **silicate compositions**

REFERENCES

1. Kodolov, V. I.; Khokhriakov, N. V.; In: Chemical physics of formation and transformation processes of nanostrcutures and nanosystems. Izhevsk: Izhevsk State Agricultural Academy, **2009,** *1,* 365 p, *2,* 415 p.
2. Patent 2393110 Russia Technique of obtaining carbon metal containing nanostructures; Kodolov, V. I.; Vasilchenko, Yu. M.; Akhmetshina, L. F.; Shklyaeva, D. A.; Trineeva, V. V.; Sharipova, A. G.; Volkova, E. G.; Ulyanov, A. L.; Kovyazina, O. A.; declared on 17.10.2008, published on 27.06.2010.
3. Patent 2337062 Russia Technique of obtaining carbon nanostructures from organic compounds and metal containing substances; Kodolov, V. I.; Kodolova, V. V. (Trineeva), Semakina, N. V.; Yakovlev, G. I.; Volkova, E. G., et al.; declared on 28.08.2006, published on 27.10.2008.
4. Platunov, E. S.; Buravoy, S. E.; Kurepin, V. V.; Petrov, G. S.; In: Heat-physical measurements and instruments. Platunov E. S., ed.; L.: Mashinostroenie, 1986.
5. Plyusnina I. I. In: IR spectra of silicates. M.: Publishing House of Moscow State University, **1967,** 190 p.

INDEX

A

Acetate fibers, 57
Adhesive strength determination, 352
Alcoholates, 325
Aluminous silicates, hydrophility, 316
Ammoniumfluorine hectorite, 333–334
Amorphous region, 25
Analysis of variance (ANOVA)
  CA, electrospun fiber mat, 86
  design, experiments and response, 84–85
ANN. *See* Artificial neural network (ANN)
Antibiotics, 423
Anti-inflammatory drugs, administration, 421–422
Antireflective coating, 409
Artificial fibers, biodamaging
  acetate fibers, 57
  bambocane, 57
  cellulose based, 57
  chemical structure and microbiological stability, 57
  modified viscose fibers, 58
Artificial neural network (ANN), 83–84
Atomic-force microscopy, 285
Attraction force, quantum-chemical modeling, 344–345
Axial pressure gradient, 190

B

Bacillus mesentericus, 364
Bast fiber
  complex fibers, 55
  definition, 54
  fiber separation methods, 55
  hemp stalks, 54
  linen complex fiber, 55
  natural biostability, 56
Bentonite, 315
Berkovich indentor, 286, 304
Bimetallic bonding, 269
Binary hookings, 7
Bioartificial (hybrid) implants, 426
Biocompatibility, OIs, 419, 422–423
Biodegradable polymers, 428
Biological nanoparticles, 278
Bisphosphonates, 424
Black body, 135
Brownian Dynamics (BD), 211–212
BSR. *See* Butadiene-stirene rubber (BSR)
Butadiene-stirene rubber (BSR), 285, 287–288, 299

C

Cahn–Hilliard–Cook (CHC) nonlinear diffusion equation, 213
Calorimeter data, 331–332
Carbon-nanoshungite-microshungite, 292
Carbon nanotubes and related structures
  capillary effects, 160–161
  carbon nanotubes, 152–154
  chirality
    graphite plane, atomic structure, 157–158
    integer positive indices, 157
    lattice vector, indices, 159
    semiconductor, 158
    single-walled carbon nanotubes, 159
  classification
    multi-walled nanotubes, 156–157
    single-walled nanotubes, 155
  effective critical inner shear stress, 201–202
  fluid flow
    average flow velocity and applied pressure gradient, 193

axial pressure gradient, 190
dry friction, 186
empty interlayer, 202–205
enhancement water flow, measured values, 188
flow enhancement, 190
Hagen-Poiseuille flow, 187
molecular modeling, 189
nanohydromecanics, modeling (*see* Nanohydromecanics)
nanoscale friction, computer model, 186
Navier-Stokes equation, 187
nonlinearity, 192
no-slip poiseuille flow, 194
pressure gradient, 189
radial velocity profile, 192
slip coefficient, 193
velocity profile, 194
viscosity and slip length, variation, 189
volumetric flow, dependence, 189
volumetric flow rate, 187, 191, 195
fullerene, 154–155
graphene, 151–152
slippage of fluid particles
cases, 172–173
classical continuum fluid dynamics, 172
collapse, energy release, 181–185
"effective slip length", 174
effective viscosity, 177–181
liquid layer, density, 174–176
"slip length", 172
time average streaming velocity profiles, 174
"zigzag" , chiral vectors, 173
specific electrical resistance (*see* Specific electrical resistance)
Cardiavascular stents
antiproliferation effect, 417–418
biodegradable polymeric materials, 419
cardiovascular pathologies, 417
coating, 418
construction, 417
eye diseases, 428
nanoporous surface structures, construction, 417–418
restenosis, 417
Caveolae-mediated endocytosis (CvME), 116, 117, 280–281
Cell behavior, biochemical mediators, 419
Cellulose
biostability, 53
degrading microorganisms, 48–49
fibrils, 46
Central composite design, 38–39
Cerezyme implant, 433
Chemical assembly, 107
Chemical fibers, 43
Chiral vector, 157
Chitosan (CHT)
chemical structures, 14
degree of deacetylation, 15–16
mechanical and electrical properties, 15
in nanofiber form, 14
CHT. See Chitosan (CHT)
Clathrin-mediated endocytosis (CME), 116, 280–281
Clusters, 196
CME. See Clathrin-mediated endocytosis (CME)
Coating stability analysis, 123
Collagen, 13–14
Collapse, energy release
"domino effect", 182–185
elastic deformation, 182
kinematic characteristics, 184
nano cannon scheme, 185
"nanogun", 184–185
single-walled carbon nanotube, 182
squashing, 181
time dependences, 184
Conduction, 134–135
Contraception, 425
Convection, 133–134
Copper/carbon nanocomposite, 344

Copper/carbon (Cu/C) nanocomposite. See Polycarbonate modification, Cu/C nanocomposite
Cotton fabric dyeing technologies, 46
Cotton fiber, 46–47
cellulose
biostability, 53
degrading microorganisms, 48–49
enzymes and metabolites, 49
external microflora, 54
first and fifth quality grade, 49–52
fungi, 49, 54
infection, 48
maturity, 48
micrographs, 53
microorganisms effect, 53
storage, 48
surface, spontaneous microflora impact, 49, 51
swellings, 54
CvME. See Caveolae-mediated endocytosis (CvME)
Cytotoxicity, 424

D

DCM. See Dichloromethane (DCM)
Dead space, 168
Diblock polymer–particle composites, 214
Dichloromethane (DCM), 15
Dielectrics, 136
Dissipative particle dynamics (DPD), 212
"Domino effect", 182–185
DPD. See Dissipative particle dynamics (DPD)
Dry friction, 186
DSSC. See Dye-sensitized solar cells (DSSC)
Dye-sensitized solar cells (DSSC)
advantages, 394
"black dye", 393
DNH2, 393–394
electrolyte, 394
high performance dye, 393
oxide semiconductors, 392
photoelectrodes, 392–393
photosynthesis process, mimic, 392
schematic, 392
Dynamic DFT method, 213–214

E

EBID process. See Electron-beam-induced deposition (EBID) process
EDS. See Energy dispersive spectroscopy (EDS)
Effective critical inner shear stress, 201–202
Effective reinforcing modulus (ERM), 222
Elastic deformation, 182
Elastomeric matrix, 295
Elastomeric nanocomposites (rubbers) nanofiller particles, 284
Elastomeric polymer materials, 284–285
Electron-beam-induced deposition (EBID) process, 256–263, 268, 272. See also Fabrication and characterization
Electron spectroscopy, chemical analysis, 107
Electrospinning process
factorial experimental design, 96
image, 13–14, 94
intrinsic viscosity and molecular weight, 95
Mark-Houwink-Sakurada equation, 95
mean fiber diameter (MFD), 98–100, 102–103
parameters, 95–96
p-values, models, 99
RSM, 97
second testing hypothesis, 99
solution
preparation, 16–17
viscosity, 94
spinning distance, 95
standard deviation of fiber diameter (StdFD), 98, 100–101, 103
statistical analysis, models, 98, 102
volume flow rate effect, 95
Electrospun fibers
ANN, 83–84, 87–88
and RSM, 91–93
weights and bias obtained, 87

ANOVA
CA, electrospun fiber mat, 86
design, experiments and response, 84–85
characteristics, 14
measurement and characterization
CA measurement set up, 82, 91
gold-sputtered, 81
parameters effect
contact angle, contour plots, 89
interfibrillar distance, 88
nanoscale fibers, 88
relative importance (RI), 90
solution concentration and spinning distance, 88
RSM
and ANN model, 91–93
ANOVA, 83
CCD, 82
experiment design, 83
regression analysis, 83
scanning electronic micrographs, 22–23, 79–80
Empty interlayer, 202–205
Endometriosis, 425
Energy dispersive spectroscopy (EDS), 261
Epicuticle, 58
Epoxy polymers, fragments interaction
absolute binding energies, 346
binding energy, 345
bond lengths, 347
Cu/C NC modification, 350–351
effective charges, 348
fine suspensions (FS), 345
hardening process, 348
initial substances, fragments, 345
magnetic properties, 349
molecular complexes, interaction energies, 347
PEPA interaction, 347
polyethylene polyamine (PEPA), 349
ERM. See Effective reinforcing modulus (ERM)
Ethylene dichloride, 379

F

Fabrication and characterization
accelerating voltage effect, crystallization
bright-field TEM micrographs, 266
electron beam, 266
HRTEM micrographs, 267–268
Lattice fringes, 268
nanodendrites, dependence of length, 267
nanometer-sized area diffraction pattern, 267
nanometer-sized W-dendritic structure, 269
electron-beam-induced deposition (EBID) process, 256
high-energy electron irradiation, 256
insulator substrates, 257
metal nanostructures, 256
nanodendrites, composite nanostructures
bimetallic bonding, 269
Bright-field TEM image, 270–271
insulator substrates, 269
Lattice fringes, 269–270
Pt-nanoparticles, 269
selected-area diffraction (SAD) pattern, 270
nanofractal-like trees, 256
nanometer-sized structures, 257
nanosized branched structures
bright-field TEM micrographs, 262
EBID process, 257
electron beam, 262
energy dispersive spectroscopy (EDS), 261
fabricated nanodendrites, 261
growth mechanism, 263–265
high resolution TEM (HRTEM) micrograph, 262–263
nanodendrite structures, 258, 262

tungsten hexacarbonyl, 257
Whisker-like deposits, 258
W-nanodendrite structures, 258–259
W nanowhisker arrays, 260
nanowire arrays, 256
transmission electron microscopes (TEMs), 257
FEM. See Finite element method (FEM)
Fine suspensions (FS), 345, 349–350
adhesion, 356
IR spectroscopic investigation, 355
polycarbonate modification, Cu/C nanocomposite, 378
thermogravimetric investigations, 356–357
ultrasound processing, optimal time period, 352–353
viscous properties, 353–354
Finite element method (FEM)
continuum theory, 221
equivalent-continuum modeling, 221
Hookian potential energy term, 221
mesh generation, 220
polymer composites, 221
subdomains, 220
uses, 220
Flammability, polymeric nanocomposites, 335
Fourier transform infrared spectra (FT-IR), 17, 19–20
FS. See Fine suspensions (FS)
Fullerene, 154–155, 233
Fullerene-containing mineral shungite, 285

G

Gelatin, 13–14
Gelatin-immobilized matrix (GIM) systems, 362–363
Gelatin-immobilized silver(I) hexacyanoferrate(II), 369
Gelatin matrix, 362, 364
Geppler principle, 446
GIs. See Gynitourinary implants (GIs)
Graphene, 151–152
Green nanofibers
biocompatibility and biodegradability, 14
chitin, 13
chitosan (CHT)
chemical structures, 14
degree of deacetylation, 15–16
mechanical and electrical properties, 15
in nanofiber form, 14
CHT-MWNTS dispersions
CHT concentration effect, 22
FT-IR spectra, 19–20
morphologies, 22–24
organic acid solution, 16
protocols, 16
scanning electronic micrographs, 20–21
concentrated organic acids, 15
dynamic light scattering (DLS), 18
electrospinning process
image, 13–14
solution preparation, 16–17
electrospun fibers, scanning electronic micrographs, 22–23
electrospun nanofibers, characteristics, 14
hydrodynamic diameter distributions, 18
measurements and characterizations method
conductivity, 17
electrospun CHT/MWNTs nanofibers, 17
Fourier transform infrared spectra (FT-IR), 17
nanofiber thin film, setup, 18
modeling
average fiber diameter, 39
central composite design, 38–39
electrospinning parameters, effect, 40
independent variables, 38

Minitab and Mathlab programs, 39
Regression analysis, 39–40
RSM, 38
runs, 38
MWNTs bundle, 19
nanofiber fabrics, 13
natural polymer, 13
polyethylene oxide (PEO)/CHT, 14–15
silk fibers (see Silk fibers)
Gynitourinary implants (GIs)
bioartificial (hybrid) implants, 426
biodegradable polymers, 428
cell technology, 426
contraception, 425
endometriosis, 425
high-molecular-weight implants, 427
hydrogel membranes, 427
nanobiomedicine, application, 425
nanotechnology, 425
nonbiodegradable materials, 427
polymer matrix, 427
polymer-metal device, 425
targeted drug delivery, 425
tissue engineering, 427
urology, tissue engineering, 426

H

Hagen-Poiseuille flow, 187
Halpin–Tsai model, 217–218
HDPE. See High-density polyethylene (HDPE)
Heat and liquid moisture transport, 132–133
High-density polyethylene (HDPE), 2
High-molecular-weight implants, 427
Hydrodynamic diameter distributions, 18
Hydrogel membranes, 427
Hygroscopic fibers, 139

I

IMD. See Implanted medical device (IMD)
Implanted medical device (IMD), 414–415
Insulator substrates, 269
Integrinbinding proteins, 420–421
IR spectrometry technique, 442, 446
IR spectroscopy, 351, 450–451

L

Lattice Boltzmann (LB)
binary fluids, 212
disadvantage, 213
feature, 213
fluid particles interactions, 213
lattice gas automaton, 212
Lattice fringes, 269–270
Layered silicates, structure
bentonite, 315
clays, bentonite breeds, 314
inorganic layer, 315
inorganic structures, 314
modification, 315–317
montmorillonite, 314–315
structure, 315
LB. See Lattice Boltzmann (LB)
Light bullets, 242, 246
Linear spatial structures, 284
Linen fiber
flax biostability, 56
microbiological damages, 56
spread straw, 56
steamed flax, 56
Lipid based nanoparticles, 277

M

Macropinocytosis, 117–118, 281–282
Macropinosomes, 117, 282
Magnetically targeted nanosystems (MNSs), 118
Magnetic nanoparticles (MNPs), 282–283
bovine serum albumin (BSA) adsorption, 119
drugs delivery, 118
MNSs, 118
proteins, nonselective linking, 119
scheme, 118
Magnetic sorbent synthesis, 120
Mass transfer, 131, 133
Maximum power point (MPP), 389
MEMS/NEMS, drug-delivery-programmed systems, 432–433

Metal-based nanoparticles, 277
Metal/carbon nanocomposite (Me/C), 241
compositions analysis
heat-physical properties, investigation, 449–450
IR spectroscopy, 450–451
spectrophotometric investigations, 448
ultrasound processing, 447–448
viscosity, relative and absolute, 451–454
fine suspension, preparation, 446–447
heat-physical characteristics, 446
IR spectrometry technique, 442, 446
nanotechnology, 442
paints characteristics, 442
silicate materials, 442
spectrophotometry, 444
structure
Fe/C nanocomposite, TEM picture, 443
globular film nanostructures, 443
nickel and iron nanocomposites, 442–443
Ni/C nanocomposite, TEM picture, 444
polymeric matrixes, 442–443
ultrasound processing of suspensions, 444–445
viscosity, relative and absolute, 446
Methylmetacrylate, 314
Metropolis method, 209
Microcrystals, 196
Miniature electromechanical implants
Cerezyme implant, 433
characteristics, 433–434
electromechanical device, 432
MEMS/NEMS, drug-delivery-programmed systems, 432–433
polymer and inorganic implants, 432
Minitab and Mathlab programs, 39
MNPs. See Magnetic nanoparticles (MNPs)
MNSs. See Magnetically targeted nanosystems (MNSs)
Modeling and simulation techniques
mechanical properties, multiscale modeling (see Multiscale modeling)
mesoscale and macroscale methods, 214–215
micromechanics
average compliance, 216–217
composites, average properties, 216
concepts, 215–217
effective bulk properties determination, 219
equivalent-continuum and self-similar approaches, 218–220
equivalent-continuum approach, 219
finite element method (FEM), 220–221
Halpin–Tsai model, 217–218
linear elasticity, 215
Mori–Tanaka model, 218
representative volume element (RVE), 215
strain–concentration, 217
SWCN nanoarray, 220
microscale methods
Brownian Dynamics (BD), 211–212
dissipative particle dynamics (DPD), 212
dynamic DFT method, 213–214
Lattice Boltzmann (LB), 212–213
Time-dependent ginzburg–landau (TDGL) method, 213
molecular scale methods
molecular dynamics (MD), 206–209
Monte Carlo (MC) technique, 209–210
Modern orthopedic implants (OIs)
antibiotics, 423

anti-inflammatory drugs, administration, 421–422
bacterial infections, 423
biocompatibility, 419, 422–423
bisphosphonates, 424
cell behavior, biochemical mediators, 419
cytotoxicity, 424
hydrophilicity, 422
integrinbinding proteins, 420–421
nanofibers matrices, 422
nanoscale implant surface topography, 421
osteoblasts, adhesion intensity, 420
polymeric matrices, 422
specific proteins, adsorption, 420
surface chemistry, 421
surface wettability, 422
titanium osteoimplant, 423
Modified viscose fibers, 58
Molecular dynamics (MD)
algorithms, 207–209
bonded and nonbonded interaction, 206–207
constituents, 206
interaction potentials, 206
simulations, 209
Molecular modeling, 189
Monomer intercalation, 324
Monte Carlo (MC) technique
atomic movement, 209
Boltzmann constant, 209–210
polymer nanocomposites, 210
simulation, 209
Montmorillonite, 314–315
Mori–Tanaka model, 218
Multiscale modeling
BEM Halpin-Tsai Equation, 226
carbon nanotube/polymer composites, 226–227
carbon nanotubes (CNTs)
agglomerates, 229
axial direction, 223
based composites, 224–225
CNT/PP composites, 229
CNT-reinforced polymer specimens, 232
CNTs-reinforced composites, 227
embedded, unit cell, 226
reinforced nanocomposites, 231
schematic, FEA, 230
computational nanomechanics, tools, 225
"critical" SWNT fiber length, 226
cylindrical RVE, loading cases, 224
3D finite element model, 222
ERM, 222
homogenization schemes, 228
homogenized RVE, 232
hybrid atomistic/continuum mechanics method, 226
local molecular structure and bonding information, 221–222
micromechanical models, 230
micromechanics, 221–222
Mori– Tanaka model, 222
multiscale homogenization (MH), 228
MWNT, intertube bridging, 231
nano-mechanical model, 231
nanoscale interface cracks, 223
nanoscale representative, 224
nanotube and polymer matrix, 227
polymer composite system, 221
Stone–Wales defects, 226
Multi-walled nanotubes, 156–157

N

Nanobiomedicine, application, 425
Nano cannon scheme, 185
Nanochemistry, 111, 235
Nanoclusters, 233
Nanocomposites PBT/organoclay, 331
Nanocomposites structures
atomic-force microscopy, 285
Berkovich indentor, 286, 304
butadiene-stirene rubber, 285, 287–288
butadiene-stirene rubber/microshungite (BSR/microshungite), 299
butadiene-stirene rubber/nanoshungite (BSR/nanoshungite), 299

butadiene-stirene rubber/technical carbon (BSR/TC), 299
carbon-nanoshungite-microshungite, 292
elasticity moduli and matrix polymer, 290
elastomeric matrix, 295
elastomeric nanocomposites (rubbers) nanofiller particles, 284
elastomeric polymer materials, 284–285
force modulation regime, 291
fullerene-containing mineral shungite, 285
initial particles diameter, 294
interfacial layer, 287–289
nanocomposites BSR/TC, 297
nanofiller
mass content, 290
particles cluster (aggregate), 293
powder, 284
nanoindentation process, 304–305
nanointentation method, 286
nanoshungite, 287
nanostructure, 285
particle-cluster (P-Cl), 291
particles sizes, 296
polymer nanocomposites multicomponentness, 283
"quadrates method", 298
reduced elasticity modulus, 301–302
SPIP, 285
statistical flexibility indicator, 288
statistical walkers diffusion constant, 295
Nanocrystal (NC), 277
Nanocrystallite, 113, 277
Nanodendrites
bimetallic bonding, 269
Bright-field TEM image, 270–271
dependence of length, 267
fabricated, 261
insulator substrates, 269
Lattice fringes, 269–270
Pt-nanoparticles, 269
selected-area diffraction (SAD) pattern, 270
Nanofibers matrices, 422
Nanofibers, TSC
in DSC solar cells
electrospun scattering layer, 404–405
encapsulated electrolyte, 405
flexible nanofiber, counter electrode, 405–406
model, 406–407
electrospun nanofiber, 401–403
electrospun nonwoven, organic solar cell, 407
organic, 408
Nanofiller mass content, 290
Nanofiller particles cluster (aggregate), 293
Nanofractal-like trees, 256
Nanogun, 184–185
Nanohydromecanics
chemical techniques, condensation, 238
clusters, 196
dispersion chemical methods, 238
flow structure, 198
fluid flow, 195, 197
iron-containing nanostructures, 241
microcrystals, 196
nanostructures, 234, 236, 239
"organized" move, 196
physical methods of dispersion, 237
powder-like nanoproducts, 240
quasi-solid phase, 197
velocity distribution, 199
viscoplastic fluids, 199–200
Nanointentation method, 286
Nanomolecular structures
biodegradable polymers, 414
gynitourinary implants (GIs)
bioartificial (hybrid) implants, 426
biodegradable polymers, 428
cell technology, 426
contraception, 425
endometriosis, 425

high-molecular-weight implants, 427
hydrogel membranes, 427
nanobiomedicine, application, 425
nanotechnology, 425
nonbiodegradable materials, 427
polymer matrix, 427
polymer-metal device, 425
stem cells, growth, 427
targeted drug delivery, 425
tissue engineering, 427
urology, tissue engineering, 426
implanted medical device (IMD), 414–415
medical implants, 414
miniature electromechanical implants
Cerezyme implant, 433
characteristics, 433–434
electromechanical device, 432
MEMS/NEMS, drug-delivery-programed systems, 432–433
polymer and inorganic implants, 432
modern orthopedic implants (OIs)
antibiotics, 423
anti-inflammatory drugs, administration, 421–422
bacterial infections, 423
biocompatibility, 419, 422–423
bisphosphonates, 424
cell behavior, biochemical mediators, 419
cytotoxicity, 424
hydrophilicity, 422
integrinbinding proteins, 420–421
nanofibers matrices, 422
nanoscale implant surface topography, 421
osteoblasts, adhesion intensity, 420
polymeric matrices, 422
specific proteins, adsorption, 420
surface chemistry, 421
surface wettability, 422
titanium osteoimplant, 423
modified cardiavascular stents
antiproliferation effect, 417–418
biodegradable polymeric materials, 419
cardiovascular pathologies, 417
coating, 418
construction, 417
nanoporous surface structures, construction, 417–418
restenosis, 417
nanoimplants, encapsulated drugs
eye diseases, 428
eye drops, 428
local drug delivery, 430
nanomedicine, 431
ocular micro and macroscale implants, installation, 429
photodynamically activated system, 431
nanotechnology products, commercialization, 416
neurological implants, 431–432
passive and active transports, 415
publications, exponential growth, 416
Nanoreactors, 111
Nanoshungite, 287
Nanostructures (NS), 344
architecture, 107, 149
Nanotechnology, 236
Nanotherapeutics
agents, 276
nanocarriers, drug delivery
liposomes, 278
non-phagocytosis pathway, 280–283
phagocytosis pathway, 278–279
reticuloendothelial system (RES), 278
nanoparticles
biological, 278

lipid based, 277
metal-based, 277
polymeric, 277–278
properties, 276
Nanotubes, 150
Nanowire arrays, 256
Natural fibers, 43
Natural hybrid nanocomposites
amorphous glassy polymers, 7
Avogadro number, 7
binary hookings, 7
elasticity modulus, dependences, 6
Euclidean space, 3
fractal dimension, values, 4–5
gas-phase HDPE, 2
high-density polyethylene (HDPE), 2
local order domains, 4
melting and testing temperatures, 4
nanoclusters, 2, 4, 5, 7
natural nanocomposites, 2
organoclay, 9
Poisson's ratio, elastic strains, 7
reinforcement degree, dependence, 8
semicrystalline polymers, 2, 4, 5, 7, 9
temperature dependence, 5
testing specimens, 3
uniaxial tension mechanical tests, 3
yield stress, dependence, 6
Navier-Stokes equation, 187
Navier–Stokes hydrodynamics, 212
Neurological implants, 431–432
Non-phagocytic pathway
caveolae-mediated endocytosis (CvME), 117, 280–281
clathrin-mediated endocytosis (CME), 116, 280
macropinocytosis, 117–118, 281–282
magnetic nanoparticles (MNPs), 118–119, 282–283
No-slip poiseuille flow, 194
Nylon-6, 326, 333, 335

O

OCTA, polymer materials simulation, 214
Opsonins, 114, 279
Optical spectroscopy, 351
Organic solar cells (OSCs), 391–392
Organoclay, 316
content, 329–330
distribution, 317
mixing, 325
OSCs. See Organic solar cells (OSCs)
Osteoblasts, adhesion intensity, 420

P

Phagocytosis pathway, nanocarriers
actin, 279
in drug delivery, 114–115
endolysosomal compartment, 279
Fc receptor (FcR), 279
opsonization, 278
phagolysosomes, 279
professional phagocyte, 278
Photovoltaic (PV) fibers, TSC
carbon nanotube (CNT), 397
electron-donating groups (EDG), 400
flexibility level, 397
hetrojunction photoactive layer, 397
ITO free hole-collecting layer, 398
ITO layer uses, 397, 399
low band gap polymers, 399
photoelectrical characteristics, 399
P3HT, 397
polymers, 396
poly (3,4-ethylenedioxythiophene):poly (stirenes ulfonate), 397–398
preparation procedure, 398
solar cell characteristics, 399
Photovoltaic solar energy, 386
Pinocytosis, 115. See also Non-phagocytic pathway
Poise, 451
Poly(3-hexyl thiophene) (P3HT), 386
Polyamide fibers
biological degradation, 71–72
capron fiber surface, 73
chemical inhomogeneity, 73
fabrics, 73
IR-spectroscopy method, 73
microorganism interaction, 72

natural nutritive and energy source, microorganisms, 72
polycaproamide fiber degradation, 74
silt microorganisms and test-bacteria, 72–73
test-bacterium complex, 72
uses, 72
Polybutylenterephtalate, 318
Polycarbonate modification, Cu/C nanocomposite
ethylene dichloride, 379
fine suspensions, 378
IR spectra, 381–382
optical density, 379–381
and polymethyl methacrylate, 379
self-organization, 382
structure formation, 381
super-molecular regularity, 378
surface and volume, 378
thermal-physical characteristics, 382
Polyethylene oxide (PEO)/CHT, 14–15
Polyethylene polyamine (PEPA), 349
Polymer and inorganic implants, 432
Polymeric clay, 314
Polymeric composite materials (PCM), 344
Polymeric matrix interaction, 314
Polymeric nanocomposites
layered silicates, structure
bentonite, 315
clays, bentonite breeds, 314
inorganic layer, 315
inorganic structures, 314
modification, 315–317
montmorillonite, 314–315
structure, 315
polymeric clay, 314
polymeric matrix interaction, 314
production, aluminous silicates
alcoholates, 325
extrusion method, 325
monomer intercalation, 324
organoclay mixing, 325
polymerization, 325
silicate nanocomposite, 325
in situ, 324
sol-gel technologies, 325
properties, 327
ammoniumfluorine hectorite, 333–334
calorimeter data, 331–332
combustion processes, 335
fire resistance, 333
flammability, 335
heat and flame resistance, 326
heat release, speed, 332, 334
nanoaddition, organoclay, 326
nanocomposites PBT/organoclay, 331
nylon-6, 326, 333, 335
optical properties, 336
organoclay content, 329–330
polybutyleneterephtalate, organoclay into, 327, 329–330
quaternary ammonium montmorillonite, 333–334
thermal stability, 326–327, 329
thermogravimetrical curves, montmorillonites, 328
structure, montmorilonite basis
clay particles, distribution, 320–324
interlayer space determination, 318–320
organoclay distribution, 317
schematic formation, 318
tactoid formation, 317
thermal stability, 314
Polymeric nanoparticles, 277–278
Polymeric nanosystems
aspects, 112
biological nanoparticles, 114
black body, 135
chemical assembly, 107
cluster, 107–109, 108, 119, 123, 127–128
coating stability analysis, 123
convective heat transfer, 133–134

copolymerization, 109
electron spectroscopy for chemical analysis, 107
enzyme activity estimation, 123
ESR spectra, 124–125
fibrinogen, 123, 130
FMR spectra, MNPs, 126–127
globules and fibrils, 108
heat transfer
coefficient, 137
conduction, 134–135
convection, 133–134
drying characteristics, 131–132
heat and liquid moisture transport, 132–133
mass transfer, 131, 133
moisture, in nanofibres, 139
porosity and pore size distribution, 137–138
radiation, 132, 135–137
thermal equilibrium and comfort, 138
water repellent fabric, 131
wet nanofabrics, thermal analysis, 131
internal channels, diameters, 110
lipid based nanoparticles, 113
magnetic sorbent synthesis, 120
metal-based nanoparticles, 113
molecular nanotechnology ideology, 111
nanocarriers
non-phagocytic pathway (see Non-phagocytic pathway)
phagocytosis pathway, 114–115
nanochemistry, 111
nanoreactors, 111
"nanoscience", 107
nanostructures, 107, 110
nanotherapeutics properties, 112
polymer-based nanoparticles, 113–114
protein coatings formation, 120–121
protein precipitates, 129
proteins adsorption on MNPs, 121–122
Rayleigh light scattering intensity, 130
relative intensity, dependence, 125
scanning tunnel microscope, 107
science of nanomaterials, 107, 108, 111
self-organization, 110–111
supernatant solutions, IR-spectra, 129
system classification, 110
thrombin adsorption, kinetics, 126
volume distributions, 128
Polymer nanocomposites multicomponentness, 283
Polymethylmetacrylate, 314, 318, 379
PowderCell program, 364
Protein coatings, 120–121
Pt-nanoparticles, 269

Q

Quantum-chemical modeling
attraction force, 344–345
copper/carbon nanocomposite, 344
epoxy polymers, fragments interaction
absolute binding energies, 346
binding energy, 345
bond lengths, 347
Cu/C NC modification, 350–351
effective charges, 348
fine suspensions (FS), 345
hardening process, 348
initial substances, fragments, 345
magnetic properties, 349
molecular complexes, interaction energies, 347
PEPA interaction, 347
polyethylene polyamine (PEPA), 349
fine suspensions, 349–350
adhesion, 356
IR spectroscopic investigation, 355
thermogravimetric investigations, 356–357
ultrasound processing, optimal time period, 352–353

viscous properties, 353–354
investigation technique
adhesive strength determination, 352
IR spectroscopy, 351
optical spectroscopy, 351
relative viscosity, determination, 351
thermogravimetric technique, 351–352
nanostructures (NS), 344
polymeric composite materials (PCM), 344
Quasi-solid phase, 197

R

Radial velocity profile, 192
Radiation, 132, 135–137
Receptor-mediated endocytosis (RME), 116
Regression analysis, 39–40
Relative viscosity, determination, 351
Reprecipitation, 362, 364, 372, 375
Representative volume element (RVE), 215
Response surface methodology (RSM), 38
ANOVA, 83
CCD, 82
experiment design, 83
regression analysis, 83
Restenosis, 417
Ripe cotton fiber, 46
RME. See Receptor-mediated endocytosis (RME)
RSM. See Response surface methodology (RSM)
Runs, 38
RVE. See Representative volume element (RVE)

S

Selected-area diffraction (SAD) pattern, 270
Shifts, 55
Short circuit current, 389
Silicate compositions, 442
Silicate nanocomposite, 325
Silicon solar cell, 409
Silk fibers
B. Mori silk, 25
concentration, effect
and applied voltage, 32–33
diameter, 28
flat and ribbon like fibers, 32
mean fiber diameter, 33–34
polymer solution, 30–31
SEM micrographs and diameter distribution, 28–30
definition, 25
electric field, effect, 34–35
electrospinning temperature, effect
bending instability, 35–36
branched jets, 36
morphology and texture, 36–38
non- uniform lateral stresses, 35
unstable jet, 36
fibroin, 25
fibrous proteins, 25
high voltage, 24–25
properties, 25
protein-based materials, 25
sericin, 25–26
silk fibroin (SF)
characterization, 27–28
electrospinning, 27
regenerated solution, preparation, 26
spinning solution, preparation, 26
stiff and strong fibers, 25
Silver nano-particles formation
Ag-GIM, 364–365
XRD-pattern, 373–375
complex forming reagents, 369
gelatin-immobilized matrix (GIM) systems, 362–363
gelatin-immobilized silver(I) hexacyanoferrate(II), 369
gelatin, intermolecular cavities, 370

maximal degree of amplification, 365
PowderCell program, 364
proteolytic enzymes, 364
"reprecipitation", 362, 364, 372, 375
"threshold" concentration complex-forming reagent, 372
Single-walled nanotubes, 155
Slip length, 172
SLN. See Solid lipid nanoparticles (SLN)
Solar cell
categories, 388
dye-sensitized solar cells (DSSC)
advantages, 394
"black dye", 393
DNH2, 393–394
electrolyte, 394
high performance dye, 393
oxide semiconductors, 392
photoelectrodes, 392–393
photosynthesis process, mimic, 392
schematic, 392
electrical measurement
maximum power point (MPP), 389
Photovoltaic devices, 389
short circuit current, 389
electron excitation, 388
inorganic
multiple junction cells, 390
P–N homojunction, 389–390
scheme, 390
semiconducting material, 390–391
silicon thin-film cells, 390
light composition, 387–388
organic solar cells (OSCs), 391–392
structure, 388
textile solar cell (TSC)
flexible, 395–396
flexible electronic devices, 395
nanofibers (see Nanofibers)
organic semiconductors, 395
photovoltaic (PVs), 395
polar bear hair, 396
puzzle, 395
PV fibers, 396–400
PV integration, 400–401
Solid lipid nanoparticles (SLN), 113, 277
Specific electrical resistance, 170–172
charges transfer, 161
continuum hypothesis, 163
convection, 161
dead space, 168
electrokinetic and hydraulic, 161
electrokinetic processes, 161
electro-osmosis, 162
electrophoresis, 162
potential of sedimentation, 162
streamy potential, 162
gases, molecular weight, 169
intermolecular interaction, potential energy chart, 165
Knudsen number, 164
laminar flow, 163
migration, 161
molecular dynamics method, 164
molecules, location scheme, 168
particles position, 167
single-walled nanotube, 161
Spectrophotometric investigations, 448
Squashing, 181
Swellings, cotton fiber, 54
Synthetic fibers, biodamages
biostability, 71
carbochain polymer, 71
chemical inhomogeneity, 70
enzymes, 70
microorganisms, energy source, 70
occurrence and biodegradation, 71
soil fungi and bacteria, 69
soil tests, 71
supermolecular structure, 70
System self-organization, 235

T

TDGL method. See Time-dependent ginzburg-landau (TDGL) method
Tents, 387

Textile materials, 43
- annual losses, 44
- bacterial degradation, 44
- biodamages, 45
- biotechnological methods, 45
- cotton fiber, 46
- degradation, 43
- fiber and fabric biodamaging, 44
- protection, against microorganisms
  - antimicrobial nonwoven materials, 78
  - antimicrobial properties, 75
  - biocides, 76
  - biostability, 75–76
  - capron polymerization, 76
  - disinfectants, 78
  - dyes, 78
  - hydrophobic finishing, 78
  - nanoparticles, 77
  - nanotechnologies, 77
  - oxyquinoline derivatives treatment, 76
  - physical modification, 77–78
  - single bath coloring and bioprotective finishing, 79
  - water repellency, 78
- temperature and humidity, 46

Textiles, 386–387

Textile solar cell (TSC)
- flexible, 395–396
- flexible electronic devices, 395
- organic semiconductors, 395
- photovoltaic (PVs), 395
- polar bear hair, 396
- puzzle, 395
- PV fibers, 396–400
- PV integration, 400–401

Thermal equilibrium and comfort, 138

Thermogravimetric technique, 351–352

Threshold concentration complex-forming reagent, 372

Time-dependent ginzburg-landau (TDGL) method, 213

Titanium osteoimplant, 423

Triflouroacetic acid (TFA), 15

U

Ultrasound processing, 447–448

Urology, tissue engineering, 426

W

Water repellent fabric, 131

Wet nanofabrics, thermal analysis, 131

Wool fiber biodamaging
- bacteria, 60
- Black color, 61
- cell membrane complex (CMC), 59
- chemical feature, 58–59
- cortical layer, 59
- cuticular cells, 58
- keratin, 58–60
- layers, 58
- microorganisms
  - average rate of strength reduction, 63
  - discoloration, 62
  - exposure, 68
  - exposure to, 63, 66
  - flaky layer destruction, 63, 65
  - resazurin ability, 62
  - thin merino and coarse caracul, yellowness, 67
  - thin merino wool, 63–64
  - visual coloration and optical density, 62
  - "yellowing", 66, 68–69
- splotches of color, 61

Y

Yellowing, 66, 68–69

Z

Zigzag, chiral vectors, 173